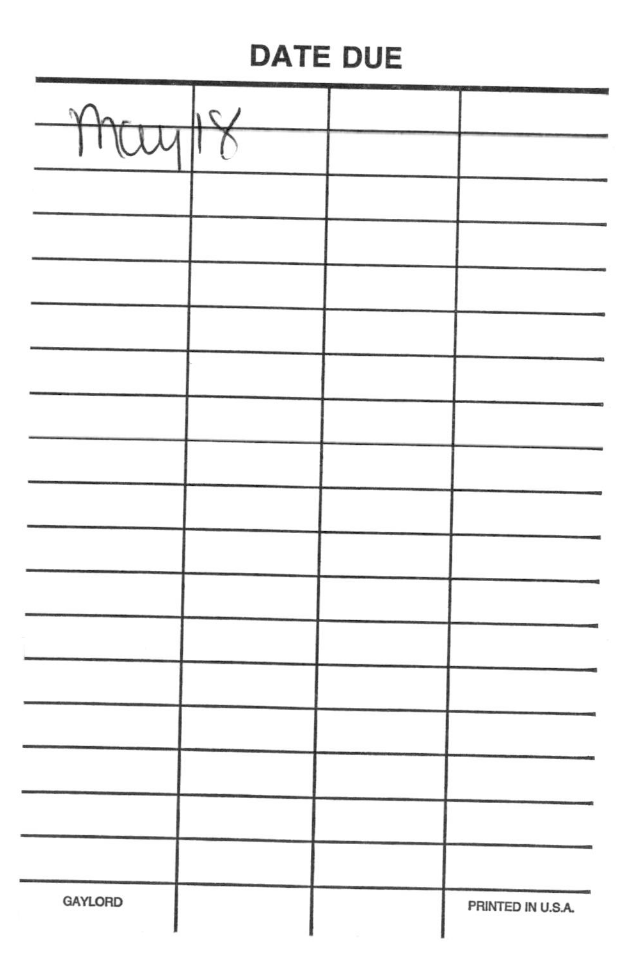

Shifting and Rearranging

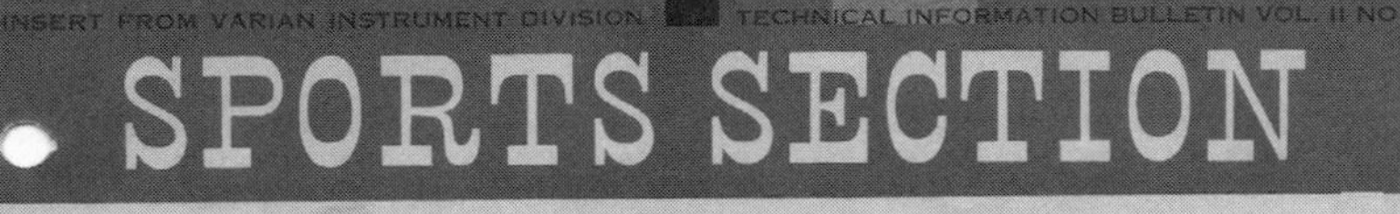

NMR WORLD'S RECORDS . . .

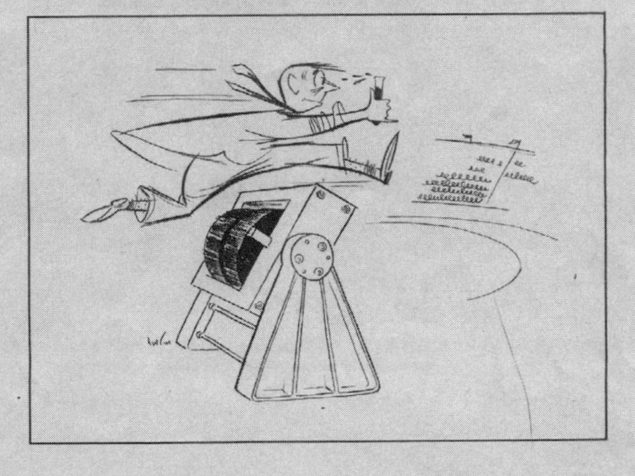

The subject of track and field records has long enjoyed a position of eminence around the cracker barrel, but we have noted with rising interest that the subject of NMR performance records has been making a bid of late, at least among our acquaintances, to displace the more time-honored topic. Possibly this is due to the fact that we are not in an Olympic year, but whatever the reason, we feel that perhaps it is time to inject a thought or two of our own into the conversation.

One of the first points that occurs to us is the fact that the performance of an NMR Spectrometer must be compared to a decathlon participant, rather than to a single-event competitor. Just as the all-around athlete must excel in running, jumping, vaulting and throwing, so must a High Resolution NMR instrument display simultaneous excellence in resolution, sensitivity, stability, and versatility.

As an example of what we mean, let us consider the case of the various ways in which one of these important criteria — resolution — can be achieved. Since magnetic field uniformity over the sample dimensions determines performance in this particular event, it is obviously possible to make an assault on the world's record by shrinking the sample down to a vanishingly small size. Another way in which extreme resolution might be sought would be to employ a lower and lower magnetic field, thereby approaching a blissful state of zero inhomogeneity at zero magnetic field. The former could be compared to a shot-putter who heaves a weightless ball an infinite distance, while the latter is best represented by a high jumper who displays perfect form as he floats effortlessly over the bar while it is lying on the ground.

The main trouble with the first approach can be traced to the fact that sensitivity falls off with the sample volume. This blunder can be fatal in the case of a sparingly soluble compound, or in the case of a substance whose spectral

1

Title page of an insert of the *Technical Information Bulletin*, 2, no. 2 (1958) of Varian Associates' Instrument Division. It shows a magnet used in NMR spectrometers, surmounted by a chemist. In the original, green paper mimicks the Sports Sections of the *San Francisco Chronicle* of the period. John D. Roberts Papers. Reprinted with permission of Varian, Inc.

Shifting and Rearranging

Physical Methods and the Transformation of Modern Chemistry

Carsten Reinhardt

Science History Publications/USA
Sagamore Beach
2006

First Published in the United States of America
by Science History Publications/USA
a division of Watson Publishing International
Post Office Box 1240, Sagamore Beach, MA 02562-1240, USA
www.shpusa.com

Printed with support of a publication grant from the Deutsche Forschungsgemeinschaft (Re 1492/5-1).

Library of Congress Cataloging-in-Publication Data

Reinhardt, Carsten.
Shifting and rearranging : physical methods and the transformation of modern chemistry / Carsten Reinhardt.
p. cm.
Includes bibliographical references and index.
ISBN 0-88135-354-X (alk. paper)
1. Chemistry—methodology. 2. Chemistry—Research—Technological innovations. I. Title.

Qd6.R43 2006
542—dc22 2006042384

Design and typesetting by Publishers' Design and Production Services, Inc.

Manufactured in the U.S.A.

Contents

Preface

In the second half of the twentieth century, chemistry underwent a profound transformation. Its object of examination, the chemical substance, was transmuted into abstract structure; its most important method, the chemical reaction, was supplemented by physical methods; and its practitioner, the chemist, was partially displaced by technical instruments. At the center of this transformation were physical methods. As much as shifts of data in nuclear magnetic resonance spectroscopy and molecular rearrangements in mass spectrometry were used to interpret the results obtained by these techniques, so their adoption in chemistry was bound to shifts and fundamental rearrangements of the discipline.

For most chemists, physical methods are now a central part of their arsenal of methods. For some, physical methods even constitute the very core of their work, their characteristic research aim being development of methods and paradigmatic applications. To ensure dissemination, such methods-oriented scientists conduct exemplary research in their preferred areas of endeavor, and they engage in wide-ranging teaching efforts. Success inevitably depends on the interplay of academic research, industrial instrument-manufacture, and governmental science-funding policy.

My intention here is to analyze the individual research programs and strategies of this crucial membership in the community of twentieth-century chemists, and not to write a complete history of physical methods in chemistry. Work on this book began in the late 1990s, at a time when studies on instruments and experiments had already been well established in the historiography of twentieth-century physics and biology. Contributing to this topic with a project on the history of chemistry seemed to me an ideal opportunity, my own background being in this science. Most of the work was done during my tenure as assistant professor at the University of Regensburg, Germany. It was the convivial, academic environment at the history of science unit that made research and writing a pleasure. The head of the unit, Christoph Meinel, backed my research with organizational advice, and supported it with intellectual input. Thomas Steinhauser was my sparring partner and sounding board in all fields, and I especially benefited from his expertise in matters chemical (and his dexterity in repairing coffee machines). Special thanks to Jörg Deschermeier and Christian Forstner, who participated as assistants in the project; to Daniel Becker, who took care of illustrations; and to Angelika Mak who prepared the index. Many thanks to Lis Brack-Bernsen, Britta Görs, Martin Kirschke, Michael Klein, Peter Konecny, Celia von Lindern, Christine Nawa, Inken Rebentrost, Martin Schneider, Sandra Wilde and Peter Zigman.

The beginning of this project was earmarked by my Edelstein International Fellowship 1998/99. I worked six months at the Chemical Heritage Foundation in Philadelphia, Pennsylvania, spending substantial amounts of time networking, and traveling to undertake interviews. In Philadelphia, I enjoyed the company of members of the historical group, most importantly Mary Ellen Bowden, Dave Brock, Rich Hamerla, and Leo Slater. The stay was made one of the most wonderful experiences in my life through the friendship of our hosts, Thelma and Francis McCarthy. To Tony Travis of the Edelstein Center in Jerusalem, Israel, where I spent the final period of the fellowship, I am indebted for much more than I can describe in a preface. I am especially grateful that he has read the whole manuscript, and corrected my English. Thanks, Tony, for so many discussions at the Beit Belgia faculty club!

Funding for archival trips to the United States was provided by a grant of the Deutsche Forschungsgemeinschaft (DFG), the project being part of the DFG-funded research unit "Interrelations of Science and Technology". The DFG has also subsidized the publication of this volume.

The manuscript has benefited tremendously from the criticisms of many colleagues. Bernhard Dick, Christoph Meinel, Mary Jo Nye, and Jürgen Teichmann agreed to serve as referees in the *Habilitation* procedure. For comments and help I thank Michael Eckert, Gerd Graßhoff, Ralf Hahn, Kristen Haring, John Heilbron, Klaus Hentschel, Martina Heßler, Jeffrey A. Johnson, Andreas Karachalios, Daniel Kevles, Ursula Klein, Peter J. T. Morris, Kärin Nickelsen, Hans-Jörg Rheinberger, Terry Shinn, Leo Slater, Friedrich Steinle, Arnold Thackray, Helmuth Trischler, and Ulrich Wengenroth. I am grateful for the help supplied by staff members of the various archives that I have visited.

Without the continuous support and advice of the scientists who were the historical actors of the events described in this book, this study could not have been made. For their willingness to allow me to interview them, and their liberality in providing access to their papers I thank all of them.

A portion of chapter four appeared in the June 2006 issue of *ISIS* entitled "A Lead User of Instruments in Science. John D. Roberts and the Adaptation of Nuclear Magnetic Resonance to Organic Chemistry, 1955–1975," and is reprinted with permission. © 2006 by The History of Science Society, Inc.

For Tony

Chapter One

Physical Methods Ante Portas

Methods are central to research. Sometimes, their adoption can even mark the boundaries of disciplines. This book describes the transfer of instrumental research methods from physics to chemistry. During this process, both the instrumentation and the practice related to it were transformed. Essential for this transition was the adjustment of physical instrumentation to chemical challenges and problems. To make instruments chemical—in devising useful methods for chemists—was research in its own right, and a strategy of chemists for creating disciplinary space.

In mid-twentieth century, chemists took up novel kinds of instrumentation that had originated in physics and high technology. Chemists integrated nuclear magnetic resonance spectroscopy (NMR), mass spectrometry, infrared and ultraviolet spectroscopy—to name only the most important techniques—in their research projects, and directed their research programs according to the opportunities and needs afforded by their instruments. Some scientists concentrated on the development of problem-solving, instrument-based methods for use in research fields they knew well. In doing so, they merged the objects of chemical inquiry—molecular structures and dynamics—with high-technological instruments. In this process, chemical substances became inextricably intertwined with physical apparatus, at the hardware level as well as in theoretical concepts. In introducing physical methods in chemical research and routine analysis, the chemical sciences and technologies underwent a major transformation. Prior to World War II, chemists determined the constitution of an unknown substance through its chemical reactions with known compounds. By the 1970s, chemists commonly obtained their results using a variety of instruments that allowed the analysis of chemical substances in terms of their physical properties. This shift from chemical separation and analysis to physical identification and elucidation altered the research practices of chemists in all fields, changed the curriculum of chemistry as a discipline, and gave further impetus to the industry of instrument makers. It greatly expanded the scale and scope of chemistry: Physical instruments enabled the search for extraterrestrial life as well as tremendously improving the ability to search for minuscule traces of substances on earth. Moreover, instrumentation affected the social context of chemistry, though the introduction of technical apparatus did not lead to a complete changeover to Big Science. In many ways, the dominance of Big Science in physics, especially as the public perceived it, served both as model and as counterpart for chemistry. High-technology instrumentation became a characteristic feature of this science.

In July 1966, the organic chemist and Nobel laureate Sir Robert Robinson opened a symposium on physical methods at New College, Oxford, quoting a phrase of a famous mountaineer:

> The late A. F. Mummery once said that all difficult peaks pass through three stages: (1) complete inaccessibility; (2) an interesting climb for experts; (3) an easy day for a lady. An analogous series applies to the introduction of new physical techniques in organic chemistry. Initially, as before the first ascent, the problems appear so difficult as to be nearly insoluble. The second stage is that early workers show the general usefulness of the techniques, which are quickly adopted for the solution of difficult problems. The third stage is that every laboratory worth its salt simply must have the necessary instruments as a matter of routine. I think we are approaching this third stage in respect of the new techniques to be discussed in this Symposium, and scientists everywhere will echo Barrie's curtain in "What Every Woman Knows": "How much do they cost, those machines?"[1]

Robinson had been among the more reluctant organic chemists in accepting physical instrumentation.[2] Often, the avowal of a convert is more telling than that of a missionary, and Robinson's sudden enthusiasm prompts some questions: When did chemists adopt physical instruments and how did they adapt their research programs? What processes had to take place to change the lack of perception into the recognition of research opportunities? Why did chemists suddenly regard physical instruments as being comprehensible and relevant for their investigations? In the search for answers, we will turn to some of the chemists who thought of physical instrumentation as being an "interesting climb." But we will not look at them as mountaineers, forcing their way up. In

---

[1] Robinson, "Introductory Remarks," 1. Albert Frederick Mummery (1856–1895) wrote *My Climbs in the Alps and Caucasus* (for the quotation by Robinson see pages 120–160). See Barrie, *What Every Woman Knows.*

[2] Robinson did not mention physical instrumentation in his 1955 *The Structural Relations of Natural Products.* On Robinson's reluctance to use physical instruments, in contrast to his main competitor in physical organic chemistry, Christopher Kelk Ingold, see Nye, *From Chemical Philosophy to Theoretical Chemistry,* 267; and Slater, "Woodward, Robinson, and Strychnine."

contrast, we will get to know them as builders of meeting grounds, constantly reshaping the boundaries and expanding the outreach of their discipline. Their work created the ease of use that Robinson alluded to when he, in his final quote, referred to James Matthew Barrie, the author of *Peter Pan*.

A Second Chemical Revolution?

In a remarkably short span of time, ultraviolet and infrared spectroscopy, NMR, and mass spectrometry changed the contents and style of chemical research.[3] The influx of instrumental "spies" who could "watch" molecules and their interactions at atomic levels strengthened and extended the emerging new intellectual core of chemistry, most importantly quantum-mechanical theories of molecular structure, and concepts of reaction mechanisms. However, the second chemical revolution, the "instrumental revolution,"[4] was not an overthrow of chemistry by physics, thus, it was not an act of reductionism. Why did the broad-scale application of methods and instruments that originated in physics, a scientific discipline with wholly different theoretical underpinnings and conceptual outlooks, not cause a complete physicalization of chemistry? First, the impact of physical instrumentation on chemistry was closely related to the inroads of quantum physics into chemistry, simply because most of the measured phenomena could be explained as quantum effects. In the 1930s, quantum mechanical concepts such as the theory of the chemical bond seemed to bolster the claims of some physicists that chemistry was nothing more than the physics of electron orbitals. But theoretical physicists were not able to calculate for molecular structures larger than the very simplest molecules, and chemical knowledge kept being crucial. Moreover, the emerging quantum chemistry explained concepts and regularities previously accepted on empirical, chemical grounds alone. This included the periodic system of chemical elements as well as structures and reactions of chemical substances. Thus, the new knowledge did not really contradict the old, although both differed substan-

[3] The techniques in routine use in organic chemistry at the end of the 1960s (either at the bench of the organic chemists themselves or in service laboratories) were ultraviolet, visible, and infrared spectroscopy, mass spectrometry, paper and gas chromatography, and to a certain extent X-ray diffraction. A large variety of other techniques stayed in the hands of specialists and was never widely disseminated in mainstream chemical research. For an excellent survey of the techniques available around 1964, their potential use, cost, and distribution, see Schwarz, *Physical Methods in Organic Chemistry*, 7–19.

[4] Tarbell and Tarbell, "The Instrumental Revolution, 1930–1955." The first chemical revolution was the introduction of the concept of oxidation, and a new nomenclature by Antoine Laurent Lavoisier and allied scientists in the late 18th century. On the diverse meanings of the concept of revolution in history of science see Cohen, *Scientific Revolution*; Cohen, *Revolution in Science*; and Kuhn, *Structure of Scientific Revolutions*.

tially. The second part of the answer lies in the consensus that scientists had to reach in order to relate instrumental data to natural phenomena, and not to technical artifacts. In this process, the data-generating instruments achieved a stage of transparency that enabled their use in an uncontroversial manner.[5] This transparency had to be negotiated again each time when the instrument moved from one disciplinary context to the other: chemists rendered the instruments unproblematic while incorporating them in their research. Instruments did indeed change chemistry; but they were themselves transformed in this process. Third, it is important to note that physical methods did not completely replace traditional chemical techniques. In organic synthesis, for example, classical chemical reactions remained the core of scientific practice. In addition, physical instrumentation did not decisively alter the social cohesion and cultural mode of the teaching of chemistry.

The argument of resilience of chemistry leads to comparing this science to Peter Galison's model of twentieth-century physics.[6] In discussing the long-term stability of physics—despite of its many revolutionary changes—Galison differentiates three layers, or levels, of this science: theory, experiment, and instrumentation. There were breaks, revolutions, either in the instrumental, experimental, or theoretical domains of physics. While one was disrupted, the structures of the other layers remained largely intact. Does this picture hold for twentieth-century chemistry? Theoretical chemistry, understood as a merger of physical chemistry, physical organic chemistry, and quantum chemistry, came into existence from the 1940s on.[7] Even today, theoretical chemistry has not reached the same maturity and influence as theoretical physics. In chemistry, theories remained largely restricted to a semi-empirical approach. It can further be argued that only during the instrumental revolution did a substantial part of chemistry become instrument-related at all. Both the theoretical and the instrumental layers of chemistry were subordinated to its third part, experimentation. This observation is valid until today to such an extent that, although there exists theoretical chemistry in the curriculum, there is no experimental chemistry: arguably, nearly the whole of chemistry counts as its experimental domain. Still, the split of chemistry according to its objects, and not its methods, is the prevalent ordering principle, and during most of the twentieth century, organic, physical, and inorganic chemistry were its most important sub-divisions. In mid-century, transformations in theory and instrumentation occurred at roughly the same time, and their interconnection and co-stabilization became crucial for their respective success. On the one hand, theoretical chemistry explained the function of the new instruments, and it supplied much of the research questions that could be tackled. On the other hand, the empir-

[5] On the notion of transparency see Schaffer, "Glass Works."

[6] Galison, *Image and Logic*, 797–803.

[7] Nye, *From Chemical Philosophy to Theoretical Chemistry*, 1–2.

ical performance of the instruments greatly supported the claims of theoretical chemistry.

The inroads of physical instrumentation and theory threatened to revolutionize the strongest level of chemistry, experimentation, in attempting to displace its core, the chemical reaction. During the 1950s and 1960s, physical instruments replaced chemical reactions to such an extent that Carl Djerassi, organic chemist and early proponent of mass spectrometry and other physical methods, once remarked that chemists "don't do any more chemistry with the natural products" as a result of this process.[8] Instruments of the novel physical type did not simply replace familiar chemical tools, such as beakers, test tubes, and stirring rods. In our context, instruments are understood as inscription devices,[9] generating meaningful data that enable direct inferences to the scientific question in mind. The most ubiquitous chemical method that existed before the advent of physical instrumentation was the chemical reaction. Reactions produced data for inferences on molecular structures which were at that time the main objective of chemists. Djerassi's remark aptly describes the events in his own domain of chemistry. In natural product chemistry, chemists strived for the isolation of previously unknown substances (most of them available only in minuscule amounts), and tried to establish their molecular structures. Spectrometers replaced the art of structural elucidation formerly done with the help of chemical degradation reactions. At face value, instrumental methods completely changed the experimental practice of chemistry. At the same time when chemical practice changed, the physical techniques were altered, too. The transfer of instrumentation from one discipline to the other involved processes of intrusion, promotion, and, most importantly, an active adjustment by its users. This adaptation occurred at several levels: in the theoretical background, the hardware, the moral economy, and the social organization of the novel research-technology. Take one example of Djerassi's own field of research, steroids. In order to enable inferences on chemical constitution and structure, the data obtained by mass spectrometers had to be interpreted in terms of the chemical theories of reaction mechanisms and electronic effects of molecular bonds and groups. Thus, theoretical concepts developed for reactions in the test tube were applied to explain molecular events in the mass spectrometer. Special inlet systems made possible the insertion of large, heat-sensitive steroid molecules into the spectrometer, which was situated at the center of an organic chemistry research group. A dozen or more chemists and technicians synthesized hundreds of compounds necessary to establish the

[8] Carl Djerassi, interview by Jeffrey L. Sturchio and Arnold Thackray, 31 July 1985, p. 39, Center for History of Chemistry, Oral History Program, Chemical Heritage Foundation, Philadelphia, USA.

[9] This follows Bruno Latour. The crucial point for Latour is that the inscriptions produced by instruments are "used as the final layer in a scientific text." Latour, *Science in Action*, 68.

interpretation rules of organic mass spectra in an empirical way, published in a stream of research articles and textbooks, and fed into computer programs for automatic retrieval and interpretation of spectra. (See Chapter 3). Chemists incorporated physical instrumentation only after they had it transformed according to their needs.

In organic chemistry—arguably the most important subdiscipline of chemistry—the effects of the instrumentation revolution were most clearly visible, if only because the new style of instrument-based analysis involved a much deeper change in experimental practice than in the neighboring discipline of physical chemistry with its basis in physical instruments and theories. Since the mid-nineteenth century, the most important aims of organic chemists consisted in the structural elucidation of unknown compounds and their synthesis. Analysis and synthesis were tightly interwoven: if lacking in structural knowledge, a chemist's attempt at synthesis was a blind shot; without synthesis, the postulated structural formula did not count as a proven fact.[10] Chemists routinely determined the structures of organic chemical compounds by chemical means, so called wet methods, and by combustion analysis. Elemental analysis by combustion of the compound and the subsequent weighing or volumetric measurement of the oxidation products yielded the relative amounts of carbon, hydrogen, and oxygen present in the substance, and thus established the empirical formula. Structural formulae were derived from chemical tests indicating the presence of certain atomic groupings, and degradation reactions that led to the carbon skeleton of the compound. The classical era of chemical structure determination lasted from the 1860s to the 1950s, when the structures of a multitude of complex natural products, such as dyes and pigments, alkaloids, vitamins, and hormones were determined in often decades-long work of research groups. Some successful determinations were found worthy of the highest awards, including the Nobel Prize, and gave rise to industrial prospects.[11] The end of structure determination by traditional chemical degradation and synthesis represented the core of intellectual and practical changes that the instrumental revolution brought about in the 1950s and 1960s.

New aims took its place, notably the art of the rational synthesis of complex molecules. In this case, synthesis has to be understood not by its role in the final confirmation of structural formulae. The novel physical methods did not need this confirmation by synthesis. The trust in them relied on other fields of science and engineering: quantum mechanics and electronics. Synthesis, after the mid 1960s, set its goal for the invention of novel synthetic pathways, reactions, and reagents. No longer did complex multi-step syntheses, piling up already known reactions, win the prizes. The establishment of elegant syn-

[10] Russell, "Changing Role of Synthesis."

[11] Slater, "Woodward, Robinson, and Strychnine"; Morris and Travis, "Role of Physical Instrumentation," 717–719; Birch, *To See the Obvious*, 57–59, 67.

thetic methodology began to be regarded as the highest peak of scholarship. Because physical instrumentation, in ample supply to organic chemists in the 1960s, was based on firm foundations outside the chemical realm, it could truly be of assistance to this modern style of synthesis.

This development began in the 1940s, with the advent of commercially available ultraviolet and infrared spectrometers. The correlation of spectral peaks with certain molecular groups gave hints to the unraveling of structures of molecules, and added a new tool to the reasoning of organic chemists, who previously had relied solely on the information obtained by chemical degradation and identification methods. Because the novel evidence could be incorporated smoothly into traditional knowledge about chemical structures, the new instruments gradually supplemented, and later replaced, chemical methods.[12] In addition, the isolation and separation of small amounts of material was greatly simplified with the innovation of paper chromatography. The prominent organic chemist Derek H. R. Barton even compared the impact of paper chromatography (later modernized to thin layer chromatography) to the role that photographic emulsions had in particle physics. Immediately after its reinvention in the early 1950s, paper chromatography showed its power by the elucidation of the constitution of the peptide hormone insulin. X-ray crystallography had a similar impact, solving the riddle of the structure of the antibiotic penicillin, and preceding its synthesis by fifteen years. In their old frame of mind, organic chemists would have regarded the synthesis of penicillin as the final proof of its structure. Now, and in the case of penicillin, the physical technique of X-ray crystallography had the last word. Soon, NMR, mass spectrometry, and gas chromatography supplemented the array of techniques available to the organic chemists.[13]

Barton described the replacement of classical chemical practice and inductive reasoning in terms of technological determinism. Already in 1963, he singled out X-ray crystallography as the instrument with the highest impact on chemistry:

> It is clear that in the future organic chemists will no longer need to solve major structural problems by degradational methods. All this can be left to x-ray crystallography. We can use the more inductive tools of NMR and of mass spectrometry for all the minor problems that arise in synthetic work . . . . Ideally, it should be possible to feed the x-ray photograph into the computer and have the printed structure emerge at the other end![14]

Barton overestimated the immediate effects of X-ray crystallography, but was right for the whole array of physical methods. With the wide use of ultraviolet

[12] Slater, "Woodward, Robinson, and Strychnine"; Slater, "Instruments and Rules."

[13] Barton, "Some Reflections," 87.

[14] Barton, "Some Reflections," 96.

and infrared spectroscopy, mass spectrometry, NMR, and other methods, the establishment of reliable instrumentation and methodology became both a science and an industry in itself. The profits of instrument manufacturing companies and the careers of innovative scientists relied on the wide use of high-technology instrumentation. Thus, structural elucidation became high technology with its own rules and special accolades for achievement.

When physical instruments pushed aside classical chemical methods, the epistemic meaning of structural formulae changed, too.[15] In the old system, structural formulae were guides for predicting what reactions could undergo, mnemonic symbols referring to chemical affinities. In the new, physical mode of chemistry, the structural representation itself became the focus of research. The material characterization nearly disappeared in this process. Other questions appeared on the chemists' agenda: conformational structure, the analysis of minute amounts of samples that could not be isolated in a pure state, and the study of short-lived reaction intermediates. No field was more important in this respect than physical organic chemistry, a successful blend of quantum chemistry and traditional chemical concepts. Physical organic chemistry supplied most of the research problems that could be tackled by physical methods, and even afforded a theoretical understanding of the instruments' working. Moreover, physical organic chemistry, being a lead-discipline in 1960s U.S. chemistry, supplied the social prestige necessary in academia to achieve success. Later, physical organic chemistry faded away when scientists moved into bioorganic and organometallic chemistry, and other novel fields. But the physical instruments were there to stay. As a result, some chemical subdisciplines experienced a complete transformation of experimental set-up and research orientation. In natural product chemistry, the intellectual goals switched to methods development, biosynthesis, and biological function. In contrast, the core of mainstream organic chemistry, synthesis, subordinated the novel physical instruments to a pure service function. This became apparent through the reactions of practitioners: Natural product chemists debated the credibility of the novel techniques. Synthetic chemists simply followed the rules of interpretation that originated during the interchange processes between physical chemists, instrument manufacturers, and chemists in the "front row," among them many natural product and physical organic chemists.

In important domains, chemistry lost its universal method, the chemical reaction, and with it a part of its unity. As a result of the impact of physical instrumentation, among other reasons, chemistry became a decentralized science: "Biochemistry, geochemistry, chemical physics, atmospheric chemistry,

[15] For a stronger view that connects the changeover in instrumentation to a change in the ontological status of structural formulae see Schummer, "Impact of Instrumentation"; Slater, "Organic Chemistry and Instrumentation." I thank Ursula Klein for a discussion on this topic.

molecular biology, and astrochemistry" were among the hybrid fields that leading chemists named in the mid-1960s to argue for the virtues and the benefits of interdisciplinarity.[16] Instrumentation made chemistry more diverse, while expanding it at the same time and connecting it to other scientific fields. According to philosopher of science Davis Baird, the instrumental revolution in chemistry was part and parcel of a big revolution in science that took place in the middle decades of the twentieth century, involving deep changes in the worldview as well as the institutions of science.[17] Baird's example, analytical chemistry, is aptly chosen to present such a view, mainly because of its roots and applications in industrial, governmental, and academic science. A large part of analytical chemistry, however, tackled the development of routine methods for undertaking tasks in industrial production and environmental control, and this is Baird's example. In this book, we focus on another sub-field of the "big scientific revolution": the development of analytical methods for chemical research. Although most of these methods were in slightly modified forms of use in many different fields, and changes in one domain could be related to those in others, modes of development and terms of use of analytical instrumentation in scientific research followed different rules of conduct, and entertained different aims, than in other fields.

It is a debatable issue as to whether the described changeover in chemistry deserves the label of a scientific revolution. The changes did not come hand-in-hand with anomalies and a crisis in order to be judged a Kuhnian revolution. Moreover, proponents of the instruments' uses argued against the perception of incommensurability, and worked hard to create connections between the old and the new style of research. However, breaks in the theoretical and instrumental domains of chemistry are clearly visible, as well as in the experimental domain of structural elucidation. Nevertheless, chemistry kept being a stable scientific discipline, and showed remarkable continuity as well as coherence. The mortar that held it all together was supplied by the strategy of a growing community of scientists: the development of scientific methods that connected established paradigms of a community of chemists to the potentials of high-technology instrumentation. The notion of an instrumental revolution places too much weight on the visible discontinuities, and neglects hidden continuities and step-by-step transition processes. Our focus is on the coincidental, and interactive, transformation of chemical science and high-technology instrumentation.

---

[16] National Academy of Sciences, *Chemistry: Opportunities and Needs*, 103. For a collection of studies of such hybrid fields in chemistry see Reinhardt, *Chemical Sciences in the 20th Century*.

[17] Baird, "Analytical Chemistry." Baird is the only author among the many users of the term revolution in this context who defines it properly. He uses the notion to describe inter- or predisciplinary changes that affect social institutions and the attitude towards the world, and relies here on work of Ian Hacking.

METHODS AND INSTRUMENTS

From the early 1980s, the experimental character of scientific work has received increasing attention from historians, sociologists, and philosophers of science.[18] Among the more important (and disputed) issues of the "New Experimentalism" are the following: Experiments are largely independent from theory, consisting of a complex interplay of theories, material things, and data. In addition, scientific facts are technically produced in the laboratory, leading to the self-vindication of the laboratory sciences. Moreover, interpretation of experimental data follows largely low-level concepts, "topical hypotheses," and not high-level theory.[19] A phrase of the philosopher Ian Hacking has become a dogma: "Experimentation has a life of its own, interacting with speculation, calculation, model building, invention and technology in numerous ways."[20] This view leaves ample space to connect the dynamics of experimental science to other scientific and social spheres, and is in accord with such influential historical work as that of Steven Shapin, Simon Schaffer, and Peter Galison.[21] Much work in the history and philosophy of experiment, however, focuses on the inner laboratory alone, emphasizing intrinsic dynamics of experimental practice.[22]

An important aim of this book is to show that despite the rightness of many of the claims of the New Experimentalism, and the richness of its historical studies, it neglects a crucial dimension in the interplay of theory, experiment, and instrument. Largely absent from the discourse are, surprisingly, methods. In this context, it is important to distinguish between *the* scientific *method* and scientific *methods*. The former is a philosophical notion in the sense of a rational pursuit of knowledge, be it the experimental, the inductive, or the hypothetico-deductive method. The latter encompasses different modes of investigations, and covers procedures of laboratory practice, paradigms serving as exemplars for future scientific research, and whole knowledge domains.[23] An explanation of the lack of a historical and philosophical discourse on methods may be the fact that many historians regard methods as the static part in

[18] Among recent surveys I found most useful: Holmes, "Do We Understand?"; Golinski, *Making Natural Knowledge*, 133–161; and Hentschel, "Historiographische Anmerkungen." A detailed case study of the interplay of instrument, theory, and experiment is Hentschel's *Zum Zusammenspiel von Instrument, Experiment und Theorie*. On spectroscopy as visual culture see Hentschel, *Mapping the Spectrum*. The most recent general overviews are: Buchwald, *Scientific Practice*; Heidelberger and Steinle, *Experimental Essays*; Meinel, *Instrument-Experiment*.

[19] A critical assessment and a plea for taking theory more seriously is Carrier, "New Experimentalism." But see Hacking, "Self-vindication of the Laboratory Sciences."

[20] Hacking, *Representing and Intervening*, xiii.

[21] Shapin and Schaffer, *Leviathan and the Air-Pump*; Galison, *Image and Logic*.

[22] On the term inner laboratory see Galison, *Image and Logic*, 4.

[23] Synonymously to methods, I use the term techniques. On the issue of the scientific method, of course, there exists a rich literature. See for a start Schuster and Yeo, *Politics and Rhetoric*.

scientific inquiries, as an established and routine necessity of research. In Hans-Jörg Rheinberger's influential scheme of the experimental system, for example, methods and instruments can be identified as "technical objects" that stand in contrast to the flexible, floating, and dynamic "epistemic things."[24] In an experimental system, knowledge is produced in the interactive process that confronts technical objects and epistemic things. Notwithstanding the historiographic value of this dialectic view of scientific research, the opposition of technical object and epistemic thing undervalues the fact that research on methods can be a constantly fruitful scientific activity. Methods are as well the result of scientific research as they constitute its origin, and they do not represent just technological routine. In the words of Gaston Bachelard:

> Toutes les méthodes scientifiques actives sont précisément en pointe. Elles ne sont pas le résumé des habitudes gagnées dans la longue pratique d'une science. Ce n'est pas de la sagesse intellectuelle acquise. La méthode est vraiment une ruse d'acquisition, un stratagème nouveau utile à la frontière du savoir. En d'autres termes, une méthode scientifique est une méthode qui cherche le risque. Sûre de son acquit, elle se risque dans une acquisition.[25]

Bachelard situated his opinion in the opening remarks of a conference in philosophy of science, and maybe intended to flatter attendants with his vision of research as an everlasting endeavor of giving up the familiar and entering the unknown. But his view already entails the strategy of a community of scientists that centers on the making of methods. Moreover, Bachelard's description of twentieth-century science as being completely foreign to common-sense knowledge, and based on the dialectics of mathematical theorems and instrumental techniques, leads to his well-known picturing of instruments as *théorèmes réifiés* and his notion of *phénoménotechnique*. According to Bachelard, scientific research realizes rational constructs with the aid of instrumentation. Although normative, and not descriptive, Bachelard's philosophy of science aids in understanding the twentieth-century impact of physical methods on other sciences exactly because of his insistence that relativity and quantum mechanics brought about a radical discontinuity not only in physical, but also in chemical and even philosophical, understanding. In chemistry, spectroscopic methods could only be of use in more than just a strictly empirical manner after the quantum-mechanical concept of the chemical bond had led to new definitions of molecular structure and reaction mechanisms. Physical methods were accepted in chemistry because they merged theories, experiments, and instruments of a new kind in a comprehensive and coherent, but also adaptable, system. Thus, the use of methods in scientific experiments was closely bound to *accepting* the theoretical underpinning of the apparatus (in terms of theories of

---

[24] Rheinberger, *Toward a History of Epistemic Things*, 28–31.

[25] Bachelard, "Le Problème Philosophique des Méthodes Scientifiques" 39.

quantum physics and electrical engineering), as well as to *understand* the theoretical framework of its respective field of application (physical organic chemistry, for example). In this sense, experiment was not independent from theory.

Since the middle of the twentieth century, chemical handbooks and introductory volumes included the term "physical methods" in their title.[26] They supplemented books of an earlier generation, describing the working modes and research uses of experimental methods largely along chemical means. Thus, physical methods became an identifiable technique for practitioners, connecting theory and instrumentation on the experimental level. Development of and investigations in physical methods of chemistry evolved into research specialties in their own right.

Five dimensions of the impact of physical methods and instruments on chemistry deserve closer scrutiny. First, the making of methods in the laboratory. This dimension has to be analyzed as the interplay of technical craft, experimental practice, and theoretical knowledge, aiming in a first step at the control of the apparatus.[27] In modern science, this was largely a matter of teamwork in- and outside the laboratory. Experimenters often regarded the apparatus to be under control if they could replicate results that were made in other laboratories.[28] In addition, scientists sought knowledge about the physical causes of the data that were produced by scientific apparatus. In the second half of the twentieth century, instrumental science was a composite system involving many diverse elements, ranging from computer programming to engineering knowledge and materials science. These constituents of scientific practice were interdependent and led through interactive stabilization to the closure of disputes about experimental results. The apparatus made the transition from an

---

[26] Among the earliest treatises that specialized in organic chemistry is Weissberger, *Physical Methods of Organic Chemistry*, in 1945–46. Houben-Weyl, the acknowledged series of experimental methods in organic chemistry published volumes on physical methods in 1955 (Müller, *Physikalische Forschungsmethoden*). Both Weissberger and Houben-Weyl include such routine techniques as determination of melting/freezing points and advanced instrumentation such as electron or X-ray diffraction. Understandably, physical chemistry much earlier had handbooks of this kind (a 1926 example is Reilly and Rae, *Physico-chemical Methods*), while inorganic chemistry was rather late (the first volume of *Technique of Inorganic Chemistry* appeared in 1963 and merged with *Technique of Organic Chemistry* to *Techniques of Chemistry* in 1971). Notoriously, prefaces emphasize the wide uses of physical methods for chemical tasks, and the large impact of the work of physicists and physical chemists. A bit different is the tone of the preface of the treatise *Physical Techniques in Biological Research*, in 1955/56. The biologists underline their own inventive capabilities (among their examples is paper chromatography, invented by Archer Martin and Richard Synge), and downplay the cooperation with physicists and chemists (p. v). In the first volume of the handbook, on optical methods, however, most articles are written by scientists employed in chemistry departments.

[27] Ravetz, *Scientific Knowledge*, 78.

[28] See Collins's study of the TEA-Laser, whose proper working mode was established by its ability to produce the right outcome. Collins, *Changing Order*, chap. 3. On the lack of studies on teams and technicians in history of science see Russell, Tansey, and Lear, "Missing Links."

object of inquiry to an instrument when it was trusted, initially locally and temporarily. The aim of the experimental scientist was to bring together in one scheme the apparatus's operation, a model of its working mode, and a conceptual understanding of the world. If this three-way coherence was found, and exported to other localities, the disputed phenomenon became an established fact and the doubted apparatus a trusted instrument.[29]

Second, adaptation to chemical concepts. The closure of disputes and the setting up of uniform standards did not prevent that instruments remained machines for producing knowledge about nature. In classical chemistry, chemical substances and reactions were at the same time natural things, technological artifacts, and instruments. For example, in physical organic chemistry, conjugated dienes served as instruments in the study of tautomerism.[30] Chemists assigned substances and reactions the status of objects of inquiry or of investigative instruments on functional grounds, and this could change rapidly during the course of an experimental project. The advent of physical instrumentation made this interplay more complex. Correlation rules connected the realms of spectra and structures. Most famous were the Woodward rules that related ultraviolet spectra of α,β-unsaturated ketones and dienes to their structures, and the Octant Rules that served the same purpose for optical rotatory dispersion and saturated ketones. The eminent organic chemist Robert Burns Woodward was the acknowledged master of synthetic organic chemistry, and with his status raised the acceptance of physical methods.[31] These rules were generalizations, and not theories in an abstract and mathematical sense. They enabled chemists to apply the novel instruments in a straightforward and semi-empirical manner. Through such rules, and correlation charts, chemists made use of spectroscopic data in the same way they used information from reaction sequences. Both kinds of information, spectra and data from chemical reactions, became directly related to molecular structure. Thus, information from spectroscopic techniques could be mixed in an unproblematic manner with information acquired by chemical means. This was an important precondition for the acceptance of the novel methods by chemists, and for its practicability at times when spectroscopic methods could only supply partial information on molecular structures. Model systems, such as steroid ketones in organic mass spectrometry, were the Drosophila of chemical experimental systems. Because they were readily available and unproblematic to use, chemists established with the help of such compounds rules that later were applied to tackle other substance classes.

---

29 Pickering, "Living in the Material World."

30 Nye, *From Chemical Philosophy to Theoretical Chemistry*, 268. See also Kohler, "Systems of Production," 128.

31 Slater, "Instruments and Rules." On Woodward see Benfey and Morris, *Robert Burns Woodward.*

Third, standardization and teaching. A good deal of stabilizing work had to be done in the chemical community in order to accept physical devices as credible instruments. Once stabilization had been achieved, the instrument was embedded in an unproblematic way in ongoing research programs, serving for the calibration of other instruments and the disciplining of individual experimenters.[32] Standardization was a crucial part in the adoption of novel instrumentation, and the different actors in instrument-based twentieth-century science understood the need to set standards as a most important feature of their respective programs. The standardization of data accumulation, presentation, and interpretation were socially determined; the same applied to the norms that governed the definition of precision and accuracy. In this regard, instrument manufacturers, scientific and industrial associations, governmental agencies, and individual scientists competed for influence, market shares, and power.[33] A related, but distinct aspect, is the teaching of the new methods. Instrument manufacturers engaged in the organization of workshops, offered assistance in dealing with research problems, and published technical guidebooks and data files. Early users held lectures and seminars, and wrote textbooks and articles for dissemination of the techniques.

Fourth, the university-industry nexus. Most routine scientific instruments were designed, constructed, and marketed by specialized instrument manufacturers. Instrument builders of the second half of the twentieth century tended to have strong roots in physics and especially electronics. Thus they had a basis in a field of technoscience that was far apart from the areas of activities of most of their customers: medicine, biology, chemistry, and many others. In order to acquire knowledge of this market of scientists, and to obtain the capability to direct the market in their interest, instrument manufacturers engaged to a limited extent in the scientific activities of their customers. They became scientific players in their own right, and they cooperated and competed with their customers in this regard. Instruments shaped the directions of research programs, and decided on their outreach. Because of their crucial importance, scientists often took part in their development, as inventors, designers, and innovative users. For those scientists who could not or did not wish to engage fully in design and construction of instrumentation, contacts to instrument manufacturers became helpful in the supply of advanced instruments. The exclusive use of novel instruments was a huge advantage in the competitive environment of citation statistics and fund raising. Access to unique instruments was gained through specifications of certain features of a novel instrument, transferred

[32] See Golinski, *Making Natural Knowledge*, 141.

[33] On the influence of RCA, the manufacturer of electron microscopes, on standards and presentational issues in biology, see Rasmussen, *Picture Control*, chap. 1.

from the scientific user to the manufacturer. This provided the company with market knowledge. For both sides, the benefits of exclusivity only counted if this exclusivity was time-limited. The manufacturer had to sell some exemplars of the same type to make a profit. The scientist's profit was in citations, and thus he was interested in the wide use of his type of instrument. By necessity, exclusivity was a transitory phenomenon. Through this, the instrument became a product of the manufacturer as well as a scientist's commodity.

Fifth, the social organization. The topography of instruments reveals important insights into the acceptance they enjoyed within the chemical community. Most advanced physical instrumentation was assembled in instrument centers, at the department, university, regional, or even national level. This served several goals: Expensive instruments were made available on a service basis to a wide community of potential users. The centers were meeting grounds where instrument experts and routine users interacted and collaborated in research projects. At the same time they created a distance between traditional chemistry and the novel, instrument-focused science. Instrument centers enabled mainstream chemists to maintain their intellectual and social autonomy, while relegating the instrumental approach to the edges of the discipline, and subordinating it to a service function.[34] Some techniques led to bench-top instruments, used by graduate students who had already been trained in their use during their studies; others were kept in departmental service laboratories, being operated on a sign-on basis or by technicians; still others formed the core of regional or national research facilities, the equivalent to Big Science in physics. Their physical location in and migration through the system of chemistry reflected their acceptance and prestige, as well as their size and cost.[35]

These were the dimensions of the method makers' science. Not all of them were important for each method-oriented scientist, and their relevance changed with time and place. Most dimensions, however, together formed the micro-context where scientists contributed to the development of scientific techniques, and they will be taken up in the chapters that follow this introduction. The general situation, however, the macro-context that confronted the first chemists interested in the application of physical methods in the late 1940s and early 1950s, is best described when we scrutinize two developments of the 1930s and 1940s: the forming of a new hybrid discipline between chemistry and physics, and the physicalization of analytical chemistry in the industrial context.

---

[34] Laszlo, "Tools, Instruments and Concepts," 83.

[35] I thank Thomas Steinhauser for an insightful discussion on this topic.

EXCHANGE AND TRANSFORMATION

Physical methods and instruments did not experience a straightforward transfer from physics to chemistry. Scientists tended to look with distrust at instruments if they yielded results that did not fit the preconceived notions of an established scientific community. Thus, often, instruments (and important improvements) that originated in the context of their later scientific use were more successful (in terms of rate of adoption) than instruments that were introduced from neighboring fields. In contrast, the latter had a higher potential of enabling revolutionary insights.[36] In periods of normal science, evoking the Kuhnian picture, such upheavals were disliked by most scientists. Given this view, physical instruments had to be smoothly incorporated into chemistry. In what manner was instrumentation that originated in physical research transferred to chemical investigations? Taking up a metaphor from evolutionary biology: Which processes of adaptation made it possible that some techniques thrived in a new environment while others did not? The first part of the answer lies in the fact that the transfer of these methods was a gradual, step-wise process, taking advantage of disciplinary niches close to the original context.

In contrast to this evolutionary account, the origins of molecular biology, the most famous hybrid field between physics, chemistry, and biology, are sometimes presented as the result of a revolutionary one-step strategy. According to this picture, Warren Weaver, the director of the division of natural sciences of the Rockefeller Foundation, funded costly research instrumentation, mostly operated by physicists and chemists, and applied it to problems of biological significance. On the one hand, this sort of physical reductionism via instrumentation was a failure, partly because biologists resisted such colonization by physicists. On the other hand, the higher prestige of physics made sure that physicists did not develop a real interest in biology, although they commanded substantial resources in the biological field.[37] How did chemists react to similar challenges by physics, the lead-discipline of twentieth-century science? The case of quantum chemistry might be a telling one. According to the opinion of some physicists, quantum chemistry would reduce chemistry to physics. Indeed, chemists at one time feared the upcoming supremacy of physics and maybe also for this reason resisted anything physical: "Like handloom weavers, drivers of steam engines and cavalrymen, chemists felt the threat of being deskilled, of finding that in the new world of physical methods all that they had learned, the very essence

[36] See Rasmussen, "Innovation in Chemical Separation," 257, referring to von Hippel, *Sources of Innovation*. Some of von Hippel's examples show that often scientific users made improvements to instrumentation and through this contributed to the adoption of the instrument in a new application field.

[37] Abir-Am, "Discourse of Physical Power."

of being a chemist was becoming irrelevant."[38] In the end, the chemists' fears were not justified: Quantum chemists were never able to account for chemical problems on purely physical terms. Consequently, they emphasized semi-empirical methods, relied on quantum chemistry's success in other chemical and biochemical disciplines, and developed a pragmatic bent.[39] Thus, the direct transfer of physical theory and practice failed. This is also in accord with Kohler's historiographical account of the origins of molecular biology, emphasizing the collaborative character of the encounter of physicists and biologists at a crucial time in the development of instrumental methods.[40] Interdisciplinary fields took over the tasks of knowledge transfer and cultural adaptation. In the case of the transfer of physical instruments to chemistry, two such interfaces of chemistry and physics come to mind: the research field of chemical physics,[41] and physicists' teams in the chemical industry,[42] both established in the 1930s.

Chemical physics merged quantum mechanical theories and spectroscopic techniques, and was rooted in a decade-long series of attempts to use quantum mechanics in the solution of problems related to chemical bonding and molecular structure. Because of the failure of quantum mechanical attempts to explain and predict accurately the chemical behavior for all but the simplest molecules, empirical methods dominated advances in theory, and molecular spectroscopy became the backbone of quantum chemistry.[43] Molecular spectroscopy attracted a considerable number of chemists who were eager to take part in the increased prestige of physical theorizing and a few physicists interested in expanding their exact methods to larger molecular systems. In order to establish their credibility, chemical physicists based their techniques both in the "hard" fields of theoretical quantum physics and electronic instruments and in the "soft" areas of chemical structures and reactions. A novel experimental culture emerged, combining theoretical, practical, and social aspects of research. The crucial feature of the experimental culture of chemical physics was its material culture, and a research style that focused on the control of the instrument.[44] Questions that come to mind in this context concern the hierarchies of

---

[38] Knight, "Then . . . and Now," 90.

[39] See Zandvoort, *Models of Scientific Development*, 234–237. For the chemical orientation of quantum chemistry in this period see Nye, *From Chemical Philosophy to Theoretical Chemistry*, chap. 9; Gavroglu and Simões, "The Americans, the Germans."

[40] Kohler, *Partners in Science*, 358–391.

[41] Kohler ascribes the innovation of many novel instruments during 1925–1935 to a "heightened activity on the border between physics and chemistry," coinciding with the origins of chemical physics, and focusing on investigations on the solid state. Kohler, *Partners in Science*, 361.

[42] Rabkin, "Technological Innovation in Science."

[43] Nye, *From Chemical Philosophy to Theoretical Chemistry*, 241.

[44] For the notion of experimental culture see Rheinberger, *Toward a History of Epistemic Things*, 28–31, 137–138; for the research style of molecular spectroscopy and the early history of NMR see Reinhardt, "Chemistry in a Physical Mode."

chemists and physicists in the new field, the interplay of theory and experiment, and the status of the instrument itself. Physics was regarded as the field of greater prestige. Consequently, theory was regarded as supreme, and the control of instrumentation was a value in itself. Nevertheless, chemical physicists were able to contribute conceptually to the uses of physical instruments in their field. Gradually, they transformed physical theory into a system that organic chemists could understand, supplementing physics with chemical concepts. Moreover, their pragmatic, empirical correlation of experimental facts and chemical reasoning made possible the uses of the instrument without a deep, quantitative foundation in physical theory. In the end, the diffusion of the instruments in the wider community of chemists had to await commercial manufacture and distribution. Chemical physicists prepared some foundations for this academic market, which developed under the influence and guidance of the chemical industry, in close cooperation with instrument manufacturers.

Coincidentally with chemical physics, but largely independent, a general movement in industry took place, when traditional chemical analytical methods were supplemented, and sometimes replaced, by methods based on physical instrumentation.[45] This had far-reaching consequences for chemistry as a whole. For qualitative and quantitative identification of compounds, analytical chemists displaced chemical separation methods with physical instrumentation. This development was heavily under way in the 1930s, especially in the United States, and World War II greatly expedited its course. During the war, through large-scale R&D programs on synthetic rubber and petroleum refining in the United States and the United Kingdom, chemical and oil companies developed infrared spectroscopy and other physical instruments into routine instruments for analytical purposes. Characteristically, for commercial production of these spectrometers, chemical firms joined forces with instrument manufacturers. After World War II, these instrument-building companies actively attempted to create a market in academic chemistry, mainly to accustom students (as future users of their equipment) to their products.[46] Due to industrial requirements, instruments had to be fool-proof, reliable, and simple to use. No doubt, such black-boxing was a crucial part in the transfer processes from physics via chemical industry to academic chemistry. Nevertheless, a black-boxed instrument was looked at with distrust by scientists who often had bad experiences with misinterpretations of experimental results due to lack of understanding of the working mechanism. Thus, research chemists worked hard to grasp a basic understanding, and to distribute their newly acquired knowledge among students and colleagues. In this process, physics and electronics had their share, but chemical theories took center stage. Simply for the

---

[45] Baird, "Analytical Chemistry."
[46] Rabkin, "Technological Innovation in Science."

reason that physical instruments had to produce data that could be converted into chemical knowledge, the "chemistry of instrumentation" was the crucial feature in chemists' strategies. The interpretation of instrumental data along chemical lines started after World War II in the chemical industry, when users realized that such chemical rationalizations could save time and money.

Soon, academic chemists took up this approach in order to place the credibility of the respective techniques on a sound chemical basis. In this process, the theoretical "wrapping" of instruments changed considerably. A mass spectrometer, even if technically unchanged, encoded different theoretical meanings for physicists measuring nuclear forces in the 1920s, industrial chemists collecting compound data in the 1940s, and natural product chemists disentangling molecular structures in the 1960s. What, then, moved with the instruments? Hardwired, inbuilt knowledge? It did, certainly, but the move of black-boxed knowledge was an implicit knowledge transfer only, if not unpacked at the other side of the transfer line. With the instruments, certain representation forms and data-handling techniques crossed the boundaries between disciplines and social spheres. In addition, institutional organizations, such as industrial service laboratories, served as models of similar developments in academic departments and regional instrumentation centers. Naturally, people switched from industrial to academic jobs, carrying with them their experience. All these issues were malleable features, and they changed together with the receiving fields. There were issues that blocked the transfer. Most important was the inability and unwillingness of most chemists to engage in construction of electronic instrumentation. This problem was taken care of largely by instrument makers (at universities, government agencies, in the chemical industry, and instrument-building firms), the number of which considerably increased during World War II. A second blockade concerned the lack of suitable methods and medium-range theories to apply and explain the novel kind of instrumentation. Here, chemists themselves, partly employed by instrument manufacturers, but mostly on the users side in academic and industrial chemistry, contributed heavily. In a way, they were forced to do so. Physical instruments threatened to destroy the methodological autonomy of chemistry. The design of novel reaction pathways had been entirely in the hands of chemists themselves. Now, engineers and physicists appeared on the scene, attempting to displace chemistry by electronics. Chemists saw a chance for themselves to develop methods for and with this novel instrumentation, paving the way for the influx of the new technology, and at the same time trying to rescue the methodological autonomy of chemistry.

The wider reasons for this process are to be found in the effects of World War II, and the subsequent decades of tremendously increased science funding. Events during World War II pushed forward the development of physical techniques, notably infrared spectroscopy, when large-scale R&D programs on aviation gasoline, synthetic rubber, and penicillin provided resources and successful applications. In the aftermath of the lead-technique of infrared spectroscopy, other physical methods intruded upon chemistry. The arrival of the age of electronics, increased governmental funding of science and technology, and the availability of user-friendly instruments made this possible. Equipped with a novel armory and increased resources, some chemists started out to adapt physical instrumentation to their experimental culture. In doing so, they had to overcome resistance by traditionalists, but most of all they had to interest the great majority of chemists who were hesitant to invest considerable intellectual, financial, and human resources in fancy electronic gadgetry.[47] One direct avenue aimed to impress peers by the spectacular successes of the new techniques. Another more mundane and secure possibility was to push the administrators of the large governmental science-funding agencies. John D. Roberts of the California Institute of Technology and William S. Johnson of Stanford University, both chairmen of their respective chemistry departments, and important protagonists of the instrumental approach, in 1964 attempted to find allies among their colleagues in their protracted battle for more funds for better instruments. Characteristically, they referred to physics, the lead-discipline of this era:

> It does not seem to be fully realized that chemistry has come out of the test-tube era in the same way as physics left "sealing wax and string" research some twenty years ago, and that many kinds of large, highly sensitive and expensive equipment are indispensable to work in modern chemistry—not just to save labor but to supply entirely new kinds of information.[48]

In order to achieve this, the new kinds of information had to be connected to established lines of chemical knowledge. Only this guaranteed the justification, credibility, and reliability of techniques that were hitherto unknown to chemists. This happened at the laboratory level, where chemists merged the diverse strings of physics, engineering, and chemistry into a novel experimental culture. Grant

---

[47] Often, such concerns were raised on pedagogical grounds. An example in analytical chemistry (one of the more progressive disciplines with regard to instrumentation) shows that even proponents of physical instrumentation hesitated to include physical instruments too early in the teaching of chemistry. The authors of a 1956 textbook on analytical chemistry stated: "If the sole aim of the course in 'Identification' were to teach methods of rapid identification of unknown compounds, major emphasis should be placed on modern instrumental methods such as infrared, Raman, and ultraviolet spectroscopy; nuclear magnetic resonance; X-ray diffraction; kinetic methods and determination of dissociation constants by potentiometric titration. Because liberal application of these technics would, in many cases, reduce the work of the student to instrumental analysis with concomitant sacrifice of attention to the chemical behavior of the unknown compounds, the use of such technics has been strictly limited." From Shriner, Fuson, and Curtin, *Systematic Identification*, v–vi.

[48] Roberts and Johnson to John D. Ferry and 52 other chemistry department chairmen and 53 chemical company presidents, 3 March 1964, copy in Terman Papers, ser. III, box 8, folder 4.

applications, research publications, and textbooks set the platform for the rhetoric of physical instrumentation in chemistry. The story of the transfer of physical instrumentation to chemistry was a story of the domestication of the instruments, and went hand-in-hand with their popularization.[49]

Method makers, acting as mediators, were crucial actors in the process. The focus of this book is on the middleman cultures of physical instrumentation in chemistry. In cultural anthropology, middleman cultures are minorities in which a disproportionately high number of individuals is engaged in trade.[50] In science, where the producers of knowledge are at the same time its consumers, the necessity of traders or middleman cultures seems odd at first glance. But in the case of knowledge and technology transfer from one group to the other, traders—or mediators—played decisive roles. Through the development of methods, they made sense of various instrumental technologies in changing environments. Method makers attached new meanings to technological objects, and their benefit rested on the wide use of them in scientific communities. For doing so, they needed standing in the importing communities as well as allies in the exporting cultures.

Overview

Although the very definition of the chemical sciences rests on laboratory methods, the issue of practice has been neglected by historians until recently. According to the editors of a recent volume on *Instruments and Experimentation in the History of Chemistry*, "the history of chemistry has been overwhelmingly a history of chemical theory, with practice little considered and with the apparatus that rendered that practice possible almost entirely ignored."[51] This applies even more so in the history of experimentation in twentieth-century chemistry, but recent efforts show that this is beginning to change.[52] Among noteworthy older studies are two books on the history of analytical chemistry that deal at least partially with the impact of instruments and give an account of important participants.[53] In 1986, Dean S. Tarbell and Ann T. Tarbell announced an "instrumental revolution" that had occurred in organic chemistry at the middle of the twentieth century.[54] In a pathbreaking study, Yakov

---

[49] For the notion of domestication see Rasmussen, *Picture Control*, 67 ff.

[50] Examples are Jews in Europe, Chinese in Southeast Asia, and Japanese and Greeks in the United States. See Bonacich, "A Theory of Middleman Minorities"; and Zenner, *Minorities in the Middle*.

[51] Levere and Holmes, "Introduction: A Practical Science," viii.

[52] Morris, *From Classical to Modern Chemistry*.

[53] Laitinen and Ewing, *History of Analytical Chemistry*; Warner, *Milestones in Analytical Chemistry*.

[54] Tarbell and Tarbell, "Instrumental Revolution, 1930–1955." See also Taylor, "Impact of Instrumentation."

Rabkin emphasized the importance of the chemical industry for the breakthrough of one technique, infrared spectroscopy, during World War II, and in a second article he warned against a straightforward transfer of results to the history of chemistry that were gained in the history of physics.[55] For Davis Baird, a "big scientific revolution" in the definitions of I. B. Cohen and Ian Hacking took place in analytical chemistry in the 1920s and 1930s; and Timothy Lenoir and Christophe Lécuyer have tackled the growth of NMR with respect to the work of physicist Felix Bloch and the instrument maker Varian Associates.[56] An excellent overview of several of the most important techniques in organic chemistry is given by Peter J. T. Morris and Anthony S. Travis, while Jeffrey K. Stine has analyzed the policy of the National Science Foundation with regard to the funding of chemical instrumentation in the 1950s and 1960s.[57] Eric von Hippel's work on user dominated innovation gives—from the viewpoint of economics—important insights in the role academic scientists had for the development of novel instruments. From a philosophical perspective, Henk Zandvoort analyzed the early history of NMR. Zandvoort's work recruited Lakatos's philosophy of science and on the basis of primary literature alone he explains the performance of the NMR research program in physics with its "extrinsic success" in chemistry, biology, and medicine.[58] Most recently, Leo Slater has written on the importance of rule-based theories in synthetic organic chemistry. This development resulted, as shown by Slater, in an increased role of synthesis, not analysis, and a change in the ontological status of chemical structures.[59] Though all the works mentioned provide important background for the present study, it either deals with the discipline of analytical chemistry alone (Baird), and thus neglects the important disciplines of organic, inorganic, and physical chemistry, or it focuses on the growth of NMR, directed by the policy of an instrument manufacturer (Lenoir/Lécuyer). Other studies center only on infrared spectroscopy (Rabkin), or give a broad overview of the topic (Morris/Travis). Most crucially, none of the works mentioned deals with the interrelation of chemical research programs with the development and the use of physical instrumentation. The question of the interplay of chemistry and physics on both the theoretical and the practical levels thus receives no treatment.

---

[55] Rabkin, "Technological Innovation in Science"; idem, "Uses and Images of Instruments in Chemistry."

[56] Baird, "Analytical Chemistry"; Lenoir and Lécuyer, "Instrument Makers and Discipline Builders."

[57] Morris and Travis, "Role of Physical Instrumentation"; Stine, "Scientific Instrumentation."

[58] Von Hippel, *Sources of Innovation*; Zandvoort, *Models of Scientific Development.*

[59] Slater, "Instruments and Rules." See also his "Organic Chemistry and Instrumentation." For a coincident change in the status of substance identity see Schummer, "Impact of Instrumentation."

The history of two experimental techniques, nuclear magnetic resonance (NMR) and mass spectrometry, forms the scaffolding of this book. Both methods originated in physical research and had a strong and lasting effect on chemistry during the three decades following World War II. Both became routine in academic, industrial, and governmental laboratories; moreover, they constituted high-technology instruments at the forefront of research in science. The first and most important applications of NMR and mass spectrometry during this period were made in the United States, and the dissemination of both techniques was the fastest there. Therefore the historical analysis will center on some of the more important and influential American research groups, supplemented by one Swiss group with strong ties to the United States. I describe some of the chemists that were affected first, and acted foremost, by and through physical instrumentation. This happened at the laboratory level, in their research, in its style, and in its presentation. In order to obtain access to their work, and to keep the book a manageable size, I restricted the study to six scientists, three each in NMR and mass spectrometry. These represent different subdisciplines of chemistry: chemical physics, physical organic chemistry, analytical chemistry, and natural product chemistry. In my view, these fields were among those most deeply and early transformed, and they contributed much to the general transformation of chemistry. The period of this study ends when the transformation was essentially complete, in the late 1970s. The case studies presented here are to be seen as exemplary, but not typical. They are detailed studies of experimental scientific work, and this method is regarded as the most appropriate to investigate the changes in the experimental practice during this period. Experimental practice means here the practical techniques necessary to perform experiments, the interpretive methods applied to the experimental results, and the rhetoric techniques used in presentation and teaching. At this level occurred the adjustment of instrumental techniques to scientific theories, and vice versa. There was the place where disciplinary boundaries blurred, and social and cognitive aspects of the research process merged to an experimental system. Thus, this book is not an all-encompassing history of NMR and mass spectrometry. Its outline follows the major evolutionary course of these methods in chemistry as much as it is designed to represent specific historiographic themes. It spans a period during which the methods presented went full circle from their beginnings in chemists' and physicists' laboratories, via their adaptation, transformation and popularization in chemistry to their replacement by more suitable successors. Readers interested in history of NMR should focus only on chapters 2, 4, and 6, while mass spectrometry is covered in chapters 3 and 5.

Each method originated in a specific experimental environment, and the first and crucial question was how newly found phenomena could be manipulated and ascertained. In Chapter 2, I emphasize the role of laboratory practice, with the example of the beginning of chemical NMR in the work of Herbert

S. Gutowsky at the University of Illinois in Urbana between 1948 and 1953. Gutowsky was the first chemist to apply NMR to chemistry, and his research group played a crucial role in the discoveries of the effects that formed the basis of chemical NMR. The main historiographical notion applied is that of research style, in this case the experimental style of chemical physics and molecular spectroscopy. NMR was tailored according to this specific style, and inner-laboratory inertia exerted considerable constraints in the course of its development. In the research style of chemical physics, the experimental control of the instrument and its relation to both physical and chemical theory and practice were central. The set-up of self-built instrumentation on the basis of a technology that was largely rooted in radar research during World War II made possible the complex process of explaining unprecedented effects. It is argued that Gutowsky's eminence relied on the fact that he connected results obtained in chemical physics to the concepts and needs of organic chemists. While there existed other active groups in this field, Gutowsky and his co-workers deserve special credit for having dealt with chemical NMR without the background of either a big research group in physics or the backing of an instrument manufacturer. Moreover, the influence of chemical physics—one of the intermediate fields in the transfer of physical methods from physics to chemistry—on the direction that NMR took in chemistry can be shown. Gutowsky's research around 1950 centered on the construction of the NMR spectrometer, the experimental characterization of chemical effects in NMR, the interplay between theoreticians and experimental scientists, and the correlation of physical phenomena to chemical concepts. With this style of control and correlation, Gutowsky, his research group, and his colleagues made possible the subsequent breakthrough of NMR in the mainstream of chemistry.

An implicit, but decisive, question of chemists who were tempted to use novel, fancy electronic gadgetry in their own research was a simple one: "Is it chemical?" In order to follow the methods leaving the chemical physics laboratory we turn to mass spectrometry, and the second intermediate field, the chemical industry. Here, we experience how scientists transformed methods during their migration through different disciplines and institutions. The extent to which the inclusion of chemical concepts into methods and instruments was successful decided on the latters' acceptance in chemistry. As was the case with NMR, mass spectrometry originated in physical research. With infrared spectroscopy, it shared its development in the chemical industry during World War II. Though sophisticated instrumentation was available in the early 1950s, it was not before 1960 that mass spectrometry proved its utility in organic chemistry and biochemistry. The crucial stepping stone was the chemical rationalization of the processes that took place in the instrument. This was achieved in terms of the concepts and theories of physical organic chemistry. Chapter 3 focuses on the work of Fred W. McLafferty (Dow Chemical, Purdue University, and Cornell University), Klaus Biemann (MIT), and Carl

Djerassi (Stanford). With the investigation of reaction mechanisms in the mass spectrometer, McLafferty, Biemann, and Djerassi established the credibility and reliability of mass spectrometric results. In the late 1950s and early 1960s, they crucially contributed to the "chemistry of the instrument" in applying chemical concepts to the interpretation of mass spectrometric data. The chapter describes the origins of organic mass spectrometry in the chemical and petroleum industries, and analyzes in detail the transformation of the method during its transfer to academic chemistry. In this process, mass spectrometry became one of the most powerful, though in the beginning disputed, techniques for analysis of minute amounts of chemical substances, most notably alkaloids, steroids, and peptides. Next to the impact of chemical concepts on an instrumental method, the role of the chemical industry on its course is shown.

The influence of users on the development of methods is portrayed through the interplay of an important early promoter of NMR, John D. Roberts at Caltech, with the first industrial manufacturer of NMR spectrometers, Varian Associates, in Chapter 4. Though Roberts could not solve the explicit, and equally decisive, problem of chemists during this period, "How can I afford it?" he contributed to another urgent question of chemists, "How can I use it?" The experimental work of the scientist, his performance in the design of creative and useful methods, and his public activities for their dissemination in a community of user-scientists form this chapter's story. Again, as with mass spectrometry, physical organic chemistry is shown to constitute the single most important sub-discipline in the bridging of instrumental technology and chemical science. In bringing forward both sides in his research, Roberts was in the position to exert considerable influence in the two spheres. Between 1955 and 1965 there began the era of the routine use of NMR in organic chemistry. Roberts, a physical organic chemist by training, entered the field early on with commercial instruments, and established important aspects of the usefulness of NMR for organic chemistry. His textbooks had a crucial role in that, but Roberts and his research group were also very innovative in the introduction of new techniques. Themes of this chapter are the role of theoretical concepts of physical organic chemistry for the breakthrough of NMR, the routine use of NMR in research, the teaching of NMR, and the cooperation of scientists with instrument builders. The latter issue is analyzed with the help of Eric von Hippel's concept of the lead user.[60] Lead users in science were characterized in two ways: they were established scientific leaders (this leadership was further enhanced by the successful use of instrumentation), and they were first movers in the application of novel instrumentation. Moreover, and precisely because of their leadership, lead users influenced the course of both their own projects and the success of the instrument manufacturers.

[60] Von Hippel, *Sources of Innovation*, 107.

Back to mass spectrometry in Chapter 5, and to its outreach towards computer science, data files, and artificial intelligence. Here again, we realize the importance of a diverse constituency, with industry and environmental agencies having the greatest interest in analysis of samples, while scientists engaged in investigations of unknown molecules. The question of doing either routine analysis by searching compounds or interpretation of spectra to unravel molecular structures was solved by a compromise; as were attempts to displace human experts by computer programs in the art of mass spectral interpretation. The chapter continues the story of Chapter 3, and keeps its focus on the careers of Biemann, Djerassi, and McLafferty. In the period covered by this part, high resolution mass spectrometry revolutionized organic mass spectrometry. In the 1960s, Biemann established a huge laboratory at MIT, the first national facility in the field of mass spectrometry, followed by McLafferty's at Purdue University. Unlike Biemann and McLafferty, Djerassi had a broad interest in the application of several physical methods, most notably mass spectrometry, optical rotatory dispersion, and NMR. He ran a big laboratory with hundreds of students passing through during his career and used mass spectrometry mainly in the field of steroids and alkaloids. The three scientists shared their interest in the development of data processing methods and computational facilities. Thus, special emphasis is given to the use of the computer as scientific tool, which even led to attempts to develop artificial intelligence with the help of mass spectrometry. From its uses in NASA missions in space to its applications in environmental control on earth, the motto is: "How far does mass spectrometry reach?"

"How does NMR change?" is the guiding phrase of Chapter 6. It leads us back to the beginning, the laboratory, but to a deeply changed one. Instrument manufacturers, government agencies, and scientists found their meeting ground there, with the skillful manipulation of experimental parameters at its center still in place. The physical chemist Richard R. Ernst invented two techniques that changed technology, representation, and application fields of NMR like no others. Ernst, who received the 1992 Nobel Prize in chemistry for his work, began his NMR studies with Hans Primas at ETH Zurich. During a stay at Varian Associates, he developed Fourier Transform NMR, and after his return to Zurich in 1968, two-dimensional NMR methods. Ernst's entire work focused on the development of basic NMR methodology, and he cooperated with molecular biologists to find high-end application fields. Like Chapter 2, this chapter tells the story of the development of methods in a chemical physics laboratory. In comparing Gutowksy and Ernst, we recognize how sweepingly the style and research modes of NMR science differed between the years 1950 and 1975. In contrast to Gutowsky, much of Ernst's work relied on his cooperation with instrument manufacturers, first Varian Associates and later Bruker-Spectrospin. In this respect, Ernst's work resembles more the work of Roberts. Nevertheless, Ernst was a physical chemist by training and predilection, a master of the quantum mechanical concepts of NMR, and its instrumental technology. Roberts, and to an even larger extent Biemann, Djerassi, and McLafferty were much more oriented to the user-side of their respective instruments. But they all were connected by their common methods, and an analysis of the community of NMR scientists is presented at the end of this chapter.

The conclusion emphasizes the main thesis of this book, in taking up the five dimensions of method-making science. During the 1950s and 1960s, a novel kind of method-oriented chemists came into existence. Their main focus was the development of problem-solving, instrument-based methods for the chemical community at large. In doing so, their work contributed to the creative interplay of physical and chemical techniques, concepts, and theories as well as it relied on scientific cooperation and academic-industrial collaboration.

Chapter Two

The Machine in the Laboratory

At a 1957 conference on radio and microwave spectroscopy at Duke University in Durham, North Carolina, Herbert S. Gutowsky of the University of Illinois presented the main features of the applications of nuclear magnetic resonance (NMR) in chemistry. For that year, he estimated the number of laboratories using NMR "for more or less routine use" as 100, while in 1952, "no more than five laboratories [were] so occupied." What struck him most was the "amazingly wide range of chemical problems" that could be tackled by NMR methods.[1] This wide range, and the availability of commercial instruments, were, in Gutowsky's opinion, the main reasons for the tremendous rate of adoption of NMR as a research tool in chemical laboratories. But the versatility of a research method had to be worked for, and Gutowsky belonged to a relatively small group that specialized in exactly this type of scientific work. Together with his colleagues, Gutowsky, around 1950, laid the foundations of chemical NMR through correlating phenomena originating in atoms and molecules—detected by electronic circuits, and explained by the laws of quantum physics—to the completely different research problems of chemists.

In late 1945 and early 1946, two research groups in physics—one gathered around Edward M. Purcell at Harvard University, the other led by Felix Bloch at Stanford University—found the nuclear magnetic resonance effect in bulk matter. The phenomenon had earlier been investigated by the molecular beam method, with considerable experimental difficulties and restrictions. The

[1] Herbert S. Gutowsky, "Chemical Applications of Nuclear Resonance," manuscript, November 1957, pp. 1–2, Gutowsky Papers, folder "Chemical appl. of NMR Duke conf. 1957." The most important NMR applications in chemistry were structural elucidation, bond characterization, reaction dynamics, and solid-state studies. See Wertz, "Nuclear and Electronic Spin Magnetic Resonance."

method found by Purcell and Bloch enabled them to study the effect in solids and liquids, and simplified the experimental setup.[2] But still in 1952, when the inventors received the Nobel Prize in physics, only a few laboratories had chosen NMR as a major focus of their activities. At Harvard, where Purcell was joined by a number of graduate students and co-workers, the focus was the explanation of unsolved problems in nuclear magnetic resonance itself, as well as its application to structural problems in the liquid and solid state. Crucial theoretical input came from other members of the physics department at Harvard. At Stanford, Bloch and his group applied NMR to the precision measurement of nuclear magnetic moments. Motivated by Russell Varian, one of the founders of the high-tech firm of Varian Associates at Palo Alto, Bloch patented his method, and members of his group (many of them later employed by Varian Associates) improved the technology and the applicability of NMR to a large range of problems. In the geographical middle area between the East coast and the West coast, at the University of Illinois in Urbana-Champaign, a new center for NMR developed in the late 1940s. There, the physicist Charles P. Slichter and the chemist Herbert S. Gutowsky were heavily influenced by the conceptual basis and instrumental methodology of the Harvard group. Independently, the physics graduate student Erwin L. Hahn developed NMR methodology and later accepted a postdoc position at Stanford, thus serving as a bridge to the West coast. In addition to these three groups at Harvard, Stanford, and Illinois, the magnet expert Francis Bitter from the MIT Laboratory of Electronics in Cambridge, Massachusetts, early on contributed to research in the field. At Oxford, England, the physicist Bernard V. Rollin in 1946 and the physical chemist Rex Richards in 1948 began to adopt NMR methods. If we neglect laboratories only interested in the more or less routine applications of NMR for nuclear physics, these laboratories we have mentioned were the major research groups that were active in the field prior to 1952.[3]

Members of these groups made the decisive observations, inventions, and theoretical explanations that greatly extended the use of NMR for the solution of research problems in the chemical sciences. In 1949–1950, three scientists independently of each other characterized the chemical shift, demonstrating the impact of chemical structure on the values measured in NMR experiments.

---

[2] The early history of NMR is treated in Rigden, "Birth of the Magnetic Resonance Method"; Rigden, "Quantum States and Precession"; Gerstein, "Purcell's Role." The history of the molecular beam method is described by Forman, "Molecular Beam Measurements"; and Ramsey, "Early History of Magnetic Resonance."

[3] For the history of chemical NMR see Andrew and Szczesniak, "Historical Account of NMR"; Becker, Fisk, and Khetrapal, "The Development of NMR"; Emsley and Feeney, "Milestones"; Freeman, "Fourier Transform Revolution"; Lenoir and Lécuyer, "Instrument Makers and Discipline Builders"; Shoolery, "NMR Spectroscopy in the Beginning"; and Waugh, "NMR Spectroscopy in Solids."

In 1951, three different groups observed the fine structure of the NMR spectrum, the so-called spin-spin coupling. The realization of these phenomena led to high resolution NMR and was a necessary, but not sufficient, condition for the subsequent use of the technique in chemistry. The conceptual basis of this development emerged in the interface between physics and chemistry, and philosopher of science Henk Zandvoort has shown that the research program in nuclear magnetic resonance relied on its "extrinsic success," thus emphasizing its role in disciplines outside the original research field of nuclear magnetic resonance itself.[4] Expanding on Zandvoort's results, this chapter shows how the connection was made between nuclear physics and chemistry in the hybrid discipline of chemical physics and its experimental part, molecular spectroscopy.[5]

The early NMR scientists worked in the tradition of the experimental style of molecular spectroscopy, a diverse area of techniques based on the interaction of radiation with matter. These techniques were rooted as well in quantum physics and electronic instruments as in chemical structures and reactions. In this context, NMR was one of several microwave and radiofrequency techniques that entered molecular spectroscopy during World War II after technological achievements in radar research. Especially one molecular spectroscopist, Herbert S. Gutowsky, played a crucial role in the realization of the chemical shift and the observation of spin-spin coupling.

Herbert Sander Gutowsky was born on 8 November 1919 in Bridgman, Michigan, as the second-youngest of seven children; his parents were immigrants from Germany. Gutowsky grew up on a farm; later the family moved to Hammond, Indiana, where his father bought a gasoline station. The Depression made the financial situation of the family difficult; Gutowsky supported himself by driving a newspaper delivery truck during the last year of high school. With the help of an older brother he was able to attend college at Indiana University in Bloomington, where he was fascinated by astronomy, which was taught by Frank Kelley Edmondson. Gutowsky became Edmondson's undergraduate assistant for three years. The enthusiasm and the kind of social interaction that astronomers engaged in while doing their research attracted him, but economical considerations led him to concentrate on chemistry. Though, with hindsight, he regarded chemistry at Indiana University as being second rank, he benefited from a course in theoretical chemistry that Fred Stitt, a former Ph.D. student of Linus Pauling, offered in Gutowsky's senior

---

[4] Zandvoort's work focuses mainly on Imre Lakatos's concept of scientific research programs. Intrinsic success is defined as explanatory and heuristic success of the theoretical content of a research program. Zandvoort, *Models of Scientific Development*; Zandvoort, "Nuclear Magnetic Resonance." Zandvoort developed his philosophical model further in "Concepts of Interdisciplinarity."

[5] For the the beginnings of NMR in chemical physics, and Gutowsky's early work at Harvard see Reinhardt, "Chemistry in a Physical Mode."

year. In 1940, Gutowsky graduated and moved on to the University of California at Berkeley. There, he considered himself a misfit, mainly for social reasons. Moreover, his research with Willard F. Libby on isotope separation, using Geiger counters and a mass spectrometer, did not proceed well. He saw the outbreak of the war as an opportunity to leave an unpleasant scene; during the next four years he worked for the Los Angeles sub-branch of the San Francisco District of the U.S. Army Chemical Warfare Service. His responsibility was the sub-contracting and the supervising of small companies that produced chemical weapons. After the war, he returned to Berkeley and finished a master's thesis under the supervision of the chemical physicist Kenneth S. Pitzer. Later on, Gutowsky decided to go to Harvard for his Ph.D. studies, and worked with E. Bright Wilson and George B. Kistiakowsky, at first on a topic in infrared spectroscopy. At Harvard, he became fully acquainted with the research style of chemical physics, and was the first chemist to use the novel technique of NMR for the elucidation of molecular structure. He graduated from Harvard in 1948 with a combined study of infrared and NMR spectroscopy, and in the same year accepted an instructorship at the chemistry department of the University of Illinois in Urbana.[6]

As one of the few chemists active in NMR around 1950, and through his experience with the already established technique of infrared spectroscopy, Gutowsky applied NMR in chemistry even before the first commercial instruments appeared on the market. He worked in the confines of the experimental culture of chemical physics, with its crucial feature of the control of the instrument, in both its theoretical and practical ramifications. Most molecular spectroscopists resisted the temptation to use commercially fabricated spectrometers and chose the harder way of designing and constructing instrumentation by themselves. Moreover, although often unsuccessfully, they tried to base their findings on quantum physical theory, mainly through collaboration with physicists. Thus, in Gutowsky's, and many others', research programs, the instrument became a machine for constructing the new. Through NMR, novel problems arose that were worth pursuing as they suddenly got in reach of answers. Gutowsky knew about the chances and dangers brought about by a novel research technique:

> At Harvard I learned if you set out to find something particular and you focused on what you are trying to find, you'll lose, you'll lose because more

---

[6] Reinhardt, "Chemistry in a Physical Mode"; Herbert S. Gutowsky, interview by Carsten Reinhardt, 1 and 2 December 1998. See also Lee, "Nuclear Magnetic Resonance," 27–37. For Gutowsky's early work see Gutowsky, "Coupling of Chemical and Nuclear Magnetic Phenomena"; Gutowsky, "Chemical Aspects"; and Gutowsky's application to the Research Corporation, without date [September 1948], copy in Papers of George B. Kistiakowsky, HUGFP 94.8, box 5, folder "Gutowsky, H. S.," Harvard University Archives.

> often than not you won't find what you think should be there. And if something is there which is projected it usually won't be that interesting.[7]

This searching for the unexpected, while looking at the expected, took place at the borderlines between chemistry and physics, where Gutowsky's small team of chemists and electronic engineers investigated the interactions of chemical substances with electromagnetic radiation. Skillfully, and based on Gutowsky's experience with service work in infrared spectroscopy, they transferred their results from the closely-knit group of molecular spectroscopists to the wide-ranging discipline of organic chemistry. Arguably, it is the team aspect of Gutowsky's work during the crucial period of 1949–52 that is the most remarkable aspect of this story. But teamwork was not a guarantee for immediate success: Chemical experience, when combined with know-how in electronics, as often prevented the recognition of novel effects as it paved the way for new insights and applications.

This chapter is a detailed study of the work of a small group of scientists during a few years only. It is neither meant to give a complete account of the early history of NMR nor does it deal with all research projects of the group in this period. It is written to give insight into some of the inner working modes of science, especially how Gutowsky and his group tailored NMR to the needs and expectations of the science of their time, and should be read as an exemplary, but not as a typical, case. It deals first with the construction of the NMR spectrometer, and Gutowsky's initial trials to tie the instrument's performance to the working modes already established in the physicists' laboratories. In the second part it is shown how the quest for precision and the examination of anomalies led to the discovery and stabilization of effects that formed the basis of chemical applications of NMR. This was achieved only through importing chemical concepts and research problems into the practice of molecular spectroscopy. Correlating NMR parameters with accepted notions of chemistry made the applicability of the technique in chemical research possible at all. The building of such correlations remained the main activity of Gutowsky throughout his career.

Assembling the Apparatus

In the late 1940s, NMR spectrometers were self-built instruments, assembled from parts that the war technology had left behind and gadgets that a booming after-war economy provided in ample supply: radio-signal generators, amplifiers, and receivers as well as probes, cryostats, and magnets. For a chemist, even if he was familiar with the use of technical equipment, the con-

---

[7] Gutowsky, interview by Reinhardt, 1 and 2 December 1998.

struction of such a complex instrument was a major, but necessary, challenge. The advice that Gutowsky's former supervisor E. Bright Wilson published a few years after Gutowsky's first encounter with the setup of a complex research instrument fits well in this context: The experimenter had to know by himself about the qualitative possibilities of the equipment considered. If he did not do so, his efforts were prone to failure.[8] Gutowsky achieved this by technology transfer in both geographical and disciplinary terms: In 1948, he moved from Harvard University in Cambridge, Massachusetts, to the University of Illinois in Urbana-Champaign. At Harvard, he had become acquainted with infrared and NMR spectroscopic methods through his collaboration with chemical physicists and physicists, and the instruments in use were built by his collaborators. In Urbana, Gutowsky joined a chemistry department that was famous for its achievements in organic chemistry, and he built an NMR spectrometer from scratch. For Gutowsky, this meant foremost that he had to become familiar with electronic circuits and magnet design, and at least once he expressed the wish to know "more about electronics and less about beakers."[9] To improve his knowledge, Gutowsky often asked for advice from his former colleagues at Harvard, especially his former close collaborator and co-author, the physicist George E. Pake, who in the meantime had moved to Washington University at St. Louis, only a few hours drive from Urbana.

The reason that Gutowsky had been offered a position at the University of Illinois, thus at one of the biggest and most successful chemistry departments in the country, was not his knowledge of NMR, but his familiarity with infrared spectroscopy. At Harvard, Gutowsky had learned to handle the custom-built infrared spectrometer of Wilson's group. During World War II, infrared spectroscopy evolved from an arcane method to a routine technique for the study of a wide range of chemical questions. This had been achieved mainly because of the needs of the chemical industry during the war years and through the efforts of commercial instrument makers.[10] Still then, an experienced technician or scientist was needed for the operation of the instrument and the interpretation of the results. Roger Adams, an influential organic chemist and chair of the department of chemistry at Illinois, early on realized the potential of the novel instrumentation. The department of chemistry had been involved in the synthetic rubber program during World War II, a project that made extensive use of infrared spectroscopy. Thus, Adams was "convinced that spectroscopy was here to stay."[11] Around 1945, the department hired a spectroscopist from the University of Minnesota, Foil Miller, who was respon-

[8] Wilson, *Introduction to Scientific Research*, 69–71.

[9] Gutowsky to George E. Pake at Washington University, 26 December 1949, Gutowsky Papers, folder "N.M. I correspondence."

[10] Rabkin, "Technological Innovation in Science."

[11] Gutowsky, interview by Reinhardt, 1 and 2 December 1998.

sible for a Perkin-Elmer model 12A, the first infrared spectrometer built by this company (soon upgraded to the model 12B). Nevertheless, the measurements were tedious: "One had to work hard to run and plot two spectra a day."[12] In 1948, Miller left Urbana for the Mellon Institute at Pittsburgh. It was understood that his successor Gutowsky was assigned to take care of this infrared spectrometer, and, with the help of a technician, to undertake the service work for the department.[13] He was supported by the help of an experienced and research-oriented technician, Elizabeth Petersen. In the years to come, the infrared work was a welcome fill-in before the NMR program led to results, and Gutowsky continued to publish in the infrared field.[14] This work provided Gutowsky with an intimate knowledge of the needs and expectations of organic chemists. Furthermore, especially the modes of representation of infrared research gave important hints for the early development of NMR. Thus, at the very beginning of his career at the University of Illinois, Gutowsky depended on the service work that he did in the field of infrared spectroscopy for the organic chemistry division, and a balance between the needs of the organic chemists and the requirements of his own research had to be found.[15] The availability of the infrared spectrometer allowed him to sustain a research program in this area, especially between 1948 and 1950, when his upcoming NMR project did not produce any publishable results.[16]

Immediately after he arrived in Urbana late in the night of 7 September 1948, Gutowsky's concern was the fund raising for his research project, a problem that he solved with astonishing rapidity. On 14 September his local mentors, among them the chemistry professors Adams and Frederick Wall, approved his research plan, and on 6 October Gutowsky received the news that the Graduate Research Board of the University of Illinois had granted $5,000 for the equipment needed.[17] Especially Wall, a well-known physical chemist and recipient of the 1945 American Chemical Society Award in Pure Chemistry, strongly backed up the young instructor, and arranged an introduction to Louis N. Ridenour, dean of the graduate college and professor of physics at the University of Illinois. This contact soon proved very valuable: Ridenour pro-

[12] Miller, "Infrastructure." See also Nowicki, "A Tribute to Foil Miller."

[13] The contact may have been provided by E. Bright Wilson at Harvard, who had exchanged research problems with Miller. Wilson to Gutowsky, 30 September 1948, Gutowsky Papers, folder "I.R. $Hg(CH_2)_2$ and Zn $(CH_2)_2$: CORSP."

[14] Of Gutowsky's four articles published in 1950 that were based on research at the University of Illinois, three concerned research in infrared spectroscopy.

[15] On 15 August 1949, he wrote to Wilson that the "new lab for our Perkin-Elmer is just being finished and with new KBr and LiF prisms I hope to get more work out of it on physical problems rather than organic." Gutowsky Papers, folder "I.R. $Hg(CH_2)_2$ and $Zn(CH_2)_2$: CORSP."

[16] See Gutowsky, "Coupling of Chemical and Nuclear Magnetic Phenomena," 363.

[17] Herbert S. Gutowsky, laboratory notebook "HSG research notes '48–'50," p. 37, Gutowsky Papers.

vided the university's grant, and gave advice with regard to suppliers of equipment. The one precondition Ridenour had was that Gutowsky should apply additionally for a grant from the Research Corporation, and in case Gutowsky got this money, he should pay back the university.[18] Gutowsky's application to the Research Corporation was successful, and his good fortune may have been rooted in the enthusiastic statement that his former Ph.D. supervisor at Harvard, George B. Kistiakowsky, noted physical chemist and later science advisor to the U.S. president, gave on the prospects of Gutowsky's research. In his letter of support to an officer of the corporation, Kistiakowsky even compared NMR with one of the most successful physical techniques of the 1930s and 1940s, electron diffraction:

> I really do believe that the application of nuclear magnetic resonance to the investigations of problems of molecular structure and the nature of solid states is one of the most promising, if not the most promising, methods that are now being developed by physical chemists. Someday it may easily rank in importance with electron diffraction.[19]

Thus, Kistiakowsky thought of NMR as a technique in the traditional field of molecular spectroscopy. In accordance with this view, Gutowsky presented his program as a continuation of already proven applications of NMR and included both a description of the technical means necessary and a plan for their subsequent use.[20] Drawing on the research style common in molecular spectroscopy, he intended to apply NMR to the unraveling of molecular structures of solids, both empirically and from theoretical perspectives. While this followed established lines of research, the investigation of nuclei other than hydrogen and fluorine was a new feature of the program. Much later, Gutowsky ascribed his success in the starting phase of his research at the University of Illinois to the lessons he had learned during his stay at the Army: The following-up of complex technical and scientific issues, the coordination of sub-contracted work, and the efficient writing of research proposals.[21] He

[18] The Research Corporation, one of the biggest nationwide private foundations that supported research in the physical sciences, had been founded in 1912 on the financial basis of Frederick Gardner Cottrell's patents on electric precipitators for the cleaning of smokestack gases. Among the most famous projects funded by Research Corporation was the cyclotron of E. O. Lawrence at the University of California at Berkeley. Research Corporation, "Research Corporation"; Bowden, *Chemical Achievers*, 162–164; Cameron, *Cottrell: Samaritan of Science*, 162–168, 279–318.

[19] Kistiakowsky to Charles H. Schauer, associate director of Research Corporation, 21 December 1948, Papers of George B. Kistiakowsky, HUGFP 94.8, box 5, folder "Gutowsky, H. S.," Harvard University Archives. For Kistiakowsky's biography see Dainton, "George Bogdan Kistiakowsky."

[20] H. S. Gutowsky, application to the Research Corporation, without date [September 1948], copy in Papers of George B. Kistiakowsky, HUGFP 94.8, box 5, folder "Gutowsky, H. S.," Harvard University Archives.

[21] Gutowsky, interview by Reinhardt, 1 and 2 December 1998.

urgently needed this experience: Though Gutowsky later characterized the NMR spectrometer as being only "moderately complex,"[22] the instrument was a quite difficult assembly of devices.

In 1948, the theoretical basis of the nuclear magnetic resonance experiment in bulk matter was well understood, but the measurement of the resonance effect was still difficult to achieve experimentally. A sample, containing atomic nuclei that possessed a magnetic moment, was placed inside a strong magnetic field. The magnetic field caused some of the nuclei to precess around the direction of the applied field with a certain angular velocity. Quantum mechanical theory led to the expectation of a splitting of the energy levels of the nuclei. The value of this energy difference depended on the types of nuclei and the strength of the magnetic field applied. If the frequency of applied electromagnetic radiation matched this energy difference, absorption or emission of energy occurred: The nuclei were said to be in resonance with the electromagnetic field. For practical reasons, the frequency was kept constant and the magnetic field was altered until resonance was reached. This "sweep" procedure of the magnetic field was achieved with separate coils that exerted a low electric current, called the field bias current. The experimental control of the phenomenon was achieved mainly through the experiences and materials of wartime radar research (nearly all of the scientists involved in the first observation of NMR in bulk matter had participated in radar research during the war).[23] Basically, an NMR spectrometer consists of a radiofrequency (rf) source, a magnet, and an rf detector. In Urbana, Gutowsky tried to emulate George Pake's design for studies on molecular structures in solids at Harvard (see Figure 2.1).

For the magnet, Gutowsky relied on established companies and his own capabilities at learning the necessary theory. From the possible suppliers of permanent magnets, Gutowsky chose Arnold Engineering Company of Marengo, Illinois. Arnold was a major contractor for military equipment during World War II, and specialized in the manufacturing of Alnico permanent magnets, with broad use in all kinds of appliances ranging from ore separators to microphone components. This was exactly the sort of magnet Pake was using in his spectrometer designed at Harvard, with pole pieces made of Alnico V by the Indiana Steel Company. Both companies were approached by Gutowsky, but they had no particular know-how in designing magnets with excellent field homogeneity, the most important feature of an NMR magnet. This was left to Gutowsky to sort out, conferring with his former colleagues at Harvard and

[22] H. S. Gutowsky to R. L. Poynter, Research outline "Nuclear magnetic resonance line-width transitions in deuterium compounds," probably October 1949, Gutowsky Papers, folder "Res. UG R. L. Poynter."

[23] See Rigden, "Quantum States and Precession"; Forman, "Swords to Ploughshares"; Ramsey, "Early History of Magnetic Resonance"; and Zandvoort, *Models of Scientific Development.*

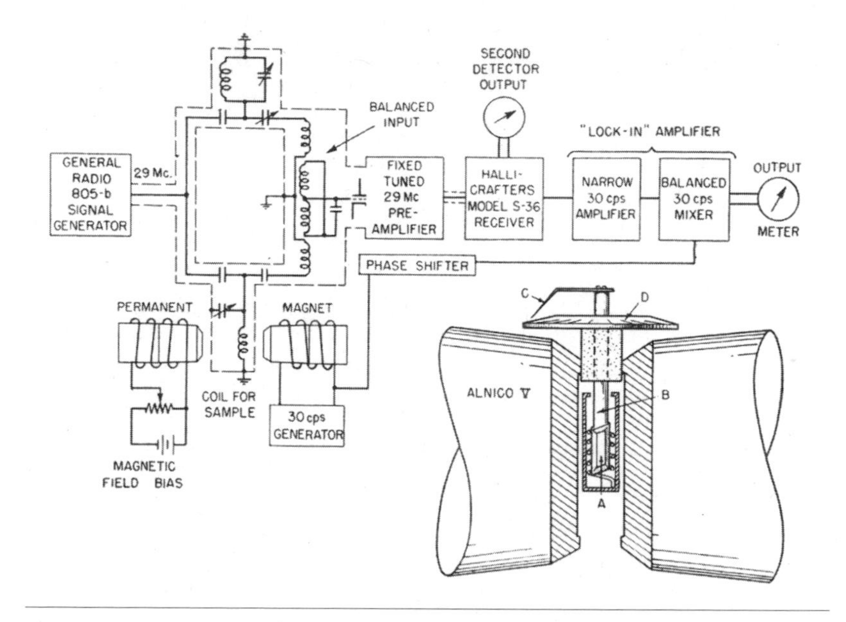

FIGURE 2.1 Block diagram of the NMR spectrometer that George Pake designed in 1947/48 at Harvard. Reprinted with permission from George E. Pake, "Nuclear Resonance Absorption in Hydrated Crystals: Fine Structure of the Proton Line," *Journal of Chemical Physics* 16 (1948), 327–336, fig. 3 on 329. Copyright 1948, American Institute of Physics.

with a local contact person that Ridenour had introduced, a former employee of Indiana Steel. "The [magnet] design wasn't that much of a chore. What was more of a chore was the radiofrequency bridge and the amplifier. Well, what I did of course was to start out with what I knew something about it, pragmatism, sure."[24] Electronic devices were delicate to handle, and sometimes hard to obtain with the specifications that were necessary for an NMR spectrometer. Well aware of this fact, Gutowsky spent a lot of time and energy exploring Urbana's university campus looking for assistance, which he found on 13 October in William Graydon Albright of the electrical engineering department: "He's the boy to cultivate. Should be able to get his group to build 30 cycle gen. and 30 cycle lock-in amplifier circuits. Can tap him also for misc.

[24] Gutowsky, interview by Reinhardt, 1 and 2 December 1998. See Arnold Engineering Company, "Group Arnold History"; and H. S. Gutowsky, laboratory notebook "HSG research notes '48–'50," pp. 21–26, passim, Gutowsky Papers. Alnico stands for an alloy made of aluminum, nickel, cobalt, and iron.

items on loan sporadically."[25] In the following months, Gutowsky often made good use of Albright's help.

The radiofrequency source, a 805-c Signal Generator (shown at far left of Figure 2.1, here the 805-b model that Pake used), was manufactured by General Radio Company of Cambridge, Massachusetts. The setup used by Gutowsky contained only a single coil for both the emission and the detection of radiation. To obtain a high signal-to-noise ratio of the small signal-voltage induced, a bridge circuit was used. This bridge balanced out the extraneous noise and allowed only the voltage change due to the absorption of energy in the sample to be amplified and detected.[26] The field of the permanent magnet could be altered moderately about a range of approximately 65 gauss with the help of the magnetic field bias coils (to the lower left). This was necessary because the radiofrequency was fixed, and resonance had to be reached through the variation of the magnetic field. The 30 cps (cycles per second, the modern unit is hertz) generator on the right side of the magnet belonged to the detection part of the system, allowing the necessary amplification of the signal through the modulation of the field. The modulation technique became a common feature among spectroscopic methods immediately after the war, and was not restricted to NMR alone.[27] Detection was achieved either with the help of an output meter, shown on the right, or an oscilloscope not shown. Robert H. Dicke at the Radiation Laboratory had designed the phase-sensitive lock-in amplifier during the war.[28] It allowed for a better signal-to-noise-ratio of the output meter, when compared with the oscilloscope. Gutowsky acquired this ingenious detection device through his own kind of technology transfer, and the help of the local engineering department, benefiting from the Ph.D. thesis of one of Purcell's early collaborators at Harvard, Nicolaas Bloembergen: "Took Bloembergen's thesis over to Albright in EE [Electrical Engineering] and asked him if he could build 30 ~ generator and lock in amplifier. Says can do and rel. soon (month or so) and for $100 for both . . . . Would be on a cost and labor basis."[29] Bloembergen, who had arrived at Cambridge in 1946 from the Netherlands, was the first student of Purcell, and made seminal contributions to the theory of NMR, most importantly the explanation of relaxation

[25] Gutowsky, laboratory notebook "HSG research notes '48–'50," p. 43, Gutowsky Papers. William Graydon Albright (1916–2000) was from 1947 on professor of electrical engineering at the University of Illinois. I thank Robert T. Chapel and William J. Maher from University of Illinois Archives for this information.

[26] See, for a detailed description, Bloembergen, Purcell, and Pound, "Relaxation Effects."

[27] See Wilson, *Introduction to Scientific Research*, 102–104.

[28] For detailed accounts on the importance of Dicke's lock-in amplifier see Forman, "Swords into Ploughshares," 435–442; Forman, "Lock-in Detection/Amplifier"; and Dicke, "Measurement of Thermal Radiation."

[29] Gutowsky, laboratory notebook "HSG research notes '48–'50," 28 October, p. 51, Gutowsky Papers.

effects and line widths. As is seen here, his experimental set-up also served as a model for subsequent work.

The preamplifier shown at the center was more difficult to obtain. On 13 October 1948, Gutowsky "spent a.m. and part of p.m. trying to choose damn pre-ampl. and also see if any of the electronic circuits might be available locally." Though he was successful with both, again with the help of Albright, he did not know that the preamplifier used in Pake's instrument at Harvard was a low-noise device that had originated at the MIT Radiation Laboratory during the war.[30] After he had become aware of this during a visit to Cambridge on 16 October,[31] he decided to try to get one of those circuits. But his sponsor, Ridenour, soon lowered his hopes in this direction because of the shortage of available funds. He suggested the modification of the amplifier available in Urbana in order to obtain an improved noise level: Ridenour "mentioned not happy at thought of my spending a lot of money and then leaving! I said perhaps if left I could take equip. along! This also didn't make him happy."[32]

On 30 November, Gutowsky went ahead and ordered the final piece of his equipment, an event he described in sober words: "Called Albright for his opinion on SX-42 receiver (good enough, some better but not enough to justify cost . . .)."[33] This receiver (it replaced the S-36 shown in the diagram of Pake) was part of the ham radio culture of the late 1940s. Gutowsky's choice fell on the new Hallicrafters SX-42, illustrated on the cover of the March 1947 issue of *Radio News* and promoted as the top product of the company. Gutowsky may have been attracted by the broad frequency range that the receiver covered, and he was not appalled by the quite high price, $275. In any case, it was a good choice from an aesthetic point of view: The SX-42 was a product of the famous industrial designer Raymond Loewy.[34]

Within three months, Gutowsky designed an NMR spectrometer and ordered the necessary parts of it. Crucial were his connections to his colleagues from Harvard, though local conditions were equally important. The contributions of the electrical engineering department of the University of Illinois and the service of the manufacturer of the magnet played a vital role in this, as did Gutowsky's ability to acquire the necessary knowledge of electronic circuits and

---

[30] An APQ-34. See Twiss and Beers, "Minimal Noise Circuits," 661–664.

[31] Gutowsky, laboratory notebook "HSG research notes '48–'50," 13 and 16 October, pp. 42-46, Gutowsky Papers.

[32] Gutowsky, laboratory notebook "HSG research notes '48–'50," 19 October, p. 47, Gutowsky Papers.

[33] Gutowsky, laboratory notebook "HSG research notes '48–'50," 30 November, p. 59, Gutowsky Papers.

[34] See the title page of the March 1947 issue of *Radio News*, where the receiver is shown in a quasi-colonial setting with a racist connotation. From http://antiqueradio.org/rn47031.htm, accessed 15 March 2004. For Loewy, see Schönberger, *Raymond Loewy*.

magnet design. Moreover, Gutowsky assembled the spectrometer from parts whose origins ranged from hand-made devices arising from war-related research at the MIT Radiation Laboratory, through high-tech gadgets available from specialized manufacturers, to commercial products generally available to the public. This underscores the importance attributed to wartime radar research, as pointed out by historian Paul Forman.[35] Certainly, a typical chemist of the time did not have the necessary knowledge of electronics to get started without access to technology from physics and electrical engineering departments. Scientists in countries that were heavily affected by the outcome of the war, and perhaps subject to research controls in the field of radar, had severe disadvantages compared with their colleagues in the United States and Great Britain. Even inside the community of American physicists, the delicate nature of the measurements occasionally made results hard to reproduce. For example, though the members of Francis Bitter's group at MIT were highly experienced in the construction of high performance magnets, they nevertheless needed the personal help of members of Purcell's group to get the NMR experiment started.[36] Also Gutowsky, though he had gained experience of how to undertake measurements during his time with George Pake, in the beginning could not repeat the results that he had achieved earlier at Harvard.

Searching for Resonance

On 29 September 1949, Herbert Gutowsky received news on the fate of the long awaited magnet, the core item of his NMR spectrometer:

> 2:30 PM Brand phoned. Saturation value for field for magnet is 6225 gauss. This [is the] consequence of their not properly estimating leakage. Brand suggested could increase diameter at base, use new tubes and coils. Estimated would require four weeks to do. Thereby could easily bring field up. Magnet should be stabilized 3% from sat[urate]d value, which means about 6050 gauss in this instance. This gives a proton res. of ~ 25.7 MC compared to 29 MC for 6820 that Pake used. F^{19} would be at 24.1 instead of 27.2. The preamp should be tuned to ~ 25, at least centered there and tunable +/–1 MC.[37]

---

[35] Forman, "Swords into Ploughshares."

[36] Nicolaas Bloembergen, interview by Joan Bromberg, Paul L. Kelley, 27 June 1983, p. 6. Niels Bohr Library, American Institute of Physics, College Park, MD, USA (AIP).

[37] Gutowsky had contracted the construction of the magnet to Arnold Engineering (see above), and had asked for a field strength of about 6.500 gauss. Thus, the news that Arnold could only reach 6.050 gauss came as a disappointment. Because the resonance frequency depends on the magnetic field strength, the values for the proton and fluorine resonances were lower than with Pake's spectrometer. Gutowsky used MC (megacycle) as unit for the frequency, which has the same number when given in megahertz. Gutowsky, laboratory notebook "HSG research notes '48–'50," p. 81, Gutowsky Papers.

This message marked the end point of a year-long period of planning, a time when Gutowsky began to set up a research program in the chemical aspects of NMR, though he lacked the most crucial part of it: the instrument. Gutowsky must have anxiously waited for the call that the engineer of Arnold Engineering, Charles Brand, made, explaining what had gone wrong in the final stage of construction. Despite the fact that the reported field strength, approximately 6,050 gauss, was lower than he originally hoped for, Gutowsky gave the order to continue. He could not afford to lose more time. A year had passed since he joined the faculty of the University of Illinois. What he needed quickly was a working instrument, a machine for producing data, ready for interpretation based on the familiar background of the theories developed at Harvard. A small group of young students had already gathered around him. Charles J. Hoffman, a graduate student who had specialized in inorganic chemistry, was working on the infrared spectrum of disilane. Robert L. Poynter, an undergraduate who had worked on electronics in the physics department, contributed to the complex electronic circuits that had to be built and repaired. In early October, Gutowsky offered him a part-time job for service work, mainly in electronics, but also other duties such as computing.[38] Electronics was also the specialty of Saul B. Yochelson, probably a student of electrical engineering, who on 6 October was hired as assistant,[39] the day before Gutowsky started a laboratory notebook dedicated to research in nuclear magnetism: "All work on and any measurement made with the nuclear magnetism equipment are to be recorded herein. Entrys [sic] are to be titled, dated and initialed. Use ink. Start page 10."[40] This was to become the handbook of the group for the set-up and the operation of the equipment. More or less, it was a manual for an apparatus under construction, and the members of the group learned some lessons the hard way in the months to come. In the beginning, mainly Poynter and Yochelson were in charge of the final set-up of the spectrometer, and this was shown to be much more complicated than Gutowsky had thought:

> I knew enough about my weakness in electronics that in the budget for the Research Board that provided funding to get the research going and operating, I allowed for a part-time graduate student skilled in electronics and the first guy I got—well there was a scarcity of people who were knowledgeable—and it took us, I hate to admit it, but it took us probably two months, maybe three months, to get our first broadline apparatus functioning.[41]

---

[38] Notes of Gutowsky, 7 and 8 October 1949, Gutowsky Papers, folder "Res. UG R. L. Poynter."

[39] Gutowsky, laboratory notebook "HSG research notes '48–'50," entry of 9 October 1949, p. 84, Gutowsky Papers.

[40] Gutowsky, laboratory notebook "Nuclear magnetism I," 7 October 1949, p. 1, Gutowsky Papers.

[41] Gutowsky, interview by Reinhardt, 1 and 2 December 1998. Broadline refers to the shape of lines in the spectrum when NMR was used for the study of solids.

As a consequence of an erroneous value given by Arnold Engineering for the magnetic field, the searches for resonance of Gutowsky and his group were in the wrong frequency region. In the beginning, the group had problems determining the exact magnetic field strength, normally measured by the flip-coil method, a complicated procedure that did not easily lead to reliable results. Only after Yochelson measured the field with a borrowed gauss meter from the electrical engineering department, showing it to have the value of ca. 6,400 gauss, did the members of the group have their first success. Gutowsky found a hydrogen resonance on 20 October, but still reported problems with the electronic equipment.[42]

In the following weeks, especially Yochelson was busy with improving the electronics. Though Gutowsky achieved a seemingly reliable calibration briefly after Christmas of 1949, there were still problems with excessive noise in the equipment. The instrument was very sensitive, and Gutowsky set careful rules for the detection of reliable data.[43] But all this search for precision, the tinkering, and the experience gained, did not help much. With the noisy electronic equipment, precision was not to be obtained. On 26 December, while in the middle of calibration efforts, Gutowsky wrote to Pake:

> We're struggling along here. We picked up a resonance shortly after I talked to you several weeks ago. Our main trouble was that the Arnold Engineering Co. claimed the field of the magnet was 6100 gauss when actually it is about 6400! Our lock-in and 30 cycle generator, using Bloembergen's thesis designs aren't working so hot. Part of the trouble seems to be that the reference voltage on the lock-in depends on the modulation used. We do have a cryostat that may be satisfactory from a temperature standpoint if the r.f. side can be worked out O.K.[44]

Unfortunately, the first months of 1950 did not bring much progress. Charles Hoffman, after being a year in Gutowsky's group, was frustrated by the fact that he had already spent two-thirds of his planned graduate research time. More and more, the infrared studies became his major topic, although from the start he had wanted to work on nuclear magnetism.[45] In May, Gutowsky had to admit his problems to the chairman of the department, Roger Adams, when he explained the change of Poynter's undergraduate research thesis with "delays in the development of essential electronic equipment."[46] But Gutowsky had

---

[42] Laboratory notebook "Nuclear magnetism I," 20 October 1949, p. 17, Gutowsky Papers.

[43] Laboratory notebook "Nuclear magnetism I," 28 December 1949, pp. 37-38, Gutowsky Papers.

[44] Gutowsky to George E. Pake at Washington University, 26 December 1949, Gutowsky Papers, folder "N.M. I correspondence."

[45] See the notes of Gutowsky of 24 January and 16 July 1949, Gutowsky Papers, folder "C. J. Hoffman 1."

[46] Gutowsky to Adams, 10 May 1950, Gutowsky Papers, folder "Res. UG R. L. Poynter."

something exciting to offer in exchange: Poynter's new thesis had the title "Chemical Shifts in the Magnetic Resonance of F^{19}."

This major change of research direction was caused by a visit to Cambridge. On 10 and 11 March, Gutowsky attended a conference on nuclear magnetic resonance, where scientists from MIT and Harvard discussed novel results in this rapidly growing field. In his notes, Gutowsky mentioned the usual discussions about high-performance equipment, and also included the work of William C. Dickinson of the MIT Research Laboratory of Electronics, published in the March issue of *Physical Review*. Dickinson had found that the resonance values of fluorine depended on the chemical type of compound in which the fluorine nucleus was located. At the same time, Warren Proctor and Fu Chun Yu in Bloch's group at Stanford found the same striking "chemical effect" in nitrogen compounds. Before that, in late 1949, Walter Knight from Brookhaven Laboratory had indicated that such shifts also might exist in phosphorus compounds.[47]

With the debates of the Cambridge meeting in mind, Gutowsky returned from the East coast to his "mid-western 'isolation'."[48] He immediately tried his hand at finding proton resonance shifts, the other three groups having found the effect with other nuclei. On 15 March, Gutowsky convinced himself that he could "determine frequency of resonance quite accurately by observing biasing current required to center line on scope." Keeping in mind the instability of his electronic equipment, he thought "that res[onance] shifts of the order of 0.10 gauss can be measured." Despite this prediction, the examination of a few hydrogen-containing compounds "gave no definitely val[uable] effects."[49]

In the following months, the group combined the complex process of improving the stability and accuracy of the instrument with measurements of shifts of various compounds containing hydrogen and fluorine nuclei. Gutowsky chose fluorine compounds because of their relative ease of examination, while hydrogen was the most important nucleus, as it was present in all organic chemicals. The improvement of apparatus and precision measurements were intertwined, and Gutowsky continuously compared his results with those of the other research groups in Cambridge, Stanford, Washington, D.C., and with the work of Erwin Hahn at the University of Illinois.[50] The existence of

[47] Knight, "Nuclear Magnetic Resonance Shifts"; Proctor and Yu, "Dependence"; and Dickinson, "Dependence." See Levine, "Short History of the Chemical Shift."

[48] For this expression see Gutowsky to Harold A. Thomas, 10 October 1950, Gutowsky Papers, folder "Magnetic shielding in H_2, H_2O, Min. O."

[49] Laboratory notebook "Nuclear magnetism I," pp. 59–61, Gutowsky Papers.

[50] Hahn was a graduate student at the physics department of the University of Illinois, and began to tackle physical aspects of NMR on his own. The first contact with Gutowsky happened to be in September 1948, when Gutowsky found out about the work of Hahn. Hahn first wanted to focus on the measurement of reaction rates, and probably because of this made contact with Gutowsky.

the shifts became an undisputed fact, for both theoretical and experimental reasons.

But at first the instabilities of the electronic equipment had to be dealt with, a problem soon taken care of by Robert McClure, a particularly able senior student in electrical engineering. Gutowsky was especially concerned about his apparent inability to reproduce the findings of Dickinson: "During past couple of months there has been an increasing amount of noise in the rig. This interfered with the attempts to measure resonance frequency shifts in compounds and generally reached the point where it had to be eliminated."[51] After painstaking efforts to establish the source of the electrostatic noise, it was found to originate in a burned-out set of contacts in a thermoregulator in a near-by room! On 25 March, Gutowsky could report that "the elimination of noise . . . is encouraging enough to try again to measure resonance frequency shifts in fluorine compounds."[52] Three runs of the experiment established measurable shifts between the reference compound perfluorolube oil (manufactured by Du Pont) and antimontrifluoride, a compound that had been already measured by Dickinson. Four days later, Gutowsky examined a sample of benzotrifluoride.[53] The average shift measured was 0.18 gauss, substantially higher than the instrument limits, estimated by Gutowsky to be ca. 0.10 gauss. Merck reagent-grade hydrogen fluoride, also measured by Dickinson, was the next target of Gutowsky's efforts. This provided the opportunity to compare directly the performance of Gutowsky's instrument with Dickinson's. In a first run, Gutowsky found the difference between the two compounds to be 0.58 gauss, much too low when compared with the value of 0.83 gauss given by Dickinson: "This seems a bit out of experimental error . . . . The general agreement, however, is encouraging."[54] Gutowsky had chosen a lower biasing current for the sweep of the magnetic field than he had used for the calibration, and he thought that this could be the cause of the difference between the two values. To the great satisfaction of Gutowsky, a second run with a higher biasing current gave an agreement "well within exptl. error." Thus, outside experimental data gave Gutowsky the necessary feedback to handle his instrumentation in the right way.

In mid April, Charles J. Hoffman started plotting the magnetic field to prepare the magnet for a better homogeneity. He used the reference compound for the fluorine shift measurements to determine the differences in magnetic field

---

He used pulsed frequencies for his work, and still in Illinois found the so-called spin echoes. See Hahn, "Pulsed NMR"; and Gutowsky to Pake, 27 September 1948, Gutowsky Papers, folder "N.M. ammonium salts—note."

[51] Laboratory notebook "Nuclear magnetism I," 25 March 1950, p. 62, Gutowsky Papers.

[52] Laboratory notebook "Nuclear magnetism I," p. 63, Gutowsky Papers.

[53] Laboratory notebook "Nuclear magnetism I," p. 66, Gutowsky Papers.

[54] Laboratory notebook "Nuclear magnetism I," 31 March 1950, p. 70, Gutowsky Papers.

strength in relation to the geometry of the magnet assembly. Thus, Gutowsky and his co-workers had already acquired enough trust in the stability of the electronic equipment to employ these data in the improvement of field homogeneity, mainly by adjustment of the geometry, and through polishing the surfaces of the magnet poles with emery paper. On the basis of these efforts, Gutowsky could report on 21 April that he was able to repeat and extend Dickinson's work on fluorine shifts. He believed that he had achieved an accuracy of +/–0.01 gauss and declared that he would try to find proton resonance shifts as well.[55] Gutowsky continued to work mainly on fluorine compounds, because he had "obtained a very wide selection of inorganic fluorine compounds from Finger—Geol. Survey as well as three new organic samples."[56] These organic samples included 2,3,5-trifluorobenzotrifluoride, whose examination two days later led to the remarkable result that two resonances could be seen, distinguishing the fluorine atoms in the same molecule (see Figure 2.2).

Gutowsky was able to see in a fluorine compound what Proctor and Yu already had reported for a substance containing nitrogen: resonance shifts of nuclei that were situated in the very same molecule. While Proctor and Yu examined a compound with ionic bonds, Gutowsky's sample contained a molecule with covalent bonds. Thus, the phenomenon of the chemical shift had been found in compounds representing all important bond types. Moreover, with the establishment of shifts in compounds containing either fluorine, nitrogen, phosphorus, or hydrogen, the general nature of the effect was proven.

Characteristically, Gutowsky checked his result with compounds that separately contained the functional groups of interest, benzotrifluoride and tetrafluorobenzene. Here, chemical reasoning entered the experiment. He found a small deviance with 2,3,5-trifluorobenzotrifluoride when compared with these, and sought for an explanation in terms of electronic influences that shaped the electron densities in the linkages of the fluorine atoms. The electronic theory of chemical valence was an important part of physical organic chemistry, and Gutowsky was well aware of its potential for the rationalization of his findings.[57] In hindsight, Gutowsky admitted that "at first we looked at every compound we could lay our hands on." But the results of 26 April "encouraged us greatly, and led me to think about how the shifts are related to molecular struc-

[55] Gutowsky to Dickinson, 21 April 1950, Gutowsky Papers, folder "Chem. shift in mag. resonance of F." Just one day later, Gutowsky made efforts to find proton resonance shifts. By taking into account the bulk diamagnetism of the sample, by careful tuning of the radio-frequency bridge, and by using the same techniques as with the fluorine compounds, he found "what may be real shifts." Laboratory notebook "Nuclear magnetism I," p. 104, Gutowsky Papers.

[56] Laboratory notebook "Nuclear magnetism I," p. 105, Gutowsky Papers. Glenn C. Finger was head of the Fluor-spar Division of the Illinois State Geological Survey.

[57] For a contemporary description of the electronic theory, mainly inductive and resonance effects, see Hammett, *Physical Organic Chemistry*, 194–196.

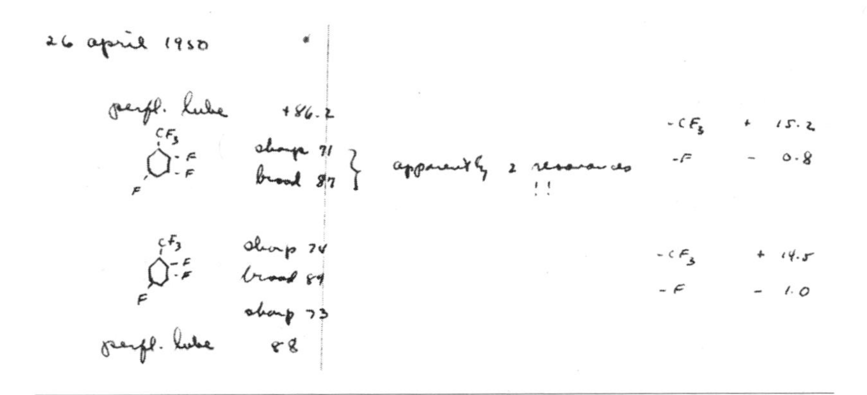

FIGURE 2.2 Entry in Gutowsky's laboratory notebook, 26 April 1950, p. 106, Gutowsky Papers. The examination of 2,3,5-trifluorobenzotrifluoride showed two resonances, representing the structurally inequivalent fluorine atoms in the molecule. A second run even led to three resonances, not commented on in the notebook, but a sign of the chemical shift differences between the three fluorine atoms bound directly to the benzene ring. From today's knowledge, four peaks should be visible (see also Figure 2.8).

ture."[58] Here, Gutowsky referred to the concept of electronic structure, i.e. the types of bonding in a molecule. The little deviation that Gutowsky found gave the first hints as to the causes, and consequences, of the observed phenomena. It seemed clear at the time that the shifts were connected to the electronic structures of molecules, in other words, that the electrons around the nucleus "shielded" the nucleus and that differences in shielding led to the shifts. The measurement of trifluorobenzotrifluoride was the embodiment of this theoretical reasoning, and it connected the phenomena to established theories and notions of physical organic chemistry.

The results achieved up to early August stimulated a "letter to the editor" that appeared in late 1950 in *Physical Review*. Knowing that "the complexity of the calculations permits their application only to the simplest molecules," the group set out to make "an experimental survey of the wide variety of existent polyatomic fluorine compounds to determine the influence of structural factors."[59] The choice of words in the article ("initial observations"), and a note written in October 1950 by Gutowsky to Harold A. Thomas of the National Bureau of Standards at Washington, D.C. revealed that the results given in the article were based on limited experimental evidence only. Thomas, as we will

[58] Gutowsky, "Coupling of Chemical and Nuclear Magnetic Phenomena," 364. Text identical in Gutowsky, "Chemical Aspects," 287.
[59] Gutowsky and Hoffman, "Chemical Shifts," 110.

see below, was interested in this matter through precision measurements of nuclear moments. It was precisely because Gutowsky saw the shifts as a valuable tool for investigating electronic structures of molecules that he could look at them with a totally different perspective. Gutowsky regarded the chemical shift not as a nuisance in the precision determinations of nuclear moments, but as an object of inquiry in its own right:

> My main interest in the matter has been in connection with the chemical shifts themselves. On the basis of some preliminary measurements on fluorine compounds, we postulated that in the simple covalent compounds (two different atoms only), the magnetic shielding of a given nucleus depends systematically on the position of the attached atom in the periodic table. In particular for atoms in a given period, the shielding decreases with z, while in a given group, the shielding increases with z.[60]

Gutowsky referred to the well-known ordering system of chemical elements, the periodic table. The atomic number z, which gave the number of protons in a nucleus, was a unique property of each element and was used for the establishment of order in the periodic table. The table was divided into groups, represented by vertical relationships of the elements, and periods, shown by horizontal relationships. While the number of electrons in the valence shells increased continuously in a given period, the elements in each group had the same number of electrons in the outer shell and thus showed a closely related chemical behavior. The linear relationship between the position of an element in the periodic table and the value of the chemical shifts of the fluorine nuclei attached to it gave hints to the refinement of the theory of the chemical shift. In addition, it indicated that the shifts were related to electronegativity, a prominent semi-empirical feature. The work on fluorine compounds, most of which Gutowsky did together with Charles Hoffman, afforded the important result that chemical shifts depended on the electronegativity of the element bound to the fluorine nucleus (see Figure 2.3). The electronegativity of an element was a measure of its ability to attract bond electrons, and consequently this observation was a proof for the postulate that the chemical shift was related to the electronic environment of the nuclei.

The search for shifts and improvements in the instrumentation were the most important and time-consuming tasks of Gutowsky, Hoffman, McClure, Yochelson, and Poynter in the spring and summer of 1950. These endeavors were interdependent. Any improvement in homogeneity meant an improvement in the accuracy of the measurements. At the same time, homogeneity was defined through the accuracy obtained. The circular argument involved in this

[60] Gutowsky to Thomas, 10 October 1950, Gutowsky Papers, folder "Magnetic shielding in H_2, H_2O, Min. O."

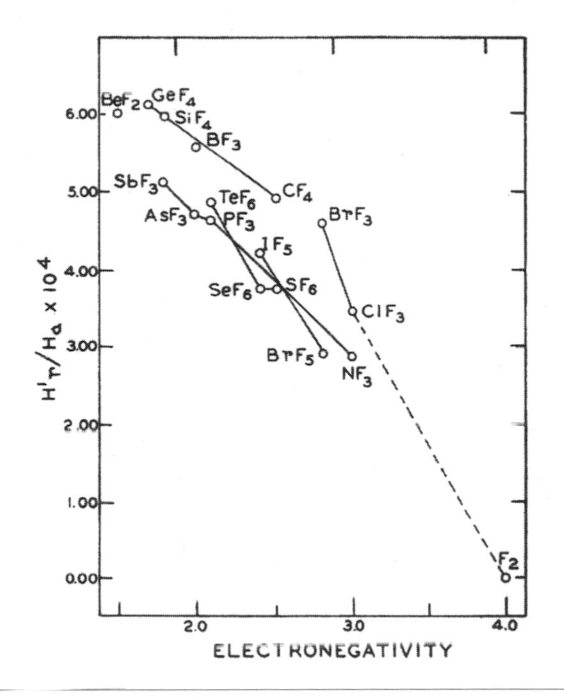

FIGURE 2.3 Correlation of fluorine resonance shifts with the electronegativity of the atoms bound to fluorine in the respective molecules. Reprinted with permission from Herbert S. Gutowsky and Charles J. Hoffman, "Nuclear Magnetic Shielding in Fluorine and Hydrogen Compounds," *Journal of Chemical Physics* 19 (1951), 1259–1267, fig. 6, on 1264. Copyright 1951, American Institute of Physics.

was broken by checks on the reliability of the instrument through correlations with outside data, mainly Dickinson's, and those obtained by Gutowsky's colleague in physics, Erwin Hahn. Moreover, Hahn used a different technique, a pulsed frequency spectrometer, and this enabled a cross check of the findings.[61] But reliance on outside data had its shortcomings, too: Gutowsky trusted Dickinson's results to such an extent that he even did not reproduce Dickinson's values for the reference compound, BeF_2 aq., until days after the submission of the manuscript for their first article on the subject.[62] Meanwhile, inside Gutowsky's group a division of labor was set in place. While Yochelson had worked mainly on the electronic equipment, his successor McClure also participated in the experiments. In addition, Gutowsky, Hoffman, and Poynter

[61] See laboratory notebook "Nuclear magnetism I," 1 May 1950, p. 125, Gutowsky Papers. See also Hahn, "Spin Echoes."

[62] Laboratory notebook "Nuclear magnetism I," entry of 8 August, p. 229, Gutowsky Papers. The manuscript was submitted on 3 August.

undertook the measurements, while Hoffman, with the help of Gutowsky, worked on the improvement of field homogeneity. In mid-1950, Poynter and especially Hoffman started with syntheses of compounds on their own. The complexity of the issues that the group was dealing with made necessary a cooperative working style, assisted by efficient means of communication. This guaranteed that the group could gain and sustain control over the instrument and the data obtained. The pages of the group's laboratory notebook show how this teamwork of specialists worked. Entries describing the synthesis of inorganic chemicals, most notably fluorine and hydrogenous compounds, were scribbled next to drawings of electronic circuits, measurement protocols showing rows of data, and plots of the geometry of the magnetic field. Building on the assumption that science is a craft-based undertaking,[63] Gutowsky's group can be seen as a team of scientists and technicians who each brought different qualifications and background skills to the task. Poynter, Yochelson, and especially McClure provided the necessary electronics expertise, and Hoffman had knowledge in the field of inorganic synthesis. Gutowsky later described a suitable assignment of tasks as a crucial feature of his research undertaking: "Along the line I developed a knack for getting people at the right job. If you put graduate students at the wrong problems they flounder and die. It takes some doing sometimes to get students in the right nitch."[64]

The entry of 5 August 1950 (Figure 2.4) shows the synthesis of phosphorustrifluoride, PF_3, a compound which would gain importance in the months to come because of the complex resonance pattern that it gave. Hoffman used the so-called Swarts reaction of phosphorustrichloride, PCl_3, and antimontrifluoride, SbF_3.[65] Antimontrifluoride was placed in the flask to the left and phosphorustrichloride was allowed to flow in through the still head funnel (which is shown in more detail to the far right). The gaseous mixture of products was cooled in the two baths shown in the middle. The first contained solid carbon dioxide, CO_2, to trap the impurities. The next bath, cooled with liquid nitrogen, collected the PF_3. Hoffman stored the product under vacuum in a glass bulb for later use in the NMR experiments. The glass apparatus drawn here is a typical example of apparatus used for organic and inorganic syntheses. Hoffman noted the boiling and melting points as well as the quantities of the raw products he used. Fluorine compounds were difficult to handle, and it made sense to describe the synthetic method in detail in case problems should arise. Nevertheless, in this case Hoffman could not prevent the loss of the larger part of the product he prepared. Moreover, because the reaction started

---

[63] Ravetz, *Scientific Knowledge*, 75–108.

[64] Gutowsky, interview by Reinhardt, 1 and 2 December 1998.

[65] Laboratory notebook "Nuclear magnetism I," entry of 5 August 1950, pp. 227-228, Gutowsky Papers. The literature cited by Hoffman is Booth and Bozarth, "Fluorination of Phosphorus Trichloride"; Yost and Anderson, "Raman Spectra."

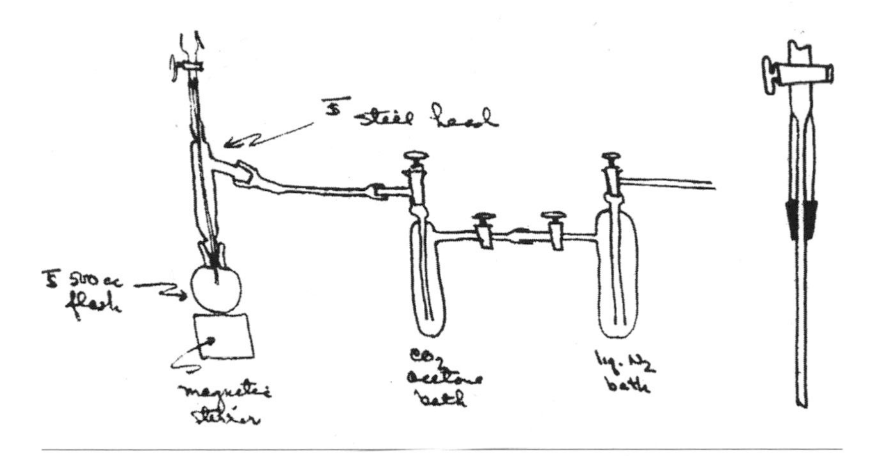

FIGURE 2.4 Entry in the laboratory notebook of Gutowsky and his group, 5 August 1950. The drawing made by Charles Hoffman shows the assembly of typical glassware apparatus for chemical synthesis, here for the compound PF_3, using the Swarts reaction. Laboratory Notebook "Nuclear Magnetism I," p. 227, Gutowsky Papers.

with phosphorustrichloride, the presence of chlorine-containing substances in the final product could not be excluded.

Twenty days later, Robert McClure constructed a preamplifier set for the resonance frequency of fluorine, 25.5 MHz, using the design of the MIT Radiation Laboratory, already familiar to the group since Gutowsky had brought news about the design used at Harvard back to Illinois in 1948.[66] McClure gave a careful description, especially of the changes that were made with respect to the published circuit. This item was probably already part of the instrument when measurements on fluorides were continued in early September. (See Figures 2.5 and 2.6.)

On 8 September, Hoffman and McClure decided to examine resonance shifts with several fluorine compounds. Because under normal conditions the substances were in the gaseous state they intended to use a cryostat to liquefy the sample, but this "was found to be useless since it prevented the detection of the resonances." As an alternative, they put the test tube, under vacuum, together with the probe coil in a small insulated tube filled with liquid nitrogen. In the case of PF_3, Hoffman and McClure made a surprising observation:

> PF_3 was condensed in probe and found to have a double resonance, when a resonance could be obtained. A steady resonance was unobtainable because of

[66] Laboratory notebook "Nuclear magnetism I," entry of 25 August 1950, p. 247, Gutowsky Papers. See Twiss and Beers, "Minimal Noise Circuits," 661–664.

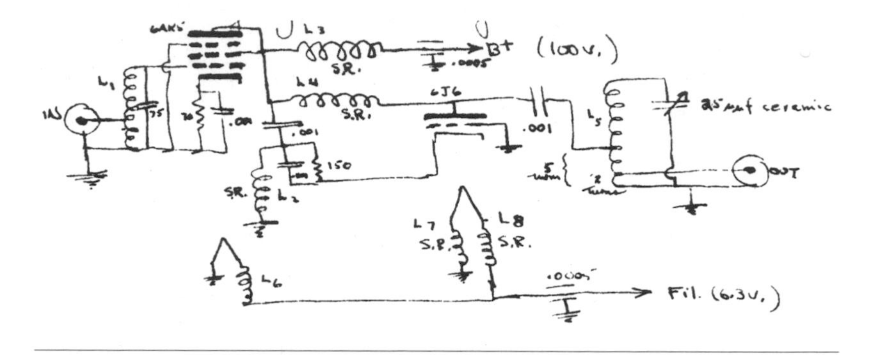

FIGURE 2.5 Robert McClure's drawing, 25 August 1950, of a preamplifier designed for the frequency of 25.5 MHz. Laboratory Notebook "Nuclear Magnetism I," p. 247, Gutowsky Papers.

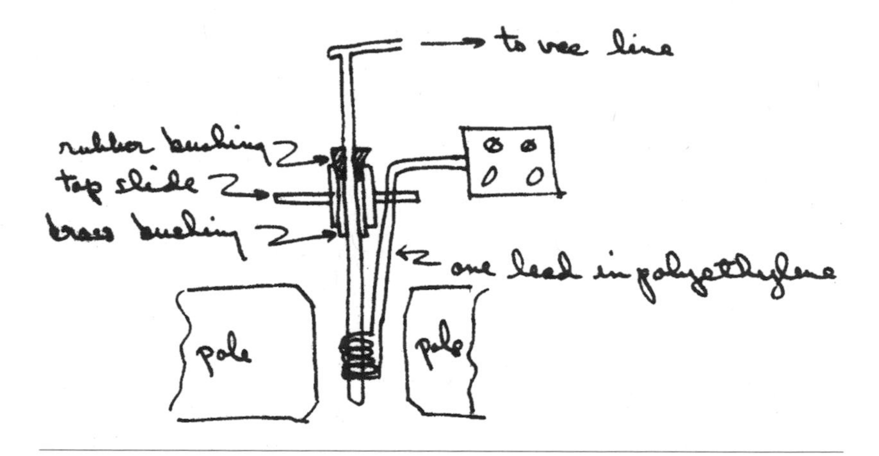

FIGURE 2.6 Hoffman's and McClure's set-up for the NMR experiment with PF_3, 8 September 1950. Laboratory Notebook "Nuclear Magnetism I," p. 277, Gutowsky Papers.

> the small quantity of sample. Double resonance indicates impurity probably $PClF_2$ or PCl_2F which are very likely to form when Swarts reaction is used.[67]

A doublet was not expected by theory and from previous experience. Because the primary product used for the synthesis was PCl_3, it was reasonable to explain the phenomenon through an incomplete reaction and the resulting

[67] Laboratory notebook "Nuclear magnetism I," entry of 8 September 1950, pp. 276, 278, Gutowsky Papers.

chemical impurity. Thus, the control the group exercised over the whole range of experiments that were leading to an actual resonance measurement seduced them to declare the double resonance found in PF_3 to be an artefact. A second trial planned three days later did not materialize because the vapor pressure of PF_3 led to the fracture of the heavy-walled glass tubes when sealed and warmed up to room temperature. A week later, Hoffman synthesized PF_3 using a recipe that excluded the presence of impurities containing chlorine.[68]

On 27 September, Hoffman and McClure found that the compound prepared in this way still gave a double resonance. This time, using a new cryostat and a larger amount of the sample, they could examine it quantitatively.[69] In addition, purer samples did not make the doublet disappear, and also phosphoruspentafluoride, PF_5, showed the clear result recorded by Hoffman and McClure on 10 November 1950: "Still two resonances."[70] While the group attributed the PF_5 double resonance to effects of its molecular structure (which proved to be wrong later), the result with PF_3 remained unexplainable. Charles Hoffman left the University of Illinois on completion of his graduate studies in early 1951 for Los Alamos and thus had no chance to follow this work up. Only following first reports by other groups of a similar phenomenon, Gutowsky became open for a different explanation. Finally, the anomaly led to the discovery of a new effect, spin-spin coupling (see below). This discovery was possible only after the group had obtained complete control over the spectrometer, and this was not achieved until the end of 1950.

Next to the integration of the group members' expertise, technical matters, such as calibration and calculation of possible errors, played an important role in gaining control. For example, when in late 1950 Gutowsky searched for a precise method to measure the proton resonances of hydrogen gas, mineral oil, and water, he very carefully considered the shape of the samples and the magnetic influences of the material of the containers. Gutowsky resolved data differences with a group at the National Bureau of Standards through an exchange of reference substances and a check on them.[71] In general, experimental data were exchanged quite frequently, and possible causes of differences were discussed.[72] Within Gutowsky's group, trust in the performance of the instrument

---

[68] Laboratory notebook "Nuclear magnetism I," entries of 11, 16, and 18 September 1950, pp. 280, 284–285, Gutowsky Papers.

[69] Laboratory notebook "NMR II. Research notes '50–'51," entry of 27 September 1950, p. 14, Gutowsky Papers.

[70] Laboratory notebook "NMR II. Research notes '50–'51," entries of 9 and 10 November 1950, pp. 100–102, Gutowsky Papers.

[71] Laboratory notebook "NMR II. Research notes '50–'51," 13 January 1951, p. 151. See also Gutowsky to Dickinson, 24 January 1951, Gutowsky Papers, folder "Magnetic shielding in H_2, H_2O, Min. O."

[72] See Gutowsky to Thomas 10 October 1950, Dickinson to Gutowsky, 21 December 1950, Gutowsky Papers, folder "Magnetic shielding in H_2, H_2O, Min. O."

was obtained also by careful observations of runs of measurements. As an example of October 1950 shows, when Gutowsky and Hoffman observed the proton shifts of arsenhydride versus mineral oil, the first run was completely discarded because of instabilities of the signal generator.[73]

Despite all these efforts, ongoing difficulties made clear that something was wrong with the calibration. Thus, in December 1950, months after the group had published a brief note describing their results, Gutowsky had to admit to Dickinson:

> This business of calibration error seems to be catching. We have discovered, finally, a ten per cent error in our calibration; previous figures of ours should be multiplied by 1.10. We have used the increment dial on our General Radio 805C signal generator to obtain the relation between change in frequency and biasing current and thereby field versus biasing current. We had also checked the signal generator versus a frequency standard set against WWV. The apparent agreement between the signal generator and the frequency standard, however, was the result of writing a 9 for an 8! I guess we were too certain the General Radio product couldn't be wrong![74]

For correcting their error, Gutowsky and Hoffman included a detailed description of design, measurement and calibration procedures in their first major publication on the subject.[75] The narration in the article by Hoffman and Gutowsky followed historical events, admitted failures, and pointed to remedies, the group wanting to regain the trust of the reader through this detailed reporting. This was achieved with block diagrams, sketches, and photographs of the magnet and the spectra, as well as detailed tables of the data, descriptions of experimental procedures and the naming of the sources of the compounds examined. Though all this certainly was normal for publications of this type, the extent and combination in this article was a sign of a determined quest to establish the new technique. Charles Slichter, Gutowsky's colleague at the physics department of the University of Illinois, Charles Hoffman, and Joseph Mayer, the editor of the *Journal of Chemical Physics*, were of the opinion that the description of the instrument should be published separately in the *Review of Scientific Instruments*, but Gutowsky chose to include this part in the article giving the results of their research, and to do so in detail: "The lengthy literary style is partially a rebellion after writing several 'Letters to the Editor.' Also, we

---

[73] Laboratory notebook "NMR II. Research notes '50–'51," 16 October 1950, pp. 39-44, Gutowsky Papers.

[74] Gutowsky to Dickinson, 11 December 1950, Gutowsky Papers, folder "Magnetic shielding in H_2, H_2O, Min. O." See the entries of 7 to 9 December 1950 in laboratory notebook "Nuclear magnetism I," pp. 139–145, Gutowsky Papers. WWV is a radiostation that provides information on time and frequency.

[75] Gutowsky and Hoffman, "Nuclear Magnetic Shielding."

are engaged in additional similar work and it was hoped that a detailed account in this first paper could serve for later publications."[76] Certainly, a detailed description helped readers to avoid repeating the same mistakes, and this was an important part and motivation of each scientific paper. But in this case, style and combination of topics made clear that more was at stake: the trust of the readers in the reliability of NMR. Gutowsky's strategy was in accordance with maneuvers of the early electron microscopists at the Radio Corporation of America (RCA). At RCA, a committee guaranteed that only valid publications would appear.[77] In Gutowsky's case he did not have the back-up of a large company, and he had to correct an earlier mistake. Thus, the calibration error, the relative novelty of NMR, and Gutowsky's precarious status as chemist and newcomer to the field provoked him to make these efforts to obtain the trust of his fellow scientists. At the same time, he took recourse to another common strategy: the correlation of his experimental findings with theoretical calculations.

Theory and Experiment

> My research is generally a combination of experimental and theoretical aspects. At the moment, our experiments are developing phenomena faster than we can interpret them theoretically.[78]

Scientists usually presented the first observations of the chemical shift as serendipitous discoveries, made possible by the advancement in precision of the measurements. But there were signs of theoretical predictions. Hints to it were first realized by Norman Ramsey, professor of physics at Harvard University, and formerly head of the physics laboratory at Brookhaven National Laboratory. As early as in December of 1949, Ramsey mentioned that effects of electrons of neighboring atoms did interfere with the measurements.[79] Ramsey discussed this on the basis of high-precision measurements of the magnetic moment of the proton by J. H. Gardner and Edward Purcell, and independently by Herbert L. Anderson of the Argonne National Laboratory.[80] Gardner and Purcell, and Anderson, respectively, used different approximations of this correction, differing by a factor of two. Ramsey was interested in calculating the exact correction factor, but this was "difficult to evaluate since it

---

[76] Gutowsky to Joseph E. Mayer, 19 July 1951, Gutowsky Papers, folder "N. Mag. shielding in F and H compounds." See also Gutowsky to Hoffman, 30 March 1951, and Hoffman to Gutowsky, 30 April 1951, ibid.

[77] Rasmussen, *Picture Control*, chap. 1.

[78] Gutowsky to Holm, 14 January 1952, Gutowsky Papers, folder "Holm, C. H."

[79] Ramsey, "Internal Diamagnetic Field Correction."

[80] Gardner and Purcell, "Precise Determination"; Anderson, "Precise Measurement."

depends on the wave function of the H_2 molecule in its various excited states."[81] Through Walter Knight, who was at Brookhaven at the same time, Ramsey acquired first-hand knowledge of experimental data. The first experimental evidence of the chemical shift indeed happened to be found during the studies of Knight on what became known as the "Knight Shift," a shift of resonance frequencies of nuclei in metals and metal salts, respectively.[82] This mangle of theoretical insight and experimental precision was also visible in the history of William C. Dickinson's discovery of the chemical shift. Dickinson, of the MIT Research Laboratory of Electronics, began his announcement of the discovery with the phrase "most unexpectedly, it has been found."[83] But he could not have been too surprised. In a later summary article he mentioned that his Ph.D. supervisor, Francis Bitter, suggested to him months before the discovery "that a second-order paramagnetic field should exist at the nucleus," and that this effect "might be observed experimentally."[84] Thus, the discoveries were all presented in an inductive manner and certainly were the consequence of the precision that scientists attained in the quest for data of nuclear magnetic moments. But from early on, theoretical interpretations also played a role in this longing for precision.

Around 1950, physicists accepted NMR as the most precise method for the measurement of nuclear magnetic moments, a thriving research program in the age of nuclear physics. Consequently, a tremendous amount of data was published by research groups all over the United States. The field was moving so fast that the author of a review article wrote: "Currently there is so much activity in the study of nuclear moments that any printed table of them is bound to be obsolescent before it reaches its readers."[85] Nearly all scientists who reported the chemical shift first were involved in routine programs for the measurement of such data. Knight was a member of the team at Brookhaven National Laboratory, whose objective was the high-precision measurement of nuclear moments, involving both the molecular beam method and NMR.[86] Dickinson, who worked with Bitter at MIT, followed up work on shifts caused by the addition of paramagnetic salts in the measurement process, and found shifts resulting from structural differences in fluorine compounds.[87] Warren

[81] Ramsey, "Internal Diamagnetic Field Correction."

[82] Knight, "Knight Shift," 432; and Knight, "Nuclear Magnetic Resonance Shifts."

[83] Dickinson, "Dependence."

[84] Dickinson, "Time Average Magnetic Field," 718.

[85] Mack, "Table of Nuclear Moments, January 1950," 64.

[86] Knight later reported that he asked his supervisor, Samuel Goudsmit, if he should follow up the effect he had found in metals. Goudsmit answered that a new effect would be more valuable than just adding new figures to a compilation of data. Knight, "Knight Shift," 432.

[87] Dickinson, "Factors."

Proctor and Fu Chun Yu, the third group reporting on chemical shifts in early 1950, were members of Bloch's team at Stanford and also were interested mainly in precise data of nuclear moments. Fortunately, the group around Bloch had abandoned an earlier plan to automate the investigation of nuclear moments. This research style was "dropped in favor of the individual creativity and labors of graduate students." Thus, and with "mild dismay," Proctor, Yu and their colleagues were able to look at the shifts found while examining a sample of ammonium nitrate. This finding later led to a new orientation of the interests of the members of Bloch's laboratory. But in the beginning, the unraveling of the exact magnitude of the shifts and its causes was simply a necessity in order to guarantee the accuracy of nuclear moment measurements.[88]

Apparently, the instrumentation had reached such a stage of precision that the small shifts caused by the electronic structure of the molecules became a matter of concern to the physicists involved. Theory had to come up with an explanation and calculation of the effect, and again it was Norman Ramsey, the key figure in the measurement of nuclear magnetic moments, who first paid attention to the problem. His first theoretical goal was the simplest molecule, hydrogen. In March of 1950, approximately one month after the problems of chemical effects had been discussed at a meeting of the American Physical Society in New York,[89] Ramsey submitted a theoretical calculation. As a member of the Harvard physics department, he was in close contact with other experts in this field, most notably Edward Purcell and John Van Vleck. After improving an older formula of Willis E. Lamb, taking into account all electrons in the molecule instead of only the atom whose nuclear moment was measured, Ramsey added a second-order paramagnetic term. This term, analogous to a term Van Vleck had used in a treatment of molecular diamagnetism in his 1932 book,[90] depended on the wave functions of all excited states of the molecule, and was very difficult to account for. Ramsey finally was able to calculate the shift for the hydrogen molecule, representing the simplest case of a diatomic molecule. On the basis of his computations, Ramsey pleaded for the relation of as many molecules as possible to the value for the magnetic moment of molecular hydrogen.[91] As a result, he argued, the numbers obtained for various nuclear moments could be based on a value that had been theoretically determined. It was agreed that this task should be tackled at two independent sites: Francis Bitter's laboratory at MIT, and the National Bureau of Standards (NBS) in Washington, D.C. Harold A. Thomas, the acting chief of the atomic

---

[88] Proctor and Yu, "Dependence." See also Packard, "Nuclear Induction at Stanford," 519; and especially Proctor, "When You and I Were Young, Magnet."

[89] See *Physical Review* 78 (1950), 339.

[90] Van Vleck, *Electric and Magnetic Susceptibilities.*

[91] Ramsey, "Magnetic Shielding," 703.

physics division of the NBS, made clear that these data were important for a number of reasons, including the exact measurement of magnetic fields.[92]

For different reasons, the hydrogen molecule was also Gutowsky's target, and he emphasized the importance of experimental data, keeping in mind that a purely theoretical treatment was not possible for molecules bigger than hydrogen. In a letter to Thomas, he revealed additional motivation: "We decided to explore the hydrogen compounds as no shifts had been unequivocally reported for them. Also, if systematic shifts were found for the proton resonance, one could be reasonably sure that the phenomenon was a general one."[93] In late August 1950, Gutowsky decided to start investigating this important question. The precision he and his team had meanwhile made possible permitted a comparison of hydrogen gas, water, and mineral oil. Although this was part of his general program of the measurement of proton chemical shifts, he decided to publish these results separately, just in time to be compared with the values of Thomas at the NBS. Their data showed a discrepancy in the value of the chemical shift of water versus hydrogen gas, but this was considered to be within the limits of experimental error.[94] Moreover, because these data did not change the previously determined values for the proton's magnetic moment, the discrepancies were never solved. At this point, according to Gutowsky, adding more precision was a waste of time. On 28 November, he wrote to Dickinson: "Actually, I think we could improve our field further and perhaps even attain the natural line width. We have not tried this because at this stage experiment is too far ahead of theory."[95]

Subsequently, Gutowsky became interested in establishing a direct link between the experimental measurements and the theoretical calculations of Ramsey. This decision was triggered by Ramsey, who, when asked by Gutowsky if his "theoretical treatment is capable of predicting the observed trends" in fluorine shifts, again expressed great interest in results that would link the values for the most often used reference compounds, water and mineral oil, to the theoretically treated hydrogen.[96] This was reason enough for

---

[92] Thomas, Driscoll, and Hipple, "Measurement of the Proton Moment," 787 and 790. For the background, and the work of Francis Bitter: Thomas to Gutowsky, 6 October 1950, Gutowsky Papers, folder "Magnetic shielding in H_2, H_2O, Min. O."

[93] Gutowsky to Thomas, 10 October 1950, Gutowsky Papers, folder "Magnetic shielding in H_2, H_2O, Min. O." Gutowsky refers here to the news given by Thomas that Bitter and Thomas had agreed at a spring meeting on atomic and nuclear constants that they would begin with precision measurements.

[94] See Dickinson to Gutowsky, 21 December 1950; Gutowsky to Dickinson, 24 January 1951 (all Gutowsky Papers, folder "Magnetic shielding in H_2, H_2O, Min. O") and Dickinson, "Time Average Magnetic Field," 731. The paper of Gutowsky is Gutowsky and McClure, "Magnetic Shielding."

[95] Gutowsky to Dickinson, 28 November 1950, Gutowsky Papers, folder "Magnetic shielding in H_2, H_2O, Min. O."

[96] Ramsey to Gutowsky, 25 October 1950, Gutowsky Papers, folder "Magnetic shielding in H_2, H_2O, Min. O." See also Gutowsky to Ramsey, 4 October 1950, ibid.

Gutowsky to send Ramsey the draft of an article and to ask for his opinion about a possible test of Ramsey's theory. In Gutowsky's opinion this should involve the measurement of the proton resonance frequencies in molecular hydrogen (H_2) and in HD (a compound containing one proton and one deuteron, the heavier isotope of hydrogen). Gutowsky thought that different nuclear spin-rotational interaction values of the molecules would be measurable. Because these data were already known for both H_2 and HD, a direct comparison of theory and experiment seemed within reach.[97] In October, after having received no response from Ramsey, Gutowsky specified his views, and predicted a value for the different resonance frequencies of HD, depending on the values of the rotational quantum number. However, at this time he added a note of caution:

> In my lack of familiarity with the physics of the situation I may be overlooking some essential aspect. However, it is the only experiment we have been able to devise which holds any hope for a direct experimental confirmation of your theoretical results. Accordingly I hope you will find time to criticize our tentative results.[98]

In a handwritten note, Gutowsky specified his doubts about the usefulness of the planned experiment, because of a possible time averaging of the different quantum states. This proved to be correct when Ramsey replied:

> I think that some of your planned experiments on H_2 and HD at different temperatures are based on a misinterpretation of my result. The asterisk comment you added to your letter in ink indicates that you were beginning to get a glimmer of the difficulty at the time the letter was written. In experiments such as yours there are many collisions with the result that the molecule is evenly averaged over all magnetic quantum numbers . . . . When this averaging is taken over all magnetic quantum numbers it makes no difference . . . whether the molecule is H_2, D_2, or HD.[99]

The wording of Ramsey's letter reflected the arrogance of the theoretician. This may be related to the fact that Gutowsky implicitly referred to work that Ramsey had done a decade earlier in his own Ph.D. thesis with molecular beam studies.[100] Though Gutowsky deferred to perform the experiment with HD

[97] Gutowsky to Ramsey, 1 August 1950, Gutowsky Papers, folder "Magnetic shielding in H_2, H_2O, Min. O."

[98] Gutowsky to Ramsey, 4 October 1950, Gutowsky Papers, folder "Magnetic shielding in H_2, H_2O, Min. O."

[99] Ramsey to Gutowsky, 25 October 1950, Gutowsky Papers, folder "Magnetic shielding in H_2, H_2O, Min. O."

[100] See Ramsey, "Early History of Magnetic Resonance"; and Ramsey, "Rotational Magnetic Moments."

after he had heard Ramsey's negative opinion, he was able to throw the ball back into the field of theory:

> I am not sure that my glimmer of understanding has increased to a full 60 watt knowledge. However, I am still interested in the possibility of obtaining a direct comparison between theoretical and experimental values for magnetic shielding. We have reasonably good experimental data for most of the binary covalent hydrides. If spin-rotation interaction data and theoretical calculations could be obtained on some molecule other than H_2, the comparison could be made. HF is perhaps the most likely candidate.[101]

The theoretician did not respond to this request. But with improvements in the experiments and progress in theory, Gutowsky hoped to contribute substantially to the sound foundation of the values of nuclear moments, as he wrote to Dickinson in November of 1950:

> My own point of view is that the 'chemical effects' may be of real value in theoretical investigations of the actual electronic distribution in molecules. At least until the problem has been attacked from such an angle, the unshielded values of the nuclear moments themselves are not obtainable.[102]

For a long time, this direction was not followed up by theoretical scientists. Gutowsky tried his best at "polling my theoreticker friends,"[103] most notably Charles Slichter, but it took them three years to come up with a simplification of Ramsey's theory. Still, even then, only a "semiquantitative" account for the experimental results for the fluorine chemical shift between F_2, as the most covalent fluorine compound, and HF, as the most ionic, was possible.[104] Crucial for this treatment was the insight of Gutowsky and Hoffman that the magnetic shielding of the fluorine nucleus depended directly on the electronegativity of the atom linked to it. Unfortunately, such a straightforward relation could not be found for proton resonance shifts.[105] Nevertheless, the treatment of Saika and Slichter allowed for a breaking-up into local contributions of the complex magnetic shielding effect of the molecular electrons and opened the possibility for a better account of the chemical shift. In general, the trials to base the chemical shift values on sound quantum mechanical calculations failed. Though the underlying theory was deemed sufficient in qualitative

---

[101] Gutowsky to Ramsey, 27 November 1950, Gutowsky Papers, folder "Magnetic shielding in H_2, H_2O, Min. O."

[102] Gutowsky to Dickinson, 28 November 1950, Gutowsky Papers, folder "Magnetic shielding in H_2, H_2O, Min. O."

[103] Gutowsky to Dickinson, 6 December 1950, Gutowsky Papers, folder "Magnetic shielding in H_2, H_2O, Min. O."

[104] Saika and Slichter, "Note on the Fluorine Resonance Shifts."

[105] Gutowsky and Hoffman, "Nuclear Magnetic Shielding."

terms, the empirically found chemical shift values could not be rigorously compared with a theoretical standard. The physicists became disinterested when they recognized that the complexity of the molecular world was an unsurmountable obstacle in this quest. Moreover, experiments showed that the accuracy of values of interest for physicists, mainly nuclear magnetic moments, was not in danger. Thus, convergence with calculations to check the experimental data was both a hopeless and a needless endeavor.

This feeling was shared by Dickinson, who moved from MIT to the cyclotron group at Los Alamos. In 1950, he was skeptical about improvements without major advances in theory. But he showed traces of optimism for an endeavor which was entirely based on empiricism:

> I hope you are right that the chemical effects will contribute to knowledge of molecular structure. I don't see how basic information will follow however until a simpler interpretation of the results is available than that of Ramsey's. I do feel that the regularities which you people have found may lead to some valuable empirical evidence as to molecular structure.[106]

Dickinson's foresight proved to be correct. The breakthrough of NMR came with its applications in structural elucidation of molecules, and this happened because Gutowsky and his colleagues connected the regularities that had been found in NMR with established theoretical entities and experimental methods of chemistry. The practical use of NMR in chemistry, and not its theoretical foundation in physics, decided on the success of the technique.

The Chemistry of the Chemical Shift

The chemist Gutowsky brought characteristic features to research in nuclear magnetic resonance. In his own words, and in historical retrospect, he described his motivation as follows:

> Chemists learn very early to look for periodicities in the chemical and physical properties of compounds, or they don't stay in chemistry very long. We deal with such a large number and wide variety of systems that we have to oversimplify their diversity to be able to remember them. Moreover, in my senior year at Indiana University I was exposed to Linus Pauling's book, 'The Nature of the Chemical Bond,' which is a masterpiece of such oversimplifications. In any case, it seemed to me that the chemical shift, as an electronic phenomenon, should be related in some way to the nature of the chemical bonds. This in turn depends upon the nature of the atoms bonded together,

---

[106] Dickinson to Gutowsky, 2 December 1950, Gutowsky Papers, folder "Magnetic shielding in H_2, H_2O, Min. O."

> so I chose to study the simple binary fluorides, which was a very happy choice.[107]

Gutowsky speculated on the causes of the chemical shift in terms of bond hybridization. Bond hybridization was an important notion in the theoretical underpinning of chemistry in quantum physics, and since the 1930s Linus Pauling most forcefully put it forward. Following the beginnings of quantum chemistry, especially the work of Robert Mulliken, the electrons of an atom were assigned specific orbitals, visualized as spaces with a high probability as to where the electron could be found. These orbitals were the outcome of quantum mechanical calculations and basically represented energy levels of the electrons. The overlapping of two orbitals led to the formation of a chemical bond. Bond hybridization refers here to the "mix" of different bond orbitals and thus to the formation of new bond types.

In 1951, Gutowsky's group changed, with the arrival of David McCall, Bruce McGarvey, and Leon ("Lee") H. Meyer.[108] Together with the expert in electronics, Robert McClure, they constituted the core of the group in 1951 and 1952, later remembered as the "4 M days."[109] McCall came from the University of Wichita, and after the completion of his graduate studies at Illinois in 1953 he left for an assignment at Bell Laboratories in Murray Hill, New Jersey.[110] Meyer, who came with an M.S. from Georgia Tech to Urbana, accepted a position at the Atomic Energy Division of Du Pont at Augusta, Georgia. Robert McClure later joined an electronics firm on Long Island, while Bruce McGarvey began an academic career, first at the University of California at Berkeley, and later at the University of Windsor, Canada.[111] Most of the funding for the group came from the Office of Naval Research.[112] The first project they tackled together, dubbed "Project M I," was the extension of the investigations of the work on binary fluorides to the more complex organic fluorides, including various substituted benzene derivatives. On 9 March 1951, McGarvey and Meyer wrote an entry in the notebook of the team describing their methods and aims:

> A few preliminary data have been taken on a series of benzene and substituted benzene comp[oun]ds with additional F's on the ring or in the -CF_3 group.

[107] Gutowsky, "Coupling of Chemical and Nuclear Magnetic Phenomena," 364. Text identical in Gutowsky, "Chemical Aspects," 287.

[108] For Meyer and his description of the early work of the group, see Meyer, "NMR Some Fifty Years Later."

[109] Gutowsky to McCall, 25 February 1992, Gutowsky Papers, folder "McCall, D. W. 2."

[110] See McCall's nomination blank for membership in the Society of Sigma Xi, not dated, and letter from Gutowsky to McCall, 30 November 1953, Gutowsky Papers, folder "McCall, D. W. 2."

[111] Gutowsky to McCall, 25 February 1992, Gutowsky Papers, folder "McCall, D. W. 2."

[112] For a judgement of the impact of military funding on post-war research in the United States see Forman, "Behind Quantum Electronics."

> It has been decided to repeat these runs . . . . The purpose of this project is, in the immediate investigation, the determination of electron density as a function of the position of the F in relation to other ring substituents.[113]

With "Project M I," NMR became firmly rooted in the organic chemistry of the day. Gutowsky assigned Lee Meyer the task of taking over "the primary responsibility of seeing that the measurements are completed on the available compounds and that as many o, m and p series as possible are obtained." The letters *o*, *m*, and *p* refer to the *ortho*, *meta*, and *para* positions of the fluorine atoms in the benzene ring, and since the nineteenth century were an important designation of structural organic chemistry. Since the 1930s, substituents were thought to influence the electronic structure of the molecule most importantly by—in the terms of physical organic chemistry—resonance and inductive effects. The resonance effect refers to situations when a substituent changes the distribution of electrons in a molecule by contributing to their delocalization. Inductive effects represent the ability of atoms in the molecule to attract or repulse electrons that constitute the bonds. Both effects were successful in the explanation and prediction of chemical reactions. In order to establish direct relations between the shifts and the positions of substituents in the molecule, Gutowsky was eager to include as many different compounds as possible that were available in a series. He even thought of a "last recourse" to write "letters to whoever has reported the syntheses." In addition, he emphasized the need for precision in the measurements and gave advice for the working procedures suitable to obtain the best results: "The signal generator drift isn't too bad now, particularly after it's been on for at least 6 hours and if there are no thermal variants in the room. Also, you might try a double probe system. As a suggestion, I think you'll find that working alone, instead of in pairs and for minimum periods of 3 to 4 hours will give the most satisfactory results." At the same time, though very precise in his suggestions, Gutowsky made clear that "the above is meant to be indicative only. There are certainly plenty of very interesting things to do."[114]

Gutowsky's most influential model was the experimental work of Charles H. Townes and B. P. Dailey on nuclear quadrupole spectroscopy, a technique closely related to NMR. With this method, Townes and Dailey had been able to give information on bond hybridization.[115] A brief remark on this work formed the beginning of the first major article on the subject by Gutowsky and his co-workers, linking their achievements and methods to those of Townes and Dailey. Though the referees, among them E. Bright Wilson, regarded the

---

[113] Laboratory notebook "NMR II. Research notes '50–'51," p. 160, Gutowsky Papers.

[114] Gutowsky to Meyer, 27 June 1951, "Memo Summer Research Program," Gutowsky Papers, folder "Meyer, L. H. 3."

[115] See Townes and Dailey, "Determination of Electronic Structure."

discussion as speculative and unduly long, an anonymous referee opined that "the speculations presented in this paper are of a kind which can stimulate much new work along a variety of lines, including classical as well as novel ones."[116] They agreed that the paper reported "an important new method of obtaining information about valence bonds,"[117] and the article appeared in the prestigious and widely read *Journal of the American Chemical Society*. There, Gutowsky considered the relation of the chemical shift to a well-known constant in physical organic chemistry, Hammett's constant σ. The latter was based on reaction rates and equilibria, and was determined by the *meta-* or *para-*positions of substituents in phenyl compounds. Hammett, in 1937 and in later publications, had reported that the σ constant represented a measure of the ability of the substituent to influence electron densities in the substituted molecules and thus the course and velocity of reactions.[118] In order to possess a parameter that gave "apparent relative electrical charges, measured with respect to the phenyl group,"[119] Gutowsky and his co-authors defined the nuclear shielding parameter δ as $10^5 \times (H_r - H_c)/H_r$. H_c was the applied magnetic field required for resonance of the compound under scrutiny, while H_r was the field strength required for resonance of a reference compound. δ later became the universally used parameter for the chemical shift.

The correlation of Hammett's σ and Gutowsky's δ provided the ideal basis and publicity for the chemical shift studies. Furthermore, Gutowsky followed a general watchword of chemical physicists that a linear correlation of two parameters helped to gain insights into the nature of both. Thus, Gutowsky's approach was another example of correlation analysis in chemistry. The systematic differences found between the two parameters provided the basis for the attribution of δ to electronic effects of the substituents, namely inductive and resonance effects. This phenomenon could not have been predicted or explained by Ramsey's physical theory of the chemical shift. Also, Gutowsky and his group used their novel method to check on how these substituent effects added to each other, an endeavor that had not before been tackled. However, they had not given up their link to theory, and sought a more secure footing in calculations of electron densities by molecular orbital methods, a project that was performed independently at the same time by Hans H. Jaffé of the Venereal Disease Experimental Laboratory at the University of

---

[116] Journal of the American Chemical Society, Comments of referee I, Gutowsky Papers, folder "Elec. dist. in molec. I. F in benzene." The article is Gutowsky, McCall, McGarvey, and Meyer, "Electron Distribution in Molecules I."

[117] Journal of the American Chemical Society, Comments of referee II (handwritten note "E. Bright Wilson"), Gutowsky Papers, folder "Elec. dist. in molec. I. F in benzene."

[118] Hammett, "Effect of Structure"; Hammett, *Physical Organic Chemistry*, 194–198; Hammett, "Physical Organic Chemistry in Retrospect." See Shorter, "Hammett Memorial Lecture"; and Shorter, "Die Hammett-Gleichung."

[119] Gutowsky, McCall, McGarvey, and Meyer, "Nuclear Magnetic Parameter," 1328.

North Carolina in Chapel Hill.[120] Nevertheless, a direct comparison proved to be impossible: "We had hoped to be able to tie our experimental observations in with calculated electron densities, but most substituted aromatic fluorides available for our measurements have not been treated theoretically."[121]

Despite this lack of direct theoretical underpinning, Gutowsky and his group were able to offer practical advantages with their method:

> The prediction of σ from δ is an attractive procedure since a δ-value can be obtained in less than a half hour with an operating high resolution spectrometer, while the evaluation of σ by the usual kinetic or equilibrium study is a much longer process.[122]

In addition, Meyer and Gutowsky connected their results to recent studies of bond lengths performed at one of the centers of molecular spectroscopy, the University of Michigan at Ann Arbor, with new and improved methods of electron diffraction.[123] When they reported their study in September 1952, at a symposium on nuclear and paramagnetic resonance at the meeting of the American Chemical Society (ACS) in Atlantic City, their results attracted the large audience of 400 people. Consequently, the meeting stimulated Gutowsky to think about plans for extending his research program.[124]

The connection to bond concepts was also a crucial feature of the work of David McCall on chemical shifts in phosphorus compounds. He correlated his results, mainly established in the spring and summer of 1951, with approaches of Linus Pauling and Kenneth Pitzer. Pitzer, the former supervisor of Gutowsky's M.S. thesis at the University of California at Berkeley, in 1948 had challenged Pauling's assignments of single-, double-, and triple bond character to certain compounds. In doing so, Pitzer opposed Pauling's concept that the distances between atoms are determined by a balance of bonding attraction and nuclear repulsion. He refined the theory by introducing the terms of "inner shell repulsion," meaning that the filled inner electronic shells of an atom interact with the bonding orbitals of another atom, and "valence shell repulsion," the interaction of p_z and p_y orbitals if p_x was the bonding orbital. While the

---

[120] Gutowsky, McCall, McGarvey, and Meyer, "Electron Distribution in Molecules I.," 4811–4812, 4815. See Jaffé, "Theoretical Considerations"; and correspondence between Lee Meyer and Jaffé (10 and 16 July, 31 August, 1951), Gutowsky Papers, folder "Elec. dist. in molec. I. F in benzene."

[121] Gutowsky to Jaffé, 6 September 1951, Gutowsky Papers, folder "Elec. dist. in molec. I. F in benzene."

[122] Meyer and Gutowsky, "Electron Distribution in Molecules II," 482.

[123] See Meyer and Gutowsky, "Electron Distribution in Molecules II," 483, table II; and correspondence between L. O. Brockway and Gutowsky, 16 and 28 August 1951, 5 and 17 January 1953, Gutowsky Papers, folder "Electron dist. II halomethanes."

[124] Gutowsky to McClure, 14 October 1952, and Gutowsky to H. L. Johnson, 14 October 1952, Gutowsky Papers, folder "Apparatus for nucl. mag. resonance."

data produced by NMR proved to be inconclusive in deciding this controversy, McCall and Gutowsky could, in the case of the phosphorus oxyhalides, propose different structures than those given by Pauling. Again, the achievement Gutowsky and McCall were looking for was to establish NMR as a method suitable for the determination of bond character and distribution of resonance structures.[125]

In 1952, Gutowsky and his co-workers had successfully established NMR in the realm of chemical physics and physical organic chemistry. Though both fields were fashionable and thriving research areas during the 1950s in American chemistry, the majority of chemists worked in the field of classical synthetic organic chemistry. What they expected from NMR (if they did expect anything at all) was a relatively simple and rugged method to determine molecular structures, along the lines of infrared spectroscopy for example. Up to this point, Gutowsky's work had nothing to offer in this respect. For that, a practical turn, and more importantly a change in perspective of interpreting the results, was necessary.

The work that led to such a result began with the usual aim: Lee Meyer, the Japanese physicist Apollo Saika, and Gutowsky were concerned with a correlation of resonance shifts with the effects of substituents on chemical reactivity. The common organic functional groups seemed to be an ideal target. For that, Meyer and Saika made the first foray into this field with a detailed investigation into the proton resonance shifts of 220 organic compounds. But soon that attempt changed its direction:

> In addition, during the course of these measurements it became clear that the proton spectrum of a molecule was distinctive enough to be of value in structural analysis, even though the proton resonance shifts are very small. Therefore, the research was extended to include a greater variety of proton types, and the emphasis was changed from observing particular groups to viewing and assigning the spectrum as a whole.[126]

The most important result of this study was the discovery that the chemical shift of protons was determined mainly by their being part of particular groups of atoms in the molecule (the functional groups), and thus could be used as a "fingerprint" signature for the presence of each functional group. In the same way, infrared spectroscopy was applied in organic chemistry. In consequence, Meyer, Saika, and Gutowsky thought that the prospects of NMR "in structural and quantitative analysis appear promising." Though the overlap of the values were a limitation to the use of the technique, "the general appearance of the

[125] David McCall, "Research Report," 15 July 1951, Gutowsky Papers, folder "McCall, D. W. 2"; Gutowsky and McCall, "Electron Distribution in Molecules IV," 162; Pauling, *Nature of the Chemical Bond*, 83; Pitzer, "Repulsive Forces."
[126] Meyer, Saika, and Gutowsky, "Electron distribution in molecules III," 4567.

proton spectrum can be predicted directly from the relative numbers of non-equivalent protons in a postulated structure. And no doubt such simplicity is one of the main advantages of this type of spectroscopy."[127] Independently, scientists of the first manufacturer of commercial NMR spectrometers, Varian Associates, had come to the same conclusion, and in July of 1953 announced "a new kind of spectroscopy: high resolution n-m-r."[128]

No wonder that at the very same time Gutowsky wrote Meyer that "we'll probably spend the next few years cleaning up the problems turned up subsequent to your work."[129] In an award address written much later, Gutowsky emphasized that their approach reflected his "experience with the Perkin-Elmer infrared spectrometer."[130] The idea of this strategy was closely related to infrared spectroscopy because in the mid-1940s, Norman B. Colthup of American Cyanamid Co., Stamford, Connecticut, had assembled an infrared spectra-structure correlations chart. Soon, it was reproduced in the instruction manuals for the Perkin-Elmer infrared spectrometers that were based on the American Cyanamid design. A much expanded chart of the same type was published by Colthup in 1950 in the *Journal of the Optical Society of America.* With the commercial success of infrared spectrometers in the 1950s, the Colthup chart became a prominent tool for organic chemists in interpreting infrared spectra, and it won tremendous acceptance for the method among organic chemists.[131]

Gutowsky modeled the chart of NMR proton chemical shifts (see Figure 2.7) after the Colthup infrared spectroscopy charts he was familiar with. This chart marked the starting point of much of routine use of NMR in virtually every chemistry department in the world. With that style of presentation, in front of the participants at the Annual Meeting of the ACS in September 1953 in Chicago, Gutowsky made clear the potential of NMR, and soon Varian Associates included a reproduction of the chart in the issue of its annual *Technical Bulletin.*[132]

This performance and the simplicity of interpretation provided a push for NMR studies, and broadened the market for instruments. Gutowsky much

---

[127] Meyer, Saika, and Gutowsky, "Electron Distribution in Molecules III," 4573.

[128] *Technical Information from the Radio-frequency Spectroscopy Laboratories of Varian Associates,* vol. 1, no. 1., July 1953, p. 1. Copy in Roberts Papers.

[129] Gutowsky to Meyer, 8 July 1953, folder "Meyer, L. H. 3," Gutowsky Papers.

[130] Gutowsky, "Coupling of Chemical and Nuclear Magnetic Phenomena," 366. Also in Gutowsky, "Chemical Aspects," 291.

[131] Colthup, "Origins of the IR Spectra-Structure Correlations Chart"; and Colthup, "Spectra-structure Correlations." I thank Peter J. T. Morris for guiding me to these sources.

[132] The meeting took place from September 6 to 11, 1953. See *Analytical Chemistry* 25 (1953), 1139, 1280-1283 for a brief description. See also *Technical Information from the Radio-frequency Spectroscopy Laboratories of Varian Associates*, vol. 1, no. 2., [not dated], p. 4. Copy in Roberts Papers.

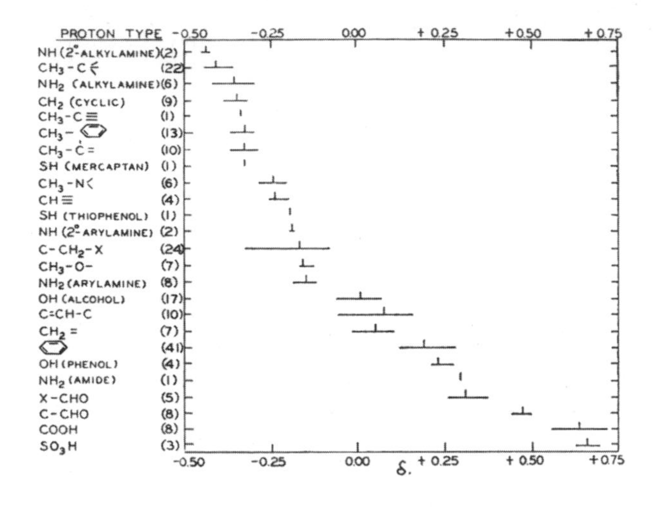

FIGURE 2.7 Chart relating the chemical shifts to groups of atoms in 220 organic compounds. Gutowsky followed here the model of charts already available for infrared spectroscopy, and emphasized the usefulness of NMR for straightforward structural research in organic chemistry. Reprinted with permission from Leon H. Meyer, Apollo Saika, and Herbert S. Gutowsky, "Electron Distribution in Molecules III. The Proton Magnetic Spectra of Simple Organic Groups," *Journal of the American Chemical Society* 75 (1953), 4567–4573, on 4568. Copyright 1953, American Chemical Society.

later reported that their method rescued the NMR program of Varian Associates, the first manufacturer of NMR spectrometers, which in 1953 marketed the first model, the HR-30. Gutowsky also made clear that Lee Meyer was a crucial figure in this endeavor:

> Lee's characterization of the proton shifts for organic functional groups was important and had an impact. A decade or so after I reported the results at the Fall ACS meeting in Chicago, I happened to meet Martin Packard (of Varian) at a meeting. He said that Varian had decided to drop their NMR program for lack of buyer interest when we reported the characterization. It greatly stimulated interest in the Varian NMR equipment, which kept them in the business![133]

---

[133] Gutowsky to McCall, 25 February 1992, Gutowsky Papers, folder "McCall, D. W. 2." The first mention of Meyer in this correspondence was made by McCall: "I always felt that Leon played a special role in those studies and did not get the unique recognition he might have. We all knew that you were the 'chief' (Roger Adams notwithstanding) and the insights that led the group through that period were all yours." McCall to Gutowsky, 17 February 1992. See also the accounts in Gutowsky, "Coupling of Chemical and Nuclear Magnetic Phenomena," 366; and in Gutowsky, "Chemical Aspects," 291, where the name of Packard is not given.

Evidently, employees of Varian Associates had first heard of the results of Meyer, Saika, and Gutowsky in June 1953, when Gutowsky reported the data at the symposium on molecular structure and spectroscopy in Columbus, Ohio.[134] They were especially interested in the chart, and James Shoolery of Varian Associates asked for permission to reproduce it for use in advertisements. At this time, and because of their own research that was going on, the company was regarded by Gutowsky as a potential, but friendly, scientific competitor:

> Dave [McCall] may have passed on my comments to the effect that life is becoming more competitive. Varian has looked at a lot of fluorine compounds for M^3 and a fair number of proton compounds for various people. They suffer however from the disadvantage of being pressed to do only the things brought in and not being able to follow up the more promising leads. In any event a few days ago I received a note from Jim Shoolery asking for a copy of the proton spectral chart with permission to make up and send out copies. As long as a testimonial wasn't also asked for, I thought the "free publicity" was OK so sent him the print.[135]

With Gutowsky's contributions, and those of his co-workers, NMR became a tool of distinct interest for chemical work. The work that related chemical shifts to accepted notions of physical organic chemistry built up the credibility of the technique, and also pointed to refinements in the theory of NMR itself. The correlation of chemical shifts to functional groups made clear the tremendous utility of NMR for analytical and structural work. It is important to note that in this process a change in viewing the same results occurred. It was the whole spectrum that constituted the basis for structural studies, while in earlier examples the picture of the spectrum was shown as a proof of the precision attained. Gutowsky and his co-authors presented such an icon of precision in their first publication in a series of articles dealing with electron distribution in molecules. The oscilloscope photograph of the fluorine magnetic resonance spectrum of 2,3,5-trifluorobenzotrifluoride (Figure 2.8) shows the very same substance that in April of 1950 first made possible the resolution of different fluorine resonances in one molecule (see Figure 2.2), in addition revealing the fine structure of the peaks of the fluorine atoms bound directly to the benzene ring. Though the lines were assigned to the respective fluorine atoms in the compound, the use of the photograph in this context was as a proof of the precision of the novel method of NMR, obtained in the laboratory of the authors: "It is apparent that the F^{19} magnetic shielding differences are readily observable."[136]

---

[134] The symposium took place from 15 to 19 June, 1953. See *Analytical Chemistry* 25 (1953), 202.
[135] Gutowsky to Meyer, 8 July 1953, folder "Meyer, L. H. 3," Gutowsky Papers. M^3 probably refers to the Minnesota Mining and Manufacturing Company.
[136] Gutowsky, McCall, McGarvey, and Meyer, "Electron Distribution in Molecules I," on 4810, fig. 2.

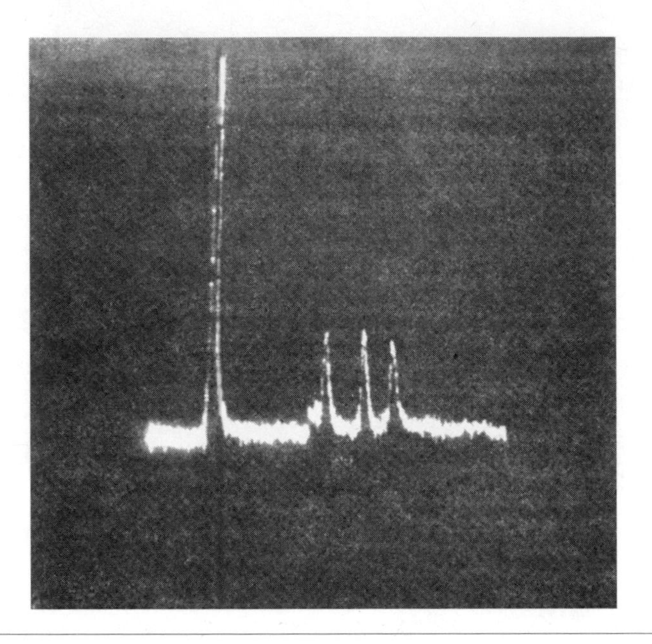

FIGURE 2.8 Fluorine resonance spectrum of 2,3,5-trifluorobenzotrifluoride. Reprinted with permission from H. S. Gutowsky, D. W. McCall, B. R. McGarvey, L. H. Meyer, "Electron Distribution in Molecules I. F^{19} Nuclear Magnetic Shielding and Substituent Effects in Some Benzene Derivatives," *Journal of the American Chemical Society* 74 (1952), 4809–4817, on 4810, fig. 2. Copyright 1952, American Chemical Society.

In 1953, the spectrum was inspected as a whole, the intensities of the absorption lines were measured, and were assigned to specific functional groups. The icon of precision had changed to a tool of structural research (see Figure 2.9).

With the conversion of the object of research from nuclear moments to electronic and molecular structures, the instrument changed, too: "The significance of nuclear magnetic resonance to chemical physics research is in the detailed measurement of particular characteristics of the resonances, and for each characteristic the instrumentation needs are somewhat different."[137] The core of Gutowsky's research program was the construction of a reliable, precise, and productive experimental system, and, at the end of 1952, he, Meyer, McClure, McCall, and McGarvey had gone a long way on this path with the correlation of the chemical shift to concepts of physical organic chemistry. At the same time, the control of the instrument that they achieved in 1951

[137] Gutowsky, Meyer, and McClure, "Apparatus for Nuclear Magnetic Resonance."

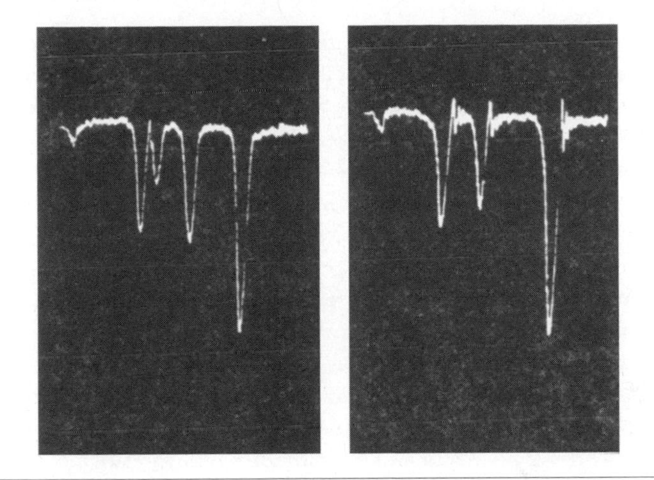

FIGURE 2.9 Oscilloscope photographs of proton resonances in *o*-toluidine and benzylamine. Here, the photographs were given to make clear the use of NMR to differentiate between the two compounds. Reprinted with permission from L. H. Meyer, A. Saika and H. S. Gutowsky, "Electron Distribution in Molecules III. The Proton Magnetic Spectra of Simple Organic Groups," *Journal of the American Chemical Society* 75 (1953), 4567–4573, on 4568, figs. 1 and 2. Copyright 1953, American Chemical Society.

enabled the experimental characterization of a second "chemical" effect in NMR: spin-spin coupling.

Constructing a New Effect

> "Errors" sometimes are more important than preconceptions as to what is to be obtained in a given experiment.[138]

In the fall of 1950, the two research groups most active in the field of high resolution NMR, Gutowsky and Hoffman in Urbana, and Proctor and Yu at Stanford, found effects that were inexplicable with the phenomenon of the chemical shift. At Stanford, the resonance of the antimony nucleus in liquid $NaSbF_6$ showed a complex pattern of five peaks (later it was shown that the resonance had seven peaks). But according to previous experience, the resonance of such a compound should be singular. Proctor and Yu resorted to the familiar, and attributed the splitting to the dipolar interaction of the fluorine nuclei with the antimony nucleus. Dipolar interaction was the basis for NMR broad

[138] H. S. Gutowsky, memo for course "Chemistry 290," September 1949, Gutowsky Papers, folder "Junior research students."

line studies of solids. In liquids, such an interaction was normally canceled out by the rapid movements of the molecules, and consequently Proctor and Yu proposed hindrance of molecular rotation. Though this theoretical understanding was so improbable that Proctor's presentation at a physics colloquium "was greeted with loud shouts more characteristic of the British parliament," it qualitatively explained the phenomenon, especially after E. Raymond Andrew proposed a mechanism for the hindrance of molecular rotation.[139]

The recourse to familiar explanations was common among the scientists dealing with the anomalies that later were attributed to the novel effect of spin-spin coupling. Also in the case of Gutowsky, splittings of resonance peaks were not regarded as due to a novel phenomenon, though they were clearly visible on the scope. In 1950, Gutowsky and his group members repeatedly made observations that later were explained with the help of spin-spin coupling. The unreliable working modus of the spectrometer and the still not completely understood theory of the chemical shift prevented an earlier realization. In addition, the complexity of the effect itself obscured interdependencies.[140]

Gutowsky and his team interpreted also the next observation, made by Charles Hoffman on 8 September 1950 with PF_3, in chemical terms, as the effect of an impurity due to the reaction pathway used for the synthesis of the substance.[141] Hoffman followed up on this on 27 September, when he and McClure still found a double resonance, although he prepared the sample by a different pathway. They made detailed measurements of the splitting: "Both resonances were of equal strength indicating about a 50-50 mixture." Thus, they still thought of a chemical effect.[142] Weeks later, the two researchers found a double resonance in the closely related compound PF_5,[143] and in November they found a double resonance in IF_5.[144] This was explained with chemical shifts between the fluorine nuclei in the asymmetrical molecule.

---

[139] The quotation is from Proctor, "When You and I Were Young, Magnet," 550. The first report of the splitting appeared in Proctor and Yu, "On the Nuclear Magnetic Moments," 27–28. See Andrew, "Nuclear Magnetic Resonance Absorption."

[140] For example, in July of 1950, during the examination of PI_3, a double resonance appeared. This was explained by a chemical reaction that the compound had undergone in the test tube. Though PI_3 later became a clear candidate for demonstrating the spin-spin coupling, this chemical explanation of a double resonance was correct, because the spin-spin splitting was suppressed in this compound by the electrical properties of the iodine nucleus. Laboratory notebook "Nuclear magnetism I," 3 July, p. 203, Gutowsky Papers; and Gutowsky, McCall, and Slichter, "Nuclear Magnetic Resonance Multiplets," 284–285.

[141] Laboratory notebook "Nuclear magnetism I," p. 278, Gutowsky Papers. See above.

[142] Laboratory notebook "NMR II. Research notes '50–'51," pp. 14-15, Gutowsky Papers.

[143] Laboratory notebook "NMR II. Research notes '50–'51," 9 November 1950, p. 100, Gutowsky Papers.

[144] See laboratory notebook, "NMR II. Research notes '50–'51," entries of 10 November 1950 and 23 March 1951, pp. 108–109, 181, Gutowsky Papers.

Following the publication of Proctor and Yu in the January issue of *Physical Review,* Gutowsky looked at the effects formerly attributed to already known causes or impurities with a different perspective. In February 1951, McClure tried to repeat the measurements of Proctor and Yu with $KSbF_6$, but could not resolve a resonance line that "may have been multiple."[145] Crucial for the discovery was the project of McCall, related to chemical shifts in compounds containing phosphorus and fluorine. With two different atomic species measurable in the same molecule and with the possibility to check on different numbers of fluorine nuclei in the same class of compounds, the range of phenomena was both so diverse and under control that alternative explanations did not make sense any more. In spring of 1951, Gutowsky finally realized that they were looking at something entirely novel:

> I did independently characterize the spin-spin coupling but we hadn't pinned it down. That was more an intuition on my part and I assigned him [McCall] a job of looking at the fluorine resonance. I think it was in compounds that I knew would have or should have the splitting. And it wasn't entirely a dry experiment, it was really the first confirmation. In other words I pinned it down in my own thoughts as a spin-spin effect and did the numerology that Bright Wilson was loading to his teaching quantum mechanics.[146]

With hindsight, the crucial step was the examination of fluoro-chloro phosphorus compounds of known purity, thus excluding impurity effects. Gutowsky much later described the "great thrill" that McCall and he experienced when a predicted triplet indeed appeared on the oscilloscope.[147] A look at the laboratory notebook reveals the "fine structure" of the discovery during late March 1951.[148] When they performed the measurements with phosphorus oxyhalides, McCall and Gutowsky on 22 March found a doublet phosphorus resonance with nearly equal intensities in $POCl_2F$, while $POCl_3$ showed a single resonance. A recheck on PI_3 gave only one resonance line. On 23 March, they examined $POClF_2$ and obtained a triplet of the phosphorus resonance, with the center resonance twice as strong as the two satellites.[149] The same day

---

[145] Entry of Robert McClure, 10 February 1951 in "NMR II. Research notes '50–'51," p. 156, Gutowsky Papers.

[146] Gutowsky, interview by Reinhardt, 1 and 2 December 1998.

[147] Gutowsky, "Chemical Aspects," 289.

[148] See laboratory notebook, "NMR II. Research notes '50–'51," entries of 22–24 March 1951, pp. 172–181, Gutowsky Papers.

[149] In the first case, the fluorine nucleus splits the signal of the phosporus nucleus into a doublet. The chlorine atoms do not have an effect. If two fluorine nuclei are in the molecule, the phosphorus signal is split into a triplet. This follows the later so-called "n+1 rule," according to which the signal of a nucleus, when coupled to n equivalent nuclei, splits into n+1 peaks. The intensities of the lines follow the binomial equation. Vice versa, also the fluorine signals are split, by the phosphorus nuclei.

they checked on the fluorine resonances of the same compounds, and found doublets. On 24 March, they looked at CH_3OPF_2, and found a doublet for the fluorine and a triplet for the phosphorus resonance. After the results with $POClF_2$, this could not be a surprise, but certainly it was confirming their ideas on the origin of the effect: "A rough calculation was made to determine the [numerical values of the] phosphorous splitting from the fluorine splitting by multiplying the fluorine splitting by 2628/1131, the ratio of the magnetic moments."[150] Thus it was clear to them that the interaction had something to do with the magnetic moments of the nuclei present in the molecule.

In their first publication on the subject, Gutowsky and McCall wrote that the intensities of the lines followed the binomial coefficients, and gave a simple rule for the prediction of the spectrum.[151] This later became the basis for the use of spin-spin coupling for structural research. On 18 April 1951, two days after submission of this paper, Gutowsky examined BrF_5, a compound of considerable interest because of the complexity of lines. Gutowsky found a doublet and a quintuplet, revealing a combination of chemical shift and the new effect.[152] In May[153] and June, Gutowsky and McClure checked the resonance multiplets in compounds containing iodine and fluorine. This time, as with BrF_5, they looked for effects of structurally non-equivalent fluorine nuclei:

> Took a detailed look at the F^{19} resonance in IF_7 . . . . The line is multiple and the components are only partially resolvable . . . . Assuming a pentagonal bipyramid structure F_2 line is sextet 1:5:10:10:5:1 F_5 line is triplet 1:2:1 or 20:40:20 using same scale as for F_2.[154]

The observations were aided by new instrumental developments following the Stanford group's model of a new modulation system of the magnetic field. The summer research program of David McCall, which included all measurable effects of nuclear magnetic resonance in phosphorus compounds, had among other things the goal of establishing the "experimental characterization"[155] of the resonance fine structure. "The outstanding experimental problems here," Gutowsky wrote, "are the T [temperature] dependence of the splitting; the line

[150] Entry of David McCall, 24 March 1951 in "NMR II. Research notes '50–'51," p. 180, Gutowsky Papers.

[151] Gutowsky and McCall, "Nuclear Magnetic Resonance Fine Structure."

[152] Entry of Gutowsky and McClure, 18 April 1951 in "NMR II. Research notes '50–'51," pp. 189, 191, Gutowsky Papers.

[153] See laboratory notebook, "NMR II. Research notes '50–'51," entry of 18 May 1951, p. 211, Gutowsky Papers.

[154] See laboratory notebook, "NMR II. Research notes '50–'51," entry of 3 June 1951, p. 221, Gutowsky Papers.

[155] McCall, "Progress report," 15 August 1951, p. 2, Gutowsky Papers, folder "McCall, D. W. 2."

widths of the components and the fact that structurally equivalent nuclei do not appear to interact."[156] In August, this had been achieved, and, together with Slichter's observations of the field independence of the splitting, the experimental characterization was "nearly complete."[157] In July, their experiments led to a differentiation of the types of splittings observed. The first system involved cases where the "nuclei of each variety are structurally equivalent," and as a consequence only the novel splitting was observed. The nuclei of the second system type were "in structurally non-equivalent sets. Main components are separated by a chemical shift. Each of these components is a multiplet with small separations of higher order interaction type."[158] The experimental data for splittings of the second type were, however, rather limited at that time. Crucial was the cooperation with theoretical and experimental physicists. For the first time, Gutowsky had found an ally in the Stanford group, with Erwin Hahn who had just moved from Urbana to Stanford. In addition, McCall's and Gutowsky's project was soon supported by the efforts of Charles P. Slichter, faculty member at the physics department at the University of Illinois. Already in their first publication, Gutowsky and McCall thanked Slichter for discussions regarding the interpretation of their results, and the collaboration with Slichter's group soon intensified.

In their second publication, submitted on 10 September, Gutowsky, McCall, and Slichter objected to their own earlier explanations. They did so on the basis of experimental evidence, presented in six statements:

1. The splitting is associated with the nuclear magnetic moments . . . .
2. Nuclei of the same species do not interact when they are in chemically equivalent positions in a molecule . . . .
3. The relative intensities and number of components of a line A are determined by the statistical weights and the number of possible spin states of the [nucleus] B which cause the splitting . . . .
4. Our recent experiments show the splittings are independent of temperature and the strength of the static field H_0 . . . .
5. Although the splittings are several tenths of a gauss, we have found in almost all cases that the components are at least as narrow as the 0.01-gauss magnet inhomogeneity over the sample . . . .
6. The ratio of these splittings is 8.2 while the calculated ratio of the magnetic dipolar fields (μ/r^3) is about 0.7.

---

[156] Gutowsky, "Memo to Dave McCall," 31 May 1951, Gutowsky Papers, folder "McCall, D. W. 2."

[157] McCall, "Progress report," 15 August 1951, p. 2, Gutowsky Papers, folder "McCall, D. W. 2."

[158] Gutowsky and McCall, "Memo to C. P. Slichter," 24 July 1951, Gutowsky Papers, folder "Coupling am. N.M. dipoles in molec."

Although these statements were strong enough to discredit the previously published explanations, it seemed questionable that they would be convincing enough to support the interpretation now favored by the authors. Thus, they put their plea in vague terms: "We wish to propose that the splittings arise from a second-order interaction between the nuclear magnetic moments and some magnetic field internal to the molecule. It seems to us that the most likely coupling is via the electrons by a mechanism analogous to the chemical shift."[159] This vagueness prompted Felix Bloch, whose reputation in matters NMR could rarely be matched, in September 1951 to afford Gutowsky the attribute of having a "lawyer's technique—a style of not sticking your neck out." But Bloch himself would have been even more cautious than the authors, at least this was reported by Hahn: "His general opinion is that the cause of this effect is yet so obscure that he would be afraid to speculate at this point." Nevertheless, in Hahn's own opinion it was "healthy that your letter introduces the problem into the literature, and that other people may have some ideas to contribute."[160] In the following weeks, Hahn's ideas contributed crucially to a change in the text of the article, whereby Gutowsky, McCall, and Slichter finally proposed an interaction of a constant independent of temperature and magnetic field, and the magnetic moments of the respective nuclei.[161]

In October, Slichter and Gutowsky reported in *Physical Review* the observation of "slow beats" in nuclear spin echoes of the compounds BrF_5 and IF_5.[162] This was achieved with an NMR apparatus using pulsed frequencies, thus differing considerably from the "continuous wave" set-up that Gutowsky and most other groups used at the time. Spin echoes had been discovered in 1949/50 by Erwin L. Hahn during a postdoc term at the physics department of the University of Illinois, in a neighborhood close to Gutowsky's laboratory. Hahn, in some of his experiments, observed a modulation of the spin echo that differed from the normally seen decay. In 1950, he attributed it to an unspecified interaction between the nuclear spin and the molecule.[163] Gutowsky later wrote about this nearly simultaneous observation, which nevertheless was not realized for a long time:

> It often happens that two independent and apparently different lines of research are found to be related. Much parallel work was done . . . before it

[159] Gutowsky, McCall, and Slichter, "Coupling Among Nuclear Magnetic Dipoles."
[160] Erwin L. Hahn reported this to Gutowsky in a letter of 11 September 1951, Gutowsky Papers, folder "The slow beat in spin echoes." Bloch referred to the article manuscript.
[161] Hahn to Gutowsky, 14 September 1951, Gutowsky Papers, folder "The slow beat in spin echoes." The change of text in Slichter to The American Institute of Physics, 1 October 1951, Gutowsky Papers, folder "Coupling am. N.M. dipoles in molec."
[162] McNeil, Slichter, and Gutowsky, "Slow Beats."
[163] Hahn, "Spin Echoes," 591–592.

> became clear that both groups were dealing with the same physical phenomena. The parallel was obscured by large differences in the types of compounds and the magnitudes of the effects observed by the two groups.[164]

After this realization, experimental evidence was strongly supported by the interrelated development at two sites. In September 1951, Hahn greeted Gutowsky's offer of simultaneous publication of the slow beat phenomenon with the remark: "There is so much yet to be understood about this business that we would be wise to avoid duplicating our researches (except for independent confirmation of important measurements) and give each other interesting dope as it arises."[165] Consequently, the two groups published their results side by side,[166] Hahn serving as contact person at Stanford for the Illinois group. At each of the sites, a group using pulsed equipment and a group using the continuous wave technique worked together to unravel the phenomenon.[167] Reproduction of the respective experiments seemed to have played a role in important experiments only, when too much was at stake to risk an error in the measurement. In other cases, the group avoided double work, in order to proceed faster, and to exclude intense competition. Thus, the reciprocal stabilization of the effects was mainly sought for by close cooperation, and not by competitive replication of experiments. It must be noted, however, that in mid-1951 the experimental characterization of spin-spin coupling was conclusive, and the effect as such not in dispute. What remained in doubt was the theoretical interpretation.

As we have seen, the interpretation of the new phenomenon relied on a new type of interaction between the nuclei.[168] In 1951, Gutowsky, McCall, and Slichter favored an interaction of electron orbitals. Though this effect certainly was present, it could not account for the magnitudes of splittings observed. A stronger coupling mechanism via electron spins was finally worked out by Ramsey and Purcell. Thus, ironically, though the Stanford and the Illinois groups made important contributions to the experimental characterization of the effect, the breakthrough in the theory was achieved by the East coast group. Again, as with the explanation of the chemical shift, Ramsey took the lead in the theoretical side of NMR. He explained the resonance fine structure as a coupling of the nuclear spins via the spins of the bonding electrons. In

[164] Gutowsky, McCall, and Slichter, "Nuclear Magnetic Resonance Multiplets," 280.

[165] Hahn to Gutowsky, 11 September 1951, Gutowsky Papers, folder "The slow beat in spin echoes."

[166] McNeil, Slichter, and Gutowsky, "Slow Beats," 1245–1246; Hahn and Maxwell, "Chemical Shift."

[167] At Stanford, the slow sweep (continuous wave) group was that of Packard and Arnold, later joined by Dharmatti, the pulsed group that of Hahn and Maxwell. In Urbana, Gutowsky and McCall were joined by Slichter and McNeil. See the literature given, and Packard and Arnold, "Fine Structure."

[168] See the summary in Slichter, "Some Scientific Contributions," 277–278.

addition to the successful explanation of the qualitative effects, i.e. number and intensities of the peaks in the multiplets, Ramsey's equations allowed for a quantitative treatment of simple molecules, such as HD.[169]

In contrast, the numerical calculations for the spin-spin coupling made by Gutowsky, McCall, and Slichter provided only a qualitative agreement with the observed splittings. "However, the approximate results do provide simple explanations for the observed general trends, and the values are certainly of the right order of magnitude. More refined calculations of this sort are of potential interest and value for investigating the dependence of bond hybridization upon molecular structure."[170] Characteristically, the approach of Gutowsky included the use of the technique for the study of molecular structure, while Ramsey's approach, though more refined and exact, did not account for this possibility. Moreover, Ramsey was satisfied with the correct calculation of the splitting in only one model molecule, namely HD.

Theoretical input continued to be important in the testing and elaboration of spin-spin coupling, especially because in some cases the splittings vanished. This was of considerable concern to Gutowsky's group because it included the danger of falsifying the new effect. Thus, it was necessary to assign the failings of observation of the splittings to explainable chemical or physical effects. On the physical side, this included relaxation and nuclear quadrupole effects. On the chemical side, impurities and the exchange of molecular groups proved to be the most important causes. The last effect, known as chemical exchange, opened a whole new avenue for the application of NMR in chemistry. It was a direction that was not immediately realized, and when it became clear that chemical exchange could be measured by NMR methods it still needed considerable improvement in instrument performance and theory before it was accepted.

The first system put under scrutiny was that of acids in aqueous solution, which showed shifts of the proton resonances depending on the concentration. Though Gutowsky and his co-author, Apollo Saika, calculated the collapse of the multiplet structure to a single line, they could not experimentally examine systems where this collapse actually occurred.[171] Much later, Gutowsky ascribed the delay in following up this promising method to the lack of suitable systems showing the exchange rates in the millisecond range, and, in addition, to the lack of mathematical knowledge among the chemists: "We, the chemists, were aware for about a while that there was something in the bushes. In other words we were aware that there was some averaging. That probably was

[169] Ramsey and Purcell, "Interactions Between Nuclear Spins"; and Ramsey, "Electron Coupled Interactions."

[170] Gutowsky, McCall, and Slichter, "Nuclear Magnetic Resonance Multiplets," 290.

[171] Gutowsky and Saika, "Dissociation, Chemical Exchange."

describable but we didn't have the mathematical confidence, tools, expertise to do it."[172] Though a system was readily at hand through Meyer's studies of the proton resonance shifts in amide compounds, this was not recognized as such, and again attributed to impurity effects or an instrumental artefact. Only after William D. Phillips at Du Pont in 1955 reported that he could resolve a doublet in amides[173] did Gutowsky see an opportunity to proceed: "The compound impressed me as a very likely candidate for a temperature-dependent rate study by NMR methods."[174] The system that Gutowsky used with his graduate student Charles Hawthorne Holm, the internal rotation of the *N,N*-dimethylamides, became a classic in the field, because this was the first case to examine exchange rates and to determine the respective activation energies.[175] However, it took years to develop the technique fully and to make it a reliable method for the study of exchange reactions and internal rotations. In this field, Gutowsky soon got competition, mostly from organic chemists interested in such conformational studies.

The Research Program of Herbert S. Gutowsky

For Gutowsky, research at the University of Illinois meant continuation of working in the experimental style of molecular spectroscopy that he had become acquainted with during his Harvard days. At first, he sought for control of the instrument, in material and intellectual terms. Nothing could serve better for this purpose than the design and construction of the apparatus. Though Gutowsky depended on the expertise of scientific and technical colleagues, most notably physicists and electrical engineers, his war-related management experience eased the difficulties and finally made the interaction successful. Nevertheless, for the crucial calibration of the instrument he relied on outside data from Harvard and MIT, and his program was stuck in spring of 1950, mostly because of the lack of qualified electronic support. At that time, Gutowsky had a stroke of luck when the publication of the discovery of the chemical shift suddenly provided him with an achievable and fruitful research topic. In this area, he was able to contribute with original observations and theoretical explanations, and his background in theoretical and physical organic chemistry allowed him to interpret the NMR data in ways not accessible to physicists. Because a quantitative interpretation of the chemical shift in purely physical terms was not possible, the pragmatic approach of the chemists was the most promising one. Moreover, his experience with infrared spec-

[172] Gutowsky, interview by Reinhardt, 1 and 2 December 1998.
[173] Phillips, "Restricted Rotation."
[174] Gutowsky, "Chemical Aspects," 292.
[175] Gutowsky and Holm, "Rate Processes."

troscopy made Gutowsky aware of the potential that NMR had for routine use in structural organic chemistry. When it became technically possible, he and Leon Meyer realized the possibility to characterize parts of organic molecules efficiently by NMR. They published this idea, emulating a model of infrared spectroscopy. This contributed to opening the market to the first commercial manufacturer of NMR spectrometers, Varian Associates.

In the early 1950s, Gutowsky was anxious to establish himself in the community of molecular spectroscopists. He emphasized the reliability of the NMR method and the precision that could be achieved. On the theoretical side, he cooperated with the physicists Ramsey, Hahn, and Slichter. His empirical vision was directed to the chemical physicists, and he light-handedly connected NMR to the established terms of this field. Gutowsky's own research field was the electronic structure of molecules; and because this approach more and more became prominent in mainstream chemistry in the United States during the 1950s, NMR became to be better known. But still, the main aim of chemical physics was the establishment of new chemical effects by the mastery of complex physical instrumentation. Nothing serves better as an example than the impeded realization of spin-spin coupling. Though Gutowsky's team commanded all necessary expertise to characterize the novel effect, six months went by before they realized that they saw something completely new. Even then, the experimental characterization took a long while, and the collaboration of four research teams was necessary, working with different techniques at Stanford and Urbana. This discovery was not a sudden event, but a slow encircling of a natural-technical phenomenon, the exclusion of competing hypotheses, and its operational characterization.

The issue of instrumental control figured prominently in the literature of the time, for example in Wilson's *Introduction to Scientific Research*.[176] But Wilson did not give an answer to the question as to when a scientist knew that he really did have the apparatus under control. This problem was raised much later by the philosopher of science Jerome Ravetz, and Gutowsky's case shows that the apparatus was thought of as being under control if the experimental scientist could replicate the results of other laboratories. In addition, and this is the argument of Ravetz, the scientist needed some knowledge about the physical causes of the data that were produced by scientific apparatus, while the standardization of data, their validity, and the norms that governed the acceptance of precision and accuracy were mainly socially determined. Moreover, data went through an interpretative stage and became information, which was presented in graphs, tables, and figures. All this led to a mutual stabilization of instrumentally produced data even before the material was published. Calculation, convergence, and calibration were important issues in the experimental

[176] Wilson, *Introduction to Scientific Research*, 137–140.

practice of twentieth-century physical and biological sciences.[177] Calculation meant a rigorous understanding of the experiment's working mode in fundamental theoretical terms. Convergence described attempts of scientists to find the same effects with different techniques. Measurements with a surrogate signal or the correlation of a new instrument's data to values obtained by older, similar ones enabled the calibration of a new instrument. These undertakings were full of pitfalls. Nevertheless, convergence and calibration largely remained inside the boundaries of the community of experimental molecular spectroscopists. In contrast, the reduction of chemical effects to fundamental physical theories involved cooperation between experimentalists and theoreticians. Here, as Gutowsky's case shows, chemists and physicists were in danger of using incompatible languages, which were related to a conflict of interests and a status gap between experimental chemists and theoretical physicists.

In his own research, Gutowsky applied at least two more strategies: experimental characterization, and correlation to chemical concepts. While the former added to the control of the method, the latter created its nexus to applications. Arguably, the correlation of NMR data to established chemical concepts decided the first successes of the method, and set it on the track that later led towards its uses in biology and medicine as well as in chemistry. But most importantly, in the case of Gutowsky's research, the process of bringing an apparatus under control was a matter of group-internal teamwork, combining chemical, physical, and electronic expertise.

Indeed, Gutowsky's criteria for an ongoing research program came pretty close to the opinion of philosopher of science Henk Zandvoort, in emphasizing the extrinsic success of the NMR research program. Especially the chart relating chemical shift data with functional groups clearly is a candidate for an extrinsic orientation, in this case towards organic chemistry. Nevertheless, it is hard to imagine that there ever existed a single, mono-layered research program in NMR. Each practitioner had his own goals and predilections, as well as specific capabilities and constraints. In addition, the focus of Gutowsky's research program remained on chemical physics and, in weaving electronegativity, bond hybridization, and inductive and resonance effects into the net of NMR concepts, he was able to firmly establish NMR inside the core concepts of chemistry. He certainly did not perceive of himself as running a spectroscopic supply program for the guide program of chemistry.[178] In addition, in Gutowsky's view, chemical theories were not subordinated to physical ones. NMR was a field where both disciplines merged.

---

[177] For a discussion see Rasmussen, *Picture Control*, 12–14.

[178] This is Zandvoort's terminology. See Zandvoort, *Models of Scientific Development*, 231–241. Gutowsky always resented becoming dependent on organic chemists' needs, despite, maybe because, of his early work on infrared spectroscopy of a service kind.

All the time, and despite his partial orientation towards chemical applications, Gutowsky used NMR as a machine for the construction of novel effects. This was the core of his research program that covered all of the important aspects of the technique. Nevertheless, Gutowsky did not think that the apparatus created the phenomena. For Gutowsky, NMR was a gate to the natural world. But the technological society he was part of, and the enormous resources the United States set aside for the unraveling of the limited secrets of nature with the help of sophisticated research instrumentation, made him doubt the future prospects:

> One of my pet sayings is that research is over-supported. That it's a natural resource we are mining, taking it out at such a rage that there won't be any left for our grandchildren. We should leave some natural discoveries for the future. I tried that one on my colleagues at academy meetings designed to get more money out of the government. That was after Frank Westheimer made a pitch for 15 percent increase per year for ever and ever. I was teaching physical chemistry then and I amused myself and the class by computing how long it would be before the national budget would be totally spent. I think more undergraduate students carried away this lesson than anything else I ever taught.[179]

Next to the development of high resolution NMR that was the focus of this chapter, Gutowsky continued at Urbana the broad line measurements after the model of the work at Harvard. With Bruce McGarvey, and beginning in April 1951, Gutowsky undertook research on metals.[180] Later, he and John G. Powles studied motion effects of the methyl group in solid organic compounds. This led to the important conclusion that a quantum mechanical tunneling effect should be invoked to explain the measured activation energies and the observed frequencies.[181] With Lee Meyer, Gutowsky approached the complicated field of NMR of polymers with a study of rubber.[182] This was all done at the time when he also pushed research in the field based on the chemical shift and spin-spin coupling. Together with Martin Karplus and David Grant, he contributed to the use of spin-spin coupling constants for the unraveling of the

---

[179] Gutowsky, interview by Reinhardt, 1 and 2 December 1998. Frank Westheimer was the chairman of the committee that prepared the so-called Westheimer report on the state of the chemical sciences in mid-1960s United States. Gutowsky was a member of this committee. See National Academy of Sciences, *Chemistry: Opportunities and Needs.*

[180] Laboratory notebook "NMR II. Research notes '50–'51," 28 April 1951, 192 ff., Gutowsky Papers, marks the beginning of the program on lithium. See among other publications: Gutowsky and McGarvey, "Nuclear Magnetic Resonance in Metals, I"; Gutowsky and McGarvey, "Nuclear Magnetic Resonance in Metals, II."

[181] See especially Powles and Gutowsky, "Proton Magnetic Resonance of the CH_3 Group, III."

[182] Gutowsky and Meyer, "Proton Magnetic Resonance in Natural Rubber."

stereochemistry of molecules.[183] With Karplus and Thomas Farrar, Gutowsky also continued to be active in the more theoretical aspects of NMR.[184] As high resolution NMR was most interesting for organic chemists and biochemists, Gutowsky occasionally collaborated with scientists from these fields, running spectra and interpreting them.[185] In 1956, when he became full professor and head of the division of physical chemistry, he increasingly concentrated on organizational and management matters. In 1967, he accepted the headship of the department of chemistry and oversaw its reorganization into a separate school in 1969–70, serving as director of the school of chemical sciences from 1970 to 1983. During that time Gutowsky continued to lead a somewhat reduced, but active research program in NMR. Beginning in the late 1970s, he added biological systems to his research, mainly through his collaboration with Eric Oldfield.[186] In 1983, after his resignation as director, he embarked on a new research field, pulsed microwave spectroscopy. He retired in 1990, but continued to work in that field until his death on 13 January 2000.

NMR scientists tend to name their methods with acronyms and initialisms. Thus, in contrast to organic chemistry, where named reactions abound, the originators of NMR techniques were easier forgotten. One of the closest competitors of Gutowsky in the 1950s and 1960s, the organic chemist John D. Roberts, was very much aware of this incorporation by obliteration. In a letter written in 1980 to a colleague, Roberts emphasized that he was convinced that Gutowsky was

> an outstanding prospect for a Nobel Prize in Chemistry for his contributions in making nuclear magnetic resonance a useful chemical tool . . . . His contributions are indeed so useful and such a part of everyday chemistry that many forget that they were not handed down from ancient Greece but by a living, breathing, still very active person.[187]

---

[183] See Gutowsky, Karplus and Grant, "Angular Dependence."

[184] Karplus, Anderson, Farrar, and Gutowsky, "Valence-bond Interpretation."

[185] For example with Melvin S. Newman of Ohio State University in 1955/56 (see Gutowsky to Newman, 1 March 1956, Gutowsky Papers, folder "NMR of hexadecilene"); with Milton Tamres of the University of Michigan and Scott Searles of Kansas State College on small ring compounds in 1954 (see correspondence in May and June 1954, Gutowsky Papers, folder "Effect of ring size on e.d. sat. het." and Gutowsky et al., "Effects of Ring Size"). Gutowsky also allowed Mildred Cohn of Washington University, St. Louis to run spectra on his machine. Mildred Cohn, interview by Carsten Reinhardt, 25 November 1998.

[186] Gutowsky, interview by Reinhardt, 1 and 2 December 1998.

[187] Roberts to Adam Allerhand (chairman of the department of chemistry of Indiana University, and a former co-worker of Gutowsky), 14 August 1980, Roberts Papers, folder "Gutowsky, H."

Chapter Three

Toward a Chemistry of Instruments

Mass spectrometry[1] developed in the first half of the twentieth century from research in cathode and canal rays to a technique for precision determination of nuclidic masses, and abundance measurements of various isotopes. During World War II, the petroleum and chemical industries provided the impetus and early context of application in organic chemistry, when they required routine and reliable methods for the control of their production processes. The data collected by industrial researchers and manufacturing chemists provided first evidence that this technique had potential in many domains of academic chemical research. Reliable instruments and suitable techniques for the handling of organic molecules were commercially available by the early 1950s. However, a huge resistance on the side of academic organic chemists had to be overcome. For them, mass spectrometry was like "black magic,"[2] and not serious science. The reason was that, to a large extent, industrial chemists used mass spectrometry for quantitative analysis of mixtures of compounds whose structures were known, mainly hydrocarbons. In contrast, academic chemists needed mass spectrometry for structural research of unknown compounds. But the manners in which long-chain hydrocarbons fragmented in the instrument did not allow spectra-structure correlations and thus did not fulfill the necessary requirement for structural research. As a consequence, mass spectrometry had a "terrible reputation" among chemists.[3] Nevertheless, an important part of the chemical community had strong inter-

---

[1] In accordance with new terminology, the term mass spectrometry includes the detection of ions by photographic plates (formerly mass spectrography), and by electronic counters (mass spectrometry).

[2] Meyerson, "From Black Magic to Chemistry."

[3] Fred W. McLafferty with reference to the 1951 Pittsburgh Analytical Conference, interview by Carsten Reinhardt, 16 and 17 December 1998.

ests in improving the situation. Industrial chemists were eager to add to their empirical and statistical arsenal new methods. Furthermore, the understanding of the fragmentation mechanisms of the sample molecules might lead to a rational interpretation of the spectra. This would contribute toward answering the urgent need for improved qualitative analysis of the fast-growing range of chemicals available in the post-war era.

During the 1960s, mass spectrometry, as did NMR, became a technique of utmost importance in organic chemistry. Even more so than in NMR, the understanding of the physical underpinning of the technique was not sufficient for the interpretation of mass spectra of most organic compounds. With its concepts of electronic molecular structure and reaction mechanisms, physical organic chemistry provided the basis for the unraveling of the fragmentation pathways leading to mass spectra. Thus, knowledge of the (physical) mechanism *of* the instrument had to be supplemented by knowledge of the (chemical reaction) mechanism *in* the instrument.[4] In the words of one of its main proponents, Klaus Biemann, mass spectrometry essentially consisted of microdegradation reactions of molecules, combined with a suitable detection system of the fragments, and explained in terms of physical organic chemistry. Enhanced by the successes of the physical organic approach in American chemistry in the 1950s and 1960s, mass spectrometry began to receive a good reputation and credibility from leading organic chemists. Also in social perspective, the recourse to physical organic chemistry was a wise choice, as mass spectrometry was first used in natural product chemistry and analytical chemistry. In this way, the minority of natural product chemists and the outsider group of industrial analytical chemists became connected to one of the dominating subdisciplines of American chemistry, physical organic chemistry.

This chapter deals first with the emergence of organic mass spectrometry in the chemical industry during World War II, and its development from a technique suitable for quantitative analysis of hydrocarbons to a method that enabled the qualitative determination of compounds in mixtures. This industrial connection of its early history parallels that of infrared spectroscopy.[5] Its afterwar development, though, was a protracted one, when compared to infrared spectroscopy. At the end of the 1950s, when infrared spectroscopy already was a routine technique in most organic chemistry departments, and NMR was on the verge of becoming the standard method in structural research, mass spectrometry practically did not exist for this purpose.

The recourse that some chemists took in order to change this situation was the rationalization of the fragmentation processes of organic ions in the mass

[4] Actually, this distinction applies only if the chemical substances injected into the mass spectrometer are thought of as not being a constitutive part of the instrument itself.

[5] Rabkin, "Technological Innovation in Science."

spectrometer in terms of accepted chemical theories, especially mechanisms of reactions and molecular rearrangements. The main part of this chapter's story centers on three scientists, Fred W. McLafferty, Klaus Biemann, and Carl Djerassi, who all contributed to the emergence of organic mass spectrometry in the period from the mid 1950s to the late 1960s. Though each of them had his unique background, specific research style and original research subjects, they shared common experiences, among them the resistance of the chemical community as well as the generous funding policy of governmental research agencies. This period and this group of scientists is chosen because the impact of chemical concepts and theories is most visible here, before, in the 1970s, progress in technology led to a change in viewing mass spectra at a time when most chemists regarded the method already a very useful addition to their arsenal. Thus, this chapter deals with a turning point in the history of mass spectrometry through just three of many participants, though influential ones. Especially with textbooks, written in the terminology of organic chemistry, Biemann, Djerassi, and McLafferty spread the novel technique in chemistry. Throughout their careers, all three of them had strong interactions with the chemical industry and instrument manufacturers. This interaction crucially shaped the history of organic mass spectrometry. Each one, although in a different vein, became an expert regarding the instrument and had an important role in its establishment.

For the analytical chemist Fred Warren McLafferty, who started out at the industrial enterprise Dow Chemical of Midland, Michigan, and in 1964 moved to Purdue University at Lafayette, Indiana, mass spectrometry always provided the crucial impetus for his work. Although analytical chemistry was in decline in the 1960s at U.S. chemistry departments, McLafferty and other industrial experts of mass spectrometry were able to reverse this trend in their respective fields and at the same time advanced their own careers. The rationalization of the fragmentation processes in a mass spectrometer along chemical lines provided the basis for both the use of the mass spectrometer in chemical analysis and the opening of a new field of studies: the chemistry of unimolecular reactions in the gaseous phase. Moreover, experts in the highly sophisticated (and expensive) field of mass spectrometry were in great demand. The use of physical instruments in chemistry in general was booming, and mass spectrometry profited from this development.

One other key person in the events making mass spectrometry a reputable scientific field was Klaus Biemann of the MIT chemistry department. He expanded mass spectrometry to the realm of natural products, namely amino acids and peptides, alkaloids, and a variety of other substances. With special interpretation methods, he proved the feasibility of mass spectrometry for the analysis of such complex compounds. The access to high-performance instruments, and the merging of instrumental know-how and chemical experience provided the basis for Biemann's rise as an acknowledged expert in the field,[6] with funding agencies and companies taking on the roles of sponsors and decision makers. Experts only fulfilled their roles if they shared and networked their expertise. Biemann connected fellow scientists, corporations, and agencies in his laboratory, and shared much of his knowledge before it was published. In the course of his career, Biemann effectively and efficiently networked his spectrometer and expertise for the community of organic chemists and biochemists, whose work he knew well from his own training. In this endeavor, Biemann participated in both scientific cooperation and academic-industrial collaboration. In return, he obtained funds to pursue his research, and procured the types of compounds he urgently needed for the development of his research program. Moreover, with his style of research he created positions for his graduate students and postdoctoral fellows. Through this, the most urgent need of 1960s mass spectrometry, the availability of trained experts, was satisfied.

In this process, the whole experimental system of natural product chemistry changed. I refer to the experimental system in the same sense that Robert Kohler has done in his work on the "systems of production" in biochemical genetics.[7] Instruments, methods, and the social organization of laboratory work were interwoven constituents of an experimental system. In Carl Djerassi's laboratory at Stanford University, mass spectrometry and steroid chemistry merged to a unique system of elucidation. In sharing Kohler's portrait of science as a productive activity in a marketplace environment, I have chosen to emphasize the unraveling of structures of unknown compounds in Carl Djerassi's research. This fits the direction of his university-based work, which is the focus here. In contrast, in his industrial activities, Djerassi embarked on the synthetic approach, the productive side of chemistry. He always underscored the divide between industry and academia in his professional life, and here we can only take a glance at the industrial part of Djerassi's work.

According to Djerassi, natural product chemistry, being a European-style chemistry, was never at the centers of academic chemical activity in the United States, these positions being taken by physical organic chemistry and, especially, by synthetic organic chemistry.[8] But natural product chemistry was important in the ways it connected the elucidation of complex chemical struc-

[6] Jan Golinski has elaborated on the consequences of the availability of an instrument, and the scientist's skills, using the early nineteenth-century example of Humphrey Davy and the voltaic pile. See Golinski, *Making Natural Knowledge*, 140–141, and Golinski, *Science as Public Culture*, chap. 7.

[7] Kohler, "Systems of Production," 88.

[8] Carl Djerassi, interview by Jeffrey L. Sturchio and Arnold Thackray, 31 July 1985, pp. 38–39, Chemical Heritage Foundation, Philadelphia, PA. Djerassi mentioned that organic chemistry in the United States developed relatively late, and thus did not have the tradition in natural products chemistry that Europeans had. The first original contributions of American chemists were in physical organic chemistry.

tures to their subsequent (partial) syntheses. Through this, and as a result of the pharmaceutical uses of many natural products, natural product chemistry was of prime interest for pharmaceutical, chemical, and later also biotechnological companies. Mass spectrometry proved to be instrumental in making natural product chemistry an American science, and in this chapter we will follow some of the strategies employed by its protagonists to make that happen.

From Physics to Chemical Industry

Mass spectrometry has its roots in the research of Joseph John Thomson on cathode and canal rays in the late nineteenth and early twentieth centuries.[9] Thomson, at the Cavendish Laboratory of the University of Cambridge, tried to determine the mass-to-charge ratio of the corpuscles that he thought were the building blocks of cathode rays and canal rays (in modern terms electrons and positively charged ions, respectively). For this, he used electric deflection of the charged particles. This is in principle the working mechanism of a mass spectrometer. In a mass spectrometer, positively charged particles (and some negative ions) are formed by electron bombardment of a compound, normally present in the vapor state. These positive ions and their fragments are then accelerated in an electric field and deflected by a magnetic one. Variation of one of these fields focuses ions with the same mass-to-charge ratio at a detector. The resulting mass spectrum is a recording of the abundance of the positive ions versus their mass.[10] In 1912, and with an apparatus called a parabola spectrograph, Thomson and Francis W. Aston were able to detect traces that led Aston to assume that the noble gas neon did consist of more than one atomic "species." Shortly before Thomson and Aston obtained these results, Frederick Soddy postulated different kinds of building blocks for radioactive elements. He named them isotopes, from having the same position in the periodic table. Isotopes cannot be distinguished by chemical behavior, but only by their differences in mass. In the case of the not radioactive neon isotopes, neon-22 and neon-20 were distinguished by Aston and Thomson. Thus, their parabola spectrograph provided the first evidence for the existence of stable (not radioactive) isotopes. With improved instrumentation and in a very short time after World War I, Aston showed that most elements were not uniform, and characterized their isotopes by mass. Moreover, in many cases he measured their relative abundances. This led to the exact establishment of many atomic mass values by

---

[9] For the history of mass spectrometry see Laitinen and Ewing, *History of Analytical Chemistry*, 216–229; Beynon and Morgan, "Development of Mass Spectrometry"; Remane, "Zur Entwicklung der Massenspektroskopie"; Sparkman, "Mass Spectrometry"; Grayson, *Measuring Mass*, 3–19.

[10] There are many other techniques for ionization, deflection, and detection. This description follows the most commonly used technology during the period of interest.

mass spectrometry. Mass spectrometers became important tools for nuclear physics, when the accurate determination of nuclidic mass enabled the calculation of inner-nuclear forces. The need for precision in these measurements led to considerable improvement in instrumental performance during the 1930s.[11]

While the accurate and precise measurement of nuclidic mass called for elaborate and expensive instrument design, isotope abundance measurements could be undertaken with a simpler spectrometer. After World War I, at the University of Chicago, Arthur J. Dempster developed a mass spectrometer for this purpose.[12] He introduced the use of electron-impact ion sources, and he applied magnetic deflection of the particles. Through the bombardment with electrons, ions could be produced in a more subtle way than in previous instruments, while the magnetic deflection allowed for directional focusing of the ion beam. Moreover, Dempster opted to detect the ions with an electrometer, and not a photographic plate. Overall, Dempster's design was the predecessor of the instruments that from the 1940s on were put to use in the chemical industry.

During World War II, at the University of Minnesota in Michigan, Alfred O. Nier developed mass spectrometry into a precise, relatively inexpensive, and versatile research technology. Nier's instruments were used in research on the separation of the uranium isotopes uranium-235 and uranium-238 in the context of the Manhattan Project for building the atomic bomb. After the war, the practical design of the simpler version of Nier's instrument allowed its widespread use in many research fields. Most important was the use of electronic parts for the control of the instrument and the detection of signals, an improved vacuum-pump, and a configuration that allowed adjustment even during operation.[13]

Up to World War II, mass spectrometry was mostly used for precision measurements of nuclidic masses and isotopic abundances. In addition, in the late 1920s, experiments on the ionization of gaseous molecules in the tradition of James Franck and Gustav Hertz were merged with the research program of mass spectrometry. Franck and Hertz investigated the ionization potential of atoms and simple molecules by measuring the energy needed to produce the ions by electron impact, but their experimental set-up was not suited to determine the nature of the ions. The mass spectrometer pointed to a solution of this problem.[14] The instrumental improvement of the ionization method at the University of Minnesota by John Tate (the first teacher of Alfred O. Nier in mass spectrometry), Walter Bleakney (later at Princeton), Wallace Lozier (later Columbia), and Phillip T. Smith, enabled the measurement of ionization potentials by mass spectrometry. Tate and his collaborators used a collimated

---

[11] Jordan and Young, "Short History of Isotopes," 527, 529–530.

[12] See Beynon and Morgan, "Development of Mass Spectrometry," 21–24.

[13] Remane, "Zur Entwicklung der Massenspektroskopie," 102.

[14] Smyth, "Products and Processes."

beam of electrons to ionize the molecules under scrutiny. As a consequence, a direct relation between the energy of the electrons and the ionization potential of the molecule could be established. Because of the small energy spread of the ions produced in this way, a relatively simple magnetic field could be used for their subsequent deflection.[15] In the late 1930s, incidentally, the determination of ionization potentials was extended to include hydrocarbons.[16]

In 1931, H. R. Stewart and A. R. Olson of the University of California at Berkeley were among the first to study simple hydrocarbons, such as propane and butane, with the aid of the mass spectrometer.[17] Around 1940, an array of simple organic chemicals had been studied with mass spectrometric methods.[18] In biochemical research, the wish to measure stable isotopes in tracer experiments made the use of a mass spectrometer mandatory. Stable isotopes had some advantages over radioactive isotopes in biochemical research, most notably the absence of further decay and thus the much longer time available for the studies.[19] Tracer techniques with stable isotopes, analysis by isotope dilution,[20] direct gas analysis, chemical kinetics (direct measurements of chemical reaction rates), and the determination of ionization potentials dominated the applications of mass spectrometry in academic chemistry at the end of World War II.[21]

In the late 1930s and early 1940s, a general movement in industry took place: the "instrumental revolution in analytical chemistry."[22] The older chemical analytical methods were replaced by methods based on physical instrumentation. Mass spectrometry was part of this development. In cases where chemists wanted to detect minute amounts of gases, to analyze mixtures, or to control a continuous chemical process, the mass spectrometer offered advantages with respect to previously used techniques.[23] The most important field of application was in petroleum refining, and this industry provided the market that instrument manufacturers catered for. While in academic circles self-built instruments were the rule, the petroleum industry used instruments that were

---

[15] Laitinen and Ewing, *History of Analytical Chemistry*, 217.
[16] A good summary of this work is Neuert, "Gasanalyse mit dem Massenspektrometer."
[17] Stewart and Olson, "Decomposition of Hydrocarbons."
[18] Examples are Smith, "Ionization and Dissociation I"; and Kusch, Hustrulid, and Tate, "Dissociation."
[19] Rittenberg, "Some Applications."
[20] Isotope dilution was an analytic method to measure minute amounts of an element in a sample. For that, the natural ratio of isotopes of that element was altered by purposeful dilution with one of the isotopes in pure form. The altered ratio could be measured to a very high degree of exactness with the mass spectrometer, and through that the amount of the element could be determined. Hintenberger, "Anwendung der Massenspektroskopie," 83–89.
[21] Stewart, "Mass Spectrometry."
[22] Baird, "Analytical Chemistry."
[23] Hipple, "Gas Analysis."

built commercially by companies such as Consolidated Engineering Corporation (CEC, from 1956 Consolidated Electrodynamics Corporation), Pasadena, California; Westinghouse Electric & Manufacturing Co., Baltimore, Maryland; and General Electric. In Great Britain, Metropolitan Vickers, and after World War II in Germany, Atlas Werke of Bremen manufactured mass spectrometers.[24] Instruments cost approximately $20,000 in 1942, with a resolution of about 100 mass units.[25]

In the United States, CEC dominated the market.[26] The firm was an outgrowth of a disputed patent-cross-licensing arrangement in the oil industry. To circumvent the enforced sharing of its patents on seismic prospecting with Texaco, the United Geophysical Company in 1938 split off its instrumentation branch as a separate company, CEC. United Geophysical Company was founded in 1935 by Herbert Hoover, Jr. (the son of the former U.S. president), and engaged mainly in exploration for the oil industry. The new company, CEC, was headed by Hoover as president, and by Harold Washburn as vice-president and head of research and development. Until 1941, the sole business of CEC remained the ownership and leasing of United Geophysical Company's instrumentation. Then, in-house developments were added, and a product line was built up. CEC became an instrument manufacturer, developing, producing, and selling instruments for the oil industry at large, with a focus on prospecting equipment. Prospecting, too, was on the mind of Washburn when soon after the founding of CEC he first encountered mass spectrometry through a contact with the California Institute of Technology (Caltech) at Pasadena. There, Daniel Dwight Taylor had just given up trials to develop a mass spectrometer to analyze hydrocarbon fractions encountered in oil refineries.[27] Taylor was hired by CEC to explore the potential of the technique for both Taylor's original purpose and soil gas analysis. Though the latter attempt failed, in 1943 Washburn could point to successful applications of the instrument in the analysis of the low-boiling constituents of refinery products.[28] At the end of 1942, their first instrument, the CEC 21-101, was installed at the Atlantic Refining Co. of Philadelphia, Pennsylvania. This had an interesting prehistory,

[24] Borman, "Brief History."
[25] This resolution allowed the generation mass spectra with unit resolution for molecules of up to a molecular weight of 100 u (atomic mass units), i.e., the analytical chemist could distinguish peaks at the values of 99 and 100.
[26] On the history of CEC see Meyerson, "Reminiscences," 198–201.
[27] In 1934, Taylor of the physics division of Caltech built a mass spectrometer using an electric radial field for velocity focusing and a magnetic field for deflection. He did so before reading Herzog's 1934 article in *Zeitschrift für Physik*, which described the same geometry necessary to obtain double-focusing. Taylor used the instrument to study the spectra of nitrogen, carbon monoxide, ammonia, and hydrazine. Taylor, "Modified Aston-type Mass Spectrometer."
[28] Washburn, Wiley, and Rock, "Mass Spectrometer"; Washburn et al., "Mass Spectrometry"; Brewer and Dibeler, "Mass Spectrometric Analyses."

as the first order originally had come from Standard Oil of New Jersey. Jersey Standard cancelled their order after they were confronted with the patent licensing requirements of CEC. Reminiscent of the trouble the mother company of CEC, United Geophysical, experienced with patent licensing from customers, CEC demanded that all improvements that were made on their instruments by the users should be licensed to CEC. Jersey Standard was unwilling to agree to this condition, and subsequently supported the mass spectrometer development at Westinghouse with expertise and manpower badly needed by the latter company. Thus, CEC's patent policy helped their major competitor in the mass spectrometer business to gain a foothold in the oil industry,[29] as both other manufacturers of mass spectrometers during the war, Westinghouse and General Electric, were large corporations lacking contacts to this valuable potential customer branch.[30]

The most important factor for CEC's success, however, as well as for Westinghouse's and General Electric's, were wartime needs. World War II had led to a "tremendous need for aviation gasoline and the new control requirements it created. Otherwise, we would never have been able to sell such an expensive and complicated instrument for refinery control analysis."[31] Before the war, hydrocarbon gas mixtures from the huge catalytic crackers were analyzed by distillation using Podbielniak fractionating columns together with infrared spectroscopy. The use of infrared spectroscopy had already resulted in a considerable saving of time, but mass spectrometry improved on that greatly. In 1943 it was shown that a nine-component mixture of C_5 and C_6 hydrocarbons could be analyzed in four hours by mass spectrometry, compared with ten days using the traditional method of refraction index measurements of fractions from the Podbielniak columns.[32] Nevertheless, complete analysis required distillation into narrow-boiling fractions and the solution of sets of large numbers of equations.

In addition to the petroleum industry, the application of mass spectrometers to organic chemical analysis was initiated by the wartime needs of the synthetic rubber program, and was made possible mostly through the work of Harold W. Washburn of CEC, John A. Hipple of Westinghouse, and their collaborators.[33] For the butadiene production program, which supplied the feedstock for synthetic rubber, mass spectrometry competed in terms of accuracy

[29] For a description of the Westinghouse instrument see Hipple, "Gas Analysis."

[30] This is the opinion of David Stevenson, an employee of Westinghouse (1940–1942) and Shell Development Company, who was intimately familiar with the development of the Westinghouse instrument. See Meyerson, "Reminiscences," 199–200.

[31] Meyerson, "Reminiscences," 201.

[32] Washburn, Wiley, and Rock, "Mass Spectrometer," 546. The authors also gave an account of how they computed the percentage of simple hydrocarbons in a mixture.

[33] Beynon and Morgan, "Development of Mass Spectrometry," 28.

and precision with infrared, ultraviolet, and distillation methods, as is shown by a report in which more than 70 laboratories in the United States were evaluated for performance in analytical procedures. When applied in a highly standardized manner (and using mainly instruments of one manufacturer, CEC), mass spectrometry made possible the exact determination of the light hydrocarbons of gas mixtures in the C_3 to the C_5 regions generally found in industry.[34] But there were deficiencies that made the use of mass spectrometry a difficult undertaking. The electronic circuitry of the instrument controlled by vacuum tubes was not always reliable. Moreover, electron multipliers had only just been developed and were not widely used for detection of fragment ions.[35] Instead, Faraday cups with vacuum tube electrometers were the basis of detection. The recording of spectra was difficult before the routine development of fast pen-operated recorders replaced the use of galvanometer and scale. Early computers facilitated the solution of equations necessary for the interpretation of spectra of complex mixtures of compounds, but still were difficult to operate.[36] It was the opinion of one of the pioneers of mass spectrometry in chemistry, John Beynon, that if the introduction of gas and liquid chromatography (around the mid-1950s) had taken place ten years earlier this would have largely prevented the development of mass spectrometry for the needs of the chemical industry.[37] Gas chromatography (GC) very well suited the needs of industry and was much less expensive. In the mid-1950s, when the technique was rapidly distributed through the communities of academic and industrial scientists, some mass spectrometrists combined both methods in the new technique of GC/MS.[38]

In the first decade of the mass spectrometer's use in the chemical and petroleum industry, the close cooperation between users and manufacturers proved to be an asset for the successful application of the instrument. Between 1944 and 1954, CEC published 108 mass spectrometer group reports, the majority of which described improvements in and trouble-shooting of instrumentation. Also from 1944, regular meetings were held, merged in 1952 with the group meetings of General Electric and scheduled to take place at the Pittsburgh Conference of Analytical Chemistry and Applied Spectroscopy. This was the predecessor of the annual meetings of the Committee E-14 of the American Society for Testing Materials (ASTM), which defined as its aim the "promotion of knowledge and advancement of the art of mass spectrometry."

---

[34] Starr and Lane, "Accuracy and Precision."

[35] For an early use at the University of Minnesota (with John Tate) see Allen, "Detection of Single Positive Ions."

[36] See King and Priestley, "Spectrometric Analysis."

[37] Beynon and Morgan, "Development of Mass Spectrometry," 26–27.

[38] Gohlke and McLafferty, "Early Gas Chromatography/Mass Spectrometry."

In 1968, these meetings were taken over by the newly-founded American Society for Mass Spectrometry.[39]

The use of mass spectrometers in the petroleum and chemical industry provided the basis for its later application in structural organic chemistry in a threefold way: First, instrument makers produced relatively large numbers of instruments and developed the mass spectrometer as a highly reliable and precise instrument suitable for organic analysis; second, many spectra and fragmentation patterns were obtained and published, mainly through the efforts of the Hydrocarbon Research Group of the American Petroleum Institute (API) with the Research Project 44;[40] and, third, some of the chemists trained in the use of the instrument for quantitative determination of hydrocarbons later applied the technique to research in other organic compounds, the most well known being Seymour Meyerson of Standard Oil of Indiana, and Fred W. McLafferty of the Dow Chemical Co. in Midland, Michigan.

Chemistry in an Instrument

When in 1951, Milburn "Jack" O'Neal and Thomas P. Wier at the Houston Manufacturing-Research Laboratory of the Shell Oil Company described a heated inlet system for the analysis of the high-boiling fractions of petroleum, the use of mass spectrometry for structural elucidation of complex molecules seemed to be close at hand.[41] Though they investigated compounds up to a molecular weight of 600, and excellent instruments were commercially available at the time, the extension of mass spectrometry to research in structural organic chemistry was not achieved immediately. Klaus Biemann, one of the pioneers of the structural approach, later explained this delay by the very success of the mass spectrometer in the petroleum industry. As an outcome, most organic chemists regarded mass spectrometers as very expensive and elaborate

---

[39] Meyerson, "Reminiscences," 202–203.

[40] The API Research project 44 was concerned with "the collection, analysis, calculation, and compilation of data on the physical, thermodynamic, and spectral properties of hydrocarbons and related compounds." Headquarters were the Carnegie Institute of Technology at Pittsburgh, Pennsylvania, with a branch at the University of California at Berkeley (Kenneth S. Pitzer). Covered were infrared, ultraviolet, and mass spectral data. For mass spectra, the most prolific contributor in 1951 was the National Bureau of Standards (NBS), Washington, D.C. with more than a half of the contributed spectra, which totaled 602. The NBS was followed by Humble Oil and Refining Co. of Baytown, Texas. See Rossini, "American Petroleum Institute."

[41] O'Neal and Wier, "Mass Spectrometry of Heavy Hydrocarbons." See, for quantitative analysis, also the contributions of scientists of Humble Oil, Baytown, Texas: Thomas and Seyfried, "Mass Spectrometer Analyses." A similar system was developed by W. S. Young, R. A. Brown, and F. W. Melpolder of Atlantic Refining Co., Philadelphia, and later, at Dow, by Victor Caldecourt: Caldecourt, "Heated Sample Inlet System." See McLafferty, "Mass Spectrometric Analysis," 307.

instruments, which were tricky to handle. This was due entirely to the quantitative industrial approach, the analysis of mixtures of compounds of known structure.[42] Furthermore, the fragmentation patterns of hydrocarbons did not show any promise for correlation with structures. This had to await systematic investigations in molecules with an aromatic moiety or particular functional groups, an endeavor that just began in the early 1950s.

In hindsight, the industrial chemist Seymour Meyerson described his work as an effort to overcome these prejudices:

> I believe that my major contribution has been to help convince myself, as well as other mass spectrometrists and chemists in general, that the things that happen to a molecule in the mass spectrometer are in fact chemistry, not voodoo; and that mass spectrometrists are, in fact, chemists and not shamans.[43]

In general, three main possibilities were available to tackle the task of making organic mass spectrometry a scientific undertaking. A theoretical approach, based on calculations of the mass spectra that were expected from quantum chemical considerations of the molecule; an empirical approach that connected certain peculiarities of mass spectra to specific molecular structures, and an orientation that rationalized the yield of molecular fragments within the terms of physical organic chemistry. It was a combination of the latter two strategies that finally proved to be the most worthwhile. But in the late 1940s and early 1950s, the majority of users certainly would not have predicted such an outcome.

At first, technical difficulties made the investigations cumbersome. Earlier attempts to use mass spectrometry for the analysis of relatively complex organic compounds were doomed, because of lack of either resolution or intensity, and because of thermal decomposition of organic molecules in the ion source.[44] Second, even when mass spectra could be measured, trials of spectra-structure correlations proved to be useless, because of the unpredictable fragmentation of molecules in the mass spectrometer. Especially rearrangements of the ionized molecules after electron impact made the task of a mass spectrometrist difficult.[45] For example in 1948, while measuring mass spectra of octanes, a group of researchers at the National Bureau of Standards in Washington, D.C. stated clearly that the major stumbling block for the use of mass spectrometry in

---

[42] Biemann, "Applications of Mass Spectrometry," 260.

[43] Meyerson, "From Black Magic to Chemistry," 960A.

[44] See McLafferty, "Billionfold Data Increase," 1–2; and Taylor, "Modified Aston-type Mass Spectrometer," 671, who stated the thermal decomposition of the hydrazine sample used. Cf. Linder, "Mass-spectrographic Study," who reported the stability of benzene compared with hydrocarbons.

[45] Stewart and Olson, "Decomposition of Hydrocarbons." For a review, see Bursey, Bursey, and Kingston, "Intramolecular Hydrogen Transfer in Mass Spectra. I."

structural research was the lack of understanding rearrangements.[46] Still, in the mid-1950s, experimental findings in the form of anomalous peaks clouded the picture of spectra-structure relation. Though some peaks were assigned to rearrangements within the molecules, these rearrangements were not thought to be specifically structure-related. For example, a 1950 paper by Alois Langer of Westinghouse described rearrangements with the statement that "all directed bonds in the molecule have vanished and the regroupings have occurred statistically." Four years later, the chemical physicist John Turkevich made a similar proposal, referred to as "sudden death theory."[47]

Most important in the endeavor to calculate for this random fragmentation was the quasi-equilibrium—or statistical—theory. Work on this, by Merrill Wallenstein, Austin L. Wahrhaftig, Henry Rosenstock, and Henry Eyring at the chemistry department of the University of Utah, started in the early 1950s and aimed to treat the decomposition of an ion in terms of reaction rates.[48] It was thought that after electron impact, the energy was distributed very rapidly and evenly over the excited ion. Thus all possible quantum states were taken, a certain fraction of which belonged to the activated states. The postulate that these activated states were in equilibrium with the products of fragmentation was the name-giving part of the theory. The correlation of the known fraction of activated states with the time the ion existed in the mass spectrometer should allow the calculation of the extent of fragmentation, i.e., the mass spectrum. The example used by Eyring and co-workers was a relatively simple molecule, propane. They based their work on contributions of James Franck, Hertha Sponer, and Edward Teller in 1932 on ion impact processes, but regarded their extension to mass spectrometry as original. Support came from molecular orbital theory using the hypothesis that the half-filled orbital left behind after initial ionization was best represented with a non-localized molecular orbital. Thus, several bonds in the molecular ion were weakened, and could be cleaved. Moreover, Eyring's work on reaction rate theory applied to isolated systems seemed a promising candidate to treat the liquid drop model of the atomic nucleus, too. Therefore, this famous concept that explained nuclear fission following neutron bombardment became the counterpart of molecular decomposition after electron impact in the mass spectrometer. Problems arose when it was later established that the original mathematical formal-

---

[46] Bloom et al., "Mass Spectra of Octanes," 133.

[47] Meyerson, "From Black Magic to Chemistry," 962A. See Langer, "Rearrangement Peaks"; and for Turkevich: Meyerson, "Cationated Cyclopropanes," 267. Turkevich was a chemical physicist at Princeton who worked on infrared spectroscopy and catalysis, among other topics. In 1948, he used the mass spectrometer in tracer analysis. See Turkevich et al., "Determination of Position."

[48] Rosenstock, Wallenstein, Wahrhaftig, and Eyring, "Absolute Rate Theory"; Wallenstein, Wahrhaftig, Rosenstock, and Eyring, "Chemical Reactions in the Gas Phase." A good review of the quasi-equilibrium theory is King and Long, "Mass Spectra of Some Simple Esters."

ism was invalid, and, furthermore, that some of the early assumptions (for example with regard to the number of degrees of freedom in the ions) could not account for the empirical data. Moreover, the basic assumption of the quasi-equilibrium theory, the equilibrium of excited states and reactant species, was in doubt. At the beginning of the 1960s, an improved form of the theory was regarded by its propagators as "neither proved nor disproved."[49] The quasi-equilibrium theory was the attempt to install an *a priori* calculation of mass spectra. As it was the case in other instances, this approach failed, and did not play a crucial role in establishing the usefulness of mass spectrometry for organic chemistry. This role was taken by the rationalization of mass spectra in terms of instrumental parameters and physical organic chemistry, an endeavor that began in industry and was transferred to the university.

In the 1940s and 1950s, several papers by scientists at Westinghouse and the Shell Development Company provided mass spectrometrists with the armory to tackle the identification of unknown compounds: John Hipple (with Edward U. Condon, who was associate director of research at Westinghouse, and after the war became the director of the National Bureau of Standards)[50] explained the origins of metastable peaks (originating from the further breakdown of an ion during its flight); and David P. Stevenson of Shell reported labeling experiments (use of stable isotopes at known positions in the molecule under scrutiny to detect fragmentation products) and appearance potentials (the voltage of the ionizing electrons necessary to make an ion's signal appear at the detector). He also published a rule later named after him, stating that the simple cleavage of a molecular ion with higher probability forms the fragment ion of lower ionization potential. Appearance potentials, investigations of metastable peaks, and labeling helped in explaining the phenomena inside the mass spectrometer. Thus, a whole arsenal for interpretation of mass spectra was at hand.[51]

In 1948, researchers at Atlantic Refining Co. reported the extension of mass spectrometry to compounds other than hydrocarbons. Using mercury-sealed inlet systems, they were able to investigate liquid samples, including long-chain hydrocarbons as well as aromatic compounds.[52] In the 1950s, using

---

[49] Rosenstock and Krauss, "Quasi-equilibrium Theory," 4.

[50] Wang, "Science, Security, and the Cold War"; and Lassman, "Government Science in Postwar America."

[51] Meyerson, "Reminiscences," 206. See Hipple and Condon, "Detection of Metastable Ions"; Hipple, Fox, and Condon, "Metastable Ions"; Hipple, "Peak Contours"; Stevenson and Wagner, "Mass Spectrometric Analysis"; Stevenson and Wagner, "Mass Spectra of $C_1 - C_4$ Monodeutero Paraffins"; Stevenson, "On the Mass Spectra of Propanes"; Stevenson, "On the Strength"; Stevenson and Hipple, "Ionization and Dissociation by Electron Impact. Normal Butane, Isobutane, and Ethane"; and Stevenson and Hipple, "Ionization and Dissociation by Electron Impact. The Methyl and Ethyl Radicals." For Stevenson's rule see Stevenson, "Ionization and Dissociation by Electronic Impact. Ionization Potentials and Energies of Formation of Sec-propyl and Tert-butyl Radicals."

[52] Taylor et al., "Mass Spectrometer in Organic Chemical Analysis."

improved instrumentation, more and more publications began to appear that covered targeted compound analysis.[53] In 1951, Sibyl Rock of CEC published an account describing the analysis of organic molecules.[54] Rock used a variety of methods for the interpretation of spectra and first demonstrated that this could be done qualitatively without reference to pre-run spectra of known compounds. The article was followed by similar investigations conducted by other researchers concerning the characterization of gaseous air pollutants,[55] and a variety of compound types in petroleum fractions.[56] Nevertheless, the reasoning of most researchers in the oil industry was bound to their specific interests. Only a few tried to bridge the gap to academic organic chemistry.

For example, together with Paul Rylander and Henry Grubb, Seymour Meyerson of Standard Oil started a research program that was characterized by four features: empirical correlation of mass spectra and molecular structure; appearance potentials; metastable peaks; and use of labeled compounds.[57] This strategy was thought to put mass spectrometry on a firmer scientific basis, and was published in a series of papers from 1956 on. While for the specialists, the footing of mass spectrometric techniques in the experimental tradition of chemical physics (e.g., with the measurement of appearance potentials) was important to gain trust in their results, the broader community of organic chemists had to be convinced by examples that fitted their domain. Especially the second paper of Meyerson's series tackled a problem that appealed much to chemists because it connected the molecular events inside the mass spectrometer to processes known in classical chemistry, with the example of the tropylium ion.[58] This compound had attracted considerable interest in the chemical community because of its unexpected aromatic behavior. Though they were aware of the preliminary character of many conclusions, Meyerson and most of his fellow mass spectrometrists regarded the relationship of spectra and structure as an established fact. Doubts regarding this were revealed by the British

---

[53] This term refers to the presence or absence of a certain compound of known structure in a sample. Gohlke and McLafferty, "Early Gas Chromatography/Mass Spectrometry," 367; Happ and Stewart, "Rearrangement Peaks"; Friedman and Long, "Mass Spectra of Six Lactones"; Long and Friedman, "Mass Spectra and Appearance Potentials"; Nicholson, "Photochemical Decomposition"; Beynon, "Qualitative Analysis"; Sharkey, Shultz, and Friedel, "Mass Spectra of Ketones"; de Mayo and Reed, "Application of the Mass Spectrometer."

[54] Rock, "Qualitative Analysis from Mass Spectra."

[55] Shepherd, Rock, Howard, and Stormes, "Isolation, Identification, and Estimation."

[56] Brown and Meyerson, "Cyclic Sulfides"; Van Meter et al., "Oxygen and Nitrogen Compounds"; Kinney and Cook, "Identification of Thiophene."

[57] Meyerson, "From Black Magic to Chemistry," 962A.

[58] Rylander, Meyerson, and Grubb, "Organic Ions in the Gas Phase. II. The Tropylium Ion." See Meyerson, "Tropylium, Chlorine Isotopic Abundances"; and Doering and Knox, "Cycloheptatrienylium (Tropylium) Ion."

chemist Rowland Ivor Reed, one of the first scientists who used mass spectrometry in organic chemistry:

> This approach . . . greatly extended our understanding of the origin of mass spectra, and as it is consonant with the intuitive arguments of organic chemists—the main practitioners—it was rapidly adopted . . . . The system has weaknesses, an obvious one being that it offers no guide as to how to convert the observed spectrum to a molecular structure, a weakness which has led to the construction of 'data' banks and the method of correlation. The latter rests on the premise that the spectra of the unknown may be interpreted by the aid of a set of spectra of associate compounds, a method which while uninspired is fairly reliable in experienced hands.[59]

In Reed's opinion, the human expert was the key in making organic mass spectrometry a credible scientific endeavor. Moreover, the analytical chemist Fred W. McLafferty pointed to the positive effects of chemical theories for mass spectrometry, implicitly relating his argument to the inference to the best explanation: "It would be surprising indeed if the success of the organic mechanistic approach in interpreting mass spectra is entirely fortuitous, and if its intuitive conclusions do not have some connection with physical principles."[60] The investigative pathway of McLafferty shows some characteristic features of mass spectrometry on its way from industry to academia.

Fred W. McLafferty at Dow Chemical Company

As did Meyerson, Fred Warren McLafferty (b. 1923) did his early work on organic mass spectrometry in an industrial environment. In 1950, he joined the ranks of the Spectroscopy Laboratory of the Dow Chemical Company at Midland, Michigan. McLafferty had obtained his B.S. and M.S. in chemistry from the University of Nebraska, concentrating on analytical chemistry with H. Armin Pagel. Pagel was a thorough analytical chemist who insisted on the proper use of the traditional techniques of analysis.[61] For his Ph.D. studies, McLafferty went to Cornell University, Ithaca, New York, and completed his dissertation in the fall of 1949. His supervisor was William Taylor Miller, who had been involved in the U.S. program for the construction of the atomic bomb. Miller, an organic chemist, developed many of the materials designed to withstand the aggressive fluorine compounds used in the enrichment process of

---

[59] Reed, "Some Problems in Organic Mass Spectrometry," 1214.

[60] McLafferty et al., "Substituent Effects in Unimolecular Ion Decompositions. XV," 6868.

[61] McLafferty, interview by Reinhardt, 16 and 17 December 1998. See also McLafferty, "Mass Spectrometry and Analytical Chemistry."

the fissionable uranium isotope. McLafferty worked on organo-fluorine chemistry with Miller. In contrast to Pagel, Miller had a very modern approach towards physical instrumentation.[62]

While pursuing a postdoctoral appointment at the University of Iowa with the famous analytical chemist Ralph Lloyd Shriner, McLafferty considered industrial appointments. At the 1949 ACS meeting in Chicago, he met a representative of Dow Chemical Company who invited him for a job interview. Though McLafferty was introduced to the achievements of Dow in the mass market fields of Latex paint and polystyrene, he was more impressed by the company's "leading spectroscopy lab in the whole world."[63] Dow, in order to control its magnesium manufacturing process, had invested heavily in the development of the first direct-reading spectrograph (independently, the competitive company Alcoa followed a similar approach). For the production of magnesium, a spark source spectrograph with phototubes—replacing the more time-consuming photoplate detection—was developed at Dow and produced and marketed by Baird Associates.[64] In addition, Dow was a leader in infrared spectroscopy, focusing on the development of a double-beam infrared spectrometer independently of American Cyanamid and Perkin-Elmer. The Dow spectrometer was also produced and sold by Baird Associates. In sum, an intensive research program on the analytical techniques of X-ray diffraction, atomic emission spectroscopy, and infrared spectroscopy tackled a wide variety of the company's problems. In addition, Norman Wright,[65] the laboratory's director, initiated research on mass spectrometry, undertaken mainly by Victor J. Caldecourt. The instruments in use in the early to mid 1950s were two Westinghouse spectrometers and a CEC 21-103B.[66] Caldecourt developed a "wonderful gadgetry to make mass spectrometry easier," improving the recording system as well as the mass calibration of the spectra, and inventing a device for the insertion of weighable amounts of samples, including solids.[67] Supplied with parts by the Westinghouse employees Russ Fox and John Hipple, he had even assembled the second Dow mass spectrometer by himself (see Figure 3.1). Dow's strategy of using mass spectrometry was similar to the style they employed with other spectroscopic methods. The main goals were process control and compound identification, and the Spectroscopy Laboratory was in charge of a considerable amount of routine analytical work. In the mid-1950s, the mass spectrometry laboratory undertook 10,000 analyses per year, and

[62] McLafferty, interview by Reinhardt, 16 and 17 December 1998.

[63] McLafferty, interview by Reinhardt, 16 and 17 December 1998.

[64] Baird, "Encapsulating Knowledge."

[65] See anon., "Norman Wright (1906–1994)."

[66] Gohlke and McLafferty, "Early Gas Chromatography/Mass Spectrometry," 367.

[67] McLafferty, interview by Reinhardt, 16 and 17 December 1998. See Caldecourt, "A Mass Indicator"; and Caldecourt, "Heated Sample Inlet System."

FIGURE 3.1 Fred W. McLafferty (right) and Herbert Woodcock at Dow Chemical Co., Midland, Michigan with an early 90° sector mass spectrometer, ca. 1951. The instrument was built by Victor Caldecourt with parts from Westinghouse supplied by Russ Fox and John Hipple. Courtesy of Fred McLafferty.

McLafferty's department employed six qualified scientists and six technicians to carry this workload.[68] Dow Chemical's product range was a broad one, encompassing thousands of chemical compounds that varied considerably in their chemistry. This variety provided the background for the work at the Spectroscopy Laboratory.

In the early 1950s, when only a few mass spectroscopists were undertaking research on organic compounds other than hydrocarbons, Dow was especially interested in halogenated organic chemicals. This was the core of the company's product range and the chemicals could be studied through mass spectrometry relatively easily by their isotopic signatures. While pursuing their pioneering work in infrared spectroscopy, members of Dow's spectroscopy laboratory had built up one of the world's best catalogues for infrared reference spectra. This proved to be the model and source for an analogous endeavor in mass spectrometry. Each time a mass spectrometer was idle, McLafferty obtained samples from the infrared spectroscopists' compound collection and analyzed them by mass spectrometry. Thus, he expanded the existing mass spectra databases on hydrocarbons with spectra of other chemicals. In addition to Dow, other major chemical companies, most notably American Cyanamid

[68] McLafferty, Application for research grant at NIH, FR 00354-01, 31 May 1966, p. 11, McLafferty Papers, folder "NIH FR 00354."

and Du Pont, were doing similar work. With the heated inlet system, samples with a lower vapor pressure came within the reach of the mass spectrometer, and with the higher molecular weight accessible, the number of investigated compounds grew dramatically. In more and more cases, no reference spectra were available. Other methods for identification had to be found, and the correlation of spectra and structure seemed the most promising analytical tool for McLafferty and his colleagues at Dow.

After more than five years of work at Dow's Midland plant, McLafferty wrote a manifesto on the mass spectrometer's usefulness in chemical research.[69] In his own research, he pointed to three directions. First, the familiar quantitative analysis of hydrocarbons could also be undertaken with other types of compounds. Second, and for targeted compound analysis, a collection of 2,500 reference spectra had been assembled at Dow, searchable with an IBM punched card system.[70] Third, and in order to find out about compounds that were not yet included in the data collection, McLafferty used a type of analysis based on the natural distribution of stable isotopes. With that input, conclusions could be drawn as to the number of each type of atoms present in the molecule, making it possible to retrieve the empirical formula.

The McLafferty Rearrangement

Though with these possibilities mass spectrometry was a powerful tool for industrial analysis, unknown structures could not be tackled with this approach. This had to await the understanding of the processes taking place in the mass spectrometer in chemical terms. McLafferty's work was based here on knowledge that had been gained in the realm of physical organic chemistry, explaining the relative stability of chemical bonds and the electron-releasing or electron-withdrawing effects of neighboring groups. In the manuscript for an article published in the first volume of *Advances in Mass Spectrometry* in 1959, McLafferty emphasized the analogy between the events going on in the mass spectrometer and "normal" chemistry:

> One can think of a MS [mass spectrum] as a quant. display of the results of a chemical rx [reaction] . . . in this case a unimolecular decom[posin]g rx caused by energetic electrons. The effect of molecular struct[ure] on the trans[ition] state of this rx should be compared with the general result found

[69] McLafferty, "Mass Spectrometric Analysis. Broad Applicability to Chemical Research"; and McLafferty's lecture manuscript, 11 May 1954, "Analytical applications of high temperature mass spectrometry to chemistry," for presentation at the ASTM E-14 committee meeting on mass spectrometry, 25 May 1954, New Orleans. McLafferty Papers, folder "Paper M.S. Anal. I 3."
[70] McLafferty and Gohlke, "Mass Spectrometric Analysis. Spectral Data File."

> to apply to a wide variety of chem rxs to see how much this higher energy level effects the results. In general, there are many striking simil[arities] between MS [mass spectrometry] and POrgChem [physical organic chemistry].[71]

But not all peaks in mass spectra could be explained by the simple cleavage of bonds. Regroupings, or rearrangements, occurred quite often. Here, McLafferty made serious efforts to correct the "terrible" reputation of mass spectrometry for analysis of organic compounds, due to what he called "random rearrangements" of hydrocarbons.[72] In contrast to the random rearrangements, the so-called "specific rearrangements" allowed insight into the mechanism involved and therefore helped to elucidate the structure of the molecules under consideration.

In the mid-1950s, no generalizations of rearrangements existed, but mechanistic studies had appeared that explained rearrangements of hydrocarbons and, especially, other molecules. It was assumed that the existence of functional groups in a molecule directed the course of an eventual rearrangement and thus made it predictable.[73] In 1953, Lewis Friedman and Franklin A. Long of the Brookhaven National Laboratory and Cornell University, respectively, found the striking result that in some cases the rearrangement peaks were the largest of the entire mass spectrum. Thus, the rearrangement had to be highly specific. Friedman and Long also concluded that "there is a large probability of rearrangement only when one of the resulting fragments is of relatively high stability,"[74] and noticed the abundance of even-electron ions as products of fragmentations and rearrangements. It was as a result of this work that rearrangements began to be seen as a tool for structural studies.

One of the most famous rearrangements was named after McLafferty, who, though he did not describe it first, did most of the work characterizing and explaining it. The mechanism of the McLafferty rearrangement involves a six-membered cyclic transition state, and includes the elimination of an olefin molecule from molecules such as alkyl esters, aldehydes, and ketones. The McLafferty rearrangement contributed a lot to explain and to predict results in organic mass spectrometry. It is a major example of the impact that mechanis-

---

[71] McLafferty, draft of "Interpretation of mass spectra of organic molecules," McLafferty Papers, folder "Rearrangements 10."

[72] McLafferty, interview by Reinhardt, 16 and 17 December 1998.

[73] McFadden and Wahrhaftig, "Mass Spectra of Four Deuterated Butanes"; Friedman and Long, "Mass Spectra of Six Lactones"; Friedman and Turkevich, "Mass Spectra of Some Deuterated Isopropyl Alcohols"; Schissler, Thompson, and Turkevich, "Behaviour of Paraffin Hydrocarbon"; Honig, "Isomerization of Hydrocarbons"; Langer, "Rearrangement Peaks"; Stevenson and Hipple, "Ionization and Dissociation by Electron Impact. Normal Butane, Isobutane, and Ethane."

[74] Friedman and Long, "Mass Spectra of Six Lactones," 2835.

tic thinking in terms of physical organic chemistry had on mass spectrometry and was used widely in textbooks and classroom teaching.

McLafferty became interested in the topic of hydrogen rearrangements in the summer of 1953, when he undertook a literature review.[75] The special reaction type he followed up had already been mentioned by Glenn P. Happ and D. W. Stewart of Eastman Kodak's research laboratories in Rochester, New York, in a 1952 publication, and A. J. C. Nicholson found the same mechanism in photochemical reactions in 1954.[76] Nevertheless, its generalization and the proof of its great utility for the interpretation of mass spectra had to await the systematic studies of McLafferty.[77] After McLafferty noticed the publication of Happ and Stewart (he also knew the two industrial scientists from meetings), he recognized other examples of such rearrangements in his own investigations: "I took it up and published on its generality."[78]

In 1955, he named some necessary conditions for the rearrangement of a hydrogen atom in a molecule. Especially, more stable fragments had to be produced through the rearrangement rather than without it, either by yielding to an "even electron ion" or a stable molecule. Moreover, McLafferty was sure that he had found a general phenomenon.[79] In the beginning, he thought of the rearrangement in terms of "lasso chemistry," i.e., the transfer of electrons and hydrogen atoms visualized in a two-dimensional manner. In the plane of a sheet of paper, the cyclic transition state of the hydrogen rearrangement looked like a resonance-stabilized benzene ring. Only later did McLafferty realize that this rearrangement was not planar, and coined it in terms of a radical site reaction. For his interpretations, McLafferty did not seek direct contact with physical organic chemists, as he was familiar with the field through his Ph.D. supervisor, William T. Miller, who taught the first course on physical organic chemistry at Cornell.[80]

---

[75] McLafferty, research notebook, entry of 19 August 1953, McLafferty Papers.

[76] Happ and Stewart, "Rearrangement Peaks." Nicholson, of the Commonwealth Scientific and Industrial Research Organisation, studied the photolytic cleavage of ketones and found that the products, acetone and propylene, could only be explained by an intramolecular rearrangement of the molecule. He used mass spectrometric methods to account for the reaction product. Nicholson, "Photochemical Decomposition." P. P. Manning, who in 1957 was independently studying photolytic reactions, proposed a cyclic transition state: Manning, "Photolysis of Saturated Aldehydes and Ketones." See Kingston, Bursey, and Bursey, "Intramolecular Hydrogen Transfer in Mass Spectra. II."

[77] McLafferty, "Mass Spectrometric Analysis. Broad Applicability to Chemical Research," 312–313; McLafferty and Hamming "Mechanism of Rearrangements in Mass Spectra"; McLafferty, "Mass Spectrometric Analysis. Molecular Rearrangements."

[78] McLafferty, interview by Reinhardt, 16 and 17 December 1998.

[79] McLafferty, notes "Conditions (necessary) for H rearr." 22 April 1955, McLafferty Papers, folder "Rearrangements 10."

[80] McLafferty, interview by Reinhardt, 16 and 17 December 1998.

Meanwhile, a change in the career path of McLafferty occurred. In 1956, he was appointed founding director of the new basic research laboratory of Dow Chemical in Framingham, Massachusetts. In this, Dow followed the example of the other major U.S. chemical companies that invested in academic-style research laboratories in the 1950s, some of them in Europe. For Dow, the step to the East coast was already a considerable one, bearing in mind the strong roots of the company in Michigan. For McLafferty's work in mass spectrometry, the move to head the Eastern Research Laboratory, as it was named, had two consequences: During the first six years, he had no mass spectrometer in Framingham.[81] In addition, McLafferty suddenly had to direct the research of around 15 scientists, and to establish his name in an academic environment. Thus, he was forced to publish without direct access to an instrument. He relied on the data acquired in six years of research at Dow, and counted on the collaboration of Dow employees back in Michigan, most notably Roland Gohlke, but also Caldecourt and Jo Ann Gilpin. Moreover, as head of a sub-committee of the ASTM-E 14 committee on mass spectrometry, he was intimately familiar with efforts to assemble a database of so-called uncertified spectra:

> I chaired what we called the uncertified mass spectral data committee. The only collection of reference mass spectra was the American Petroleum Institute. And they not only took mostly hydrocarbon-kind of compounds, but they also really thought they ought to be done under these really carefully controlled conditions of the CEC-103. It wasn't that they wouldn't take things, but they were sort of second class spectra and so forth and people didn't want to bother sending them. Really wonderful spectra. So we said, let's start a collection that's uncertified. If anybody finds a mistake in it that doesn't matter, if it's run under very different conditions. And since I started it, people started sending them in to me . . . and I promised to do it. Well, I was away and I couldn't do it . . . . I finally got Roland [Gohlke] to finish publishing this thing and I got the lab pay for it . . . . But it was actually the collection that I started way back in the early 1950s. This I think has a couple of thousand spectra in it and it essentially doubled the number of spectra available at the time. Those were the data that I actually took to the Boston lab that I could look at and do these correlations of spectra.[82]

On the basis of this experience, in 1957 McLafferty expanded his work on rearrangements to the aliphatic aldehydes, in cooperation with Jo Ann Gilpin of Dow. Citing geometrical factors in the process of ion formation, and inves-

---

[81] In 1962, Roland Gohlke installed a Bendix time-of-flight mass spectrometer. McLafferty had ordered a high-resolution instrument (the SM-1) from Atlas Werke, Bremen, around 1961. The company did not deliver the instrument before McLafferty left Dow in 1964. McLafferty, interview by Reinhardt, 16 and 17 December 1998.

[82] McLafferty, interview by Reinhardt, 16 and 17 December 1998.

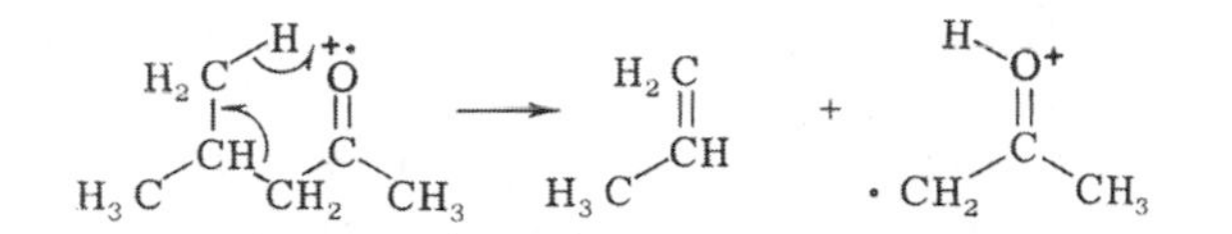

FIGURE 3.2 On top the McLafferty rearrangement, shown with the example of a ketone (4-methyl-2-pentanone). From Fred W. McLafferty, *Interpretation of Mass Spectra. An Introduction*, New York: Benjamin 1967, fig. 8-41 on 125. Below: Fred McLafferty teaching on rearrangements at a lecture in Frankfurt am Main, in 1982. Courtesy Fred McLafferty.

tigations of appearance potentials by other scientists, McLafferty and Gilpin argued in favor of their supposed transfer mechanism (see Figure 3.2, top).[83]

In July 1957, McLafferty attempted to report a general account of the findings on rearrangements in the *Journal of Chemical Physics*. The paper was rejected by the editor, but shows the development of McLafferty's ideas on the nature of the rearrangements he was dealing with. In his manuscript, he set out to classify hydrogen rearrangements in "exchange" and "replacement" rearrangements. Only in the later, published, version of the same manuscript, McLafferty coined the phrases "random" and "specific" rearrangements. This may well have been to satisfy the quite different audience of *Analytical Chemistry*, in which the article finally appeared.[84] While the title of the manuscript

[83] Gilpin and McLafferty, "Mass Spectrometric Analysis. Aliphatic Aldehydes."
[84] McLafferty, "Mass Spectrometric Analysis. Molecular Rearrangements."

submitted to the *Journal of Chemical Physics* avoided mentioning mass spectrometry, that in *Analytical Chemistry* explicitly referred to the technique. Certainly, the readership of the *Journal of Chemical Physics* was more interested in mechanistic studies than in analytical performance. The concept of exchange, with its meaning of "a higher energy reshuffling process in which various atoms or groups are equilibrated in the molecule," implied a greater emphasis on the groups that were exchanged within the molecule. The concept of "randomized" rearrangements pointed to the relatively little differences in energy and entropy of the various pathways leading to the fragmentation products in the ion source.[85] Specific rearrangements, in contrast, were guided by functional groups or electronic effects in the molecule, and by the stability of the products of such rearrangements. The term "specific" implied a certain reliability in interpreting them; and the contrast to "random" was meant to give the readers of *Analytical Chemistry* the signal that the results were reliable.

Nevertheless, the proposed rearrangement mechanism was not proven yet. Labeling studies could have definitively solved this problem, but though McLafferty and Gilpin already had started investigations with deuterated compounds, they did not report this in their published article. Although the reviewers strongly recommended this, Gilpin preferred to publish a separate article on this topic, based on her own work alone.[86] This work was done with alcohols, and the similarities to aldehydes were not large enough to warrant a direct comparison.[87] McLafferty was eager to follow this up thoroughly,[88] though deuterated compounds of the right type were hard to come by.[89] In the summer of 1958, McLafferty collaborated with Mynard C. Hamming (a former colleague at Dow, then at Koppers Company, Verona, Pennsylvania) on this issue. Hamming had obtained deuterated samples from Herbert C. Brown and Benjamin Repka of Purdue University, in the course of checking the list of uncertified spectra of ASTM. He undertook the mass spectra measurements because McLafferty, who at that time was already at the Dow Eastern Research Laboratory, did not have the necessary instrumentation.[90] Further investiga-

[85] Manuscript "Molecular rearrangement induced by electron impact. A generalized concept," 1, McLafferty Papers, folder "Rearrangements 10"; McLafferty, "Mass Spectrometric Analysis. Molecular Rearrangements," 82–83.

[86] Gilpin, "Mass Spectra Rearrangements of 2-phenyl Alcohols."

[87] See McLafferty to Gilpin, 15 February 1957; Gilpin and McLafferty, "Mass Spectra of Saturated Aliphatic Aldehydes," Dow report SL 78859, 8 August 1956; Glenn P. Happ to L. T. Hallett (editor of *Analytical Chemistry*), 26 October 1956, all McLafferty Papers, folder "Aliphatic aldehydes, 5."

[88] McLafferty to Gilpin, 5 February 1957, McLafferty Papers, folder "Aliphatic aldehydes, 5."

[89] See McLafferty to R. M. Guedin of Celanese Corporation of America, 3 April 1958, and Guedin to McLafferty, 12 May 1958, McLafferty Papers, folder "Rearrangements 10."

[90] See Hamming to McLafferty, 23 June 1958; McLafferty to H. C. Brown, 10 July 1958, all McLafferty Papers, folder "Rearrangements 10." The paper is McLafferty and Hamming, "Mechanism of Rearrangements in Mass Spectra."

tions with deuterated compounds by Klaus Biemann's group at MIT and Einar Stenhagen's in Sweden in the early 1960s showed clearly that the proposed mechanism was correct.[91] Moreover, the McLafferty rearrangement proved to be a very general one, not limited to ketones and esters, but involving many compounds with C, N, or O content, and including the possibility that the double bond was part of an olefinic or aromatic system. The reasoning in terms of physical organic chemistry was that the hydrogen rearrangement yielded a stable olefinic neutral product, not a radical species; though the cation now contained an unpaired electron, it was resonance stabilized; and the more favorable enthalpy more than compensated for the unfavorable entropy.

The history of the McLafferty rearrangement shows interconnections with photochemistry. The mechanism of the mass spectrometric rearrangement was seen as analogous to a photochemical rearrangement of the Norrish type II, and analogies were also drawn to thermolytic and radiolytic reactions.[92] But the study of the rearrangement mechanism soon developed independently from most other areas of chemistry. This drew mainly on the use of the McLafferty rearrangement in structural elucidation by mass spectrometry. The first use of the name in the open literature was made in 1964 by Gerhard Spiteller, in German.[93] At the same time Spiteller, who had been a member of Biemann's group and was now at Vienna, developed his own ideas of how the mechanism of the rearrangement worked. These ideas concurred with McLafferty's thoughts at the same time.[94] In 1965, Carl Djerassi adopted Spiteller's coinage in a study on steroids, emphasizing the role of the distances between the atoms for the occurrence of the rearrangement.[95] The naming of this rearrangement after a person was the only such instance, and was in conflict with guidelines agreed on by the community of mass spectrometrists. Seemingly, both Spiteller and Djerassi were at that time not aware of this.[96] In the 1967 edition of their book on interpretation of mass spectra, Djerassi, Williams, and Budzikiewicz dedicated a whole subchapter to the McLafferty rearrangement.[97] At the beginning of the 1970s, such a vast amount of investigation had been done on the McLafferty rearrangement that the authors of a review article on this topic concluded

---

[91] Biemann, *Mass Spectrometry*, 119–120.

[92] Kingston, Bursey, and Bursey, "Intramolecular Hydrogen Transfer in Mass Spectra. II," 231–232.

[93] Spiteller and Spiteller-Friedmann, "Zur Umlagerung aliphatischer Verbindungen." The term "McLafferty-Umlagerung" on p. 258.

[94] McLafferty to Spiteller, 20 May 1964, McLafferty Papers, folder "Genl M.S. Talk."

[95] Djerassi, von Mutzenbecher, Fajkos, Williams, and Budzikiewicz, "Mass Spectrometry in Structural and Stereochemical Problems. LXV."

[96] Klaus Biemann, personal communication, 6 May 2003.

[97] Budzikiewicz, Djerassi, and Williams, *Mass Spectrometry of Organic Compounds*, 155–162. Cf. the 1964 textbook, Budzikiewicz, Djerassi, and Williams, *Interpretation of Mass Spectra of Organic Compounds*, 3–4.

that "the *absence* of rearrangement is also good evidence that an appropriate molecular structure does not exist in the compound under investigation."[98] The ease with which such rearrangements could be recognized in the mass spectrum (rearrangements produce mostly odd-electron fragments, in contrast to normal fragmentation that yields to even-electron fragments), and the wide variety of the compounds undergoing the McLafferty rearrangement, guaranteed its adoption in the chemical community. From 1965 to 1978, McLafferty was the most cited analytical chemist in the United States.[99]

Despite these successes, and though McLafferty was head of a Dow laboratory focusing on fundamental research, he had to argue inside the company for the values of scientific standing. Thus, in the early 1960s, the widespread acceptance of the McLafferty rearrangement in the scientific community was a welcome argument. With respect to his planned editorship of the book *Mass Spectrometry of Organic Ions*, he tried to convince a member of the research management of Dow that this would benefit the company, too:

> I think that the company does stand to gain from such a project, although these gains may be difficult to evaluate. A creditable publication of this sort should enhance the company's scientific reputation generally and call attention to the reputation already established by the Chemical Physics Laboratory for leadership in the field of mass spectrometry. It may be just a pet theory of mine, but I believe it is a real help to the effectiveness of a research supervisor to maintain his scientific reputation among his colleagues outside the company, and especially within the company at his particular laboratory.[100]

To improve on the scientific output of the Eastern Research Laboratory, Roland Gohlke joined McLafferty in Framingham around 1962, installing a Bendix time-of-flight spectrometer (see Figures 3.3 and 3.4).

McLafferty tried to guide basic research in his laboratory in an academic style. His role model was Melvin Calvin, the famous Berkeley chemist who was a Dow consultant and from time to time visited the Eastern Research Laboratory. Along these lines, McLafferty arranged for conferences on basic research, which gathered a selection of Dow scientists and took place at the laboratory in Framingham. Moreover, among others he employed the later Nobel Prize laureate George A. Olah. Olah had come from the Canadian Dow laboratory at Sarnia, and joined McLafferty's laboratory shortly before McLafferty left.[101] Another important task he had was interviewing candidates for research posts

---

[98] Kingston, Bursey, and Bursey, "Intramolecular Hydrogen Transfer in Mass Spectra. II," 232. Emphasis in the original.

[99] M. Bursey, "F. W. McLafferty: An Appreciation."

[100] McLafferty to R. H. Boundy, 15 July 1960. McLafferty Papers, folder "Book on mass spectrometry of organic ions, FWM editor."

[101] McLafferty, interview by Reinhardt, 16 and 17 December 1998.

FIGURE 3.3 Fred W. McLafferty (left) and Roland S. Gohlke around 1962 at the Dow Eastern Research Laboratory. McLafferty was inserting a microgram sample of an unknown compound for molecular structure determination into the special sample introduction system. With this, the sample could be vaporized directly in the ion source of this special time-of-flight mass spectrometer. Courtesy of Fred McLafferty.

and the Framingham laboratory attracted chemists from Ivy League universities. Many later moved to Dow's headquarters in Midland, Michigan. Thereby, McLafferty got to know some of the best chemistry departments in the country.[102] This later proved to be his own entrance gate into academic chemistry, where organic mass spectrometry had gained a foothold through the efforts of a few entrepreneurial organic chemists.

McLafferty's work while at the Eastern Research Laboratory of Dow stood midway between the academic and the industrial type of scientific work, as did his institution. The restricted access to advanced instrumentation, and the availability of a huge collection of mass spectral data, led him to emphasize the rational, interpretative style that at the same time became prevalent in academic chemistry. The compounds that he tackled, though, were mostly well known, and some even manufactured on an industrial scale. Thus, he was able to develop the interpretation of mass spectra with the concepts of physical organic

[102] McLafferty, interview by Reinhardt, 16 and 17 December 1998.

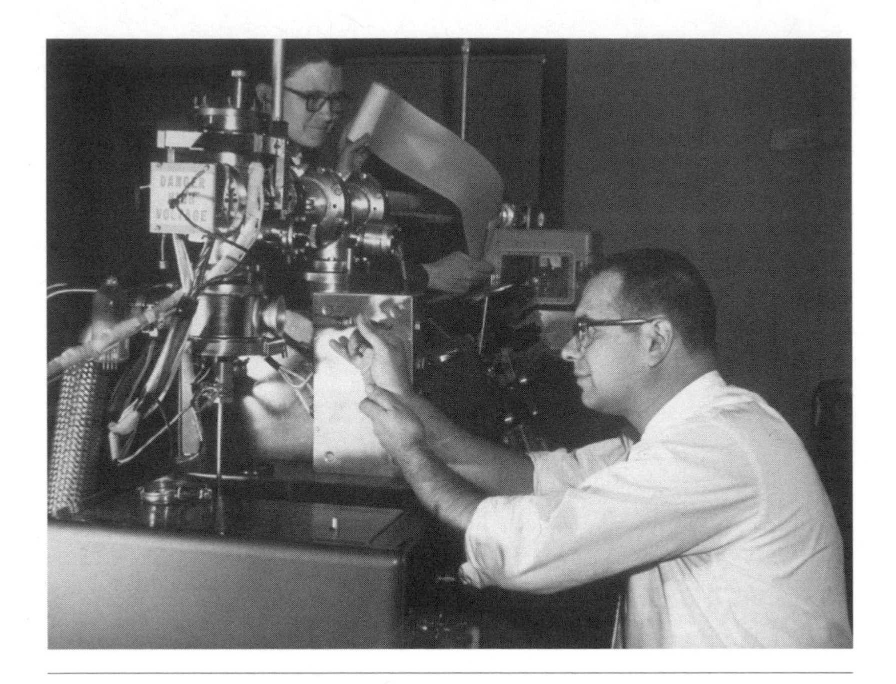

FIGURE 3.4 Roland Gohlke (right) and McLafferty with the Bendix time-of-flight spectrometer at the Eastern Research Laboratory, ca. 1962. Through the doubling of the length of the instrument, Gohlke improved the time-of-flight spectrometer considerably. In the opinion of McLafferty, Gohlke had a great pair of hands. Gohlke also built the probe for direct insertion into the ion source, shown here. Courtesy of Fred McLafferty.

chemistry on prepared ground. Soon, the first chemists sought to apply the technique for the elucidation of substances whose structures were still unknown.

Klaus Biemann at MIT

> It is, in part, due to the excellent quantitative results of mass spectrometry [in analytical chemistry] that the qualitative potentialities of the technique were until very recently very much neglected by the organic chemist. The technique is used for structure determination thus far only in a few research laboratories. This may change if the mass spectrometer is regarded not only as a complex instrument but rather as a new micro-degradation reaction, the mechanism of which is already elucidated to a certain extent and in which the

> products are separated automatically, and at the same time subjected to a molecular weight determination and even to elemental analysis.[103]

Klaus Biemann's opinion on the future of mass spectrometry in organic chemistry reflected the work of chemists in industry who undertook the pathbreaking work to establish the credibility of the technique in chemical research. But in 1962, the presentation of mass spectrometry as a combination of chemical reactions and elemental analysis was more the expression of hope for the future than the description of a present reality. Moreover, Biemann had already experienced the resistance of chemists who clung to their traditional methods. Was a compound that the chemist could not isolate in pure form a "real" chemical substance? With mass spectrometry, the sample nearly disappeared, and minuscule amounts could be detected and studied. For this reason, it was necessary to emphasize the chemical character of the molecular events in the mass spectrometer: The micro-degradation reaction in the mass spectrometer was seen as the analog to the degradation reaction in the test tube.

Klaus Biemann (b. 1926) received his education in chemistry at the University of Innsbruck in Austria. Originally, and influenced by family tradition (his father was a pharmacist), he began to study pharmacy. But his talents and interests soon led to a preference for chemistry, and he changed his subject accordingly in 1948. His Ph.D. supervisor was the synthetic organic chemist Hermann Bretschneider, also at Innsbruck. Bretschneider had a strong interest in products that were of potential use as drugs, and Biemann became his first graduate student. After receiving his Ph.D. in 1951, Biemann stayed on in Bretschneider's laboratory as instructor, in 1954 receiving a six-month Fulbright fellowship at MIT in Cambridge, Massachusetts. The fellowship was part of a program known as the "Foreign Students Summer Project," and was organized by students at MIT who had been in Europe during and after World War II. These students realized that the situation in Europe did not allow sophisticated work, and provided means for the much needed short-term exchange of postdoctoral workers. At MIT, Biemann was assigned to George Büchi,[104] mainly because it was thought that Biemann should study with somebody who knew German. Büchi, a Swiss chemist, specialized in natural products. Thus, by chance, Biemann turned from synthetic organic chemistry to natural product chemistry. After his fellowship had expired, Biemann returned to Innsbruck, but the hierarchical Austrian university system was a disappointment to him because it did not allow independent work. In 1955, Büchi offered Biemann a postdoctoral position in his laboratory, financed by the

---

[103] Biemann, "Application of Mass Spectrometry in Organic Chemistry. Determination of the Structure of Natural Products," 111.

[104] For Büchi, see Berchtold and Foley, "George Hermann Büchi."

Swiss company Firmenich & Cie. that cooperated with Büchi.[105] Back in Büchi's group, Biemann became acquainted with physical methods, mainly ultraviolet and infrared spectroscopy, while the interests of Firmenich & Cie. were behind his first encounter with mass spectrometry.

In late 1956, and on behalf of Firmenich & Cie., Biemann attended a conference on food flavors in Chicago. The talk that attracted Biemann's greatest interest was given by William H. Stahl of the U.S. Quartermaster Research and Engineering Center in Natick, Massachusetts. Stahl described the identification of fruit flavor components with the help of a mass spectrometer. Biemann, who at the time was already aware of the usefulness of infrared and ultraviolet spectroscopy in organic chemistry, pondered the possibility of using mass spectrometry along the same lines in his own research. A quick literature survey made him familiar with previous work done in industry on the quantitative determination of hydrocarbons and with the early attempts to extend mass spectrometry to hetero atom-containing compounds with the help of empirical correlations of spectra and structures. But, with very few exceptions, the method had not yet been applied to natural products. In the late 1950s, most mass spectrometrists were still concerned with mass spectra of compounds of known structure. Exceptions included Rowland Ivor Reed at Glasgow University, who did some work on steroids and terpenoids, and the group around Einar Stenhagen at the University of Gothenburg (Sweden). Stenhagen, who was mainly interested in tuberculosis research, used the mass spectrometers designed and built by Ragnar Ryhage from the Karolinska Institute in Stockholm[106] mostly for elucidation of the structures of long-chain fatty acids and esters (the fatty acid mycolic acid is an important part of the cell wall of *Mycobacterium tuberculosis*).

Biemann's first acquaintance with mass spectrometry might well have led to nothing, had not Arthur C. Cope, head of the department of chemistry at MIT, offered Biemann a position in the analytical division of the department in September 1957. Cope, one of the most influential organic chemists in the United States, wanted to strengthen organic chemistry in the analytical division. Biemann, who had some experience in the teaching of organic analysis, was deemed to be the right man for the job. The position as instructor provided the impetus for Biemann to combine organic chemistry with mass spectrometry:

> I now faced the dilemma that I could not base a research career in an analytical chemistry setting on what I had done previously, which was mainly organic synthesis and the determination of the structure of natural products

---

[105] The contact of Büchi to Firmenich & Cie. originated at the ETH Zurich, where Büchi took his Ph.D. with Leopold Ruzicka. Ruzicka was a very important consultant for Firmenich & Cie. Biemann, interview by Reinhardt, 10 December 1998.

[106] See Ryhage, "Mass Spectrometry Laboratory at the Karolinska Institute 1944–1987."

> in the conventional sense. However, it did not take long to realize that I could rearrange these fields around mass spectrometry, which would be a legitimate analytical centerpiece. The fact that I had no practical experience was an advantage, as it did not deter me, contrary to contemporary wisdom, from planning to put comparatively large and polar molecules, such as alkaloids and derivatives of amino acids and peptides, into the mass spectrometer.[107]

The first problem was the price tag of such an instrument, around $50–60,000; the second arose when Cope had to be convinced that such an expensive analytical instrument could be run by chemists, without the need for full-time technical assistance.[108] Cope had been chosen by MIT in 1945–46 to reorganize the chemistry department.[109] In the following decade, he mostly hired young research-oriented scientists, and, though this was successful in terms of the ranking among leading universities, it posed a continuous challenge when it came to keeping this aspiring faculty at MIT. Cope, aware of the attractions that the expanding chemistry departments at Harvard, Stanford, and Berkeley could offer, emphasized the need for sufficient research space and equipment.[110] Thus, Cope tried to raise additional MIT funds for the sponsoring of chemical research. When, in 1956, he was offered the appointment as director of the Mellon Institute in Pittsburgh, Pennsylvania, Cope argued for an increase of the MIT commitment to the department. He succeeded, and the largest concession of MIT president James R. Killian was the provision of a fund of $225,000, which would be available to the department of chemistry over the next decade.[111] As a result, Cope was able to invest in equipment, even if it was risky. With the acquisition of a mass spectrometer, Cope also hoped for positive effects on the teaching in the analytical division. In 1958, analytical chemistry at MIT became an advanced course in undergraduate chemistry education, placing much value on instrumental analysis. The new mass spectrometer was expected to play a role in this course.[112] For Biemann, this was a stroke of luck. At that time, governmental programs for instrumentation were just about to come into existence, and he did not fit in with the left-over programs from the war.[113]

---

[107] Biemann, "Massachusetts Institute of Technology Mass Spectrometry School," 333.

[108] Biemann, "Massachusetts Institute of Technology Mass Spectrometry School," 333; Biemann, interview by Reinhardt, 10 December 1998.

[109] For the transformation of MIT at large during this period see Lécuyer, "The Making of a Science Based Technology University," and the literature given therein.

[110] Cope to Dean George R. Harrison, 27 August 1959, MIT Archives, AC 134, box 33, folder 9.

[111] James R. Killian, "Memorandum of conversation with Professor Cope," 19 July 1956. See also Killian to Crawford H. Greenewalt, president of Du Pont, 30 July 1956, MIT Archives, AC 4, box 51, folder 1.

[112] Cope to Arnold J. Zurcher, executive director of the Sloan Foundation, 22 May 1958, MIT Archives, AC 134, box 33, folder 9.

[113] Biemann, interview by Reinhardt, 10 December 1998.

Ten years later, Cope stated that the purchase of Biemann's mass spectrometer was one of the best uses made of the fund:

> Subsequently, he [Biemann] has become recognized as the foremost person in the world in the application of mass spectrometry to the determination of structures of complex organic compounds. He now has three additional mass spectrometers purchased with government funds. He could not have begun work in this field or reached the position that he now has attained without the initial instrument purchased under this fund.[114]

In addition to the MIT funds, Biemann's project was supported with a $10,000 grant from Firmenich & Cie. for the initial purchase of the instrument and an additional support for a postdoc in Biemann's group. Together with a grant from the National Institutes of Health (NIH), which paid for a second postdoc, this put the program on a firm footing. The support of the Swiss company was crucial, as Biemann stated in hindsight:

> This was an important event, because, by myself, I could not fully utilize the instrument, interpret the data, carry out the chemistry that needed to be done, and still fulfill my departmental duties in the analytical course cycle. Needless-to-say, the chance for an instructor working in an unknown field to get good graduate students right away was nil.[115]

Ongoing contacts with Bretschneider's group in Innsbruck helped Biemann out of the dilemma. With Josef Seibl and Fritz Gapp, he was able to hire two experienced organic chemists for the postdoctoral positions funded by Firmenich & Cie., and by NIH, respectively. Both Seibl and Gapp were known to Biemann through their times at Bretschneider's laboratory. Since the Austrian chemists did not have any experience with spectroscopic methods yet, Biemann was optimistic about their prospects for finding jobs in industry after the postdoctoral fellowships, exactly because they would have made themselves familiar with the newest methods of use in organic chemistry.[116]

In 1957, Biemann ordered the CEC 21-103C mass spectrometer, aware of the fact that his plan to apply the mass spectrometer for complex research prob-

---

[114] Cope to J. A. Stratton, president of MIT, 22 July 1964, MIT Archives, AC 134, box 33, folder 9. Cope wrote this letter to argue for the continuation of the grant, which was endangered because Biemann's mass spectrometer was the only major investment coming out of these funds.

[115] Biemann, "Massachusetts Institute of Technology Mass Spectrometry School," 333.

[116] Biemann, "Massachusetts Institute of Technology Mass Spectrometry School," 333. See Biemann to Seibl, 6 and 19 February, 17 March 1958, and Seibl to Biemann, 12 February 1958, Biemann Papers, folder "Seibl, J."; Gapp to Biemann, 11 February and 20 March 1958, Biemann to Gapp, 22 February, 17 and 28 March 1958, Biemann Papers, folder "Gapp, F." Seibl later joined the faculty of the organic chemistry division at ETH Zurich, took his *Habilitation* in mass spectrometry there in 1967 and became professor of analytical chemistry. See Seibl, *Massenspektrometrie*, and Klaus Biemann, personal communication, 17 March 2003.

lems in organic chemistry required one of the more elaborate models of the company,[117] as well as some changes to the design of its sample inlet system. The requirements of the chemical industry were such that mass spectra had to be highly reproducible, also in their quantitative aspects. In the CEC instruments, a sample reservoir guaranteed a continuous flow of the sample into the ion source of the spectrometer over the recording time of the spectrum. The disadvantages were obvious: rather large amounts of sample were needed, and it had to be quite volatile. Biemann knew that for his requirements the main problem was in the handling and insertion of the samples, and this would "always decide on failure or success, not so much the ultimate sensitivity of the instrument."[118] Thus, Biemann included an inlet system whereby the sample was injected with a hypodermic needle through a silicone-rubber diaphragm, based on experiences of American Cyanamid and Du Pont. The sample was allowed to enter the needle by capillary force, one of the open ends of the needle was closed with a rubber stopper, and the other end pushed through the silicone-rubber plug (see Figure 3.5). Though this resulted in slightly higher background noise of water and other impurities in the mass spectra, it was a relatively simple and reliable method that suited Biemann's needs. In order to have it attached to his instrument, he designed an adapter in which a diaphragm was used instead of the gallium sintered disk of the original introduction system.[119] Other modifications allowed for better performance, versatility, and user-friendliness of the spectrometer.[120]

Later, with another major modification, Biemann again neglected the traditions of CEC in the design and use of mass spectrometers. Another special inlet system became necessary, allowing the direct introduction of samples into the ion source. In this way, a much lower partial vapor pressure was required of the samples. Here, Biemann and his group clearly parted from the ways in which the company was handling the operation and service of their instruments:

> Of course that was something that CEC would have never done. Because . . . when the instrument was delivered we were not supposed to take out the ion source and clean it ourselves. We were supposed to take out the ion source and send it to Pasadena to have it cleaned . . . . After the second try we didn't do

---

[117] Biemann to W. C. Cameron of CEC, 4 November 1957, Biemann Papers, folder "Instrument manufacturers, CEC." See, for a description of this instrument, and how it was used in Biemann's laboratory, Biemann, "Four Decades of Structure Determination," 1261–1263.

[118] Biemann to Max Stoll, 6 May 1960, Biemann Papers, folder "Firmenich et Cie."

[119] Biemann to Kenneth B. Wiberg, 16 July 1959, Biemann Papers, folder "Requests for mass spectra, 1959–60."

[120] Biemann, memorandum "Experiences with the CEC 21-103C mass spectrometer," written for Firmenich & Cie., n. d. [September 1958], Biemann Papers, folder CEC. See Biemann to Stoll, 9 September 1958, Biemann Papers, folder "Firmenich et Cie."

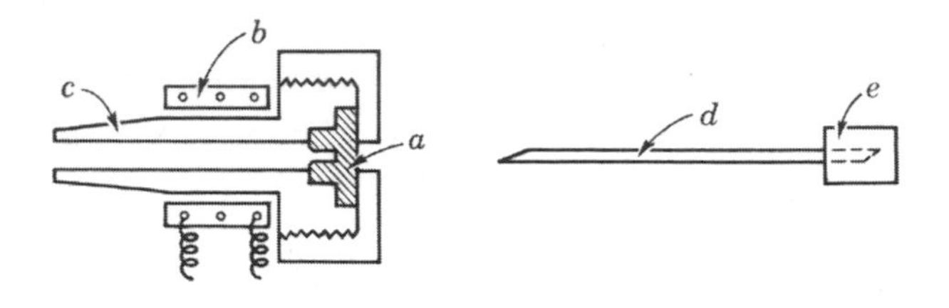

FIGURE 3.5 Biemann's inlet system for liquids. (a) Silicone-rubber stopper, (b) heater, (c) tapered joint, (d) hypodermic needle, (e) piece of silicone rubber. From Klaus Biemann, *Mass Spectrometry. Organic Chemical Applications*, New York: McGraw-Hill 1962, 27, fig. 2.5.

> [that] anymore because we could clean it ourselves just as well. But . . . they were so concerned about their reputation for accuracy of the instrument . . . . Of course any contamination in the ion source screwed it up . . . . That ion source was stabilized in temperature to 0.1 degree Celsius . . . [over a range of] 250 degrees. There was never another mass spectrometer built with that kind of temperature stabilization. But you needed that in order to get that highly reproducible mass spectrum today and a year from now. Because you had to use that library for quantitative analysis. But since we were not interested in quantitative analysis it didn't matter. That was one of the real modifications we made to the 21-103C, to put the sample directly into the ion source.[121]

The direct introduction probe was developed by Biemann's first graduate student, James A. McCloskey,[122] adapted from a time-of-flight mass spectrometer, which was financed by NASA. This instrument had a pyrolysis unit that employed a vacuum lock with a ball valve.[123] Based on these instrumental changes, Biemann applied mass spectrometry to tackle the structures of peptides, amino acids, and alkaloids.

Referring to Biemann's work on alkaloids, one of his first graduate students stated that Biemann "was possibly the first to put mass spectrometry into language that an organic chemist could really understand."[124] His book on mass spectrometry, published in 1962, abounded "with mechanisms and thus represents the diametric opposite of [the physicist's] . . . presentation of mass spectral fragmentation processes." These mechanistic studies connected mass spectrometry to physical organic chemistry, and made it a scientific field in its own right. Moreover, the rationalization of the fragmentation processes in the mass spectrometer counteracted the previously favored method of representing

---

[121] Biemann, interview by Reinhardt, 10 December 1998.

[122] Biemann, "Massachusetts Institute of Technology Mass Spectrometry School," 334.

[123] Biemann, interview by Reinhardt, 10 December 1998.

[124] Alma L. Burlingame, quoted by Henahan, "Klaus Biemann," 52.

molecular fissions with a "wiggly line," with no indication of how and why a particular cleavage occurred.[125] To achieve this, the mass spectra had to be interpreted in terms of chemical structure:

> In fact, for a mechanistically trained organic chemist who just came up at that time [this] was not that difficult. It was, I think, mainly the expense and to have somebody who did it as a fulltime job. And of course, at that time, money, even though it was a comparatively large amount for chemistry, was easier to come by than it is now . . . . If a department chairman felt that we should have that capability it wasn't so difficult to actually set it up.[126]

Alkaloids and Peptides

Biemann's choice of substances that he wanted to investigate was an expansion of the traditional field of organic mass spectrometry pursued in industry, and he concentrated on molecules whose structural formulae were still unknown. One of the most successful endeavors in this respect concerned alkaloids. Indirectly, the tremendous success of reserpine, an alkaloid used as a sedative and antihypertensive agent, lured Biemann, and many others, into the chemistry of alkaloids. For many centuries, leaves and roots of a plant called Indian snake root were applied to treat the effects of snake bites and to calm excited patients. In the 1930s, researchers in New Delhi worked at the isolation of the active substances, and after World War II, *Rauwolfia* alkaloids appeared also in Europe for medical treatment. The structure of reserpine, isolated from the Indian plant *Rauwolfia serpentina*, was established by chemists at the Swiss company CIBA in the mid-1950s. Soon, the search for a synthetic pathway to the coveted alkaloid began, no mean task because of the importance of introducing the stereochemistry of the natural molecule in the synthesis. Robert Burns Woodward achieved the synthesis, and commercial production was taken up by the French pharmaceutical company Roussel-Uclaf in 1958. The desire to repeat the reserpine story was behind many industrial and academic activities in the alkaloid field from the mid-1950s.[127]

Of considerable interest to both pharmacologists (because of their biological activity) and chemists (because of their complicated structures), alkaloids were an ideal target for mass spectrometric methods. Their use was enhanced by the fact that many alkaloids were available only in minute amounts, and mostly in mixtures of series with closely related structures. Before the advent of mass spectrometry, the elucidation of the structure of a specific alkaloid had

---

[125] Djerassi, book review "Mass Spectrometry. Organic Chemical Applications," 2190.
[126] Biemann, interview by Reinhardt, 10 December 1998.
[127] See Benfey and Morris, *Robert Burns Woodward,* 178–181.

been achieved either by its chemical degradation into smaller identifiable molecules or by conversion into other alkaloids of known structure. This required laborious chemical reactions and often considerable amounts of starting material. The identity of these degradation products with pure samples was then established by infrared spectroscopy or melting point measurements. With gas chromatography, the separation of complex alkaloid mixtures became much easier; and mass spectrometry was the ideal technique to tackle the identification, and, if necessary, the structural elucidation of relatively large numbers of alkaloids.[128]

Despite their great complexity, alkaloids showed some properties that eased the burden of unraveling the riddle of their structures: They contained aromatic and alicyclic moieties, and in a group of the same structural type many members differed only in the substituents in the aromatic part. If the scientists ignored more complicated cases, this peculiar property gave rise to mass spectra with peaks at identical mass numbers and similar intensities for the fragments containing the alicyclic parts. Those fragments that retained the aromatic moiety showed peaks that differed in the mass values of their substituents. Biemann developed this technique, known as the "mass spectrometric shift," (see Figure 3.6), soon put to great use also by other alkaloid chemists. Biemann, for example, in his investigations of indole and dihydroindole alkaloids, compared the conversion products of ajmaline and sarpagine, whose mass spectra were extremely similar but differed by sixteen mass units. Because the structure of ajmaline was already known, Biemann and co-workers were able to confirm a predicted structure of sarpagine by mass spectrometric methods. This was just before much more laborious chemical conversion made possible the direct comparison of the conversion products of ajmaline and sarpagine by melting point and infrared measurements.[129]

In order to prove the validity of his newly found method, Biemann, in collaboration with the group of William Taylor at CIBA Pharmaceutical Products in Summit, New Jersey, investigated a number of related alkaloids from *Tabernanthe iboga*, namely ibogamine, ibogaine, and ibogaline. This investigation began as a comparative study to the one related to sarpagine, and Biemann looked for alkaloids whose carbon skeleton was isomeric, but definitely different, from the one present in sarpagine.[130] The CIBA chemists were able to provide Biemann with the compounds he asked for. With this help, he built up trust in his method of the mass spectrometric shift and it led to a further application of Biemann's method to determine an unknown structure. Though the

[128] For an excellent account of Biemann's ventures into alkaloid chemistry see his "Four Decades of Structure Determination," 1254–1259.

[129] Biemann, "Determination of Carbon Skeleton of Sarpagine"; and Bartlett, Sklar, and Taylor, "Rauwolfia Alkaloids XXXIII."

[130] Biemann to William I. Taylor, 21 March 1960. Biemann Papers, folder "CIBA."

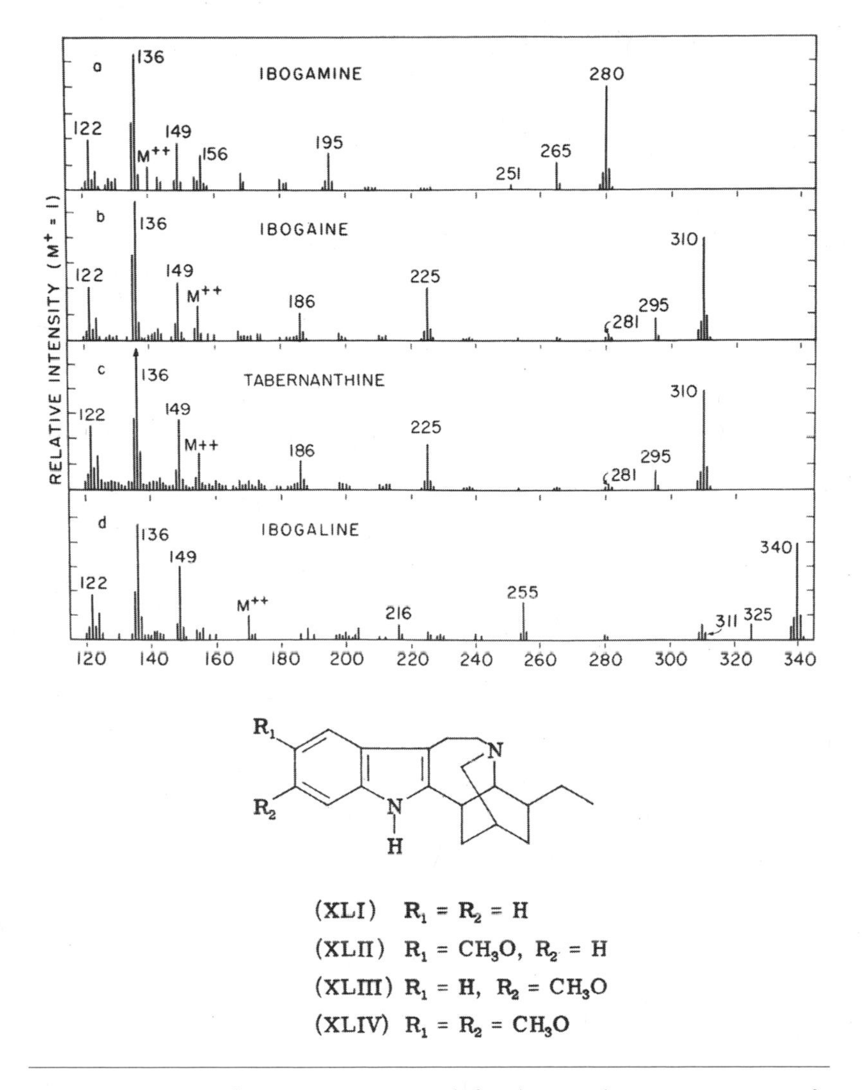

FIGURE 3.6, Top, the mass spectrometric shift technique, showing mass spectra of ibogamine (a), ibogaine (b), tabernanthine (c), and ibogaline (d). Important peaks (in (a) at masses 156, 195, 251, 265, and 280) differ by 30 atomic mass units (u), representing the molecular weight of a methoxyl group. Thus, ibogaine was found to be a methoxyibogamine, and ibogaline a dimethoxyibogamine. Tabernanthine shows nearly the identical mass spectrum as ibogaine, its molecular structure differing only in the position of the methoxyl group. Bottom, the structural formulae of the alkaloids ibogamine (XLI), ibogaine (XLII), tabernanthine (XLIII), and ibogaline (XLIV). From Klaus Biemann, "Application to Natural Products and Other Problems in Organic Chemistry," in Fred W. McLafferty, ed., *Mass Spectrometry of Organic Ions*, New York: Academic Press 1963, 573, fig. 19, 574.

structures of the first two alkaloids, ibogaine and ibogamine, had already been established, the small amount available for ibogaline was not enough to allow classical proof of its structure. Ultraviolet and infrared spectra indicated that ibogaline contained a 5,6-dimethoxy-indole moiety. The use of the mass spectrometric shift technique enabled Biemann to prove the correctness of this proposed structure.

Another technique, deuterium labeling, enabled the structural elucidation of iboxygaine, a hydroxyl derivative of ibogaine. It indicated a fragmentation scheme that allowed a rational correlation of many of the peaks with fragments of known structure. Biemann's reasoning was based on the assumption that fragments stabilized by electrons of an adjacent heteroatom or a π-bond showed stronger peaks than fragments which were not. Moreover, he proposed that multiple cleavages of bonds "should occur in a concerted fashion and lead wherever possible to neutral molecules rather than fragments with a number of unpaired electrons."[131] This followed a generalization of "Stevenson's rule,"[132] according to which the positive charge after cleavage will reside on that fragment which has the lower ionization potential. In most cases this was caused by delocalization due to resonance or the inductive effects of substituents. Thus, Biemann used and refined rules and generalizations developed in the industrial period of organic mass spectrometry, rationalized in the language of physical organic chemistry.

The responses to the advances in alkaloid chemistry that were possible by mass spectrometric methods were disparate. Some referees were cautious, thinking that the technique did not live up to the promises that had been made or suggested. Such was the case in a comment on a paper by Biemann and Margot Friedmann-Spiteller:

> Since in fact the "potentialities" of mass spectrometry have been generally known for some time, although the instances of its actual applications to the rigorous determination of structure of complicated molecules are exceedingly rare, if the author has really obtained unambiguous proof of the ibogaline constitution in this way he should certainly be encouraged to publish it.[133]

Other colleagues were excited. When Biemann established the structure of quebrachamine, the specialist in the field, Bernhard Witkop, of NIH, congratulated him on this success. Witkop was glad that a twenty-year-long struggle had come to an end:

---

[131] Biemann, "Determination of the Structure of Alkaloids by Mass Spectrometry," 412.

[132] Stevenson, "Ionization and Dissociation by Electronic Impact. Ionization Potentials and Energies of Formation of Sec-propyl and Tert-butyl Radicals."

[133] Comments of referee I, in Biemann Papers, folder "Early manuscr. submissions and reviewers comments." The article is Biemann and Friedmann-Spiteller, "Application of Mass Spectrometry to Structure Problems. V. Iboga Alkaloids."

> Your correlation of aspidospermine and quebrachamine via a common degradation product differing only in the methoxy group has been indeed most gratifying. When I started work on these alkaloids almost 20 years ago I had no way of knowing that aspidospermine and quebrachamine were of a structure accessible more to the methods of mass spectrophotometry and Fourier analysis (sic) than the classical chemical methods . . . . Thanks to your investigations the case seems to be air-tight and shut by now."[134]

Biemann's method was most suited to determination of the structures of large numbers of closely related alkaloids. Here, he could make best use of his mass spectrometric shift technique. The minute amounts of alkaloids often prevented their isolation and crystallization using classical means. For Biemann, with his new sophisticated methods, this presented no problem.

But the breakthrough of mass spectrometry did not come automatically. This is most visible in the case of the determination of elemental compositions and molecular weights. Who could guarantee the accuracy of the new methods that endangered the supremacy of old and well-established procedures? One of the first conflicts in the vicinity of Biemann emerged in relation to a paper by George Büchi on the structure of patchouli alcohol, submitted to the *Journal of the American Chemical Society*.[135] Büchi had used Biemann's services to establish the molecular weight of this compound solely by mass spectrometry. In answering the criticism of the journal editors, Büchi asked Biemann to support his view. Biemann was glad to do that, "because I also think that with the coming of age of organic mass spectrometry, this problem will arise more frequently in the near future,"[136] he wrote in November 1960 to Marshall Gates, editor of the prestigious journal. He was sure that the determination of molecular weights by mass spectrometry afforded more reliable information than the traditional methods.

Biemann carefully indicated the criteria for justifying the substitution of elemental analyses by mass spectrometry. Because of the considerable complexity and diversity of mass spectra of organic compounds, the novel technique asked for more personal experience and skills of the interpreter than did the traditional methodology of combustion analysis: "The determination of the mol. wt. [molecular weight] by mass spectrometry requires a certain amount of interpretation and experience and puts more responsibility on the investigator (or the person determining the spectrum) than does the use of a C, H value."[137] Biemann did not mention here that microanalytical methods in elemental

---

[134] Witkop to Biemann, 28 August 1961, Biemann Papers, folder "Witkop."

[135] See Büchi, Erickson, and Wakabayashi, "Terpenes. XVI"; and Büchi and MacLeod, "Synthesis of Patchouli Alcohol."

[136] Biemann to Marshall Gates, 4 November 1960, Biemann Papers.

[137] Biemann to Gates, 4 November 1960, Biemann Papers.

analysis needed considerable skill on the part of the experimenter, usually a technician. Thus, the use of mass spectrometry shifted the needs of expertise to the higher educated (Ph.D.) mass spectrometrists. Furthermore, he regretted that mass spectrometry could lead to a situation in which chemists would not bother any longer to prepare their compounds in a pure state. Thus, he suggested the acceptance of mass spectrometric molecular weights if

> a. there is an indication of the purity of the material associated with the description of the experiment;
> b. the compound belongs to a class expected to give reasonably intense peaks for the molecular weight;
> c. the type of instrument used (including operating conditions like temperature and ionizing potential), and perhaps even the laboratory in which the spectrum was determined, is indicated; and
> d. the compound was derived in a reasonably clear-cut reaction from another completely characterized substance.

Biemann closed by stating that he tried to objectively compare the respective merits of the two major possibilities for establishing the molecular weight of an unknown compound. Mass spectrometric methods, in his opinion, "should be permitted, where applicable, to substitute for the classical elemental analysis to which we subconsciously attach more significance than it deserves in these times of advanced instrumental methods."[138]

Gates appreciated the balance of Biemann's proposal, and knew of possible inaccuracies in conventional analysis of carbon and hydrogen. But Biemann's letter did not cause more than a lukewarm response, and a vague reference to the future. Gates offered to consider the acceptance of mass spectrometric data, but only in favorable cases:

> It seems likely that as techniques improve and as knowledge of the versatility of the method becomes more widespread that we will have increasing use of mass spectrometric measurements of this sort for characterization, and the time may well come when in many cases it will be proper to use such data in place of the traditional analyses. I do not think, however, that the time has come.[139]

Two years after this incident, the time had not come yet. A colleague of Gates, Robert L. Autrey, wrote, on the basis of two referee reports, and with reference to a manuscript submitted by Biemann and his collaborators Margot Friedmann-Spiteller and Gerhard Spiteller:

---

[138] Biemann to Gates, 4 November 1960, Biemann Papers.

[139] Gates to Biemann, 16 November 1960, Biemann Papers.

> We realize that mass spectrometry is a powerful method, and we are confident of the correctness of your conclusions. However, its use is still in its infancy, and the samples to which you have applied it are open, perhaps, to question as to their homogeneity. You have detected 20 compounds, but we question that you have "isolated" 16 of them. There is no demonstration, other than by gas chromatography, of their purity. None of the usual criteria of purity have been applied; none of the compounds have been characterized by any of the classical means: there are no melting points, no analyses, few ultraviolet and no infrared spectra. The chemist or chemical taxonomist discovering one of these compounds in another plant may be quite unable to ascertain that he has in hand a known compound, as few institutions have available a mass spectrometer of the caliber required to obtain these data.[140]

Biemann replied that he was concerned about the possibility that "other workers in the field of alkaloid chemistry may use these mass spectrometric techniques too freely or may draw unjustified conclusions." He explained that in order to make clear the applicability of mass spectrometry he has written his book, which appeared in print at the same time.[141] Furthermore, he questioned the critique concerning the isolation of these alkaloids, claiming that this would be a matter of definition: "One might say we did even if we have in some cases merely collected it from a gas chromatograph and carried it (in a glass tube) to another room to run the mass spectrum."[142] However, he altered the text of the relevant table in the final manuscript, now stating only that the alkaloids are listed in the order of elution from an alumina column. From the beginning, and in contrast to the critique that no melting points were reported, Biemann and co-authors Friedmann-Spiteller and Spiteller had given these data in cases where the compounds had been obtained in crystalline form. With reference to the lack of elemental analyses, he stated in the final publication that this

> is balanced by the presentation of mass spectrometric molecular weights (and the entire spectrum), which are very reliable for this type of compound . . . and leave as little ambiguity (albeit of a different kind) regarding the true empirical formula as does a conventional elemental analysis. Furthermore, we feel that the mass spectrum is at least as good a method for the characterization of organic compounds as other spectra at present accepted for this purpose.[143]

---

[140] Robert L. Autrey to Klaus Biemann, 9 October 1962, Biemann Papers, folder "Aspidosp." The publication referred to is Biemann, Friedmann-Spiteller, and Spiteller, "Application of Mass Spectrometry to Structure Problems. X."

[141] Biemann to Autrey, 18 October 1962, Biemann Papers, folder "Aspidosp." The book is Biemann's *Mass Spectrometry. Organic Chemical Applications*, New York: McGraw-Hill, 1962.

[142] Biemann to Autrey, 18 October 1962, Biemann Papers, folder "Aspidosp."

[143] Biemann, Friedmann-Spiteller, and Spiteller, "Application of Mass Spectrometry to Structure Problems. X," 637.

Biemann's discussions with the editors of the *Journal of the American Chemical Society* reveals a momentary snapshot in the struggle of getting physical instrumentation to supplement traditional methods. His was not a solitary case. Carl Djerassi had similar experiences with the editors of the same journal.[144] This conflict is to be seen more as a continuum than as a short, revolutionary strike. Most mass spectrometrists in the late 1960s regarded the traditional microanalysis as a "rather crude method and one that has been overtaken by recent developments in characterization by physical means."[145] Consequently, mass spectrometry was recommended by organic mass spectrometrists to establish the purity and constitutional formula of a compound. This was part of a long-term development from 1950 on replacing traditional methods with novel, spectroscopic data. Characteristically, Biemann emphasized the responsibility that mass spectrometrists had for the correct application of the technique and the interpretation of its results. Thus, in Biemann's view, traditional, labor-intensive methods of analysis would not be replaced by automated procedures. Still, a human interpreter would rule the process. In Biemann's system, the academic mass spectroscopist should fill this role. In the long run, the organic chemists countered this challenge to their status by taming mass spectrometry (and other spectroscopic techniques) in the form of service laboratories run by subordinated technicians. In response, experts, such as Biemann, focused on the development of methods for structural elucidation of unknown molecules. This more demanding intellectual field guaranteed their independence from the organic chemists, and enabled them to establish a scientific specialty in its own right, organic mass spectrometry. However, they depended on the applicability of their methods by biochemists and natural product chemists.

While Biemann rapidly gained acceptance among alkaloid chemists, the community of peptide chemists showed much more reluctance. In the 1950s, the sequencing of peptides and their long-chain homologues, proteins, was a very fruitful research field. The famous biochemist Frederick Sanger had developed a method to determine the amino terminals of peptides,[146] and Biemann planned to develop a counterpart for the carboxyl group end of the chain. This was to be achieved with purely chemical means. With this in mind, he applied for an NIH grant, which was approved. Fortunately, NIH rules enabled changes in the approach to a chosen topic, and after Biemann had gotten access

---

[144] Carl Djerassi, interview by Carsten Reinhardt, 20 January 1999.

[145] Allan Maccoll, editor in chief of the journal *Organic Mass Spectrometry*, to the regional editors and members of the editorial advisory board, 20 February 1968, and the enclosed policy for confirmation of a molecular formula of a compound, McLafferty Papers, folder "Organic Mass Spectrometry."

[146] See Sanger, "Free Amino Groups of Insulin." In 1958, Sanger received the Nobel Prize in chemistry for this work.

to the CEC instrument, he switched strategies in order to tackle the problem by mass spectrometric methods.[147]

With mass spectrometry, Biemann planned to replace the commonly used chromatographic and electrophoretic separation techniques of peptides and proteins, followed by the identification of the degradation products via unspecific color reactions. These methods were the workhorses of biochemists, and generally highly successful, as Biemann conceded in 1960 before an assembly of leading European peptide scientists. But he promised a solution for their most urgent problem, the "reverse salient" of protein chemistry:

> Over the last decade these techniques have led to a dramatic advance of the knowledge of protein structure. For the solution of the many more structural problems in the peptide field there seems to be only one obstacle: the amount of time and manpower required due to the slowness of the methods of separation employed, the tedious procedures used for the determination of the amino acid sequence in the degradation-peptides, and the sometimes not very reliable identification of the products."[148]

Moreover, peptides eluded physical methods such as infrared and ultraviolet spectroscopy because these methods were not sufficiently sensitive to the subtle differences in the sequences of the amino acid chains constituting peptides and proteins. Biemann was convinced that mass spectrometry could fill this gap.

Biemann thus proposed to tackle one of the major themes of the biochemistry of that time, the determination of the amino acid sequences of proteins. It was thought that the data gained by the mass spectrometer could be fitted back into a structure of the whole protein. But Biemann's work on the sequencing of peptides took a long time to be accepted by biochemists. The problematic features of Biemann's early methods were rooted in the necessity for converting the peptides chemically before they could be inserted into the mass spectrometer. Thus, for his investigations in the peptide field, Biemann had to make full use of his chemical skills and those of his collaborators. The "zwitterionic" character of peptides, causing their high polarity, had to be removed, and for this Biemann applied acetylation of the amino groups and esterification of the carboxyl groups. Finally, to increase volatility, the molecule was converted into a polyamino alcohol. Thus, only substantial experience in micro-methods of organic chemistry could make the mass spectrometry of peptides viable at all. Though Biemann, Gapp, and Seibl obtained generally useful results, this was a

---

[147] Biemann, "Massachusetts Institute of Technology Mass Spectrometry School," 333. See Biemann, Gapp, and Seibl, "Application of Mass Spectrometry to Structure Problems. I. Amino Acid Sequence in Peptides."

[148] Biemann, "Application of Mass Spectrometry in Amino Acid and Peptide Chemistry," 393.

chemical "tour-de-force on a microscale."[149] Thus, the potential of the mass spectrometer, especially its speed and accuracy, could not be used to its full extent. Indeed, the technique proved to be too labor-intensive and complex to compete with the newly improved Edman sequencing methods.

Consequently, Biemann changed his approach. First, he included the qualitative and quantitative determination of amino acids in mixtures. The increased use of chromatographic methods in natural product chemistry had resulted in the findings and structural assignments of a large number of previously unknown natural amino acids. Their separation and identification became more and more difficult. Because amino acids showed highly characteristic fragmentation patterns, their qualitative identification in complex mixtures was relatively straightforward. Moreover, and for quantitative purposes, mass spectrometry was an efficient tool, making unnecessary the very time-consuming separation of the components via chromatography. This was reminiscent of the analytical procedures in the petroleum industry,[150] the reasoning behind quantitative mass spectrometric determination of amino acids being enhanced speed and efficiency. Emphasizing the industrialized style of many procedures used in biochemistry, Biemann recommended his method to laboratories where it was necessary "to analyze accurately a large number of samples in a relatively short time, which might be necessary in connection with the structure determination of a protein, studies of metabolism, multiple analyses to increase the accuracy by statistical treatment of the results, etc."[151]

In contrast to these analytical methods, the structural elucidation of unknown amino acids was a truly novel approach, and required all the knowledge gained from fragmentation patterns of organic molecules in the mass spectrometer. Biemann soon became a master in this field, and his abilities were acknowledged by his colleagues and peers. The referee's comments on a paper of Biemann, Seibl, and Gapp on the mass spectra of amino acids contained the remark that the paper established a valuable new technique for the structural determination of amino acids. In addition, the referee was of the opinion that "it is one of the finest discussions of the mechanisms of molecular degradation under electron impact for a particular class of compounds that I have seen."[152] But despite all the individual performances, mass spectrometry alone did not suffice to establish the structure of complex natural products, mainly because it could not unequivocally prove the presence or absence of certain functional groups in the molecule. What could be shown by mass spectrometry, however,

---

[149] Biemann, "The Coming of Age," 1921. See Biemann, Gapp, and Seibl, "Application of Mass Spectrometry to Structure Problems. I. Amino Acid Sequence in Peptides," 2274–2275.
[150] Biemann and Vetter, "Quantitative Amino Acid Analysis by Mass Spectrometry," 93.
[151] Biemann, "Application of Mass Spectrometry in Amino Acid and Peptide Chemistry," 397.
[152] Comments of referee I, in Biemann Papers, folder "Early manuscr. submissions and reviewers comments." The paper is Biemann, Seibl, and Gapp, "Mass Spectra of Organic Molecules. I."

was the arrangement of atomic groupings. The first complex natural product where mass spectrometry decisively contributed to structure elucidation was the amino acid lysopine. This compound was of substantial biological interest, and was isolated from tumors found in leguminous plants. It was an ideal compound to be tackled with mass spectrometry as it was available only in very small amounts and further purification through crystallization was made impossible by retention of the solvent.[153] In this project, Biemann collaborated with the French biochemist Edgar Lederer, whom he knew through his contacts to Firmenich & Cie. Lederer was also a consultant for the Swiss firm, and had read about Biemann's work on peptides. Biemann was glad to get the opportunity to tackle an unknown amino acid with mass spectrometry. When he reported his preliminary results on the compound in a letter to Lederer, he offered his cooperation in the final proof of its structure in case Lederer did not have other means at his disposal:

> I have tried to get somewhere a sample of an unknown amino acid to determine its structure using mass spectrometry. But they are not easy to come by. Your compound would be an interesting example and if, because of lack of material, you should not be able to deduce its structure, I would try it with more mass spectrometry on some additional derivatives.[154]

Biemann had, as it was later shown, immediately established the correct structure (he originally assumed the possibility of some isomers of the compound). Lederer was excited about Biemann's results, and took measures for their confirmation via synthesis. Additionally, he provided Biemann with the names of scientists who would have additional unknown amino acids.[155] The structure was finally proved by comparison of the synthetic with the natural compound. The contact with Lederer was important for Biemann and his group in the long run. Three of Biemann's associates, Walter Vetter, James McCloskey, and Bhupesh Das, in the early 1960s joined Lederer's group at the Institut de chimie des substances naturelles in Gif-sur-Yvette. All of them did mainly service work in mass spectrometry there. Moreover, Biemann received the honorary title of an "Inspector General" of this mass spectrometry laboratory.[156] Lederer's institute belonged to the prestigious Centre National de la Recherche Scientifique, and was—in Biemann's eyes—the most advanced institution in this field in France.

---

[153] Henahan, "Klaus Biemann," 51–52; Biemann, "Application of Mass Spectrometry in Amino Acid and Peptide Chemistry," 396; Biemann et al., "Structure of Lysopine."

[154] Biemann to Lederer, 20 October 1959. See also Lederer to Biemann, 24 June 1959, Biemann Papers, folder "Lederer, Prof. E."

[155] He named Artturi Virtanen of Helsinki and Marcel Renard of Gembloux in Belgium. Lederer to Biemann, 24 October 1959, Biemann Papers, folder "Lederer, Prof. E."

[156] See Biemann to Lederer, 23 May 1961, Biemann Papers, folder "Lederer, Prof. E."

In 1960, Biemann felt that of the various applications of mass spectrometry to protein and peptide chemistry, "the most important contribution appears to be the one for the direct determination of the amino acid sequence in small peptides . . . since it is much faster and at least as reliable as conventional methods."[157] But the attack of "real problems," i.e. structures of proteins, took a long period of refinement of the methodology. Thus, in 1969, one of the peers of protein chemists, Stanford Moore, prophesied that Biemann's plans "will never work" because of the huge differences in the cleavages of peptide bonds in proteins. As a result, there would not be enough overlap of the pieces, and no possibility to reconstruct the primary structure of the protein. Indeed, it took more than fifteen years before Biemann successfully contributed to the structural elucidation of a protein and thus showed that mass spectrometry could complement the established Edman procedures.[158]

Networks of University and Industry

One of the difficulties that Biemann encountered was of having chemical problems at hand that could be solved with the new method. For structural studies, Biemann's intention was to start out with alicyclic compounds, such as cyclohexane derivatives and mono- and bicyclic terpenes. Because he knew of the interest of Firmenich & Cie., of Geneva, in this field, he suggested that they send him samples:

> [My] advantage was that I had access to interesting unknown compounds. I never was very much in favor of running large numbers of known compounds and interpreting their structure which people at that time did, petroleum people, and chemical industry people . . . . As I said, I came from the other side, namely having an unknown compound and trying to find out about it, and Firmenich was one source for such compounds.[159]

Though not very well known in the public domain, the Swiss company Firmenich & Cie. was one of the most dynamic and research-intensive companies in the manufacturing of flavors and fragrances.[160] Founded in 1895 by the chemist Philippe Chuit, the company originally known as *Chuit & Naef* successfully engaged in the production of vanillin, and soon added other flavor compounds to its product range. The scientific development of the company from the 1920s to the 1950s was decisively shaped by Leopold Ruzicka, who

---

[157] Biemann, "Application of Mass Spectrometry in Amino Acid and Peptide Chemistry," 401.
[158] Biemann, "The Coming of Age," 1923.
[159] Biemann, interview by Reinhardt, 10 December 1998.
[160] Ohloff, "In Place of a Foreword"; Ohloff, "Ein fragmentarischer Beitrag"; Ruzicka, "Rolle der Riechstoffe."

headed the chemistry department of ETH Zurich, and in 1939 received the Nobel Prize. Ruzicka's research areas belonged to the chemistry of terpenes and macrocyclic compounds, and both substance classes were of utmost importance as odors. He was a major scientific consultant for Firmenich & Cie. (the company's name was changed several times, in 1956 to Firmenich & Cie., successeurs de Chuit, Naef & Cie.). In the mid-1930s, the firm embarked on a research project on the perfume ingredient ambergris. The first real breakthrough came when Ruzicka, and, independently of him, Edgar Lederer in Paris, in 1946 were able to elucidate the structure of the triterpene ambreine. Subsequently, research in this field was undertaken at four different geographical sites: Zurich (Ruzicka, later Albert Eschenmoser, both at ETH), near Paris (Lederer, Institut de chimie des substances naturelles in Gif-sur-Yvette), Geneva (Max Stoll, Firmenich & Cie.), and from 1955 Cambridge, Massachusetts, with George Büchi at MIT who had taken his Ph.D. with Ruzicka in Zurich. The center of coordination for research into ambergris constituents was Geneva, with Stoll's research department.[161]

Ambergris was not by far the only research topic of scientists working for Firmenich & Cie. To the topics that involved chemists at MIT belonged investigations of cocoa and coffee constituents, and jasmine oil. Josef Seibl of Biemann's group was actively engaged in the unraveling of the constitutions of pyrazine compounds that form a part of cocoa components. The synthetic work in this area was done by Irving M. Goldman, a graduate student of Büchi, during 1958–1959.[162] In addition, a decade-long research effort enabled the analysis of 226 odorant and gustatory constituents of coffee.[163] An important ingredient of jasmine oil, methyl jasmonate, was tackled by Büchi, while the structure of another constituent, jasmine lactone, had been established by the Firmenich chemist Max Winter.[164] Winter was at MIT for a while in 1957 and 1958, and often acted as mediator between Biemann and Firmenich & Cie.

Thus, Firmenich & Cie. had considerable experience in the coordination of complex research projects undertaken at different places. Furthermore, in 1955 they engaged in modern instrumental methods of analysis. Most important for Firmenich & Cie. was the novel technique of gas chromatography (GC).[165] For the natural product chemists at Geneva, Zurich, and Cambridge, the gas chromatograph soon became an indispensable tool for the separation of compounds, both for preparative and analytical purposes.[166] The peculiar per-

---

[161] Ohloff, "In Place of a Foreword," ix–x.

[162] Ohloff, "In Place of a Foreword," x. Irving M. Goldman was mentioned in connection with work on coffee in Biemann to Stoll, 20 April 1959, Biemann Papers, folder "Firmenich et Cie."

[163] Ohloff, "In Place of a Foreword," xiii.

[164] Ohloff, "In Place of a Foreword," x–xi.

[165] See Ettre, "Gas Chromatography."

[166] Ohloff, "In Place of a Foreword," xi.

formance of the mass spectrometer, allowing detection and identification of minute quantities of a sample, consequently showed high promise to tackle the small amounts available after their isolation by gas chromatography. But instead of building up a mass spectrometry group in Geneva, Max Stoll of Firmenich & Cie. decided to cooperate with Biemann, who already was in close contact with the firm through his research with Büchi.[167] Biemann's expertise in a field that was of interest to Firmenich & Cie., and his ambitious drive into an analytical method of high potential for natural product chemistry were ideal conditions for the beginning of their cooperation. Biemann hoped to receive samples from the Swiss company, and in exchange expressed his willingness to make himself familiar with other compound classes if this would suit Firmenich & Cie.[168]

In addition, Biemann provided Stoll with first-hand knowledge of the working conditions of his CEC spectrometer. When he sent a report in September 1958, describing three months of his own experience with the instrument, Biemann showed great optimism for the future of mass spectrometry: "In my personal opinion, mass spectrometry will become at least as useful for structure work as NMR or even IR in some respects."[169] In the early period of their collaboration, a free flow of information and samples in both directions took place. On the basis of published literature, Winter provided Biemann with a fourteen-page summary of the interpretation of mass spectra, dubbed the "general rules" in subsequent correspondence.[170] Both partners also exchanged reports on conferences that the other side was not able to attend.[171] Soon, this was about to change, when Biemann was increasingly directed into a service role.

Many of the samples that Biemann received from Geneva were effluents of gas chromatographs, available only in tiny amounts and diluted with water from the carrier gas. This complicated the insertion of the samples into the mass spectrometer. Even more time-consuming was the interpretation of mass spectra, "because the compounds apparently do not belong to the simple types discussed in the literature."[172] Thus, Biemann proposed to start a wide-ranging

---

[167] In 1956 he had co-authored with Stoll and Büchi a mechanistic paper on the cyclization of a derivative of α-ionone, one of the classic products of the company. Another paper of Büchi and Biemann published in 1957 was related to the production of substances reconstituting the odors of ambergris. In addition, a synthetic method of Biemann and Büchi was later adopted by Firmenich for the synthesis of macrocyclic compounds. Büchi, Biemann, Vittimberga, and Stoll, "Terpenes IV"; Biemann, Büchi, and Walker, "Structure and Synthesis of Muscopyridine"; Büchi and Biemann, "Conversion of Sclareol to Manool." See also Ohloff, "In Place of a Foreword," xiv.

[168] Biemann to Stoll, 22 April 1958, Biemann Papers, folder "Firmenich et Cie."

[169] Biemann to Winter, 12 September 1958, Biemann Papers, folder "Firmenich et Cie."

[170] Max Winter, memorandum "Mass spectrometry," Geneva, 18 June 1958; Biemann to Winter, 12 September 1958, Biemann Papers, folder "Firmenich et Cie."

[171] See, for example, Winter to Biemann, 14 October 1958, Biemann Papers, "folder CEC."

[172] Biemann to Stoll, 3 November 1958, Biemann Papers, folder "Firmenich et Cie."

project in the mass spectrometry of furanes (contained in coffee), and terpene alcohols (menthol is an example of this class). His aim was to establish reference spectra to ease the interpretation of substances that would arise from the research at Firmenich & Cie.[173] This was possible only with access to the rich sample collection of reference compounds at Firmenich & Cie. Because the interpretation of mass spectra was an art requiring a lot of experience and knowledge "about the history and source of the compounds," Biemann appreciated Stoll's assistance and critical questions.[174]

Though he obtained a variety of interesting samples for establishing mass spectrometry in the area of terpenes and other odorific compounds, he could not publish his results, except in some very few cases. One such exception was his work on *cis* rose oxide, a monoterpene ether discovered in 1959 by Casimir F. Seidel and Max Stoll, and found to be an important ingredient of Bulgarian rose oil. Its structure was elucidated through a combined research effort of the chemists at Firmenich, the group of Albert Eschenmoser at Zurich, and Klaus Biemann's group in Cambridge.[175] In this case, the publication speeded up because a competitor of Firmenich & Cie. was known to be on the verge of finding the same results.[176] But in general, the publication restrictions of Firmenich & Cie. were a serious scientific disadvantage to Biemann.

In 1959, the cooperation became more defined and formalized. To "facilitate the exchange of ideas between Dr. Biemann and Geneva and to make MS-application as efficient as possible," Stoll and Biemann reached an agreement on the organization of mass spectrometry research. Most importantly, Biemann obtained the assurance that only a fixed number of samples (15 spectra per month) had to be run by him. A biweekly conference in Geneva decided which samples would be sent to Biemann, considering "all the different reasons in favor and against MS analysis . . . when making this selection."[177] From then on, in addition, all correspondence and the forms containing the information on the compounds had to go through the hands of Stoll on the side of Firmenich & Cie.,[178] stating, among others, physical data (boiling and melting points), state of purity, purpose of mass spectrometric analysis, history of the sample, and available information on the structure of the compound. Bie-

---

[173] Biemann to Stoll, 18 November 1958, 20 April 1959, Biemann Papers, folder "Firmenich et Cie."

[174] Stoll to Biemann, 4 December 1958, 6 January 1959, and Biemann to Stoll, 17 December 1958 (quote), 15 January 1959, Biemann Papers, folder "Firmenich et Cie."

[175] Seidel, Felix, Eschenmoser, Biemann, Palluy, and Stoll, "Zur Kenntnis des Rosenöls. 2. Mitteilung."

[176] Stoll to Biemann, 15 November 1960, Biemann Papers, folder "Firmenich et Cie."

[177] Memorandum "Organization of MS-Research," 9 April 1959, agreed on by Stoll, Biemann, and Winter, Biemann Papers, folder "Firmenich et Cie."

[178] Memorandum "Organization of MS-Research," 9 April 1959, Biemann Papers, folder "Firmenich et Cie."

mann's part was to make information available about the mass spectrometer runs, including attempts at interpretation.[179]

In May 1959, at a meeting between Biemann and Stoll in New York, Stoll assured Biemann that he would receive an annual remuneration of $5,000 if Biemann agreed to get extra help. Furthermore, Stoll expected that

> he will continue to do our mass spectrometric work without any limitation, but it is clearly understood that we will give him only those problems which we can absolutely not solve otherwise in a reasonable time. We must make sure that Klaus has sufficient time to do his own research work. When we send him new substances of unknown constitution, or for which there are no comparison spectra available, we will try to synthesize substances according to Klaus' suggestions. We will avoid as much as possible all service work, i.e. work on substances where Klaus has only to take the spectrum without being able to identify it. If we send Klaus a substance of unknown constitution, he is fully authorized to handle this substance according to his ideas about the best way to perform its identification, so that he is really a co-worker of ours and not merely a service-man.[180]

Although the purpose of this meeting was to further clarify the manner of cooperation, Stoll's wording of the report injected the causes of what turned out to be future misunderstandings. The mentioning of unlimited support by Biemann, especially, was at odds with the explicit statement in the April agreement that only up to 15 spectra per month had to be run by Biemann. Moreover, with regard to the latest developments in mass spectrometry instrumentation and Biemann's recommendation that Firmenich & Cie. should purchase a mass spectrometer for the Geneva-based research laboratories, Stoll came to entirely different conclusions:

> There is no doubt that this kind of apparatus is in a fast development period, so that in my opinion it would be unwise to purchase at the present time an instrument for Geneva. Klaus, however, thinks that when waiting for a better instrument, we are delaying our research work. If our competitors are working with such a tool, they will advance more rapidly and win time, as well as experience over us. But this, I hope, Klaus will overcome. So, whenever we are purchasing ourselves an instrument, all his experience will be available to us, a fact that Klaus confirmed.[181]

---

[179] "Comments regarding form M for MS-research," and "Comments regarding form G for MS-research," attached to letter of Winter to Biemann, 21 April 1959, Biemann Papers, folder "Firmenich et Cie."

[180] Stoll, "Report on my discussions in USA," May 1959, Biemann Papers, folder "Firmenich et Cie." See letter of Stoll to Biemann, 10 June 1959, ibid.

[181] Stoll, "Report on my discussions in USA," May 1959, Biemann Papers, folder "Firmenich et Cie."

The burden of service spectra increased considerably in the fall and winter of 1959, at a time when Biemann desperately needed to publish. To circumvent the non-publication rules of Firmenich & Cie., and in the case of the ingredients of "base noisette," he proposed a strategy that would mask the connection of his work to the coffee-related research problem of Firmenich & Cie:

> Since we probably should not publish anything about the "base noisette" in the near future, we would like to write a paper on the spectra of the synthetic pyrazines and the trick with the piperazines, which I think is very neat. Since it would contain also pyrazines not present in 'base noisette,' it would not give any clue to other investigators.[182]

Nevertheless, no publication resulted of this work, probably for reasons of secrecy. To supply Biemann with topics that would be publishable, Stoll tried to hand over to Biemann more independent research problems, in addition to service spectra.[183] When Stoll acknowledged that their investigations of the components of strawberries and raspberries made necessary a large amount of spectra, he hastened to assure Biemann that he was regarded to be a close co-worker, and that he would participate in eventual profits if the project would be a success.[184] Stoll was in a classical dilemma: He decided to rely on outside help, while the nature of the work (a backlog of a huge amount of samples after the harvest season and the fractionating work was over) asked for in-house, dependent service work. In December 1959, Biemann wrote:

> As to the number of samples which we received since the beginning of November I must tell you that their processing has stopped my own research problems almost completely and has taken up all my spare time (including Sundays) . . . . We have tried to do the job considering it one of the "emergency cases" which we agreed upon . . . [but] it is technically impossible to exceed a reasonable amount of samples.[185]

---

[182] The piperazines are hydrogenated products of pyrazines, and the strategy Biemann mentioned was related to the additional information that the study of the mass spectra of the piperazines made it possible to gain about the pyrazines. Mass spectra of pyrazines could not distinguish between the different positions of the substituents of the aromatic ring, while hydrogenated compounds yielded better evidence with regard to these positions. See report of J. Seibl, TR 691, bases "A," Biemann Papers, folder "Firmenich et Cie." For a very good example of Seibl's reasoning in establishing components of base noisette see J. Seibl, report TRL 15, base noisette, Biemann Papers, folder "Firmenich et Cie."

[183] The topics were related to an anti-oxidant substance and tobacco. Stoll to Biemann, 12 November 1959, and Biemann to Stoll, 14 December 1959, Biemann Papers, folder "Firmenich et Cie."

[184] Stoll to Biemann, 20 November 1959, and Winter to Biemann, 20 November 1959, Biemann Papers, folder "Firmenich et Cie."

[185] Biemann to Stoll, 14 December 1959, Biemann Papers, folder "Firmenich et Cie."

Since November, the pile of samples had run up to 96, and since June more than 130 spectra had been recorded and interpreted for Firmenich & Cie. Still, Stoll tried to acquire at minimal cost Biemann's high-level scientific expertise in mass spectrometry, while Biemann tried to talk the Firmenich & Cie. scientists into the acquisition of a mass spectrometer that resembled his own. Through this, Biemann saw the chance of removing a large part of the burden the Firmenich connection had become for him.

When discussing the potential of the Bendix time-of-flight mass spectrometer (Biemann opined that its resolution was not good enough for this type of work), and the new high resolution mass spectrometers that were in the process of development at the instrument manufacturers (too precise and time-consuming for the needs of Firmenich & Cie.), Biemann argued on a sound technical basis for the imitation of his own type of instrumentation in Geneva. In contrast, Stoll proposed that Firmenich & Cie. acquire a time-of-flight spectrometer for the more routine type of work, while Biemann continued to do the more precise analyses.[186] But Biemann envisioned a situation where all the type of work recently done in Cambridge, especially the identification of GC fractions, would be accomplished in Geneva,

> whereas we would do some basic work on general methods . . . and occasionally work on the structure determination of certain compounds of special interest for Firmenich, in addition to supplying information in the field of mass spectrometry or help in the interpretation of spectra. You might say, these are the things which your boys would want to do as a refreshing change from the daily routine but I can assure you that there will be rather too many than too few of these problems.[187]

As a result, Stoll concurred with Biemann's views. In sum, Stoll saw no other possibility than to incorporate mass spectrometry in his research laboratory.[188] Nevertheless, the Swiss partners of Biemann were convinced that they could not dispense with Biemann's counseling. His advice was urgently needed to make a rational choice prior to the purchase of a mass spectrometer. Stoll made clear to Biemann how valuable Biemann's help had been in the previous two years, asking Biemann to participate in early stages of important research projects, in order to get directions as to how to proceed and to make the best use of mass spectrometry.[189] Biemann's rather negative answer[190] led to a change in the nature of Biemann's cooperation with Firmenich & Cie., and Stoll strengthened his efforts to have sufficient mass spectrometry expertise in-house.

---

[186] Stoll to Biemann, 24 December 1959, Biemann Papers, folder "Firmenich et Cie."
[187] Biemann to Stoll, 9 January 1960, Biemann Papers, folder "Firmenich et Cie."
[188] Stoll to Biemann, 16 February 1960, Biemann Papers, folder "Firmenich et Cie."
[189] Stoll to Biemann, 14 April 1960, Biemann Papers, folder "Firmenich et Cie."
[190] Biemann to Stoll, 30 May 1960, Biemann Papers, folder "Firmenich et Cie."

In addition, he tried to arrange for other means of knowledge transfer from Biemann's laboratory than just to rely on Biemann himself. Josef Seibl, who had been in charge of the samples run for Firmenich & Cie. in Biemann's group, in 1960 accepted a position at ETH Zurich. With Seibl's supervisor in Zurich, Vladimir Prelog, Stoll arranged for Seibl to be allowed to serve as consultant for Firmenich & Cie.[191] He further expected that Seibl would take over an "important part of our MS. work in Zurich," using the instruments available there,[192] and that he would help with the interpretation tasks in Geneva. With regard to the foundation of an in-house mass spectrometry group at Geneva, Stoll proposed to send Bruno Willhalm to Cambridge for a few months. Willhalm was supposed to lead the new mass spectrometry department and "should learn as rapidly as possible the whole MS. technique."[193] Though Biemann in principle agreed to the latter proposal, he was hesitant to have Willhalm over at MIT in the summer of 1960, because Biemann had only a limited amount of time to spare. He remarked that it might be good for Willhalm to "learn the art by investigating a basic problem of mutual interest," or that Seibl could come to Geneva to teach Willhalm in the interpretation of spectra.[194] In the outcome, the direct knowledge transfer from Cambridge to Geneva failed for these reasons, as Stoll withdrew his request to send Willhalm to MIT. Nevertheless, Biemann continued to assist Firmenich & Cie. He proposed that Seibl showed the Geneva scientists the design of his special inlet system once their instrument was delivered. To round off the knowledge transfer he also proposed to inform the Swiss chemists on papers in the literature that were published in journals to which they normally did not subscribe.[195]

For the Swiss industrial researchers, it was important to acknowledge Biemann's contributions in the rare case that they published their findings at all. In the case of *cis* rose oxide, Stoll was convinced that Biemann had contributed considerably to the elucidation of its structure and thus urged the MIT chemist to become co-author of the publication. Though Biemann would have preferred to be mentioned in a footnote only, he finally agreed to his co-author status.[196] While in this case Biemann was able to control the text that referred to his work (he actually wrote it), soon thereafter another situation arose in which he was not happy at all with the way his work was presented by Firmenich sci-

[191] Vladimir Prelog was the head of ETH's chemistry department at that time.

[192] Seibl had a Metropolitan Vickers MS-2 and a Bendix time-of-flight mass spectrometer at Zurich. Stoll to Biemann, 17 May 1960, Biemann Papers, folder "Firmenich et Cie."

[193] Stoll to Biemann, 14 April 1960, Biemann Papers, folder "Firmenich et Cie."

[194] Biemann to Stoll, 6 May 1960, Biemann Papers, folder "Firmenich et Cie."

[195] Stoll to Biemann, 17 May 1960, Biemann to Stoll, 30 May 1960, Biemann Papers, folder "Firmenich et Cie."

[196] Stoll to Biemann, 28 November 1960, Biemann to Stoll, 9 December 1960, Biemann Papers, folder "Firmenich et Cie."

entists. Referring to a publication in *Helvetica Chimica Acta*, the most prestigious Swiss chemical journal, Biemann criticized the wrong interpretation the author gave to the mass spectra that had been recorded in Biemann's laboratory. This publication did "not add to the prestige of my group, and mass spectrometry in general." Consequently, he demanded the right to control the publications wherein his results were discussed, or to omit his name and the names of his associates in the published paper, a demand that was agreed to by Firmenich & Cie.[197]

Biemann's collaboration with Firmenich & Cie. is the story of an attempted knowledge transfer in two directions. First, the Firmenich resources in money, manpower, and samples helped to get Biemann's project started. Then, mainly because of the special needs of the company, the flow of information became one-sided, though the cooperation continued to be very friendly. After Stoll had set up his own mass spectrometry unit, the sharing of the instrument by two groups on different sides of the Atlantic came to an end, although Biemann continued to consult for Firmenich & Cie.

In general, technology transfer from Biemann's group to other laboratories was not confined to Firmenich & Cie., but extended to pharmaceutical companies as well. As with Firmenich & Cie., the need to obtain interesting samples to establish mass spectrometry was behind Biemann's initiatives to make contact. One such example is Biemann's connection to the Lilly Research Laboratories of Eli Lilly & Company of Indianapolis, Indiana. Eli Lilly was a major manufacturer of pharmaceuticals, and their product range included peptide-related antibiotics, and alkaloids. Thus, the interests of Eli Lilly were close to the scientific fields that Biemann tackled with mass spectrometry. It was Biemann who started the first, though unsuccessful, attempt to establish a contact with the Lilly Research Laboratories, when, in March of 1959, he wrote to Koert Gerzon, a Lilly chemist whom he knew from a 1956 Gordon Research Conference on natural products.[198] Biemann hoped that Gerzon would provide further contact to scientists working on peptides within the company, and, in particular, was seeking suitable compounds for his newly developed method to establish the amino acid sequences in small peptides:

> Since many antibiotics are found to be peptides, I wonder whether perhaps at Eli Lilly some compound of this type is around whose structure is not known and on which nobody is working (or everybody has given up). I

---

[197] Biemann to Stoll, 1 January 1961, Biemann Papers, folder "Firmenich et Cie." See Stadler, "Gas-chromatographische Untersuchung"; Stoll to Biemann, 28 December 1961; A. F. Thomas to Biemann, 13 December 1963; Biemann to Thomas, 6 January 1964; and Thomas to Biemann, 14 January 1964, all in Biemann Papers, folder "Firmenich et Cie."

[198] Klaus Biemann, personal communication, April 2000.

> would be very interested in obtaining such a compound to determine its structure.[199]

In addition, Biemann mentioned that he would also like to obtain compounds other than peptides. One of the major bottlenecks in the early development of organic mass spectrometry was the lack of availability of suitable compounds. In the isolation of natural products, human labor was as important as it was in the synthesis of artificial chemicals. Biemann's task could only begin after many stages of chemical work using traditional methods of distillation, separation, and chemical conversions. Moreover, because mass spectrometry was in the process of just being established, only compounds of closely related structures suited the needs of the mass spectrometrist. These compounds were available mostly in industry, with enough manpower to undertake the necessary work, and with rather limited individual scientific interests. But while Biemann was flooded with compounds by Firmenich & Cie. in which he was not interested, Gerzon from Eli Lilly did not see any way to fulfill his wishes:

> I have inquired here about the availability of such materials and have found that we have a long waiting list of investigators which have requested these and other antibiotics over the past few years . . . . Of antibiotics in the research stage we usually have too little even for our own limited needs.[200]

At that time, Gerzon did not understand the potential benefit that mass spectrometry could have for the purposes of Eli Lilly. This reflects as much the difficult diffusion of mass spectrometry among the community of peptide scientists as it points to the special requirements of a pharmaceutical company. For Biemann, as so often, this created a dilemma. To obtain compounds, he had to show that his technique could solve problems. In order to do so, he needed samples. As Gerzon hastened to add in his letter: "If you have any published data I would be glad to circulate a copy or reprint amongst my colleagues so that we would know when and where your technique could be called upon."[201]

Biemann's other attempt to obtain samples from a pharmaceutical company, CIBA Pharmaceutical Products Inc. of Summit, New Jersey, was more successful. In October of 1958, Biemann received an interesting CIBA sample from Büchi, who was unable to proceed with a planned NMR investigation. Biemann tackled the problem with mass spectrometry, and proposed a structure. Although data obtained later by NMR were not in accord with it,[202] this

---

[199] Biemann to Gerzon, 3 March 1959, Biemann Papers, folder "Lilly & Company."
[200] Gerzon to Biemann, 10 March 1959, Biemann Papers, folder "Lilly & Company."
[201] Gerzon to Biemann, 10 March 1959, Biemann Papers, folder "Lilly & Company."
[202] Biemann to William I. Taylor, 10 October 1958, 17 January 1959, Biemann Papers, folder "CIBA."

work led the well-known natural product chemist William I. Taylor of CIBA to come to Cambridge in order to check out how these experiments were run.[203] Moreover, CIBA was willing to undertake something for Biemann that Eli Lilly would not do: the isolation and purification of a peptide suitable for Biemann's studies.[204] Taylor âlso helped in the alkaloid field: When Biemann needed some indole compounds to undertake a comparative study, he asked Taylor to send compounds. Taylor immediately made available the alkaloids ibogaine, ibogamine, and tabernanthine, and in the same vein asked Biemann to help with another, related problem.[205]

In 1961, CIBA gave Biemann a research grant. In his application, Biemann emphasized that his previous work had been related to a few alkaloids only, "because they seemed to be attractive problems to test the method." He regarded it necessary to begin a systematic investigation of mass spectra of alkaloids of known structures, and,

> once this has been accomplished, it is hoped that by a combination of mainly ultraviolet and mass spectrometric methods we will be able to elucidate the structures of trace constituents present in plants, particularly of minor alkaloids. This knowledge we expect to lead to a much better understanding of the biogenetic relations between those alkaloids and between the different species of a botanical genus.[206]

Moreover, Biemann planned to work with alkaloid types that, because of their lack of functional groups, were very difficult to study by traditional chemical methods. Due to their higher volatility, they were particularly suitable for mass spectrometric studies. In addition, and in order to solve the difficulties encountered with substances of low vapor pressure, Biemann proposed that CIBA contribute to the rental of a Bendix time-of-flight spectrometer. The Bendix was supplied with a direct introduction system, and Biemann hoped for results that he could not achieve with his magnetic sector instrument. In order to do this work, he asked CIBA for the annual salary of a research assistant (Ph.D. candidate), and a part of the rental fee for the spectrometer, totaling $5,000.[207]

---

[203] Taylor to Biemann, 27 January 1959, Biemann Papers, folder "CIBA."

[204] Vinactin, identical with a product of Parke Davis & Co., Viomycin. See Taylor to Biemann, 24 February 1959, Biemann Papers, folder "CIBA."

[205] Biemann to Taylor, 21 March, and Taylor to Biemann, 24 March 1960, Biemann Papers, folder "CIBA."

[206] Biemann, application for research project "Mass spectra of alkaloids," 18 May 1961, Biemann Papers, folder "CIBA."

[207] Biemann, application for research project "Mass spectra of alkaloids," 18 May 1961. See also Taylor to Biemann, 10 April 1961, Biemann Papers, folder "CIBA." The Bendix time-of-flight instrument, model 12-101, was rented from May 1961 to November 1961, when Biemann bought it. The rental fee was not paid by CIBA, but shared by Upjohn, Petroleum Research Fund, possibly Eli Lilly, and Firmenich & Cie. See Memo from L. F. Hamilton to Purchasing Agent of MIT, 12 May 1961 and D. B. Harrington to Biemann, 24 November 1961, Biemann Papers, folder "Bendix Corporation."

Taylor was important for Biemann because he was a well-known specialist in the field of Biemann's main interest, indole alkaloids. Thus, Taylor's use and appreciation of mass spectrometry to solve problems that he encountered would show other natural product chemists that this was a valuable new method. He also provided the contact to Norbert Neuss at Eli Lilly, who possessed a sample of ibogaline, another member of the interesting series. In a letter to Neuss, Biemann emphasized that he needed only a very small amount of material, thus taking advantage of the performance of mass spectrometry when compared with traditional methods: "Your remark in the paper in the Journal of Organic Chemistry that there was not enough material for a chemical structure proof would make it an even more attractive little problem for mass spectrometry."[208] Because Neuss had heard through Taylor of Biemann's "elegant method," he was glad to supply Biemann with the requested sample. Furthermore, he announced that Eli Lilly "would like to take advantage" of Biemann's method and promised to get back to Biemann about this topic in the near future.[209] Thus, Biemann's contact to Taylor, and the established performance in alkaloid chemistry finally opened the door to the Eli Lilly research laboratories.

In early 1961, Biemann's cooperation with Eli Lilly was formalized in the form of a research grant to MIT covering the expenses of a postdoctoral fellow and chemicals for one year.[210] Pointing in his proposal to already established expertise in the elucidation of structures of certain alkaloids, Biemann was convinced that mass spectrometry could supplement the information gained by ultraviolet spectroscopy:

> Thus, mass spectrometry may eventually serve the same purpose ultraviolet spectra serve, but for the recognition of the alicyclic and not of the aromatic part of molecules. It should eventually be possible to gain considerable insight into the structure of newly isolated alkaloids from only these two spectra, and this possibility illustrates very well the point that mass spectrometry is a valuable addition to the array of physical methods at the disposal of the organic chemist.[211]

But this point had not been reached yet. Although mass spectrometry had the great advantage of requiring only small amounts of samples, the correlation of spectra to structural features of alkaloids was still in its infancy. Thus Biemann proposed to go ahead with labeling studies and investigations of a series of

---

[208] Biemann to Neuss, 2 June 1960, Biemann Papers, folder "Lilly & Company." See also Taylor to Biemann, 5 May 1960, Biemann Papers, folder "CIBA"; and Neuss, "Alkaloids from Apocyanaceae. II."

[209] Neuss to Biemann, 10 June 1960, Biemann Papers, folder "Lilly & Company."

[210] Totalling $6,400. See F. W. McClelland to Biemann, 23 March 1961, Biemann Papers, folder "Lilly & Company."

[211] Biemann, proposal of research project, "Mass spectra of alkaloids," pp. 1–2, enclosed with Biemann to Neuss, 7 February 1961, Biemann Papers, folder "Lilly & Company."

closely related compounds, as he did in his research proposal to CIBA. This work would constitute the basis for a collection of mass spectral data of alkaloids, and through this put the mass spectrometry of this class of compounds on a sound foundation.[212] Biemann achieved this goal, with the support of Eli Lilly, CIBA, and other funding institutions. More efficiently, as with Firmenich & Cie., he transferred his knowledge of the technique into the pharmaceutical company. Already in the summer of 1961, less than half a year after the cooperation had been formalized, managers of Eli Lilly were fully aware of the potential mass spectrometry had for their own work. Reuben Jones, director of the organic chemical division, was convinced that details of Biemann's work should be circulated as widely as possible inside Eli Lilly, and invited Biemann to give a talk in Indianapolis. This talk was one in a series of lectures aimed at making an audience of around 100 Lilly scientists familiar "with the newest developments and knowledge in the various fields of science."[213] The lecture, and Biemann's visit, contributed towards pushing Lilly scientists into a "program of structure investigation by mass spectrometry . . . . The enthusiasm which Dr. Neuss had caught from you has certainly infected our research generally since you were here."[214]

In early 1962, Biemann agreed to accept a consultancy. This included the interpretation of mass spectra that were run at MIT for Eli Lilly,[215] the centerpiece of Biemann's consulting being indole alkaloids. But he also looked into penicillin and cephalosporin antibiotics,[216] and provided expertise on instrument design.[217] Naturally, he was involved in the structural elucidation of molecules for patent matters,[218] for which he was reminded occasionally that he should be "as discreet as possible."[219] In acting as consultant, Biemann solved many of Eli Lilly's problems, as the newly appointed director of the chemical

---

[212] Biemann, proposal of research project, "Mass spectra of alkaloids," enclosed with Biemann to Neuss, 7 February 1961, Biemann Papers, folder "Lilly & Company."

[213] Jones to Biemann, 2 August 1961, Biemann Papers, folder "Lilly & Company."

[214] William W. Davis, director of the physicochemical research division, to Biemann, 10 November 1961, Biemann Papers, folder "Lilly & Company."

[215] Biemann to Jones, 2 January 1962, Biemann Papers, folder "Lilly & Company."

[216] See R. B. Morin and B. G. Jackson, Organic Chemicals Division, to Biemann, 8 November 1961 and 9 March 1962, Biemann Papers, folder "Lilly & Company."

[217] Biemann to R. J. Harley, 16 May 1962, Biemann Papers, folder "Lilly & Company."

[218] In the case of an alkaloid that probably was of the sarpagine type, Neuss wrote Biemann that "it would be nice to have this alkaloid straightened out for patent purposes, which will arise due to some interesting pharmacological properties." Neuss to Biemann, 7 December, and Biemann to Neuss, 28 November 1962, Biemann Papers, folder "Lilly & Company."

[219] In the case of the cephalosporin and 6-chloropenicillin antibiotics. As this was a competitive field, and Biemann's MIT colleague John Sheehan, a penicillin specialist, was consultant for another company, Eli Lilly scientists were anxious that Biemann should keep his work confidential. R. B. Morin and B. G. Jackson, Organic Chemicals Division, to Biemann, 8 November 1961, Biemann Papers, folder "Lilly & Company."

research division of Eli Lilly, Frederick R. Van Abeele, acknowledged in 1963. In this capacity, Biemann also agreed to have Lilly scientists for short visits and research stays at MIT in order to learn more about mass spectrometry, and to give them "an idea of the light and dark sides of experimental mass spectrometry."[220]

In late 1963, Van Abeele decided to invest in high resolution mass spectrometry, emphasizing the role of Biemann in the decision making process:

> We have tentative approval to proceed with plans to acquire a high-resolution mass spectrometer. Needless to say, your influence and recommendations were prime factors in this decision. The acquisition of this instrument will make your role of consultant even more indispensable.[221]

Biemann agreed to continue with the consulting, expressing his opinion that Eli Lilly's venture into high resolution mass spectrometry, which had just been established in Biemann's laboratory, would make the cooperation "even more stimulating."[222] In early 1964, Biemann had the Eli Lilly scientist William M. Hargrove at MIT for a three-month training period on high resolution mass spectrometry. This proved to be a long-term project, since in October 1964 Van Abeele asked Biemann to extend his consultancy for another year: "Perhaps by that time you will have taught us enough to stand on our own feet in mass spectrometry."[223] Also after the CEC 21-110 high resolution mass spectrometer had been installed at Eli Lilly, Biemann continued to counsel Hargrove in its operation.[224] For Eli Lilly, the contact to Biemann was fruitful also in product-related problems. In the 1960s, Eli Lilly undertook a large-scale research program on *vinca* alkaloids, and indeed from this work was able to market an anticancer drug, vinblastine. Biemann established the molecular weight of the alkaloid,[225] which was impossible to determine with other methods. In the late 1960s, Eli Lilly hired a former postdoctoral fellow of Biemann, John Occolowitz. Occolowitz built up a research-oriented mass spectrometry laboratory, and Biemann continued to consult for him until the late 1980s.[226]

---

[220] Frederick R. Van Abeele to Biemann, 1 April 1963, and Biemann to Van Abeele, 12 May 1964 (there is located the quote with respect to the visit of William Hargrove), Biemann Papers, folder "Lilly & Company." See *The Lilly News*, 11 July 1964, copy in Biemann Papers, folder "Lilly current."

[221] Van Abeele to Biemann, 12 December 1963, Biemann Papers, folder "Lilly & Company."

[222] Biemann to Van Abeele, 23 December 1963, Biemann Papers, folder "Lilly & Company."

[223] Van Abeele to Biemann, 15 October 1964, Biemann Papers, folder "Lilly & Company."

[224] Hargrove to Biemann, 3 December, and Biemann to Hargrove, 22 December 1964, Biemann Papers, folder "Lilly & Company."

[225] See Bommer, McMurray, and Biemann, "High Resolution Mass Spectra of Natural Products. Vinblastine and Derivatives"; Neuss et al., "Vinca Alkaloids. XXI."

[226] Biemann, interview by Reinhardt, 10 December 1998.

Biemann's contacts to CIBA and especially Eli Lilly were quite successful in terms of a bi-directional technology transfer, and Biemann was able to profit scientifically. In addition, Eli Lilly pushed forward to transfer Biemann's mass spectrometric know-how into the company. In contrast to Firmenich & Cie., CIBA and Eli Lilly were interested in becoming independent from Biemann's expertise as soon as possible. Even then, it took years before Eli Lilly had an independent group in high resolution mass spectrometry. The delay in this taking place was manageable because the scientific interests of Biemann and the Eli Lilly researchers were close enough to become the basis for a long-term collaboration. In addition, Eli Lilly, CIBA, and other big pharmaceutical firms had a tradition of patenting their inventions. This strategy allowed subsequent publication of the findings in the scientific literature. Their prime interests being patents and products thus allowed the academic consultant to acquire the scientific credit.

Somewhat different from Biemann's collaborative ventures with corporate enterprises were his cooperations with academic chemists. Biemann felt obliged to share his expertise with academic colleagues, in this way spreading the technique in the scientific community. In general, with his expertise in mass spectrometry, Biemann was a highly sought-for partner in academic collaborations. From early on, he received many requests to run mass spectra.[227] Although instruments were often used on behalf of outside academic collaborators, this did not become so prevalent an activity as it did with Firmenich & Cie. But there were similarities. On-site training periods were customary features of all cooperations. In one important case, Biemann even agreed to teach an agile competitor, Carl Djerassi, at his own university. Biemann stayed six weeks in early 1961, enabling Djerassi's group to gain a quick start in organic mass spectrometry. This group could not afford to depend on an outside consultant for a long time, mainly for scientific reasons: Mass spectrometry became a key technology for Djerassi's research program, a technology that he had to master inside his group.

Carl Djerassi at Stanford

In 1960, Carl Djerassi (b. 1923) was already an established natural product chemist who had worked up the career ladder from industrial chemist at CIBA to full professor at Stanford University. Working for Syntex S.A., an unknown pharmaceutical company in Mexico City, Djerassi made his name with the partial synthesis of cortisone. Even more important was his work leading to the

[227] See examples in Biemann Papers, folder "Requests for mass spectra, 1959–60."

synthesis of the first steroid hormone later used in the pill.[228] As early as 1957, when at Wayne State University in Detroit, Djerassi had used mass spectrometry for the determination of the empirical formula of the cactus sterol lophenol, which made possible its structure elucidation. The mass spectrometric measurements that established the molecular weight of the compound were done by Paul de Mayo in the laboratory of Rowland Ivor Reed in Glasgow. Lophenol was interesting for the reason that its structure pointed to new biogenetic pathways, linking the biogenesis of sterols in plants to those in animals.[229] In 1958, Djerassi became familiar with Einar Stenhagen's attempts to tackle natural products with mass spectrometry. Stenhagen ran spectra for Djerassi's group at Wayne State, and Djerassi provided samples of some steroids and related compounds for study at Stenhagen's laboratory in Sweden.[230] While still at Wayne State, Djerassi had made an important foray into the field of physical instrumentation with optical rotatory dispersion (ORD), which he first used to investigate steroid ketones.[231] Later, he was convinced that it was his work on steroids that made the subsequent use of ORD in the whole field of organic chemistry possible. He showed confidence that the same would happen with mass spectrometry, and that the work on steroids would lead to results of much wider significance.[232] His interest in mass spectrometry was stimulated by the work of de Mayo and Reed, Stenhagen and Ryhage, and particularly by the investigations on alkaloids by Biemann.[233] Due to Biemann's early work, Djerassi realized that mass spectrometry could solve many of his research problems:

> The thing that made a real impression on me in his lectures, when he was here, was aspidospermine. It was a very famous paper, it was his first structure elucidation paper with Spiteller. He really was able to write a mechanism for the fragmentation of aspidospermine rather than the mambo jambo, that's the wavy lines that all the people before did, it fragments here, it fragments there, so what . . . . In fact we then moved rather quickly also into alkaloids, all kinds of alkaloids, because we did so much alkaloid research at that time. It isn't that we moved into alkaloid chemistry because of mass spectrometry, it is the other way around.[234]

---

[228] Djerassi has written two book-length autobiographies: *Steroids Made it Possible* and *The Pill, Pygmy Chimps, and Degas' Horse*.

[229] Djerassi, Mills, and Villotti, "Structure of the Cactus Sterol Lophenol"; Djerassi et al., "Natural Constituents."

[230] See E. J. Eisenbraun to Stenhagen, 6 October 1958, Djerassi to Stenhagen, 9 January 1959, Stenhagen to Djerassi, 31 January 1959, Djerassi Papers, ser. VII, box 15, folder 7.

[231] Djerassi, *Steroids Made it Possible*, 53–56.

[232] Carl Djerassi, NIH grant application, "Steroid studies," 1 October 1963, 8, Djerassi Papers, box 22a.

[233] Djerassi, *Steroids Made it Possible*, 77.

[234] Djerassi, interview by Reinhardt, 20 January 1999.

With the ORD experiences in the background, and the strong indication that mass spectrometry showed great potential for the problems Djerassi was interested in, the decision was quickly taken to tackle mass spectrometry in his own laboratory. The grant application for NIH funds was probably made some time in 1959 or early 1960. News that the money had come through reached Djerassi in September of 1960, shortly after he arrived at Stanford. In December 1960, Herbert Budzikiewicz, an Austrian postdoc, joined Djerassi's group and became the main investigator of mass spectrometry in Djerassi's laboratory until the late 1960s.

The transfer of information and knowledge was central for the quick start that Djerassi had in mass spectrometry. On-site training was a customary feature of many cooperations between academic scientists, but mainly involved exchange of postdocs. In this case, a direct contact between researchers was especially important, because there was little published on the method. With the exception of *Tetrahedron Letters*, publication was quite slow, while the field developed fast. The manufacturers of mass spectrometers offered manuals and training, but these covered quantitative analysis only. In contrast, most important for Biemann and his soon-to-be competitor Djerassi were natural products of unknown structures. Nearly nothing along these lines was published in 1960. In August of that year, Djerassi heard Biemann talk on mass spectrometry and natural products at a conference in Australia,[235] and immediately invited the MIT chemist to come to Stanford to give a training course. In September 1960, Djerassi asked Biemann to run a sample of the alkaloid pyrifolidine. The structural elucidation of this molecule was well advanced, and Biemann's mass spectrometric shift method could be used to decide if the proposed structure was correct. This work led to a joint publication.[236] Biemann also made other measurements on behalf of Djerassi in October and November 1960.[237] At that time, a quite intensive discussion between Biemann and Djerassi on the correct structure of deacetylpyrifolidine took place, and Biemann convinced Djerassi of the correct assignment based on mass spectral data that were not in agreement with an assumed structure of Djerassi.[238] In addition, Biemann also gave advice on special features of Djerassi's instrument, the CEC 21-103C, of the same type as his own.

---

[235] The international symposium on the chemistry of natural products, Melbourne, Canberra, Sydney, 15–25 August 1960, organized by the Australian academy of science and the section of organic chemistry of IUPAC. Biemann gave a talk on "The use of mass spectrometry in structure determination." Abstracts of Papers in Biemann Papers, folder "IUPAC 1960 (Australia)."

[236] Djerassi to Biemann, 23 September 1960, Biemann to Djerassi, 26 October 1960, Biemann Papers, folder "Dr. Carl Djerassi"; Djerassi, Gilbert, Shoolery, Johnson, and Biemann, "Alkaloid Studies XXVI."

[237] Djerassi to Biemann, 19 October, and Biemann to Djerassi, 21 November 1960, Biemann Papers, folder "Dr. Carl Djerassi."

[238] Djerassi to Biemann, 1 December, and Biemann to Djerassi, 9 December 1960, Biemann Papers, folder "Dr. Carl Djerassi."

From late January to early March 1961, Biemann taught organic mass spectrometry in Djerassi's laboratory. In addition, he was willing to have Herbert Budzikiewicz, Djerassi's newly hired postdoc from Vienna, make a short stay at Cambridge to become acquainted with a mass spectrometry laboratory during the time before Djerassi's instrument was delivered. Djerassi emphasized the value of this stay:

> I consider this visit to MIT very important because . . . Professor Biemann's laboratory is one of the most active in the field of mass spectrometry of organic chemical compounds. Furthermore, he is using precisely the same type of instrument that we are installing at Stanford.[239]

Djerassi offered Biemann the use of his instrument during his stay at Stanford for advising students, and even of bringing his own postdoctoral fellows. Incidentally, the idea was to have Biemann's whole group work at Stanford for a few months.[240] While Biemann was willing to help Djerassi, he did not want to disrupt his own teaching and research projects going on in Cambridge. Thus, he proposed to come for a shorter period, and not to bring his associates with him:

> I would think that the primary object of my visit to make the organic group at Stanford, and particularly you and your associates, acquainted with past, present, and possibly future work in mass spectrometry as related to problems of organic chemistry—might be accomplished just as well.[241]

Though their collaboration continued for a little while after, it soon became clear that they were competitors. Biemann's colleagues pointed out to him that his generosity might have been a mistake: "Many people told me: 'How could you have been so stupid to teach Carl Djerassi—who was known to be a very aggressive person—the technique which you developed, because he will beat you over the head.' I wasn't used to think in that way."[242] In Biemann's opinion, however, he should assist his colleagues in working with his methods. Djerassi indeed used the methodology aggressively in scientific terms and published several hundred papers in the first years. As Djerassi put it in April 1962, "it is amazing to what extent we seem to be picking the same alkaloids,"[243] hastening to remark that his own way to this compound was independent from Biemann's.

---

[239] Djerassi to Budzikiewicz, 17 October 1960, Biemann Papers, folder "Dr. Carl Djerassi."
[240] Djerassi to Biemann, 20 September 1960, Biemann Papers, folder "Dr. Carl Djerassi."
[241] Biemann to Djerassi, 11 October 1960, Biemann Papers, folder "Dr. Carl Djerassi."
[242] Biemann, interview by Reinhardt, 10 December 1998. The term aggressive person refers not to Djerassi's personality, but to his reputation of publishing fast and in many scientific fields, irrespective of claims of others.
[243] Djerassi to Biemann, 27 April 1962, Biemann Papers, folder "Dr. Carl Djerassi."

Even after he had left Stanford, Biemann did not hesitate to impress on Djerassi his style of interpreting mass spectra. When he was asked to review the first paper of Djerassi and Budzikiewicz on mass spectrometry, which was concerned with steroids, he sent—in addition to the review—his private comments directly to Djerassi. Biemann challenged Djerassi that he lacked the sophistication for advanced interpretation of mass spectra used by physical organic chemists. In Biemann's opinion, Djerassi presented a purely descriptive paper, based on empirical considerations only. Although the quality of the content of this article equaled that of many others in organic mass spectrometry, far more was expected of Djerassi, due to his status:

> While many of the proposed fragmentations may be correct, I think we should not present the subject in a way which makes it look like magic and gives the impression that matching of related spectra is all one can do in this field. First, because one can do much more; secondly, matching is highly dangerous. And with a well known name like yours as one of the authors, the average organic chemist will automatically assume this is the most advanced method of the interpretation of mass spectra.

Biemann added that he felt very strongly against the use "of wiggled lines for the explanation of complex fragmentations," and enclosed copies of manuscript pages from the draft of his book that explained his views. On a personal level, he did not want to annoy Djerassi,

> but I personally feel that if we organic chemists enter this field as a major occupation, we have the obligation to come up with something clearly superior to the past method of discussing mass spectra, which was often more confusing than enlightening for the organic chemist.

Biemann strongly recommended the use of labeled compounds, and the interpretation of spectra with the help of established reaction mechanisms of physical organic chemistry.[244] Djerassi obviously took this to heart, and indeed had planned from the very start to include mechanistic studies of the type Biemann had in mind. But before he published them, he wanted to have results of studies with deuterated compounds, in order to put the conclusions on a sound empirical basis. Maybe in response to Biemann's critique, Budzikiewicz and Djerassi mentioned in the published article that they "purposely refrained from possible mechanistic proposals," before they could use the more conclusive results of these studies.[245] Not only did Djerassi undertake one of the most extensive programs in labeling of compounds, he was also very innovative in

---

[244] Biemann to Djerassi, 22 September 1961, Biemann Papers, folder "Dr. Carl Djerassi."
[245] See Budzikiewicz and Djerassi, "Mass Spectrometry in Structural and Stereochemical Problems. I," 1431.

the use of mechanistic pathways for the rationalization of mass spectra. Moreover, exactly the same criticism that Biemann showed for Djerassi's early work was returned when Djerassi reviewed Biemann's book.[246] In the following years, both scientists became competitors, and Biemann showed some bad feelings towards Djerassi after the Stanford chemist in one of his articles did not acknowledge that the method he used had originated with Biemann:

> I am probably very subjective regarding the other points, but the novelty of this work is definitely the structure of the alkaloid and not the mass spectrometry. I am certainly proud that our technique is so useful to others but, frankly, you seem unconsciously to push the publicity in this particular area a bit further than necessary for someone as well known as you.[247]

Nevertheless, they later reciprocated by acknowledging the achievements of each other. Biemann's cooperation with Djerassi certainly is an example of how far scientific cooperation occasionally went.

Djerassi's style in using mass spectrometry fitted Stanford University and its industrial environment. In the 1950s, Stanford University had subjected itself to deep and critical changes in organization and outlook regarding research and teaching. As with many other American universities after World War II, Stanford officials placed the output of research and graduate students above the teaching of undergraduates. This was made possible by a tremendous increase in federal funding of science, mainly to satisfy the military needs of the Cold War era. Stanford's dean of engineering, Frederick E. Terman, sometimes called the "father of Silicon Valley,"[248] had since the end of World War II creatively and ruthlessly brought Stanford to the top five of American research universities through making use of governmental and industrial funding to build academic repute. In Terman's view, outside patronage was necessary and benign for development of the university, provided that the sponsored projects could be integrated into the academic program.[249]

Terman was not much concerned about topics that were to be followed for their own sake. His strategy of building "steeples of excellence" was that of an opportunist. Fields of interest were mostly defined by the pull of funding agencies and industrial companies. Furthermore, in order to be able to integrate outside interests to the university, departmental authority had to be substantially eroded. In the late 1940s and early 1950s, Terman efficiently transformed the school of engineering along these sponsor-directed lines. In 1954, on his

---

[246] Carl Djerassi's book review in *Journal of the American Chemical Society* 85 (1963), 2190–2192.
[247] Biemann to Djerassi, 23 May 1961, Biemann Papers, folder "Dr. Carl Djerassi." The paper is Djerassi et al., "Mass Spectrometry in Structural and Stereochemical Problems. XIII. Echitamidine." Biemann was referee for this article.
[248] See Leslie and Kargon, "Selling Silicon Valley."
[249] Lowen, *Creating the Cold War University*, 103–109, passim.

appointment as provost of Stanford, he was provided with the opportunity to impress his policy upon the whole university.[250] The department of chemistry, as it existed in the middle 1950s, apparently did not live up to Terman's expectations. For example, though Philip Leighton, former chair of the department and wartime director of the Chemical Warfare Service, had secured a grant from this agency, the resulting work did not enhance the standing of the chemistry department since it was classified and separately organized, and thus did not contribute to research papers and the training of students.[251] In physics, meantime, the novel Stanford Linear Accelerator was initiating Big Science at Stanford; the medical center had just moved from downtown San Francisco to the main campus; and George Beadle, Edward L. Tatum, and Joshua Lederberg, three biologists related to Stanford at some point in their careers, were awarded the Nobel Prize. Chemistry, the "hub of the wheel," or the central science between all these disciplines, fell behind. Terman's vision for the chemistry department was to hire "great" chemists who brought along federal and industrial support with less restrictive publication and research policies. The hiring of William S. Johnson as new department head in 1959 was thought to set this plan in motion. It had not been easy for Stanford to secure a new head; three candidates had refused offers since the search began in 1956, among them Herbert S. Gutowsky from the University of Illinois.[252] In early 1959, Djerassi received word that Johnson, a steroid chemist whom he knew when at the University of Wisconsin, received an offer from Stanford University to become head of the chemistry department. Johnson asked Djerassi to join him at Stanford if he accepted the offer. In April 1959, the two steroid specialists agreed to join Stanford, where they formed the nucleus that brought about the subsequent transformation of academic chemistry there.[253]

In order to bring Johnson and Djerassi to Stanford, Terman granted privileges, such as relatively high salaries and exemption from administrative tasks and undergraduate teaching. In Terman's view, the experiences and expertise of Djerassi, as those of Johnson, justified such measures. Although the acting chair of the department, Eric Hutchinson, strongly disliked Terman, he willingly assisted in absorbing Johnson and Djerassi.[254] Johnson asked for considerable support on the side of Stanford's administration to ensure that he and Djerassi could "assist in making the Chemistry Department one of the best in

---

[250] Lowen, *Creating the Cold War University*, 147–190.

[251] Lowen, *Creating the Cold War University*, 101–102. For the insiders' public accounts of these events see Hutchinson, *Department of Chemistry, Stanford University, 1891–1976*, 27–35; and Johnson, *Fifty-Year Love Affair with Organic Chemistry*, 83–98.

[252] Memorandum, "Chemistry at Stanford," 3 November 1958, Terman Papers, ser. III, box 8, folder 9; notes, "Chemistry search," n.d., Terman Papers, ser. III, box 8, folder 4.

[253] Djerassi, *Steroids Made it Possible*, 66–67.

[254] Lowen, *Creating the Cold War University*, 188–189.

the world." His list of requirements would bring, as a first step, Stanford to the level of the University of Wisconsin. For example, Wisconsin held second place after Harvard in obtaining National Science Foundation (NSF) predoctoral fellowships. In contrast, Stanford, in Johnson's view, was "not even in the running with respect to attracting graduate students in chemistry." University funding to provide sufficient teaching stipends for graduate students, at least for the period up to 1965, was deemed absolutely necessary. Also, Stanford should commit funds to hire three physical chemists to strengthen that side of chemistry. As for instrumentation, Wisconsin could boast of excellent service facilities (two NMR spectrometers and four infrared spectrometers for the organic group), and the university as well as the Wisconsin Alumni Research Foundation afforded substantial research support on a recurring basis. Johnson was aware that a privately endowed university could not invest to a comparable extent in the research programs of faculty members, but asked for a "starter kit" that should provide for the upgrading of Stanford's aging NMR spectrometer (at a projected cost of $16,000), other pieces of instrumentation, and $40,000 for the initial supply of Djerassi's and Johnson's laboratories with glassware, special chemicals, and electrical equipment. A final point compared the building situation of Wisconsin and Stanford. Stanford had a completely run-down building, with only a fraction of the funds at hand that Wisconsin promised to invest in a new chemistry building. Thus, Johnson requested renovation of the old building, and, if possible, the acquisition of funds for a new one. Johnson, who saw himself serving in a managerial position as department head, and proposed to leave the administrative business to Hutchinson, ended his long letter alluding to the space age, with "anticipation to participating in placing its [Stanford's] chemistry-satellite in orbit."[255] Terman was obviously fascinated by such prospects, since he promptly agreed to most of Johnson's requests.

If Terman thought that this orbit would be an entirely steroid one, he had erred. In June of 1959, Djerassi informed Terman on his research plans, and how he wanted to bring about clear separation between his ongoing interests in Syntex research and his commitment to Stanford:

> While I have published well over 100 papers in the steroid field, relatively little of this work emanated from Wayne and equally little will probably emanate from Stanford. I plan to teach a very effective steroid course, as I am probably in a position in this field that nobody in an American university can equal because of my intimate knowledge of industrial aspects of this highly specialized area. However, I am doing so much steroid research at Syntex, and

[255] Johnson to Terman, 15 April 1959, Terman Papers, ser. III, box 9, folder 7. For a list of the costs of the special commitments made to Johnson and Djerassi see Hutchinson to Terman, 17 August 1959, Terman Papers, ser. III, box 9, folder 3.

> I shall continue to direct a great deal of it by long distance when I am at Stanford, that I intend to dedicate most of my direct research activities at Stanford in other fields. Furthermore, under no conditions do I want do give even the slightest impression that I am doing any research for Syntex in a university and I avoided any such implication while on the Wayne staff.[256]

In this respect, Djerassi was cautious, and for good reason. He was very well aware of the fact that the type of consulting that he undertook for Syntex was of a different nature than that performed by most other university professors. He planned to continue taking part in business decisions and having a "major voice in initiating important research projects." He emphasized the academic style of Syntex research, and wanted to keep the title of vice president of Syntex Corporation (the Panama-based parent company of both Syntex S.A. and a New York branch) in order to be able to make management decisions. Djerassi explicitly mentioned that he did not plan to invest more time in his involvement with Syntex than was the case with other academic consultants. But the retention of his connection to Syntex was very important:

> I am largely responsible for having established in Mexico at Syntex one of the largest and most effective "academic" research organizations in the steroid field; as far as Latin America is concerned, it is probably unique in the entire area of scientific research.[257]

Djerassi also made clear what his research plans at Stanford were. He planned to continue the projects he had begun at Wayne State in the fields of antibiotics, alkaloids, and terpenoids. As for ORD studies, he envisaged the inclusion of steroids. Moreover, the first investigations of biogenetic pathways of sterols and terpenes using radioactive tracers were under way. Thus, Djerassi offered a well-rounded program in natural product chemistry, one that had already met with success. Six of his graduate students at Wayne State University were already assistant professors, and he expected to improve on this record at Stanford. With regard to the seeming similarities between his and Johnson's program, Djerassi emphasized the areas where they complemented each other. While Johnson clearly focused on the synthetic side, Djerassi was a specialist in degradative work. Both were interested in physical methods: Johnson centered on NMR and planned to add Raman spectroscopy; Djerassi had worked intensively on ORD and "would like to start a serious project on mass spectrographic applications among polycyclic molecules."[258]

It should be understood, however, that Djerassi also worked on the synthetic part of natural product chemistry. But this belonged to the industrial side

---

[256] Djerassi to Terman, 16 June 1959, Terman Papers, ser. III, box 9, folder 2.
[257] Djerassi to Terman, 26 April 1959, Terman Papers, ser. III, box 9, folder 2.
[258] Djerassi to Terman, 16 June 1959, Terman Papers, ser. III, box 9, folder 2.

of the academic-industrial barrier that he, over-cautiously or not, tried to establish. Physical methods were a potential connection of the two sides. Though Djerassi's work at Stanford University was not directly exploited by Syntex, the creation of a huge knowledge base from the interpretation of mass spectral and ORD data of steroids and other natural products certainly benefited Syntex (and all other companies) in the field. Djerassi's industrial management posts led to occasional suspicion among many of his colleagues. But not so to Stanford provost Terman, who in accord with his plans for the Stanford Industrial Park was delighted with the prospect that biomedical and chemical corporations might join the array of computer and electronic companies present there: "In his eyes, my industrial connection with Syntex made me attractive, not suspect."[259] In contrast, and in subsequent years, the federal fund-granting agencies were suspicious of the extent and nature of Djerassi's connection with Syntex. He signed, and respected, the patent policy of NIH that reserved the decision about the patentability and disposition of inventions made with agency support to the surgeon general. However, NIH had to be reassured repeatedly that Djerassi's industrial commitments did not exceed the extent of normal consultancies.[260]

Terman's hopes in connection with Djerassi's arrival proved to be justified. Among the first issues that Djerassi dealt with upon his arrival at Stanford was the establishment of a Syntex research institute of molecular biology. This was done in collaboration with the Nobel Prize winner in the emerging field of molecular biology, Joshua Lederberg, whom Stanford had hired briefly before Djerassi came to Stanford. While Djerassi, as a vice president of Syntex, assumed directorship of the institute, Lederberg, in an advisory role, chose the research topics as well as the scientists to pursue them. Three years later, in 1963, Syntex moved its main operations to the Stanford Industrial Park, with Djerassi as executive vice president of research. For the successful marketing of pharmaceuticals, Syntex shifted the operations center from Mexico to the United States. Stanford provided ideal opportunities at its industrial park, and Syntex moved with a large part of the manufacturing and research facilities to Palo Alto. The Syntex research center was organized along the lines of an assembly of relatively independent research institutes. Next to molecular biology, the other institutes comprised the fields of steroid chemistry, hormone biology, and clinical medicine.[261] Thus, alongside his professorship at Stanford, Djerassi was able to maintain close relationships with the pharmaceutical industry. Three spin-offs emerged out of Syntex by the late 1960s: ALZA,

---

[259] Djerassi, *The Pill, Pygmy Chimps, and Degas' Horse*, 96.

[260] Djerassi to Stuart Sessoms, 30 July 1963, William Johnson to R. G. Meader and Johnson to Carl R. Brewer (all of NIH), 30 April 1964, Djerassi Papers, box 22a.

[261] Crabbé, "Dynamic Philosophy in Industrial Research," 23; and Zaffaroni, "From Paper Chromatography to Drug Discovery," 646.

Syva, and Zoecon, Djerassi being involved in the latter two. The main business of Svya, a collaborative venture of Syntex and Varian Associates, was medical diagnosis, while Zoecon focused on insect control using insect hormones. In 1968, Djerassi switched to a part-time professorship at Stanford in order to take over larger parts of the Syntex management. Four years later, he concentrated most of his active management activities on Zoecon, as president and chairman of the board. In 1983, Zoecon was bought by Sandoz of Basle, which meant the end of Djerassi's role as manager of this company.[262]

Djerassi did not expect that Stanford would support his forthcoming venture into mass spectrometry. The first instrument was acquired with NIH funds. This was a very welcome addition to existing equipment at Stanford's chemistry department, and Hutchinson, the executive chair in charge of the department, expressed the urgent need for such an instrument.[263] As early as in May of 1959, Djerassi had indicated to Johnson that he planned to write a proposal to NIH about such a project. Additionally, he applied for industrial grants that would facilitate his start at Stanford, including Merck, Eli Lilly, Parke-Davis, Pfizer, and Schering.[264] Moreover, Djerassi took care that his ongoing grants from federal agencies would be transferred to Stanford. This included an NIH grant for research into alkaloids and related natural products. In this case he applied for an expansion of the size of the grant as well as an extended support period.[265] Although he stayed in Mexico until the fall of 1960, his Wayne State research group under Pete Eisenbraun had already moved to Stanford in the summer of 1959.

The arrival of the two research groups of Djerassi and Johnson with a total 40 members at Stanford constituted de facto a doubling of the previous size of the department of chemistry. Such an expansion called for a new building. Stanford University was able to raise funds from the heirs of Stauffer Chemical Company, an enterprise in the bay area. While this brought the larger part of the necessary means, the balance was provided by an NIH grant and Stanford University. The building was ready for use when Djerassi arrived at Stanford in early September of 1960. The official ceremony was held in early March 1961, with a symposium of internationally known natural product chemists.[266]

---

[262] Djerassi, *The Pill, Pygmy Chimps, and Degas' Horse*, 95–96.

[263] Hutchinson to Terman, 5 May 1959, Terman Papers, ser. III, box 9, folder 2.

[264] Djerassi to Johnson, 21 May 1959, Terman Papers, ser. III, box 9, folder 2.

[265] NIH grant H 2574. See Djerassi to the division of research grants of NIH, 12 June 1959, Terman Papers, ser. III, box 9, folder 2.

[266] Brochure, "The John Stauffer Chemistry Building at Stanford University and its dedication: An international chemistry symposium," March 1st, 2nd, and 3rd, 1961, copy in Terman Papers, ser. III, box 8, folder 8; application for NIH health research facilities grant, 8 July 1959, Terman Papers, ser. III, box 8, folder 7. The Stauffer gift was a sum of $600,000, NIH contributed with $210,000. Note "New funds received by chemistry as of Oct. 1, 1960," Terman Papers, ser. III, box 8, folder 4.

From the beginning onward, after he accepted Terman's offer, Djerassi took part in the planning of the new chemistry building. He regarded the organic chemical research laboratories of Cambridge University as the best in the world, and tried to convince the Stanford management to emulate this style, even with the limited budget available. Djerassi argued for a large laboratory room, divided by lab benches only, and with a central room that was closed and provided insulation from noise. This was supposed to be the instrument room.[267] The design reflected his expectations of academic-style work, and he

> had asked for one lab with twenty benches, twenty sinks, and twenty desks, because I wanted no barriers: everyone should know what everybody else was doing; equipment should be shared; cooperation should flourish. I had visions of a quasi-socialist, intellectual enterprise, presided over by a benevolent dictator.[268]

Although Stanford certainly belonged to the wealthier institutions, virtually the entire graduate and research program had to be financed by outside sources. The university provided faculty salaries, building facilities, and some fellowships, but the finances for the more than 150 pre- and postdoctoral researchers working at Stanford's chemistry department had to be applied for at grant-providing agencies. Of the 1964 budget requirements, 69 percent was obtained from federal agencies, and 6.4 percent and 5.6 percent were provided by industrial and other funds, respectively. The remainder of 19 percent was paid for by the university, mainly by faculty salaries.[269] Of the federal part, NIH paid by far the largest amount. Johnson was very well aware of this dependence: "Without NIH help, this program would largely collapse because we have nowhere else to turn."[270]

Djerassi's research group typically had around 20 predoctoral and postdoctoral workers, and two technicians. In the early 1960s, this meant an annual budget of more than $250,000, half spent on salaries, one third for consumable supplies, and one tenth for permanent equipment. Virtually all of these funds came from federal funds, with NIH representing more than 80 percent of it, and NSF covering only 15 percent. NSF had a much smaller amount available for chemistry, and in addition to its greater level of funding, NIH was willing to commit itself to grants for up to seven years, a policy that was not shared by

[267] Djerassi to Terman, 26 April 1959 (2), Hutchinson to Djerassi, 30 April 1959, Terman Papers, ser. III, box 9, folder 2. See also Djerassi, interview by Reinhardt.
[268] Djerassi, *The Pill, Pygmy Chimps, and Degas' Horse*, 100.
[269] William Johnson to Donald F. Hornig (science advisor to the U.S. president), 6 March 1964, Terman Papers, ser. III, box 8, folder 4.
[270] Johnson to Carl R. Brewer of NIH, 30 March 1964, Djerassi Papers, box 22a.

NSF.[271] Djerassi was aware that these figures were slightly larger than those at most other universities. He attributed this to the fact that his group was "spending a substantial amount of time, money and manpower on instrumentation."[272] His greatest concern was that NSF was unprepared to deal with the substantial costs for high technology instruments that had become essential for his research program. Here, for once, Djerassi was successful in obtaining a very substantial contribution from NSF. In late 1963, the agency provided the funds for a high resolution mass spectrometer, at a cost of $130,000. But NSF did not pay for the maintenance of the instrument, that is, not for supplies and manpower.[273]

In the short time span between 1959 and 1964, the research costs of Djerassi's projects skyrocketed because of the use of expensive instrumentation.[274] But NIH was not willing to fund this change in research direction with a generous increase of grant money. One of their points of critique was that the training part of his research to a large extent benefited foreign countries, and not the United States, because of a supposedly relatively large percentage of foreign postdocs among Djerassi's group. In the opinion of the associate director who was responsible for grants and training of the NIH National Cancer Institute, this dependence on foreign manpower led to critical effects:

> The real crux of your situation appears to be that there are not enough United States citizens who wish to receive training in your laboratory to fill all the billets available and that, by your method of conducting research and on-the-job research training, your research program cannot be accomplished unless all of the training or research employee billets are filled.[275]

Indeed, all four postdocs paid by the NIH training grant in the first half of 1963 were non-U.S. citizens, while the six graduate students were Americans.[276] Djerassi, in turn, was convinced that the relatively great number of foreign researchers was a phenomenon in organic chemistry all over the United States, and not restricted to his laboratory.[277] But the crucial part of his research

---

[271] The exact numbers (all in $) are: 126,000 for salaries, 24,000 for permanent equipment, 39,000 for expendable equipment and supplies, 2,500 for travel, 8,500 for publication costs, 6,000 other direct costs, and 51,500 (25 percent) indirect costs, a total of 257,500. Djerassi to Walter R. Kirner, head of NSF chemistry section, 15 February 1963, Djerassi Papers, ACCN 1999-021, box 12, folder "NSF miscellaneous."

[272] Djerassi to Kirner, 15 February 1963, Djerassi Papers, ACCN 1999-021, box 12, folder "NSF miscellaneous."

[273] Johnson to Brewer, 30 March 1964, Djerassi Papers, box 22a.

[274] Djerassi to R. G. Meader, 12 May 1964, Meader to Djerassi, 13 April 1964, Djerassi Papers, box 22a.

[275] Meader to Djerassi, 13 April 1964, Djerassi Papers, box 22a.

[276] Djerassi to Helen L. Jeffrey, 24 January 1964, Djerassi Papers, box 22a.

[277] Djerassi to Meader, 12 May 1964, Djerassi Papers, box 22a.

was the mass spectrometry facilities. If not enough money was available to staff and run them, virtually all of Djerassi's research would have to be stopped. Again, he pointed to the fact that this did not apply to his laboratory only:

> Organic chemical research used to be the cheapest research among the physical sciences. This has changed, principally due to the advent of sophisticated instrumentation, and the introduction of mass spectrometry into the everyday organic laboratory practice has been one of the most important contributing factors. What is frequently ignored is that while the instruments and ancillary activities such as maintenance and computer costs are indeed expensive when compared to costs of laboratory supplies and chemicals, they are cheap if one considers the tremendous saving in time and the nature of the problems that can be attacked with such techniques.[278]

Next to the saving of time and labor, it was the increased complexity of problems that could be attacked with the novel tools that became the incentive to start with methodological work in the field of mass spectrometry. Djerassi decided not to wait, but instead exploit the opportunity in and for his own research.

The Chemical Approach

By the late 1950s, Djerassi's projects on triterpenes and alkaloids of giant cacti were tapering off due to lack of novel substances found in these plants. He decided to concentrate on a new family of alkaloids from another plant family, the *Apocyanaceae*, and here mainly on alkaloids from the *Aspidosperma* genus. This project led him into further investigations of the chemistry of indole alkaloids (indole is the common building block of this type of alkaloids). Together with Biemann, and many other researchers at universities and companies, Djerassi added momentum to the "indole alkaloid explosion" of the 1960s. For him, in retrospect, this explosion had quantitative and qualitative dimensions. In a very short time period, hundreds of structures of new alkaloids were established. The fast pace of this research was only possible because the full scale of modern instrumentation could be used in the endeavor. Djerassi reported that the main beneficiaries of this development were the scientists working on the biosynthesis of plant materials, because the "almost bi-weekly offered superb material" provided ample opportunity for the design of biosynthetic pathways and their subsequent proof or disproof by tracer experiments.[279]

---

[278] Carl Djerassi, NSF grant application, "Mass spectrometry in organic and biochemistry," 1969, 5, Djerassi Papers, box 22a.
[279] Djerassi, "Natural Products Chemistry 1950 to 1980," 128.

Most of the new alkaloids were minor ones, differing only in minute detail from already known compounds. This made them accessible to mass spectrometry, especially with the mass spectrometric shift method of Biemann. Biemann cooperated with Djerassi in the definitive determination of one of Djerassi's *Aspidosperma* alkaloids, pyrifolidine, through the use of this technique before Djerassi got his own instrument. In addition, NMR measurements, run by James Shoolery and LeRoy Johnson at Varian Associates, were crucial for this work.[280] This connection of NMR and mass spectrometry in the determination of alkaloid structures followed a tradition established in Djerassi's group. Before he was able to rely on in-house instrumentation of any kind, Djerassi cooperated with outside experts. In the case of skytanthine, an alkaloid of relatively low molecular weight, it took from 1957 to 1961 before the constitution could be settled. For NMR, Djerassi cooperated with such highly regarded specialists as Aksel Bothner-By of the Mellon Institute at Pittsburgh and Harold J. Bernstein of the National Research Council in Ottawa, Canada. For support in mass spectrometry, Djerassi asked the analytical service of CEC in Pasadena to run a spectrum. Although CEC agreed to run this unusual compound, for a fee, they could not offer any assistance in the interpretation of the spectrum. Finally, NMR spectra supplied by Varian Associates and classical chemical work settled the problem.[281]

When Djerassi and Biemann began to work on indole alkaloids, only five types of *Aspidosperma* alkaloids were known. From the 1950s to the 1970s, they identified and established the structures of more than one hundred indole alkaloids. Of crucial importance in Djerassi's endeavor was his cooperation with a group of chemists in Brazil. This emulated the model of the Syntex agreement with the National University of Mexico, whereby students of the university's chemistry institute could do their thesis work at Syntex. At Wayne State, Djerassi continued to cooperate with the Mexican university on studies into natural products of indigenous plants. After Walter Mors, a Brazilian graduate student of Djerassi, returned to the *Instituto de Quimica Agricola* of the botanical garden in Rio de Janeiro, he proposed to Djerassi a cooperation along similar lines.[282] The principal local partner in Brazil for Djerassi became the English-born chemist Benjamin Gilbert, who had been a member of Djerassi's group at Wayne State, later having several functions at universities and private institutions in Brazil.[283] Long-term support came from the Rockefeller Foun-

---

[280] See Djerassi to Biemann, 23 September 1960, Djerassi Papers, ser. VII, box 1, folder 6, and Djerassi, Gilbert, Shoolery, Johnson, and Biemann, "Alkaloid Studies XXVI. The Constitution of Pyrifolidine," 162–163.

[281] Djerassi to Bernstein, 22 July 1959, 30 May 1958, Djerassi to Bothner-By, 28 May 1958, C. E. Johannsen (CEC) to Djerassi, 27 August 1957, and related correspondence in Djerassi Papers, ser. VII, box 3, folder 10; Djerassi, et al., "Alkaloid Studies. XXVII."

[282] Djerassi, *The Pill, Pygmy Chimps, and Degas' Horse*, 91–93.

[283] See Mors, "Editorial."

dation, funding Djerassi's trips to Brazil as well as stipends for postdoctoral fellows in residence there. Gilbert, and other postdoctoral fellows participated in the collection trips.[284] Beginning in the late 1960s, and in his function as chairman of the Latin America Science Board of the U.S. National Academy of Sciences, Djerassi spurred a much broader program in different areas of chemistry in which professors at four U.S. and two Brazilian universities were involved.[285] The work done in Brazil was related not only to collecting the plant material and the isolation of the crude alkaloids. Chemical work done in Rio de Janeiro was of a very high standard. In addition to isolation and chemical degradation, infrared measurements were mostly done in Brazil. For NMR and mass spectrometry, the samples were sent to Stanford. In some cases, Gilbert also wrote up drafts of research articles.[286]

In addition to the Brazilian chemists, the Eli Lilly Research Laboratories in Indianapolis and Jean LeMen and M.-M. Janot of the Institut de chimie des substances naturelles in Gif-sur-Yvette in France were important partners in the alkaloid studies of Djerassi. Thus, both Biemann and Djerassi independently cooperated with the very same institutions (and in the case of Lilly even had the same contacts, Marvin Gorman and Norbert Neuss).[287]

Mass spectrometry was of tremendous use in alkaloid chemistry. Low resolution mass spectrometry established the molecular weight of the unknown substance and together with combustion microanalysis this normally led to an unequivocal empirical formula. In this way, mass spectrometric methods decided on empirical formulae, where these had been previously in doubt, or corrected erroneous ones. This happened quite often, because of the intrinsic lack of precision of elemental analysis of high molecular weight samples. High resolution mass spectrometry distinguished between isobaric fragments (to one atomic mass unit resolution). This made possible the determination of the empirical formula of a substance without recourse to any traditional microanalytical work. Next, establishing the absence or presence of certain functional groups was within reach of the experienced interpreter of mass spectra. The ultimate goal of each alkaloid chemist, however, the carbon skeleton structure of the unknown molecule, was possible only with reference to compounds of known constitution.[288]

---

[284] For a typical report see Gilbert to Djerassi, 6 September 1961, Djerassi Papers, ser. VII, box 1, folder 5.

[285] Djerassi, *The Pill, Pygmy Chimps, and Degas' Horse*, 91–93.

[286] An example is a paper on *Aspidosperma marcgravianum*, submitted to the *Journal of Organic Chemistry*. See Djerassi to Gilbert, 26 April 1962, Djerassi Papers, ser. VII, box 1, folder 5.

[287] For an example, the alkaloid vindolinine, see Neuss to Djerassi, 21 July 1961, Djerassi to LeMen, 19 May 1961. See also Djerassi, "Mass Spectrometric Investigations in the Steroid, Terpenoid and Alkaloid fields," 597.

[288] Budzikiewicz, Djerassi, and Williams, *Structure Elucidation of Natural Products by Mass Spectrometry*, vol. 1, *Alkaloids*, 6–10.

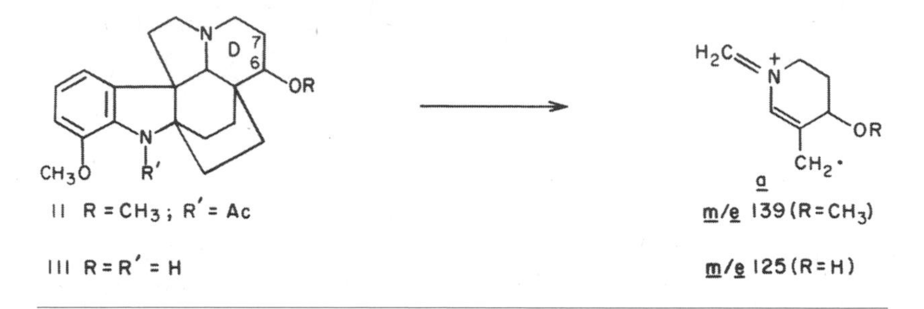

FIGURE 3.7 Use of chemical alterations in the mass spectrometric elucidation of the structure of pyrifoline (number II in the figure). An important peak at 139 was assigned to the fragment shown on the right, with R=CH_3. Demethylation (substitution of the CH_3-group with a molecular weight of 15 by a hydrogen atom) of the alkaloid led to a peak shift to a value of 125. Thus, it was proven that the methyl group was included in the particular fragment ion shown. From Herbert Budzikiewicz, Carl Djerassi, Dudley H. Williams, *Structure Elucidation of Natural Products by Mass Spectrometry*, vol. 1, *Alkaloids*, San Francisco: Holden-Day 1964, 11.

Together with chemical, usually degradative, reactions, the mass spectrometry of alkaloids led to a treasure trove of useful information (see Figure 3.7). In the case of pyrifoline, "none of the carbon atoms was 'isolated' in the classical sense in terms of known degradation compounds."[289] The localization of chemical alterations of the molecule could be detected with the help of the mass spectrometer. Moreover, the kind of reaction used was adjusted to the needs of the mass spectrometrist. The results of chemical reactions shifted characteristic peaks in a predictable manner and thus led to conclusions about structural features. Together with other physical methods, such as ultraviolet, infrared, and NMR, mass spectrometry could thus yield a complete structure determination.

Despite his announcement not to work on steroids at Stanford, it was the field of mass spectrometry of steroids where Djerassi's group took the lead. Djerassi's laboratory was the ideal place to do this type of research, his group having ample experience in the field because of the research at Syntex. Moreover, because steroids were relatively complex substances, they could be used as test cases to establish whether the general rules found with simpler model compounds could be validated. For the first systematic investigation with mass spectrometry, Djerassi chose the particular class of compounds which had been of great use in the ORD work: steroid ketones. In this endeavor, he was "motivated by the belief that a semi-empirical study of the mass spectrometric frag-

[289] Budzikiewicz, Djerassi, and Williams, *Structure Elucidation of Natural Products by Mass Spectrometry*, vol. 1, *Alkaloids*, 11.

mentation patterns of a group of closely related substances (together with certain deuterated analogs) would lead to generalizations, which might prove very fruitful in structural and stereochemical investigations of natural products currently under way in our laboratory."[290]

Steroid is the chemical notation for a class of compounds that are based on a common type of carbon skeleton, a molecule made of fused rings made up of carbon and hydrogen atoms arranged in three six-membered rings and one five-membered ring in the fashion shown in Figure 3.8. Minute variations of this molecule, especially the introduction of oxygen atoms at various positions, and dehydrogenation leading to double bonds between two carbon atoms, lead to dramatic changes in the biological activity of the steroid. Thus, the male and female sex hormones, cortisone, some vitamins, and many other substances are produced by small variations of the steroid skeleton. But these slight chemical variations did not come easy, at least not in the test tube of the organic chemist. For example, in the synthesis of cortisone, the most difficult step was the introduction of the oxygen atom at the position "11" of the steroid skeleton.[291]

The variations in the position of the oxygen atom (called keto group) in steroids were of highest interest for Djerassi. The accepted dogma of organic mass spectrometry was that functional groups such as the keto group would direct the fragmentation of the molecule, and thus lead to different mass spectra depending on the position of the functional group. Consequently, Djerassi and his co-workers systematically investigated all steroid molecules with keto groups in as many different positions as they could retrieve. If established, they believed that these ground rules of steroid mass spectrometry would have proved invaluable for the steroid chemist. But for this particular class of compounds, Djerassi's hopes were too optimistic. Extensive correlations between the mass spectra of saturated steroid ketones and their structures were not feasible. They came to realize that "the fragmentation pathways are quite complex and do not lend themselves to a generally predictable behavior."[292] Nevertheless, the labeling with deuterium made possible the investigation of dissociations and rearrangements.[293] Fortunately, Djerassi and his co-workers found other classes of steroid derivatives that showed a more predictable mass spectral pattern. Among the most important were steroid ethylene ketals, structurally loosely related to diosgenin, the plant feedstock of the Syntex steroid syntheses. The ketal group contains two oxygen atoms, and therefore can better stabilize

---

[290] Budzikiewicz and Djerassi, "Mass Spectrometry in Structural and Stereochemical Problems. I. Steroid Ketones," 1431.

[291] For a brief, but excellent, introduction to steroid chemistry for lay persons see Djerassi, *The Pill, Pygmy Chimps, and Degas' Horse*, 34–38.

[292] See Shapiro and Djerassi, "Mass Spectrometry in Structural and Stereochemical Problems. L," 2825.

[293] Budzikiewicz, Djerassi, and Williams, *Structure Elucidation of Natural Products by Mass Spectrometry*, vol. 2, 64.

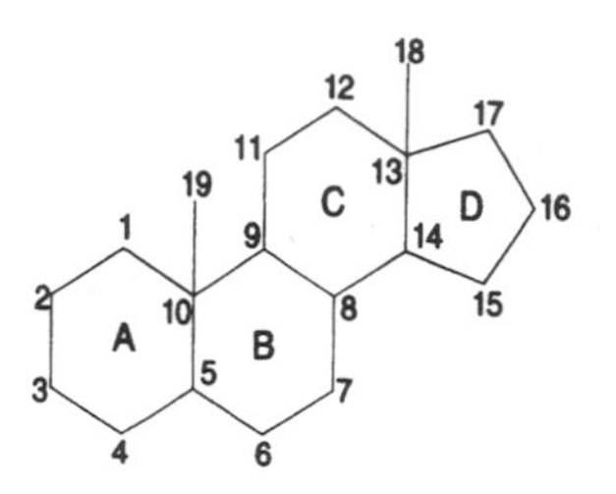

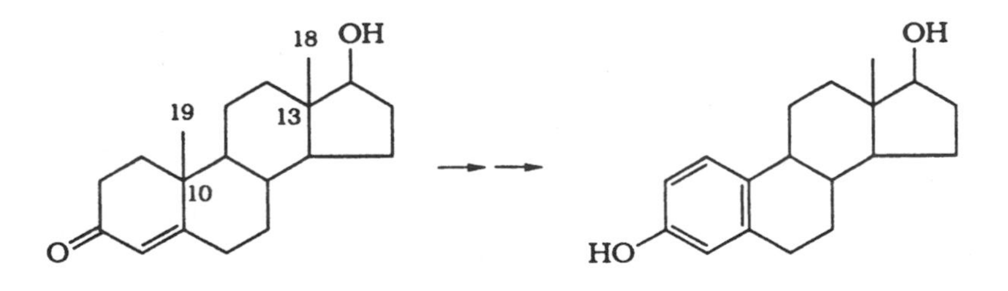

FIGURE 3.8 Chemical formulae of the basic steroid skeleton (top), and the formation of estradiol from testosterone (bottom). Top graphic formula from Carl Djerassi, *The Pill, Pygmy Chimps, and Degas' Horse. The Autobiography of Carl Djerassi*, New York: Basic Books 1992, 35, and bottom graphic formula reprinted with permission from Carl Djerassi, *Steroids Made it Possible*, Washington, DC: American Chemical Society 1990, 21. Copyright 1990, American Chemical Society.

the positive charge. This leads to enhancement of the directing influences of the functional group on the mass spectral fragmentation pattern.[294] Studies of other functional groups with similar effects strengthened these results. Djerassi's laboratory was the only one in a U.S. university that engaged systematically in the mass spectrometry of steroids. This happened at a time when the market introduction of the birth control pill led to considerable publicity. Djerassi's research helped to lay the groundwork for the control of steroid hormones in human bodies, and thus supported the acceptance of the pill in society at large. In 1963, for Djerassi it was obvious that efforts to advance mass spectrometry were central for further progress in the science of steroids in particular and the medical field in general: "The development of new methods which would permit the recognition of a steroid on a microscale is of the utmost importance and mass spectrometry is very likely to play a crucial role."[295]

---

[294] Von Mutzenbecher, Pelah, Williams, Budzikiewicz, and Djerassi, "Mass Spectrometry in Structural and Stereochemical Problems. XLVI."

[295] Djerassi, NIH grant application, "Steroid studies," 1 October 1963, 7, Djerassi Papers, box 22a. Later, Djerassi reflected on the social implications of the pill. See Djerassi, *The Politics of Contraception*.

In the mid 1960s, the structural determination of unknown steroids of natural origin lost its momentum. Lack of novel material was the probable cause. But in the 1970s, steroid, or at least sterol (sterols are solid alcohols with a steroid carbon skeleton), mass spectrometry experienced a renaissance. The reason was the sudden discovery of unknown sterols of marine origin, some of them with totally unexpected structural features. In contrast to the classical period of steroid chemistry from the 1930s to the 1950s, physical instruments were now in place. This greatly enhanced and speeded up the determination of the multitude of marine sterols that were isolated in the 1970s.[296]

Soon, mass spectrometric studies and investigations that applied mass spectrometry as a major tool comprised most of the work going on in Djerassi's group. He emphasized the cooperative style of his work, mentioning in a 1963 grant application to NIH that "nearly 30% of all our measurements are being performed for outside academic investigators."[297] Increasingly, methodological research became the focus of Djerassi's research program. He carefully separated analytical work (the determination of known substances in complex mixtures) from the structural deductions that were possible with mass spectrometry, and considered the latter the "fundamental basis" of his work.[298]

Not working on improvements of the instrumentation, Djerassi and his group concentrated on the chemical output and foundations of mass spectrometry. They did so by looking at the same functional grouping in different hydrocarbon skeletons as well as taking into account the effects of combinations of functional groups on the mass spectral fragmentation processes. It was important to state the conclusions "in a fashion that is meaningful to the organic chemist."[299] This ensured subsequent success, and the methods they established were used by a huge number of practitioners in their respective fields. Moreover, the meaningful application of the theories and rules of physical organic chemistry in mass spectrometry strengthened the position of the former even more. Thus, mass spectrometry added to the framework of physical organic chemistry itself. The focus on ion reactions, the conventions of electron movement, and finally the sophisticated proofs of mechanistic pathways—for example, showing the importance of interatomic distances—were all important contributions from mass spectrometry.[300]

With many of his colleagues, Djerassi shared the assumption that the relatively high ionizing energy in the ion source of the mass spectrometer (normally 70 eV) was sufficient to remove an electron from any bond in the

---

[296] See Djerassi, "Recent Advances."

[297] Djerassi, NIH grant application, "Steroid studies," 1 October 1963, 8, Djerassi Papers, box 22a.

[298] Djerassi, NIH grant application, "Studies on steroids and related compounds," 31 July 1967, 7, Djerassi Papers, box 22b.

[299] Djerassi, "Isotope Labelling," 159.

[300] Djerassi, interview by Reinhardt, 20 January 1999.

molecule, and thus resulted in an electronically excited positive ion, which did not immediately decompose. Its excess electronic energy could be transferred via vibrations or oscillations to yield lower-lying electronic states, and in cases where the molecules contained heteroatoms or π-bonds, it was possible to localize the charge prior to further decomposition. Therefore, while most compounds of interest to the organic chemist did not decompose statistically, it was possible to identify the molecular ion according to the canonical rules of physical organic chemistry, and, on this basis, to rationalize the mechanism of its fragmentation. This could be shown for example by the introduction of functional groups in compounds that had given mass spectra with many peaks (and therefore many fragments), which simplified the spectrum in favorable cases to such an extent that only one or two fragment ions carried the bulk of the ion current.[301] Reasoning along these lines allowed the prediction of bond fissions, assumed to be homolytic,[302] though not based alone on analogies with the accepted rules of modern organic chemistry. It was assisted by the use of additional techniques, such as the measurement of the appearance potential of a molecular ion, which is equal to the dissociation energy of a compound plus the ionization potential. Thus, the appearance potential could provide hints enabling prediction of which bonds were broken, and therefore which ions were formed. Equally important was the investigation of so-called metastable peaks, which originated when positive ions further decomposed after their acceleration, but before they reached the analyzer region of the mass spectrometer. With both methods, and together with kinetic studies, it was possible to indicate the structures of the ions formed.

Thus, the first point of dogma of this school considered the positive charge of a molecular ion after ionization (through the loss of an electron) as being localized. Preferred positions of this localized charge were heteroatoms and tertiary carbonium ions. These molecular entities were known to be more stable than others, and thus it was rational to assume that after electron impact such structures would be formed. The second dogma concerned the kind of bond fission that followed this localization of charge. Again in agreement with theories of physical organic chemistry, Djerassi preferred to assume that homolytic bond fission would take place. As a result, the transfer of single electrons or radicals had to be taken into account. To clarify this, Djerassi and his co-authors even used a special notation to symbolize the transfer of only one electron, the fishhook (see Figure 3.9). This symbol made it easily possible to distinguish one-electron transfers from two-electron movements, the latter symbolized by

[301] Budzikiewicz, Djerassi, and Williams, *Mass Spectrometry of Organic Compounds*, 9–12.

[302] In homolytic cleavage of a bond, each of the two resulting fragments keeps one of the two bonding electrons.

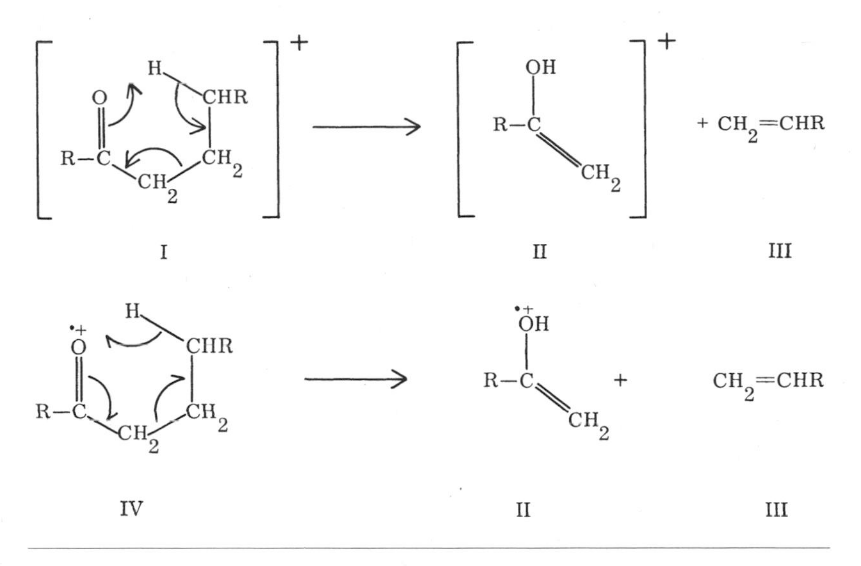

FIGURE 3.9 The conventional representation of the McLafferty rearrangement (top), with arrows and delocalized charge, and the style adopted by Budzikiewicz, Djerassi, and Williams (bottom) in their book. The latter gives one-electron movements (symbolized by the "fishhook") and the positive charge is localized at the oxygen atom. From Herbert Budzikiewicz, Carl Djerassi, Dudley H. Williams, *Interpretation of Mass Spectra of Organic Compounds*, San Francisco: Holden-Day 1965, xi and xii.

an arrow. These two dogmas were introduced in Budzikiewicz's, Djerassi's, and Williams's first book on the subject, as a special "note to the reader."[303]

In the mid-1960s, however, it was not certain that the positive charges really were localized at specific centers of the ions. For example, Gerhard Spiteller, in response to Djerassi's "note to the reader," challenged the views of the Californian group. Spiteller, who had been a postdoctoral fellow with Biemann at MIT and was now at the University of Vienna, was not convinced that charge localization was a valid concept in large molecules. In the case of the McLafferty rearrangement, which was the example used by Djerassi and his coauthors, Spiteller favored the transfer of an electron pair, and thus the movement of a proton. Djerassi and his group presented counter evidence, based on the opinions of experts in the more physical aspects of mass spectrometry, including McLafferty himself. However, in the end, Djerassi expressed that this

[303] Budzikiewicz, Djerassi, and Williams, *Interpretation of Mass Spectra of Organic Compounds*, xi–xiii.

was "certainly an open question at this point and we simply are going to retain our present system of representation for the sake of simplicity rather than to imply that it may be the gospel truth."[304]

In general, Djerassi was well aware of the possible pitfalls involved in assigning specific "mechanisms" to fragmentations of molecules. At the end of the 1960s, Djerassi's investigations, which then were based mainly on high resolution work and isotopic labeling techniques, had either strengthened or revised many of the earlier proposed fragmentation mechanisms. But Djerassi, and his first collaborators in mass spectrometry, Herbert Budzikiewicz and Dudley H. Williams, stated in 1967:

> The quotation marks around the term "mechanism" are still well deserved and the term is hardly used throughout our present book; "rationalization" is a much better substitute. Since the fragment ions are not isolated, only indirect support can be presented to describe their nature and the evidence is by no means as rigorous as in many other organic chemical reaction mechanisms. Nevertheless, the circumstantial evidence is now overwhelmingly in favor of the approach used in our first book—namely that much of organic mass spectrometry can be discussed in terms of the standard and really oversimplified language of the organic chemist. It is largely for that reason that mass spectrometry has found such rapid acceptance by organic chemists during the past few years, and it is precisely through the use of such oversimplified concepts and generalizations that the more detailed and refined knowledge of the future will be derived.[305]

Djerassi and his group at Stanford did much to establish mass spectrometry as a tool for the organic chemist, based on the assumption that most of them would not measure the spectra by themselves but would interpret them. Therefore they regarded the "mechanistic" approach as the best from a pedagogical perspective. To achieve their goal they published a series of books, starting with relatively simple organic molecules and then dealing with natural products such as alkaloids, steroids, terpenoids and sugars.[306] The structure of the first book, *Interpretation of Mass Spectra of Organic Compounds*, was well suited to the needs of the general organic chemist. The chapters were organized according to the most important compound classes of organic chemistry, and thus followed the lines of most textbooks in classical organic chemistry. Moreover, because the

---

[304] Djerassi to Spiteller, 21 May 1964. See also Spiteller to Djerassi, 9 and 26 March, 23 June; Djerassi to Spiteller, 18 March; and Budzikiewicz to Spiteller, 6 July 1964, all Djerassi Papers, ser. VII, box 15, folder 5.

[305] Budzikiewicz, Djerassi, and Williams, *Mass Spectrometry of Organic Compounds*, vi.

[306] Budzikiewicz, Djerassi, and Williams, *Interpretation of Mass Spectra of Organic Compounds*; Budzikiewicz, Djerassi, and Williams, *Structure Elucidation of Natural Products by Mass Spectrometry.*

technical issues of the instruments were tackled in other treatises,[307] Budzikiewicz, Djerassi, and Williams did not include such practical matters. Thus, they concentrated on the chemistry, which was their main concern. Two succeeding volumes showed the applications of mass spectrometry mainly in the fields of steroids and alkaloids. Prefiguring perhaps Djerassi's later move into the field of *belles lettres*, they compared their approach to the field of literature:

> Our general approach has been essentially of the type that one might employ in teaching the appreciation of poetry. In that sense, our first book . . . was written to lay down the metric and ground rules of mass spectrometry in the language of the organic chemist, while the actual poetry is represented by the two volumes . . . which demonstrate that the relatively simple ground rules of the first book are applicable to quite complex organic molecules.[308]

The success of this strategy may be well illuminated by the fact that a few years later, a second, enlarged and revised edition appeared under the title *Mass Spectrometry of Organic Compounds*. This was the last of the four joint volumes written by Budzikiewicz, Djerassi and Williams. Budzikiewicz went on to become professor of chemistry at the University of Cologne in Germany, and Williams moved to the University of Cambridge. While Budzikiewicz mainly took care of the instrument and ran the measurements, Williams was, as a more classical organic chemist, interested in the chemistry itself. He also wrote, together with Norman S. Bhacca of Varian Associates, a book on NMR applications in steroid chemistry.[309] For a long period in the 1960s, Budzikiewicz and Williams were at the core of Djerassi's venture into mass spectrometry. In Djerassi's group, it was always the senior postdoctoral fellow who was in charge of mass spectrometry. In this, he was assisted by a technician who operated the instrument. With few exceptions, no one else was allowed to run the spectrometer.[310] The danger of contamination of the ion source was great, but in addition the factory-like style of work in Djerassi's group did not permit that each researcher would be allowed to operate the high performance instruments. Thus, Djerassi organized his team of some twenty researchers around the instruments. The large majority was busy with chemical work: classical degradation, and especially synthesis of the deuterated model compounds needed for the establishment of organic mass spectrometry. This availability of a huge number of experienced synthetic chemists enabled Djerassi to tackle organic mass spectrometry on a very broad front, and yielded a mass of detailed data ready for interpretation. Occasionally, the synthetic methods used were worthy

---

[307] Among them Biemann's and McLafferty's monographs.

[308] Budzikiewicz, Djerassi, and Williams, *Structure Elucidation of Natural Products by Mass Spectrometry*, vol. 2, *Steroids, Terpenoids, Sugars, and Miscellaneous Classes*, v.

[309] Bhacca and Williams, *Applications of NMR Spectroscopy in Organic Chemistry.*

[310] Djerassi, interview by Reinhardt, 20 January 1999.

of publication in their own right; but mostly they were developed to obtain results in organic mass spectrometry.[311] The synthetic chemical part of Djerassi's factory of structures was as important as the machine that brought about detection and identification. At the head stood the human interpreter, soon assisted by computer programs for data handling and interpretation. Before computer handling of data became commonplace, the wives of the graduate students did all the plotting of the spectra. Highest in the hierarchy, immediately below Djerassi, were those with access to the mass spectrometer and the ability to interpret the spectra in a sophisticated manner. While the interpreters were academic chemists, technicians served as the operators of the instruments.

Djerassi used the whole assembly of physical instrumentation available at the time: infrared, NMR, ORD, and mass spectrometry were the most important and most often used techniques. Together with diverse chromatographic methods for separation, and microanalysis, these physical methods were used on a routine basis in nearly every single attempt to elucidate the structure of an unknown compound. His research program, as did the projects of his professorial colleagues all over the world, heavily relied on the availability of sophisticated instrumentation, and a professional workforce to operate it, often assembled in a service laboratory of the department.

CONCLUSION

> The position of organic mass spectrometry can probably best be gauged by the observation that in 1960 one had to search far and wide in the organic chemical literature to encounter even an occasional reference to the use of mass spectrometry in structural organic chemistry, while in 1969 it is probably impossible to find a journal . . . dealing with organic chemistry in which there are not numerous references to the use of this technique and several specific articles dealing with research in organic mass spectrometry per se.[312]

Arguably, it was the coinage of the fragmentation mechanisms of organic compounds in the mass spectrometer in terms of physical organic chemistry that was the single most important issue in this fast adoption process. But, as we have seen, other features proved to be important as well. The need of access to suitable compounds by mass spectrometrists, and the urgent desire of natural product chemists to gain more information about the molecules they studied

---

[311] An example of a combined article reporting synthesis and mass spectrometry of a particular class of compounds is Beard, Wilson, Budzikiewicz, and Djerassi, "Mass Spectrometry in Structural and Stereochemical Problems. XXXV."

[312] Djerassi, NSF grant application, "Mass spectrometry in organic and biochemistry," 1969, 4, Djerassi Papers, box 22a.

led to an alliance between the two groups. This included the chemical industry, and as a consequence, mass spectrometry became part of the established academic-industrial relationship. Originating in the chemical industry, further developed in academic departments, and introduced again into industrial companies, organic mass spectrometry crossed many boundaries. The networking of the instrument between research groups was a crucial feature in establishing its usefulness.

In this process, the level of adaptation of the mass spectrometer to the experimental culture of chemists decided its success. For example, although Djerassi considered steroids as his scientific specialty, he and his group very broadly and systematically studied mass spectrometric mechanisms in all fields of organic chemistry. Much of Djerassi's output consisted in detailed studies making use of isotopic labeling. In this, he put mechanistic studies on a firm empirical basis, and he made full use of the substantial human workforce in his laboratory. Thus, the mechanization of chemical practice was closely connected to its handicraft side: Without syntheses of labeled compounds, the mass spectrometer would have stood idle.

The mass spectrometric work on alkaloids, steroids, and peptides shows that the technique competed with other methods, and was not at all immediately welcomed in all parts of organic chemistry. The elucidation of alkaloid structures, after initial reservations, was a great success for mass spectrometry, but the work on peptides took a much longer time before it could yield results not achievable with other techniques. While mass spectrometry was employed as a tool to make forays in the chemistry of alkaloid, peptides, and steroids, this very chemistry yielded the necessary insights to advance the understanding of mass spectrometry itself. Thus, mass spectrometry and natural product chemistry were at the same time both instruments and objectives of inquiry. Next to knowledge in mass spectrometry and natural product chemistry, human expertise constituted the third kind of output that this system of elucidation yielded. Especially in the 1960s and 1970s, the demand for trained personnel was high. Experience, that is, the "ability to interpret in a sophisticated manner mass spectral data,"[313] was the qualification of the expert that was most urgently sought for.

Chemists aimed at getting support by mass spectrometric methods, but they did not want to be replaced in their endeavors. Many organic chemists were recalcitrant when they were confronted with spectroscopic methods, especially X-ray analysis:

> In my opinion a number of people didn't use it because it was an insult . . . . If anyone can prove a structure with an X-ray analysis, we are nothing. The

[313] Djerassi, NSF grant application, "Mass spectrometry in organic and biochemistry," 1969, 10, Djerassi Papers, box 22a.

> organic chemist is nothing but a little technician who crystallizes the compound and gives it to someone who sticks it in an X-ray machine, and even the rest is computerized. So what's your function? I would like to see if I can prove a structure without X-ray, but otherwise use everything else. Well, some other people may have said I want to prove a structure without mass spectrometry, or I want to prove a structure without NMR, but then it becomes pretty pointless after a while. I think there was [a strong tendency] . . . to show 'how clever I am using intellectual techniques.'[314]

Mass spectrometry did not experience the same fate as X-ray analysis, and kept being an "intellectual technique." In the end, intellectual pride did not prevent the breakthrough of novel instrumentation, including X-ray methods. Due to strong competition, natural product chemists either adapted to the new methods or they quit the field.

With their strong interconnections with industry, while holding academic positions, McLafferty, Biemann, and Djerassi preceded the style of biotechnology that intertwined the university with industry in this field from the 1970s on. Although they all carefully divided the two spheres, a considerable flow of personnel from the university to the firms connected the two sides. Up to the early 1960s, the chemical and petroleum industries provided most of the material and intellectual resources for organic mass spectrometry. Here, as was shown with the examples of Seymour Meyerson at Standard Oil of Indiana, and Fred W. McLafferty of Dow Chemical, organic mass spectrometry first emerged from a routine analytical technique to a technology enabling the research chemist to perform investigations into molecular structures of unknown compounds. Nevertheless, despite all this support, the industrial context showed severe setbacks. Although McLafferty, from the late 1950s on, directed a Dow basic research laboratory, he lacked the resources to undertake fundamental studies on a full-time basis. Mass spectrometry, for Dow, still was mainly an analytical method best organized in large service laboratories near the main manufacturing plants. Thus, to fully apply the potential that mass spectrometry showed for chemical research, McLafferty moved from industry to academia, accepting posts first at Purdue and then at Cornell University.

In the career path of Klaus Biemann, we recognize familiar features of an instrument-based research program. The first purchase of novel instrumentation was made possible by the institution he worked for, the Massachusetts Institute of Technology. Understandably, scientists who were not established faced problems in obtaining sums from federal agencies large enough to start an expensive research project. The "venture capital" of MIT was a fruitful investment, in scientific terms. Biemann multiplied the original capital many times in his subsequent fund-raising. With this development, chemistry changed

[314] Djerassi, interview by Reinhardt, 20 January 1999.

from a labor-intensive to a capital-intensive endeavor. The contributions of Biemann, and others, expanded mass spectrometry to such an extent that the technique tackled the most minuscule amounts of substances found in nature and made by chemical technology. He connected natural product chemistry, physical organic chemistry, and mass spectrometry in a unique way. In his research projects, the development of novel mass spectrometry methods took center stage.

Though Biemann developed crucial parts of new instrumentation, he did so only when forced by necessity. Most technological inventions, and their subsequent diffusion, originated with the instrument manufacturers. Biemann invented new methods to adapt this instrumentation to the areas he was most interested in: amino acids, peptides, and alkaloids. He efficiently connected his fellow scientists with the instrument manufacturers, making both sides aware of their respective needs and the shortcomings. In this respect, Biemann was a mediator. Instrument manufacturers guaranteed that Biemann's methods could be used around the globe (and beyond), provided that sufficient funds and trained experts were available. For the latter, his laboratory was an important source. Cooperations of mass spectrometrists with chemists at companies and universities had their focus in the instrument itself. Cooperation meant transfer of information and knowledge. The latter was the expert knowledge in mass spectrometry, comprising the experience necessary for interpretation of spectra, and the skills for running the instrument successfully. Often, the transfer of knowledge was preceded by the sharing of the instrument, as the case of Klaus Biemann's cooperation with the Swiss Firmenich & Cie. has shown. After the flow of information was supplemented by a flow of expertise, the Swiss company was empowered and encouraged to build up its own mass spectrometry center. The same thing happened with academic chemists. The sharing of the instrument actually became transfer of knowledge, and not just information, through the training that accompanied its joint use.

The knowledge transfer in Biemann's network of chemists and companies was a transfer of samples, instruments, and people. Through the network of his mentor George Büchi, an experienced natural product chemist at MIT, Biemann gained access to the world of natural product chemistry, both in the material and social perspectives. This academic and industrial network provided funds, compounds, and enthusiastic collaborators for Biemann's developing expertise in the novel field of organic mass spectrometry. The instrument was at the center of Biemann's network, and the control over it a necessary and important part of Biemann's standing in the community. He shared data from it, when necessary, and he networked it wherever possible. His strategic aim was the expansion of mass spectrometry to novel fields of chemistry as well as to institutions of all kinds. This can be seen as a bi-directional exchange of resources. Technology transfer from mass spectrometry groups to other, often industrial, laboratories were strongly shaped by older networks of academic-industrial collaboration. Through connections of this kind, Klaus Biemann got access to firms such as Firmenich & Cie. in Geneva, CIBA, of Summit, New Jersey, and Eli Lilly of Indianapolis. While Biemann's direct cooperations with firms and scientists were the synchronic part of his influence, the diachronic component came through his students. Through his students and postdoctoral fellows, many of them coming from his native country, Austria, Biemann established a flow of knowledge from university to industry.

In many ways, Biemann was the founder of a school of mass spectrometry, and he regarded himself as such.[315] As with many other scientific schools, Biemann's was centered around a method. Because of the large demand in university, industry and governmental agencies, members of the first generation of Biemann's graduate students and postdoctoral fellows were highly successful in obtaining influential positions. In the 1960s, governmental funds were still relatively easy to come by, and the broad applicability of mass spectrometry in many areas of science, industry, and society ensured that Biemann's students had good prospects. Biemann estimated the size of the group of graduate students, postdoctoral fellows, visiting scientists, and their own students of subsequent generations, to comprise nearly 1,000 members in the early 1990s. This was made possible by the accidental entry into a new and largely unexplored field, and by increasing federal funds that were available during the 1960s, reflected in the fact that nearly one half of his graduate students took their degrees before the mid-1970s. Many of them obtained academic positions. Later, the pharmaceutical and biotechnological industry provided most of the employment opportunities. In the early years of his independent research at MIT, he desperately needed personnel to pursue the tasks he was confronted with. He found them with the help of his connections to his native Austria and to the department where he had taken his Ph.D. before he moved to the United States. Most of Biemann's early postdoctoral fellows came from Austria and Germany, and many of them returned with the expertise gained at MIT. This was a partial reversal of the brain drain from Europe to the United States. At the same time, it was a sign of the importance of on-site training, especially when new instruments played the most important role. This feature of Biemann's research continued in the mid-1960s and 1970s, when most of his students came from and remained in the United States. In general, it was the huge demand for experts in mass spectrometry in an enormous variety of fields that made this development possible. Chemistry expanded with the advent of mass spectrometry and spectroscopic methods. Only with their help, previously inaccessible areas in science, industry, and the environment came within reach.

The environment of Stanford University in the 1960s was amenable to Carl Djerassi's style of doing chemistry in both the industrial and the academic

[315] Biemann, "Massachusetts Institute of Technology Mass Spectrometry School."

spheres. At other places he might have felt less comfortable. With his industrial interests, and his successes in fund-raising with governmental agencies, he matched very well the needs of a research university of the Cold War era. Stanford certainly was a model university in this respect. This environment greatly fostered the mechanization and industrialization of chemical research. High performance mass spectrometers were large-scale instruments that could be handled only with a considerable investment and highly specialized personnel. In order to make full use of them, a large constituency had to be assembled. The private research universities Stanford and MIT provided an ideal environment for such entrepreneurship.

Djerassi and his group concentrated on the foundations and the chemical output of mass spectrometry. They systematically searched for fragmentation processes that were triggered and directed by the presence of functional groups and heteroatoms in certain molecules. In Djerassi's system of elucidation, mass spectrometry and steroid chemistry constituted the two sides of one coin. While mass spectrometry was employed as a tool to make forays into the chemistry of steroids (and natural products in general), steroid chemistry yielded the necessary insights to advance the understanding of mass spectrometry itself. A characteristic issue was the division of labor in Djerassi's laboratory. Only very few members of his group were allowed to operate the mass spectrometer, usually the most senior postdoctoral fellow and the technician in charge. Djerassi was convinced that organic chemists would use the mass spectrometer as a black box, through simply delivering samples to the mass spectrometrist and receiving spectra in return. In his own laboratory, the output of the assembly line of twenty graduate students and postdocs, synthesizing model compounds, was fed into the instrument. The result were methods and structures, published with an enormous rapidity.

Despite some differences in style and content of their research, McLafferty, Biemann, and Djerassi stood for the first generation of organic mass spectrometrists. Together, they directed the foundations of this field towards the prevailing paradigm of the time, the physical organic approach to organic chemistry. In the years to come, they shared also the focus on computer-based methods of mass spectral data processing (see Chapter 5).

Chapter Four

A Lead User at the University

In 1959, Herbert S. Gutowsky wrote about the "boon and boom" of NMR in chemistry. NMR, according to Gutowsky, belonged to the "most important events in the past 50 years for the advancement of organic chemistry." He vigorously advocated this view in a review of John D. Roberts's book, *Nuclear Magnetic Resonance: Applications to Organic Chemistry*, a book that like no other brought NMR methods into mainstream organic chemistry:

> Initially, it was thought that nuclear physics would benefit most from the NMR techniques . . . . On the other hand, the detailed characteristics of the nuclear magnetic phenomena are influenced by many features of the nuclear environment. Because of the diversity of phenomena and the sensitivity of the methods, it has proved possible to use stable nuclei of known and favorable magnetic properties as probes to investigate the structure of matter. In fact, such "chemical" applications have occupied more than 1000 publications in the past 10 years and have overshadowed the original intent of the physicist.[1]

In his review, Gutowsky addressed the availability of commercial instruments and the massive increase in funding, mostly provided by government agencies, when he looked for reasons for this "boom." Curiously, he neglected the role of scientists in this process. Chemists were not only the users of the novel techniques, often they were innovators and made important improvements. NMR, as a technique with a theoretical underpinning in quantum physics, and its roots in high-technology fields such as magnet design and electronics, was a difficult field in which organic chemists could contribute to the development of innovative hardware. Consequently, innovation in novel NMR technology was not so heavily "user-dominated" as it is described in the literature, if we focus

[1] Gutowsky, "NMR Boon and Boom."

on the development of the hardware itself.[2] But at the same time, the introduction of new methods became imperative. Chemistry and instrument technology had to be closely linked. For this, a detailed knowledge of both the working modes of the instrument and of its utility in chemistry was mandatory. Partially, this knowledge rested with chemists who were employed by the instrument manufacturers and who were in charge of bridging the realms of electronics and organic chemistry. Equally important were chemists at universities and in the chemical industry. As innovative users, they specified novel technology for prototype construction and later, perhaps, serial production by the instrument manufacturers.[3] As fund raisers for their projects and equipment, they supplied a large part of the funds needed for the innovation of new technology. Thus, they connected the instrument industry and government science agencies through their research needs. In their roles as teachers, peers, referees, and consultants for government and industry, they paved the way for the diffusion of physical instrumentation in the scientific community. As active researchers in their respective fields, they felt the need for exclusive access to unique instrumentation. The modus of the scientific credit system assured their quest for the dissemination of the technology, after the proof of its utility had been achieved. The diffusion of the instrument itself was tied to the use of their innovative methods, and bound to the citation of the pioneers' work.

This concept is closely related to Eric von Hippel's notion of the lead user. He defines lead users as users who "face needs that will be general in a marketplace, but they face them months or years before the bulk of that marketplace encounters them, and . . . are positioned to benefit significantly by obtaining a solution to those needs."[4] Von Hippel distinguishes lead users from his concept of user-dominated innovation. The latter affords important insights into the roles academic scientists played in the development of novel instruments, among other products. The former concept, the lead user, in contrast describes the effects that users had on the later stages of innovation and the marketing of

---

[2] See von Hippel, *Sources of Innovation*, 11–27, 133–163. Von Hippel's book contains innovation histories of gas chromatography, ultraviolet spectroscopy, NMR, and electron microscopy. His assignment of the user as innovator is often hampered by the fact that von Hippel neglected the strong ties of academic scientists to instrument manufacturers. For example, one of the inventors of NMR, the Stanford physicist Felix Bloch, was a paid consultant of the pioneering manufacturer of NMR spectrometers, Varian Associates. It is doubtful to consider his innovations as user dominated, as von Hippel did.

[3] An earlier example is given by the chemist Alexander Todd, who described the innovation process during the 1930s and 1940s as follows: A prototype of equipment was designed by a member of his laboratory, and licensed to an instrument manufacturer. The proviso was that Todd's laboratory would have priority in all orders, and thus "became perhaps the best equipped laboratory in Britain." Todd, *A Time to Remember*, 78–79.

[4] Von Hippel, *Sources of Innovation*, 107.

high technology.[5] Expanding on von Hippel's concept, lead users of scientific instruments can be seen to fulfill their functions best if they lead the actual uses of novel instrumentation (i.e., are "first movers"), and, moreover, if they are already established scientists (their leadership being further enhanced by the successful use of instrumentation). In this position, scientific lead users are enabled to create and to direct the market for scientific instruments. Thus, precisely because of their scientific authority, lead users influence the success of the instrument manufacturer; and their participation in academic administration and funding agencies allows them to lobby for the new techniques from the inside of the scientific establishment. Moreover, through specifications of novel custom-built instruments, and their eventual success in the scientific marketplace, lead users supply the instrument manufacturers with crucial ideas for improvements. On the other hand, the exclusive ownership of up-to-date instrumentation is a huge advantage in one's own research, when done in a competitive environment. While design and construction of the instruments is mainly the responsibility of a company, the marketing is a shared activity. The course of this cooperative strategy could not be thoroughly planned, neither by the company nor by the scientist. But it was recognized and acknowledged as such on both sides.

In this chapter, we investigate such an alliance of a research scientist, John D. Roberts (b. 1918) of the California Institute of Technology (Caltech) in Pasadena, and an instrument manufacturer, Varian Associates of Palo Alto, and we will encounter the story from the scientist's point of view. In his career, Roberts climbed many steps of the academic system. He became full professor at a major research university at the age of 35, dean of one of the most prestigious chemistry departments in the United States ten years later, and accepted the office of provost in 1980. Influential positions in the National Science Foundation and the U.S. Academy of Sciences round up a picture that set the stage for his role in introducing NMR into organic chemistry. Of course, he was not the only scientist who had a role in this. But his example shows how instruments and science were interwoven through the active input of scientists, influencing innovation and diffusion of new instruments and methods.

First, it is argued that the subdiscipline of physical organic chemistry, and Roberts's role in this area, prepared the ground for his later successful involvement in NMR in both cognitive and social respects. Physical organic chemistry, with its key concepts of the electron bond and mechanisms of reactions,

---

[5] The borderline between user-innovators and lead users is fluid. Because the focus here is on the development of methods, their illustrative use, and subsequent diffusion in science, the term lead user seems to be adequate. The concept became widely known in economics through von Hippel's efforts (partially with the companies 3M and Business Genetics) to establish it as a development and marketing tool. This even led to the founding of a company named Lead User Concepts Inc. (LUCI). See Anon., "About Lead User Concepts."

was very well suited for the early diffusion of NMR. New types of mechanisms, entirely inaccessible by chemical means, could be investigated by physical methods. Moreover, the modern quantum chemical concepts made their first inroads into organic chemistry through proponents of the physical organic approach. As NMR, and other physical techniques, were based on closely related concepts, the diffusion of both the theory and the laboratory technique went hand in hand. From the 1940s to the 1960s, physical organic chemistry boomed, especially in the United States. Thus, NMR found a prepared and fertile ground for early growth, and contributed to opening up a new scientific field, molecular dynamics.

Second, in the process of cooperation, academic and industrial participants exchanged some of their original functions, if only to a limited extent and for a short period. Here, the focus is on these changes of functions, and the conflicts and benefits that they created. Employees of Varian Associates engaged in chemical research and issues of standardization, while for Roberts the contact to the company proved to be the single most important precondition of his research. It is not unusual for chemists to have close and longstanding connections to industrial enterprises. But in this case, Roberts depended on the company for access to his major research instrumentation. He counterbalanced this with his role as promoter of NMR instrumentation. This informal, but close, cooperation enabled Roberts and Varian Associates to successfully channel their respective experiences and needs towards the design and development of new instruments and methods.

Third, teaching comprised the central function of the scientist's part in diffusion of instrumentation in a scientific community. The writing of textbooks and the holding of lectures not only educated a new generation of NMR users, but also set the tone of how to employ the instrumentation and to what ends. In this endeavor, Roberts distributed Varian Associates' commercial instruments and his own scientific methods at the same time.

Fourth, and perhaps most crucially, Roberts himself engaged in the design and planning of novel instrumentation. Here it is most clear that NMR became a scientific specialty, and its further development constituted the major part in the research program of a whole community of scientists. When the growth of physical organic chemistry slowed down in the 1970s, NMR practitioners headed to the newly booming fields of bioorganic chemistry, and the biomedical sciences in general. Roberts took the same path in his own career. In doing so, Roberts established himself as an instrument-driven scientist in the innovation and pioneering application of a complex research technology.

John D. Roberts and Physical Organic Chemistry

> And then as techniques developed, we began to use less material. We developed things that we could use on droplets instead of with hundreds of cc's.

> So the style has changed, so that we're getting by using much less stuff. When we wanted to find out what a structure of a compound was when I was doing my PhD thesis, I would make ten or fifteen grams of stuff in a run, and then do some chemistry on it, make some derivatives and cut the molecules up and try to find out what kind of pieces they were made of. Now we can usually do that stuff in ten minutes by just running . . . [an] NMR spectrum. We have to worry about the structures, still, in some cases, but most of the time, structure determination is done by physical methods. Very, very powerful, and in a much, much shorter time.[6]

With this quote, Roberts referred to changes in structure determination that the broad use of high-technological instrumentation brought about in the 1950s and 1960s. He might well have added the consequences for the idea of molecular structure itself. Spectroscopy strengthened the novel theoretical understanding of the chemical bond, and paved the way for its acceptance in the different subdisciplines of chemistry. Bonds, the crucial concept of chemistry, acquired a new epistemological status, and were attributed for example with length and heats of formation. Groups of atoms rotated and vibrated, electrons had various possible energetic states. These parameters provided the foundation of spectroscopic methods. Their use by organic chemists affected the status of chemical structures and bonds in this important subdiscipline. Another, and closely related, shift was the changeover from research into structure to research into function and mechanism. It is here that physical organic chemistry, a new subfield of organic chemistry, had the most to offer.[7] During the time of its emergence from the 1920s to the 1940s, the field still was overshadowed by traditional synthetic organic chemistry. It gained prominence in the 1950s, at least in the United States, Japan, and most Western European countries. As such, it became the new center of gravity of the discipline of organic chemistry, supplying methods and research problems that were seen as worthwhile to follow up. Thus, physical organic chemistry is different from most other sub- or borderline fields of organic chemistry, i.e. polymer, bioorganic, and organometallic chemistry. While the latter tackle special classes of substances (either because of their function or their structure), physical organic chemistry deals with the whole of chemistry. The term physical organic chemistry was coined in 1940 by the Columbia chemist Louis P. Hammett. Hammett founded his announcement of a new subdiscipline on two decades of work by several research groups, many of them based in Great Britain. In his

---

[6] John D. Roberts, interview by Rachel Prud'homme, February to May 1985, Caltech Archives, Oral History Collection, 76.

[7] Morris, Travis, and Reinhardt, "Research Fields and Boundaries," 14–20; Nye, *From Chemical Philosophy to Theoretical Chemistry*, 196–223; Brock, *Fontana History of Chemistry*, 506–569; Saltzman, "Development of Physical Organic Chemistry"; and Gortler, "Physical Organic Chemistry Community."

view, the mission of physical organic chemistry was to make organic chemistry scientific.[8] Referring to the opinion of the later Harvard president and influential science manager James Bryant Conant, who once had called organic chemistry a fascinating art slowly changing to an exact science, Hammett announced that organic chemistry indeed was in the process to become a science: "It must be the task of science to replace the qualitative judgments with quantitative statements of reaction rate and reaction equilibrium."[9] As with chemical physics, the higher prestige of physics in the world-view of academics and the public alike helped to make this name popular: physical organic chemistry was a "scientific" organic chemistry. "The great systematizer" of organic chemistry, Christopher Kelk Ingold,[10] in the 1920 and 1930s performed the task of interpreting the jungle of organic structures and reactions with the guidance of mechanistic schemes and a classification in basic reaction types. The main achievements of Ingold were in his ability to connect three-dimensional chemical structures and physical forces. In doing so, he was able to investigate the role of electrical effects for the direction of chemical reactions, always having in mind the classical chemical concepts of valence and structure. With this approach, he made electrical interaction the major topic of theoretical reasoning on reactivity.[11] As many principal revolutions in chemistry did, the advent of physical organic chemistry went hand in hand with the pushing-ahead of a new nomenclature. The novel notions were related to a physical rationalization of molecular structure and reaction mechanisms. Ingold's systematization of physical organic chemistry was conceived before Linus Pauling, Robert Mulliken, and others achieved the foundation of chemistry in quantum mechanics. The latter led to the emergence of another interstitial discipline between chemistry and physics: chemical physics. In the following decades, most theoretical concepts of the two systems merged, and physical organic chemistry began to receive a firm place in the curriculum of organic chemistry.[12] In addition, and parallel to the developments in chemical physics, the practitioners of physical organic chemistry increasingly relied on the use of physical instrumentation. In the beginning, mostly electric dipole measurements and refraction indices were used, supplemented by spectroscopic meth-

---

[8] See Hammett, *Physical Organic Chemistry*. In this respect, Hammett was influenced by Percy Bridgman's operationalism. See Benfey, "Teaching Chemistry," 19.
[9] Hammett, *Physical Organic Chemistry*, 2. Conant forcefully argued in favor of including physicochemical studies in organic chemistry, but ended his article with the sentence: "We may rest confident, moreover, that the fascinating art of organic chemistry will yield only slowly to the devastating inroads of an exact science." Conant, "Equilibria and Rates," 472.
[10] Brock, *Fontana History of Chemistry*, 507.
[11] Nye, *From Chemical Philosophy to Theoretical Chemistry*, 199, 222.
[12] Brock, *Fontana History of Chemistry*, 549.

ods. The determination of dipole constants, for example, in 1933, led the Ingolds to their concept of mesomerism.[13]

In the United States, physical organic chemistry advanced first in a few schools only: Harvard, Columbia, Chicago, Caltech, and the University of California at Los Angeles (UCLA).[14] In the 1940s, physical organic chemistry began to spread out, first to Berkeley, MIT, the University of Illinois, and Notre Dame.[15] The pinnacle was reached during the 1960s and 1970s. Among the positive effects that physical organic chemistry had for organic chemistry as a whole were the revitalizing of synthesis by the postulation of interesting compounds, and a deeper understanding of synthetic reactions.[16] Later, physical organic chemistry faded away as it moved into bioorganic and organometallic chemistry, and other novel fields.

It was not just the "push" of instrumentation that led organic chemists to think of new areas of inquiry. Physical organic chemistry also exerted a "pull," because it was attractive in its own right and gave instrumental methods the background to unfold their capacities. Most important in this respect was the fact that the topics of highest interest in physical organic chemistry concerned intermediate molecular states that could not be isolated, and thus were not approachable by classical means. Roberts himself worked intensively on such issues, and here NMR had the most to offer. With its help, these short-lived reaction intermediates suddenly became observable. Moreover, physical organic chemistry could rationalize synthetic methods; it stimulated the development of synthetic methods, and the synthesis of novel compounds for the testing of its hypotheses. As an outcome, physical organic chemistry transformed traditional organic chemistry, and the new organic chemistry can be considered a hybrid of traditional organic synthesis, physical organic reasoning, and physical instrumentation.[17] In principle, the history of physical organic chemistry resembles the history of physical instrumentation. Both fields transformed the practice and theory of organic chemistry, and together they achieved the eclipse of traditional methods. In doing so, they interacted and often depended on each other, but should be seen as separate units.

Roberts's contributions to physical organic chemistry were far-reaching. With the term "nonclassical carbocation" he pushed forward a notion that was controversially discussed during three decades. His concept of benzyne (dehydrogenated benzene) explained the intermediate states of important reactions.

---

[13] Christopher Kelk Ingold and his wife Edith Hilda Ingold. Brock, *Fontana History of Chemistry*, 533–534.

[14] Gortler, "Physical Organic Chemistry Community," 753.

[15] Gortler, "Physical Organic Chemistry Community," 753, 756–757.

[16] Roberts, "Beginnings of Physical Organic Chemistry (1)," 53.

[17] Morris, Travis, and Reinhardt, "Research Fields and Boundaries," 20.

He was a pioneering user of the radiotracer method with carbon-14, and applied early on the mathematically difficult method of molecular orbitals for the calculation of possible organic structures. Roberts combined the knowledge of traditional chemical methods with a curiosity-driven adaptation of innovative methods and concepts. His research style was described by his Caltech colleague George S. Hammond, in the foreword of Roberts's *Collected Works* in 1969:

> Roberts fastened onto carbon-14 when most chemists still feared that radioactivity would sterilize them; he learned molecular orbital theory when most organic chemists regarded an operator as a bogeyman; he set a style now copied by hundreds of young men for the study of small carbocyclic compounds; he applied nuclear magnetic resonance to organic compounds when most of us were still pondering spin flipping in ice; and he emerged from a violent romance with computer science as a truly triumphant lover.[18]

In his autobiography, Roberts gave a more sober account, presenting his life as being at "the right place at the right time." But Roberts's inclination for exploring unknown ground certainly helped in the "demystification" of NMR, an achievement "that opened the doors for organic chemists."[19] Some traditional chemists perceived his contributions as shocking, but many of his results, opinions, and novel methods survived the test of time. Hammond ascribed the high state of sophistication that was achieved in NMR to Roberts's determination to "do things right," and for a counterexample he pointed to the still empirical state of methodology in infrared and ultraviolet spectroscopy. Above all, in the opinion of Hammond, Roberts kept being an organic chemist, while he contributed to the redefinition of the boundaries of the discipline.

Already at a relatively young age, in the early 1950s, Roberts became one of the best-known and most influential physical organic chemists. He achieved this with a series of investigations into the reaction behavior and structure of small molecules, and their unusual interpretation. Although his views were not always immediately unanimously accepted, he installed path-breaking modern, physical thinking in organic chemistry. Most American chemists accepted the novel view, and some of Roberts's results became included in textbooks: His work belonged to the mainstream of the chemistry of the day. Achieving these successes already with the help of instrumentation, among others refractometry, infrared spectroscopy, and radioactive tracer experiments, Roberts was eager to add novel methods.

John D. Roberts had received his early education in chemistry at the University of California at Los Angeles. UCLA, in the 1930s, did not have a Ph.D.

[18] Hammond, "Foreword," vi.

[19] Dervan, "John D. Roberts," 75.

program, and Roberts in 1941 went to Penn State University to begin his Ph.D. studies there with Frank Whitmore. The beginning of World War II brought him back to UCLA, and after the end of his war-related research and the inauguration of a Ph.D. program there, Roberts was able to commence his project in Los Angeles. Most influential during his early career were William G. Young and Saul Winstein. Both had been trained in physical organic chemistry by Howard Lucas of nearby Caltech. When he graduated in 1944, Roberts had very solid training in physical organic chemistry, a lot of research under his belt, and hands-on experience with the building of laboratory equipment. He received a stipend of the National Research Council, and went to Harvard University for postdoctoral studies. In 1946, Roberts was offered the position of instructor at the department of chemistry of MIT at Cambridge, Massachusetts, where Arthur C. Cope was in the process of reforming the organic chemistry group.[20] Roberts was one of three young organic chemists (the others were Gardner Swain, who was also a physical organic chemist, and John Sheehan, a synthetic organic chemist who later became famous for his work on penicillin), who were chosen by Cope.

His first encounter with developing new methods happened to be at MIT, and was triggered by the availability of nearly inexhaustible funds. Cope made available the money through the new MIT Laboratory for Nuclear Science and Engineering. For the first year, Roberts obtained the princely sum of $44,000, in the following years half that.[21] Its use was bound to application of the radioactive isotope carbon-14, and Roberts had to come up with ideas for it. In the following years, he used carbon-14 mostly for studies of reaction mechanisms, his method of "isotope-position" rearrangements having a model in work of researchers at Shell Development Company in Emeryville, California.[22] In the spring of 1950, Roberts began to use the radiotracer method for the elucidation of the pathways leading to the concept of the nonclassical carbonium ion (or, in modern terminology, carbocation), one of his greatest achievements.

Still at UCLA, Roberts had the idea of studying small ring compounds,[23] and on his move to the East, carried one kilogram of his favorite raw product, cyclopropylmethyl ketone.[24] Small ring molecules were a favorite playground

---

[20] See Chapter 3.

[21] Roberts, *The Right Place at the Right Time*, 60.

[22] Roberts, *The Right Place at the Right Time*, 75.

[23] He described his rationale for doing research on small rings in Roberts, *The Right Place at the Right Time*, 39, as reminiscent of reading Frank Whitmore's book (*Organic Chemistry*, New York: D. van Nostrand 1937), with the dictum that "cyclopropanol apparently does not exist." Moreover, his work relating to the nonclassical carbonium ion came into being at Harvard through a suggestion of Paul D. Bartlett. Roberts, interview by Prud'homme, 51–52.

[24] Commercially available through wartime research on atabrine, an antimalarial drug. Roberts, *The Right Place at the Right Time*, 44.

for organic chemists, who were investigating reactivity and the bonding structure of molecules. At MIT, Roberts, with his graduate student Robert H. Mazur, studied the unusual reactivity of another small ring compound, cyclobutyl chloride, and its connection to the cyclopropyl compounds.[25] To explain the unusual behavior of the molecule, and building on work of other physical organic chemists (including Ingold), Roberts chose an equally unusual interpretation of its structure, and dubbed it "non-classical carbonium ion."[26] With that, Roberts gave one of the most enduring controversies in organic chemistry its name; and he installed himself as one of the central figures in the rather one-sided debates that would last for the next three decades.[27] While the majority of chemists agreed with the experimental evidence and theoretically convincing calculations provided by Roberts, Winstein, Michael J. S. Dewar, and others, only Herbert C. Brown of Purdue University did not admit defeat. With this controversy, a fundamental issue of organic chemistry was at stake. In the view of Roberts, his and Mazur's work was

> especially important as the opening of the Pandora's box of an extraordinarily difficult and subtle problem—a problem concerned in an important way with what we mean when we write chemical structures on paper.[28]

According to historian Stephen Weininger, the nonclassical ion "was a hook on which to hang a much larger agenda."[29] For chemists, matters of representation were as important as matters of fact (actually they were matters of fact).[30] With the coinage of the name nonclassical cation, and more so with his extremely successful investigations of the field, Roberts made his name in the community

---

[25] This follows Dervan, "John D. Roberts," 74. See also Roberts, *The Right Place at the Right Time*, 64–74, for a description of Mazur's work.

[26] Roberts and Lee, "Nature of the Intermediate." A carbonium ion (carbocation) is an ion with a positive carbon atom, and, in chemical reactions, represents a reactive intermediate. The controversy on "nonclassical" cations was about the structures of the intermediates involved. In classical carbocations, the positive charge is either localized, or it is delocalized by involving an unshared pair of electrons or double/triple bonds in neighboring position. In nonclassical cations, the positive charge is delocalized by a single bond, or by double/triple bonds not in the neighboring position. In the traditional view, a single bond had been thought of as being connected to the two bonded atoms only, thus not being delocalized and not contributing to the electronic states of other parts in the molecule. The term nonclassical ion was used rather loosely, and referred to many kinds of unusual reaction intermediates in organic chemistry. See Bartlett, *Nonclassical Ions*, and, for a modern treatment March, *Advanced Organic Chemistry*, 312–327. Roberts chose the name in a discussion with Robert Burns Woodward at Harvard. Roberts, *The Right Place at the Right Time*, 80–82.

[27] Weininger, "What's in a Name?" There is a rich literature on this controversy. See for a start, and additional literature, Brock, *Fontana History of Chemistry*, 558–569; Bartlett, *Nonclassical Ions*.

[28] Roberts, *The Right Place at the Right Time*, 64.

[29] Weininger, "What's in a Name?" 128.

[30] See Weininger, "What's in a Name?" 128–129.

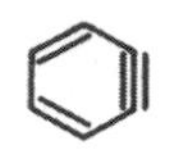

FIGURE 4.1 Structural formula of benzyne (dehydrobenzene). According to Roberts, the choice to draw the additional bond outside the ring caused a visually perceived increase in its plausibility. From John D. Roberts, "Benzyne as an Intermediate in Nucleophilic Aromatic Substitution Reactions," *Chemical Society Special Publication*, no. 12 (1958), 115–128, on 115. Adapted from John D. Roberts, *Collected Works*, New York: W. A. Benjamin 1970, 535. Reproduced by permission of the Royal Society of Chemistry.

of physical organic chemists at the age of thirty-three: "That research program on carbocations really went along like a house afire."[31]

A similar success that Roberts had with nonclassical ions was achieved with another of his concepts, benzyne. Here, he recognized the intermediate product of certain reactions as dehydrogenated benzene. For such a molecule, Roberts chose a presentation with two double bonds and one triple bond (see Figure 4.1). While the majority of chemists agreed with his proposal, a minority of old-style organic chemists could not believe the reality of such an unusually looking structure.

Roberts's investigations of reaction intermediates that are most often too unstable to be isolated by classical chemical means prepared the ground for his later entry into NMR, a technique that was ideally suited to tackle such problems.

Along with his research, Roberts entered the area of theory in physical organic chemistry with his teaching. In the early 1950s, Linus Pauling's coinage of quantum chemistry, the valence bond theory, dominated the field, ably popularized by the notion of resonance. In classical physics, resonance is often introduced with a system of coupled pendulums, exchanging energy. In quantum physics, and in a very general meaning, resonance is the interaction of wave functions. In Pauling's view, resonance was based on several possible solutions of the wave equation for a molecular structure. Each solution represented a resonance structure, the actual molecular structure being determined by their average. In the beginning, it was thought that the molecule resonated between these possible structures, but soon it was recognized that only one structure existed, the resonance hybrid. The popularity of Pauling's concept partially rested on the fact that most organic chemists misunderstood the quantum mechanical background of resonance. The most common mistake of a chemist was to equate resonance with the purely chemical concept of tautomerism,

[31] Roberts, interview by Prud'homme, 53.

involving an actual, intra-molecular transfer of atomic groupings and thus meaning an equilibrium of different, "real" molecules.[32]

> Resonance theory was well entrenched, generally highly satisfactory in a qualitative way, even if you could not claim to have the faintest understanding of the quantum mechanics involved . . . . To be sure, there were books such as Eyring, Walter, and Kimball's *Quantum Chemistry* or Pauling and Wilson's *Introduction to Quantum Mechanics* that covered matters such as these, but precious little of it was either understood by or useful to organic chemists. Most of us were resigned to explaining resonance to students by some variation on Wheland's rather tortuous qualitative descriptions of pendulums and similar analogs in *The Theory of Resonance*, with the hope that, in the repetition, familiarity could mimic understanding.[33]

The success of resonance theory in organic chemistry rested in its ability to systematize an immense quantity of organic phenomena. Moreover, it explained and even predicted reactivity. But there were some anomalies ("cracks in the façade") that bothered the specialists. Meanwhile, a competitor for the concept of resonance gained popularity: the molecular orbital theory of Robert S. Mulliken. Molecular orbital theory was considered to be a better approximation to quantum mechanics. After Pauling's valence bond (resonance) method had prepared chemists to accept quantum mechanics as a theoretical basis, molecular orbital theory took over as the better method of the two.[34] "The winds of change had started to blow"[35] with an article of the British quantum chemist Charles A. Coulson in 1947,[36] and especially with the book *The Electronic Theory of Organic Chemistry* of Michael J. S. Dewar. Dewar, who had the support of the British doyen of physical organic chemistry, Robert Robinson, recast the older electronic theory of organic chemistry in terms of the molecular orbital approach. In doing so, he tried hard to connect to the experience of organic chemists.[37] Roberts was intrigued by the "siren song" of Dewar and decided to shift from resonance to molecular orbital theory in the fall semester course of 1951. Roberts was impressed by the simplicity and elegance of the method in treating effects quantitatively: "I was learning how to go to a variety of sources and look for anything at all I could understand, then try to transport that back to solving the problem that really interested me."[38]

---

[32] For a warning against this misunderstanding see Wheland, *Resonance in Organic Chemistry*, 3.
[33] Roberts, *The Right Place at the Right Time*, 120.
[34] Brush, "Dynamics of Theory Change," 291.
[35] Roberts, *The Right Place at the Right Time*, 122.
[36] Coulson, "Representation of Simple Molecules."
[37] Dewar, *Electronic Theory of Organic Chemistry*, x. Robinson first considered co-authoring the book.
[38] Roberts, *The Right Place at the Right Time*, 124.

Characteristically, Roberts did not stop at simplified understanding. Together with Andrew Streitwieser, he started out "to calculate everything we could think of that might be of interest to organic chemists and did not offer too great computational difficulties."[39] In the early 1960s, Roberts (as did Streitwieser) even wrote a book on molecular orbital theory.[40] Designed to "help people to get over the activation barrier," it had 13 printings by 1990.[41] Though "it certainly wasn't widely admired among the physical chemists, and a lot of it wasn't really right . . . students loved it . . . . So that got me started in a lot of new directions and I got some reputation for doing that job, and helping to bring organic chemists more generally into the molecular orbital age. I didn't make any basic contribution except as sort of a teaching function."[42] Also in the case of NMR, this role as facilitator and teacher proved to be important.

In 1950, Cope proposed and promoted Roberts for a consultancy at Du Pont. Consulting for industry was a competitive business, and Roberts tied with the Illinois chemist Elliot R. Alexander, the author of the first undergraduate textbook on reaction mechanisms, and Roberts's former laboratory partner at Harvard.[43] Roberts received the position, and the consultancy for Du Pont proved to be a constant important source of ideas about novel techniques and research problems for him. In 1952, while giving lectures on molecular orbital theory at Du Pont, Roberts met a Du Pont scientist, Rudolph Pariser, and a theoretical chemist at Johns Hopkins University, Robert G. Parr. Pariser and Parr were developing molecular orbital methods for the calculation of properties of synthetic dyestuffs. Roberts used their method to calculate the unknown organic molecule pentalene during the greater part of his 1952 sabbatical, which he spent on a Guggenheim Fellowship at Caltech.[44] Though the tedious calculations of Roberts did not lead to anything publishable, this was an ideal project for his research at Caltech, which had a strong tradition in theoretical chemistry.

From his days at UCLA, Roberts knew the small, but elite division of chemistry at Caltech very well. He kept in touch with Verner Schomaker (an expert in electron diffraction methods and collaborator of the division head Linus Pauling), and Edwin J. Buchman, a senior research associate who was doing small-ring chemistry along the same lines as Roberts did at MIT. "And of course, I had a tenuous connection with Yost."[45] Don Yost was a versatile

---

[39] Roberts, *The Right Place at the Right Time*, 125.
[40] Roberts, *Notes on Molecular Orbital Calculations*; and Streitwieser, *Molecular Orbital Theory for Organic Chemists*.
[41] Roberts, *The Right Place at the Right Time*, 126.
[42] Roberts, interview by Prud'homme, 51.
[43] Roberts, interview by Prud'homme, 54.
[44] Roberts, *The Right Place at the Right Time*, 128–129.
[45] Roberts, interview by Prud'homme, 55.

physical chemist who early on ventured into NMR with his graduate student John S. Waugh (later at MIT and a specialist on solid-state NMR) and with James N. Shoolery (later at Varian Associates). In addition, Caltech professor Howard J. Lucas was one of the early proponents of physical organic chemistry in the United States, and best known through the publication of a 1935 textbook in organic chemistry that included physical organic ideas.[46] At Caltech, he had built a small but influential research program. Thus, Caltech was an attractive place, and Roberts knew that Lucas was going to retire in 1953. If physical organic chemistry were to stay at Caltech, Roberts was a serious candidate. Indeed, in late 1952 or early 1953, Pauling offered Roberts a professorship, and Roberts accepted.[47] In Roberts's opinion, Pauling dominated Caltech chemistry, and many of Pauling's colleagues, faculty members and postdocs alike, were doing research that was of interest to Pauling himself.[48] In the early 1950s, Pauling shifted his focus to biological applications of chemistry, and just had published his alpha-helix model of proteins.[49] Moreover, he worked on the structure of the hereditary material, DNA. Thus, at Caltech strong interdependencies between chemistry and biology existed, but not as many between physics and chemistry. Nevertheless, Pauling's program centered heavily on the use of physical instrumentation, and he relied on X-ray measurements and electron diffraction methods for his theoretical work in chemistry and biology. As a consequence of Pauling's domination, the organic department had only three full professors, Lucas, Laszlo Zechmeister, and Carl Niemann. Zechmeister concentrated on the use of paper (column) chromatography for the study of colored plant materials, Niemann investigated enzymatic activities, and although he used synthetic methods, he was more a biochemist than an organic chemist.[50]

When Roberts succeeded Lucas in 1953, he changed from a large department where he represented only a fraction of organic chemistry to a small division with the responsibility for a main part of it. In his last year at MIT, 19 people worked in his group. At Caltech, Roberts decided to work with no more than eight or ten people.[51] With this change, Roberts switched to a different system of directing the research of his group. More or less, he began to act as a sort of consultant, and let his group members develop their own ideas. Though he demanded monthly reports, and later on held weekly research seminars, this

---

[46] Lucas, *Organic Chemistry*. For Lucas, and his succession by Roberts see California Institute of Technology, *Chemistry and Chemical Engineering*, 95–99.
[47] Roberts, *The Right Place at the Right Time*, 141.
[48] Roberts, interview by Prud'homme, 64.
[49] Pauling and Corey, "Atomic Coordinates."
[50] For Niemann and Zechmeister see Roberts, *The Right Place at the Right Time*, 146.
[51] Roberts, *The Right Place at the Right Time*, 145, and Roberts, interview by Prud'homme, 75.

resulted in quite a lot of freedom for his graduate students and postdocs.[52] When Roberts arrived in Pasadena, Pauling had already arranged for some research support, including funds for a postdoctoral position for the first years. In addition, though he naturally lost the money from MIT, Roberts had a grant from the National Science Foundation for the study of small ring compounds.

The Move into NMR

Roberts was the first organic chemist at a university who became fully acquainted with NMR. In his view, an abundance of questions of high importance in physical organic chemistry could be tackled by the method. Chemical structures, reaction rates, mechanisms, and internal rotations were the riddles that had to be solved, and NMR offered an excellent potential for doing so. Thus, NMR opened new research areas for the organic chemist and promised to answer questions they had not before dared to ask. This applied especially to the investigation of interconvertible three-dimensional structures of organic molecules, conformational analysis. NMR enabled chemists to study the dynamics of molecules:

> Conformational analysis used to be not very actively pursued in physical organic chemistry because there wasn't very much you could do. That all changed when NMR came along, and I decided at the early stages that I wanted to work on NMR and conformational analysis. We did some of the very early experiments to show that you could stop rotation around a carbon-carbon single bond.[53]

The departmental structure at Caltech fostered the use of advanced physical instrumentation by individual scientists. For Roberts, this proved to be the opportunity to advance his own research program to areas unforeseeable before:

> At that point, I didn't care whether I would ever know how NMR worked, I just knew it would solve problems that I was interested in, and, with the help of Linus Pauling, the Caltech administration, bless them, came up with the funds to buy the first commercial NMR installation in a university. And so I was able to ride the early crest of the NMR wave which has swept along through chemistry and biochemistry.[54]

---

[52] Roberts, interview by Prud'homme, 72–73.

[53] Roberts, "Interview," 31.

[54] Roberts's Priestley Address at the National Meeting of the ACS, Denver, CO, April 1987, quoted after Dervan, "John D. Roberts," 74.

Still in Cambridge, Roberts had missed several opportunities to get involved with NMR even earlier. In late 1949 or early 1950, Richard Ogg, a physical chemist from Stanford University who was visiting Harvard, described in great detail to Roberts the potential of NMR in chemistry. At that time, the effects of NMR that later proved to be of such importance in organic chemistry were not yet known, and the unavailability of commercial equipment excluded its routine use by organic chemists. Thus, Roberts understandably did not join in with Ogg's excitement. Moreover, Roberts at that time simply did not understand the physical principles of NMR, a fact that he later frankly admitted. It also seems that Roberts did not have any contact with the group of Edward Purcell at the physics department of Harvard that was so active in the early study of structures of solids with the help of NMR.[55] Clearly, NMR in 1950 still belonged to the physicists and physical chemists. In the next four years, this was to change rapidly.

In 1954, while Roberts was on a consultancy visit to Du Pont, William D. Phillips introduced him to the applications of NMR in organic chemistry. Du Pont scientists and their questions led Roberts to consider many different areas of chemistry, and he became acquainted with a huge variety of instruments. For Roberts, this exposure was a continuous source for novel ideas and techniques.[56] Du Pont had bought one of the first commercial NMR spectrometers, manufactured by the Palo Alto-based firm of Varian Associates. The first high resolution spectrometers of Varian Associates were all delivered to petroleum and chemical companies. These companies could afford the price tag of $25,000, had the necessary infrastructure for operation and maintenance, and they recognized the possibilities of NMR for analytical uses in chemical research. At Du Pont, Phillips informed Roberts not only about the impact that NMR had on the structural elucidation of chemical compounds, but also talked about the measurement of rotation rates of atomic groups in molecules. Phillips chose the example of *N,N*-dimethylformamide to study internal rotation. One year later, after the publication by Phillips in the *Journal of Chemical Physics*, the same example prompted Gutowsky and Charles H. Holm to resume their studies on this topic.[57] Thus, through his consultancy for Du Pont, Roberts gained first-hand knowledge about a novel development in NMR that was clearly of crucial importance for physical organic concepts. He had every reason to "hyperventilate with excitement."[58]

After his return to Pasadena, Roberts tried to convince Pauling to provide funds for an NMR spectrometer. Though Pauling liked the idea, he had the

---

[55] Roberts, interview by Prud'homme, 99–100; Roberts, *The Right Place at the Right Time*, 149–151.

[56] John D. Roberts, interview by Carsten Reinhardt, 28 January 1999.

[57] Phillips, "Restricted Rotation." See Chapter 2.

[58] Roberts, *The Right Place at the Right Time*, 152.

opinion that NMR should be under the auspices of an expert, preferably somebody from chemical physics. At Caltech, Don Yost had brought forward a research program in NMR since 1950. Together with John Waugh, James Shoolery, and other students, Yost undertook investigations of the structures of solids along the lines of Purcell's group at Harvard, and the early work of Gutowsky there. Yost, who had embarked also on microwave studies immediately after 1945, built his own equipment, financed mainly by funds from Newmont Exploration Ltd., New York. Strangely enough, there seemed to have been no direct contact between Roberts and Yost in the field of NMR, and Yost—though he continued his investigations of NMR until his retirement in 1964—began to focus his main efforts on a purely mathematical study.[59] Pauling's opinion that he should rely on expertise in chemical physics was exactly the opposite of what Roberts had in mind of how to use NMR:

> I saw the thing differently, because when I was at MIT they had put in a new infrared spectrometer at Harvard, and while they had infrared experts around—Wilson and people like that—it was set up for use by the organic chemists. They could go in and put their samples in; they didn't have to ask Wilson how to do it. They then took the spectra away, and they could talk to people about the spectra as their own thing. I saw NMR in terms like that, not as the kind of thing that you had to have an expert to carry out for you, and then tell you what the results meant, in your context. I wanted results much faster than I knew they could be gotten that way . . . . So I saw NMR as a technique for organic chemists or chemists in general, to use in their own way.[60]

Roberts intensively used infrared spectroscopy for his research and already supervised an infrared spectrometer at the division level. He envisaged the same for NMR, and convinced some of his colleagues, among others Norman Davidson and Carl Niemann, to assign funds from their ongoing research to the purchase of the instrument.[61] In a memorandum to Caltech's president Lee A. DuBridge, Niemann emphasized the potential of NMR as well as the relatively early stage at which Caltech would enter the field:

> From the results that have been obtained with current installations, which with but two exceptions are in industrial laboratories, it is clear that n-m-r

---

[59] For the beginning of Yost's work on NMR see his annual report to Linus Pauling, 7 July 1950, Yost papers, Caltech Archives, folder 10.6; for the grant of Newmont Exploration see Fred Searls, Jr. to Lee A. DuBridge, 14 September 1954, ibid., folder 2.2.4; for Yost's interest in mathematics see his annual report to Linus Pauling, 7 June 1954, ibid., folder 10.6.

[60] Roberts, interview by Prud'homme, 101. For the chemical physicist E. Bright Wilson, see Reinhardt, "Chemistry in a Physical Mode"; and Kistiakowsky, "Edgar Bright Wilson."

[61] Roberts, interview by Reinhardt, 28 January 1999. Among the colleagues of Roberts, R. M. Badger (physical chemistry), W. Corcoran (chemical engineering), Davidson, Schomaker, and Niemann expressed their interest in the use of NMR.

> spectroscopy, either alone or when coupled with infrared spectroscopy, is capable of solving in an elegant way many of the problems of structural chemistry and chemical reactivity.[62]

Roberts collected $17,500, which was nearly enough for the spectrometer. With reference to a research problem of interest for Pauling (the borderline of resonance and tautomerism), Roberts finally convinced Pauling to apply for extra funds at the board of trustees of Caltech. Surprisingly, the board agreed to fund the full price of the spectrometer, and the extra money could be used for additional equipment and the installation of the instrument.[63]

Varian Associates delivered the instrument in the summer of 1955. The small instruction manual was of no real help in getting the spectrometer started, as was the brief introduction provided by James Shoolery, who installed the machine.[64] By reading Felix Bloch's original articles, George Pake's overview articles in the *American Journal of Physics*, and by listening to a series of talks that Purcell gave at Caltech, Roberts managed to understand the basic principles of NMR and the working of the spectrometer. But because the machine shop of the Caltech chemistry division at that time was not able to give support in electronics, engineers of Varian Associates came down to Pasadena every time a major problem emerged with the machine. In contrast to some other instrument manufacturers, Varian Associates was willing to give this kind of service. For example, Roberts had bought a machine from Applied Physics, founded by Howard Cary, to do carbon-14 analysis. When the machine showed an urgent problem, Cary just informed Roberts that he should work out the problem with the help of the instruction manual:

> He was rather incensed that somebody would use his instrument who really didn't know what it was all about. That kind of attitude was slow in changing. But now people are not expected to really know all about each part of the instrument.[65]

Soon, the service of Varian Associates enabled mainstream organic chemists to participate in NMR studies. While competition with physicists did not play a role for Roberts, because their interests were different from those of the chemists, the chemists entering the new field saw an increased competition among themselves. This had been noted by Gutowksy before, and it proved to be a disadvantage to chemical physicists because of their smaller number and

---

[62] Carl Niemann to L. A. DuBridge, memorandum 24 February 1955, Roberts Papers, folder "Varian correspondence through 1969."
[63] Roberts, *The Right Place at the Right Time*, 153; Roberts, interview by Prud'homme, 103.
[64] For a detailed description of the Varian V-4300b NMR spectrometer and the way of its use see Roberts, *The Right Place at the Right Time*, 156–158.
[65] Roberts, interview by Prud'homme, 103.

lower profile in the area of chemical NMR. With reference to Gutowsky, Roberts remarked:

> We talked occasionally. We were competitors, but friendly competitors. This was an area where there was so much new stuff that we found that everybody was pretty sharing of what they found. But after a while it became pretty competitive. Gutowsky, as I've said got sort of rolled away by the wave of organic chemists rushing into this field.[66]

Roberts knew first hand what he was talking about, because he was one of the main competitors of Gutowsky. The first research problem Roberts was interested in, the structure of diketene, was solved by Gutowsky before Roberts could get started.[67] Roberts heard about Gutowsky's work through the natural product chemist Martin Ettlinger of the Rice Institute at Houston.[68] Ettlinger was a close friend of Roberts's Caltech colleague and friend Buchman. The contact to Ettlinger also proved to be important for the first two problems that Roberts tackled with the help of NMR, the structure of Feist's acid, and the inversion of nitrogen in cyclic imines.

Feist's acid had been synthesized in 1893 by Franz Feist at ETH Zurich.[69] The detailed structure of this dibasic carbon acid with a three-membered ring was a long-disputed question. In the 1920s, the compound attracted the attention of leaders in physical organic chemistry like Christopher Kelk Ingold and Jocelyn Field Thorpe. Their concept of the semi-aromaticity of the compound led to a controversial reply by Feist.[70] Essentially, the unsolved structural problem centered on the position of the double bond, which could be situated inside the ring or at the methylene group adjacent to it (see Figure 4.2). Because of the lability of the derivatives of Feist's acid, chemical methods could not settle this question. When, in the 1950s, physical methods became available, Feist's acid was an ideal target to show the utility of the novel instrumentation. Thus suddenly, in 1956, studies by X-ray diffraction, infrared spectroscopy and NMR appeared in chemical journals. A 1952 article by

[66] Roberts, interview by Reinhardt, 28 January 1999.

[67] Roberts, interview by Reinhardt, 28 January 1999. The structure of diketene was a subject of dispute because the information gained by several instrumental techniques had never been unequivocal. Five or more possible structures were discussed. Foster, "Application of Nuclear Magnetic Resonance," 272–274.

[68] Ettlinger to Roberts, 20 January 1956, Roberts Papers, folder "Ettlinger correspondence." Martin Grossman Ettlinger (b. 1925) took his Ph.D. at Harvard in 1946, and was Frank B. Jewett Fellow at Caltech in 1946–47. He worked on steroids and naphthoquinones. From 1946, Ettlinger had been a member of the society of fellows at Harvard University and it is very well possible that Roberts and Ettlinger met there.

[69] Feist, "Ueber den Abbau des Cumalinringes."

[70] Goss, Ingold, and Thorpe, "Chemistry of the Glutaconic Acids, Part XIV"; and Feist, "Über 3-Methyl-cyclo-propen-1,2-dicarbonsäure."

FIGURE 4.2 Possible structural formulae of Feist's acid. Reprinted with permission from Martin G. Ettlinger and Flynt Kennedy, "The Structure of Feist's Acid and Esters," *Chemistry and Industry* (1956), 166–67, on 167. Copyright *Chemistry and Industry*, 10 March 1956.

Ettlinger preceded this series, strongly arguing in favor of the cyclopropane structure (left in Figure 4.2), which he based on chemical and infrared evidence. Investigations with X-ray and electron diffraction by Douglas Lloyd, T. C. Downie and J. C. Speakman of the University of St. Andrews supported this view.[71]

The cyclopropene structure (right in Figure 4.2) was again put forward by Frank R. Goss (who in the 1920s had co-authored an article with Ingold and Thorpe). But his interpretation of the infrared spectra was regarded by both Roberts and Ettlinger as "foolish."[72] In addition to Roberts and Ettlinger, D. R. Petersen of Caltech and Andrew S. Kende of Harvard University were interested in the problem.[73] Petersen used mainly X-ray diffraction methods, and in 1955 he proved the structure of Feist's acid to be a methylenecyclopropanedicarboxylic acid, i.e. the double bond was outside the ring.

Though the accumulated evidence of Ettlinger and Petersen seemed to be convincing, Roberts—in accordance with his general interest in small ring compounds—set out to establish an unequivocal assignment of the structure with the help of NMR. He asked Ettlinger to supply him with a sample of the compound.[74] Ettlinger agreed to support Roberts, assuming that Roberts would publish only a general account of his NMR work on small ring molecules. Feist's acid itself was regarded by Ettlinger as belonging to his own turf. Roberts and his graduate student Albert Bottini were able to solve the structural

[71] Ettlinger, "Structure of Feist's Methylcyclopropenedicarboxylic Acid"; and Lloyd, Downie and Speakman, "Structure of Feist's Acid."

[72] Ettlinger to Roberts, 14 November 1955, and Roberts to Ettlinger, 22 November 1955, Roberts Papers, folder "Ettlinger correspondence."

[73] D. R. Petersen, Ph.D. thesis Caltech 1955 (worked with J. D. Sturdivant); Boreham, Goss and Minkoff, "Structure of Feist's acid"; Ettlinger and Kennedy, "Structure of Feist's Acid and Esters"; Kende, "Nuclear Magnetic Resonance Spectrum of Feist's Acid"; Petersen, "X-ray Investigation"; Bottini and Roberts, "Nuclear Magnetic Resonance Spectrum of Feist's acid." Ettlinger used NMR equipment at Shell Oil Company in Houston, Texas. Kende had a Varian high resolution spectrometer at his disposal at Harvard. For the use of NMR by Petersen see Roberts to Petersen, 3 January 1956, Roberts Papers, folder "Ettlinger correspondence."

[74] See Ettlinger to Roberts, 14 November and 22 December 1955, Roberts Papers, folder "Ettlinger correspondence."

question decisively in favor of the cyclopropane structure. Moreover, they could show by NMR and infrared spectroscopy that an exchange of hydrogen atoms of the compound with deuterium atoms of the solvent took place. In the opinion of Ettlinger, this was an unexpected and original discovery, and he became especially interested in the mechanism of the exchange process.[75] In addition to NMR, Roberts also presented evidence established by the classical method of ozonolysis of the compound. The mechanism of this degradation reaction that was often used to establish the position of double bonds in molecules was a matter of lengthy discussion between Roberts and Ettlinger in their correspondence.[76] Thus, Feist's acid was a nice little problem for NMR, confirming the already quite clear structure, and leading to new evidence with regard to exchange reactions. It also gave the opportunity to combine "classical" methods, such as ozonolysis, with "modern" ones, e.g. NMR.

In connecting NMR to established modes and procedures in chemistry, Roberts and Ettlinger also conferred about the standardization of NMR. Gutowsky, in 1952, had established the δ values for the numbers of the chemical shifts. The δ values allowed one to compare results independently of the strength of the magnetic fields used.[77] While Ettlinger defended the use of the values in this way, Roberts opposed that and chose to give absolute values of the chemical shifts. His reasoning was that solvent effects would influence the chemical shifts, making a direct comparison of the δ values senseless.[78] More or less, Roberts challenged the validity of a δ value to give a correct judgment about the absence or presence of a functional group in a molecule. Thus, their use for fingerprinting—analogous to wave lengths in infrared spectra—seemed doubtful. Chemical effects had to be taken into account. In February 1956, Roberts wrote to Ettlinger:

> I still feel that delta-values can be red herrings in nuclear magnetic resonance structure determination unless determined by extrapolation to infinite dilution in a common solvent. This is not as important in the type of compounds you are working with—they are all pretty much of a kind. However, we have seen some rather outlandish shifts with more diversified substances and I hesitate to use a delta-value the way one uses a wavelength of absorption in infrared measurements. Excuse this note of caution; it reflects my basic conservatism regarding many instrumental methods, and extends, naturally, to infrared as well.[79]

[75] Ettlinger to Roberts, 22 December 1955 and 20 January 1956, Roberts Papers, folder "Ettlinger correspondence."
[76] See Ettlinger to Roberts, 22 December 1955 and 20 January 1956, Roberts Papers, folder "Ettlinger correspondence." On ozonolysis see Morris, "From Basel to Austin."
[77] See Chapter 2.
[78] Roberts to Ettlinger, 1 February 1956, Ettlinger to Roberts, 20 February 1956, Roberts Papers, folder "Ettlinger correspondence."
[79] Roberts to Ettlinger, 27 February 1956, Roberts Papers, folder "Ettlinger correspondence."

Though Roberts certainly was right in his cautious dealing with the fingerprint use of NMR data, most organic chemists adopted this style of working with the technique. But Roberts himself, as a specialist interested in NMR in its own right, attempted to gain deeper insights into the potential pitfalls and opportunities of NMR.

Also Roberts's "earliest real triumph"[80] in NMR originated in a hint given by Ettlinger. Referring to Roberts's interest in methylene derivatives of small ring compounds, Ettlinger reported on anomalous evidence obtained by infrared spectroscopy about the structure of 1-ethyl-2-methyleneaziridine. In contrast to the published structure,[81] Ettlinger proposed a ring structure containing one nitrogen and two carbon atoms as the basic constituent of the molecule (see Figure 4.3). He encouraged Roberts to investigate the compound by NMR, stating that he by himself would not work on it in the near future.[82] Because of the two possible (cis- or trans-) positions of the two ring-hydrogen atoms with respect to the *N*-ethyl-group, a chemical shift between them should be discernible. At room temperature, only one resonance peak was visible, and Roberts and his graduate student Albert Bottini were able to attribute this to the inversion of the nitrogen-bound ethyl-group, quickly flipping between the two possible positions in space. At minus 80 degrees Celsius, when the inversion was hindered, two resonance lines could be seen.[83] This work was the beginning of a series of studies by Roberts and his collaborators on imine nitrogen inversion.[84] Bottini, who was an NSF predoctoral fellow at Caltech from 1954 to 1957, went for postdoc research to the University of Basle, Switzerland, and later joined the faculty of the department of chemistry of the University of California at Davis.

Together with the Indian scientist P. Madhavan Nair, who had obtained his Ph.D. from the University of Arkansas and had an Arthur A. Noyes Fellowship at Caltech in 1956–57, Roberts began to investigate the kind of problems he later regarded as the most important of his early work in NMR: Rotation about single bonds, and the effects of molecular asymmetry on the spectra observed.[85] He first did so with fluorine derivatives of relatively simple

---

[80] Roberts, *The Right Place at the Right Time*, 159.

[81] Pollard and Parcell, "Synthesis of N-allylidene-alkylamines."

[82] Ettlinger to Roberts, 14 November and 22 December 1955, Roberts Papers, folder "Ettlinger correspondence."

[83] In the beginning, both Roberts and Ettlinger erroneously had expected a splitting of the double bond methylene protons, and not the two ring protons as it turned out to be. Ettlinger to Roberts, 20 January 1956 and Roberts to Ettlinger, 27 February 1956, Roberts Papers, folder "Ettlinger correspondence."

[84] The first studies were Bottini and Roberts, "Nitrogen Inversion"; and Bottini and Roberts, "Nuclear Magnetic Resonance Spectra."

[85] For a detailed description of this work see Roberts, *The Right Place at the Right Time*, 162–167; Roberts, interview by Prud'homme, 105–106; and Nair and Roberts, "Nuclear Magnetic Resonance Spectra."

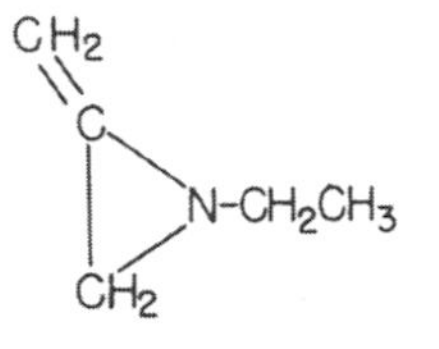

FIGURE 4.3 Structural formula of 1-ethyl-2-methyleneaziridine. Reprinted with permission from Albert T. Bottini and John D. Roberts, "The Nitrogen Inversion Frequency in Cyclic Amines," *Journal of the American Chemical Society* 78 (1956), 5126. Copyright, American Chemical Society.

alkanes, the fluorohaloethanes, mainly because fluorine chemical shifts were easier to observe than proton chemical shifts. Fluorine derivatives were difficult to handle, but after Roberts—through his consultancy at Du Pont—had heard of an efficient method to introduce two fluorine atoms at one carbon atom (using sulfur tetrafluoride), he had the possibility of exploring this area at hand.[86] The main results of these studies, which Roberts's group continued until the late 1960s, were that the populations of the several possible conformers were not equal and that NMR could be a suitable method for determining both the rates of rotation and the population distributions. Thus, Roberts connected NMR to a burgeoning field of research in organic chemistry: conformational analysis. Doubts arose because of his use of fluorine compounds, and many organic chemists questioned the validity of the results for hydrogen-containing substances (the main target for organic chemists): "A lot of what we did, then, I guess wasn't popular with the organic chemists, because we were using the fluorine as a tracer to find out what would happen if the fluorines weren't there."[87]

For the studies on nitrogen inversion, a device for temperature control would have been preferable, but the Varian Associates V 4300 spectrometer only came with a Dewar that could be heated or cooled outside the instrument. Already in his letter of inquiry about the cost of the Varian Associates instrument, Roberts had asked about the possibilities of including a more elaborate temperature control.[88] In the months after delivery, Roberts set out for his first improvement of the apparatus. Together with James Shoolery of the applications laboratory of Varian Associates, he designed a temperature control device

[86] Roberts, interview by Prud'homme, 106. John J. Drysdale and Phillips of the Experimental Station of Du Pont at Wilmington, Delaware, embarked on very similar studies. See Drysdale and Phillips, "Restricted Rotation." Drysdale had been a student of Roberts at MIT.
[87] Roberts, interview by Prud'homme, 106.
[88] See Emery Rogers to Roberts, 18 January 1955, Roberts Papers, folder "Varian correspondence through 1969."

(first only for heating). Though he co-authored the publication,[89] Varian Associates did not offer to share the patent that the company filed. The contribution of Roberts mainly consisted of the design of the glassware that was part of the device.[90] Nevertheless, for routine service, even electronics did not constitute an insurmountable hurdle for Roberts. Carl Niemann had previously introduced him to the use of Heathkits, produced by the Heath Company for the building of electronic instruments.[91] As both Niemann and Roberts were interested in manual work, but found it hard to take interruptions from laboratory work, they switched to the building of electronic equipment. For Roberts, this was a helpful exercise in the light of his later experience with NMR, as the early Heathkits consisted of an arrangement of vacuum tubes in steel chassis, and this gave Roberts some confidence for doing simple repairs on his first NMR machines.[92]

An important asset of Roberts was his ability to synthesize the necessary compounds for innovative and meaningful research. Moreover, he was carefully considering the complex chemical issues that played a role in the interpretation of the spectra. Because of this, and of the fact that he was the first university-based organic chemist who used Varian Associates NMR equipment, he became a preferred customer. A symbiotic relationship that lasted over more than a decade was at its beginning.

User and Manufacturer Intertwined

Varian Associates needed this kind of partner in academic chemistry. Founded in 1948 by the brothers Russell and Sigurd Varian, and—among others—the Stanford physicists William Hansen, Edward Ginzton, and Leonard Schiff, the company's expertise heavily centered on physics and electronic engineering. In many ways, Varian Associates can be regarded to be an off-shoot of the physics department of Stanford University. But, as historians Christophe Lécuyer and Timothy Lenoir have shown, the technology transfer between university and corporate enterprise was not unidirectional. Varian Associates contributed tremendously to the academic culture at Stanford. Many Stanford

---

[89] Shoolery and Roberts, "High Resolution Nuclear Magnetic Resonance Spectroscopy."

[90] Roberts, interview by Reinhardt, 28 January 1999. See Roberts to Shoolery, 11 January 1956, Roberts Papers, folder "Varian correspondence through 1969."

[91] Founded in the 1930s, the Heath Company first manufactured kits for airplanes. During World War II, they moved into electronics, and established after the war a business in electronic kits, based on war surplus equipment. The company's products were relatively inexpensive, and the manuals gave detailed and reliable descriptions for assembling the kits. I thank Kristen Haring for information on this topic.

[92] Roberts, *The Right Place at the Right Time*, 147–148.

physicists aimed to escape the pressures of nuclear particle physics, a direction of research that was enforced by the military-funded style of research at leading American universities. Varian Associates promised to retain the goal-oriented, but nevertheless creative, research atmosphere that many of its founders had experienced during the war. Moreover, the microwave physicists at Stanford University had gotten more and more into a dependent service position with regard to high-energy physics. For the upkeep of their intellectual and social independence, they either had to convert to nuclear physics research itself or to build up a new environment of their own. It was the corporation of Varian Associates, and not the academic institution of Stanford, that allowed them to achieve this goal.[93] Key products for the company in its early years were klystrons, vacuum tubes for the generation of microwave radiation that were based on a design that Hansen and Russell Varian had invented in 1937. These tubes were used for navigation and guidance of aircraft and missiles. Essentially, in its first twenty years, Varian Associates was a research and development company that relied heavily on orders from the military. With excellent links to Stanford University, Varian Associates was a prime example of the plans of Stanford provost Frederick Terman to establish a closely knit relationship between academic and industrial cultures in Stanford's vicinity.

Though he did not belong to the circle of founders of Varian Associates, Felix Bloch, co-inventor of NMR and a theoretical physicist at Stanford, supported the goals of the company. On the advice of Hansen (who joined Bloch for the work leading to the first observation of NMR in bulk matter in early 1946), Bloch filed a patent. He licensed it exclusively to Varian Associates after the company was founded. In the first years, Varian Associates developed NMR as a tool in geology and military surveillance, the technique proving its utility in exploration and the finding of submarines.

The title of Bloch's patent already suggested chemical applications, but these were different from the direction that NMR later took in organic chemistry. In 1946, Russell Varian considered NMR a method to identify trace elements and isotopes in minute amounts of material. Crucial for development was funding from the Office of Naval Research (ONR), supporting Varian Associates' investigations as well as research undertaken at universities. Thus, government money was indispensable in two ways: Scientists at Varian Associates were paid through government contracts, and the grants to universities "allowed normal market forces to shape the development of NMR."[94]

For Bloch, the laboratories of Varian Associates became an extension of his own sphere. Many of his students wanted to stay in California and—lacking

---

[93] Lenoir and Lécuyer, "Instrument Makers and Discipline Builders," 287.

[94] Martin Packard, "The Varian story. As presented at the 1980 Pittsburgh conference on analytical chemistry," p. 9, Varian Associates Papers, box 4, folder 23.

academic opportunities—welcomed the opportunity to continue their work in the setting of a favorable corporate environment. Initially, Bloch had followed up the discovery of NMR for use in the precise measurement of nuclear magnetic moments, with substantial funding from the Research Corporation and (from 1947 on) from the ONR. Members of his laboratory were among the first to explore the application of NMR for organic chemistry, tackling the phenomena of chemical shift and spin-spin coupling. Thus, through contacts with Bloch, Varian Associates got access to first-hand experience. Along with Martin Packard, Weston Anderson, and James Arnold, important members of Bloch's staff joined Varian Associates in the early-to-mid-1950s.

For Varian Associates, chemical applications of NMR provided a market that had the potential to counterbalance the firm's one-sided dependence on government funding. In 1953, the company was still quite small, with a turnover of a little more than $5 million. R&D efforts comprised three percent of sales, a little less than $160,000. In 1968, after the introduction of several product lines in NMR and EPR spectroscopy, and the acquisition of other instrument manufacturers,[95] sales were more than $170 million, while corporate R&D rose to eight and a half percent (14.5 million).[96] Already in 1957, with regard to the diffusion of infrared spectroscopy in hundreds of chemical laboratories, the commercial future of NMR seemed promising. An auditing company praised the prospects of Varian Associates shares, as NMR complemented existing techniques and thus would find a prepared market.[97] Early customers of Varian Associates' high resolution NMR equipment were oil companies; the first was either Magnolia Petroleum of Dallas, Texas (Magnolia was a predecessor of Mobil Oil Co.) or Phillips Petroleum, joined in 1953 by Humble Oil in Baytown, Texas (later Exxon), Shell Development Company in Emeryville, California, and the chemical company Du Pont of Wilmington, Delaware.[98] The early sales of NMR spectrometers thus depended on the attitudes of the instrumentation laboratories of large enterprises:

---

[95] Varian Associates bought the business of Aerograph in liquid and gas chromatography in 1965, Cary's line of spectrophotometers in 1966, MAT (Mess- und Analysentechnik, Bremen, Germany, formerly Atlas-Werke, a specialist in mass spectrometry) in 1967, and Techtron Pzy., Ltd. (Melbourne, Australia) with atomic absorption spectrometers in 1967. For a while, Varian Associates manufactured a nearly complete product line of analytical instrumentation. *Varian 25 years,* special issue of the Varian Associates Magazine, April 1973, pp. 6, 39, 41. Varian Associates Papers, box 14, folder 4.

[96] Numbers from Lenoir and Lécuyer, "Instrument Makers and Instrument Builders," 304, table 1.

[97] Lenoir and Lécuyer, "Instrument Makers and Instrument Builders," 310–311.

[98] Ralph Kane, interview by Sharon Mercer, November 1989, Varian Associates Oral History, SC M 708, box 1; Woessner, "Early Days of NMR"; and Ferguson, "William D. Phillips."

> It was adventuresome for a fledgling company to try to create a viable business from a concept so esoteric and untried as NMR. And it would not have come off had it not been for the backing of some equally adventuresome chemists at such places as Shell Development Company, Du Pont, Humble Oil, and Bayer-Leverkusen in Germany. For the early instruments really did not produce that much valuable data—"you could smell the cork on the bottle and almost make the same analysis," says Martin Packard.[99]

At that time, Varian Associates created an application laboratory, with the Caltech Ph.D. James N. Shoolery as its head. Shoolery had been a student of Norman Davidson, a chemist who used physical instrumentation for unraveling questions in molecular biology. Moreover, Shoolery had worked with Yost on NMR and microwave spectroscopy, and thus was intimately familiar with the technology. Shoolery was fascinated by the prospects of NMR, and he proposed to Varian Associates to build up a laboratory for the development and demonstration of chemical applications of NMR.[100] As a chemist with research experience, Shoolery knew about the needs of the chemical community. The application laboratory was essentially a marketing tool, creating and fostering potential applications of the technique, and connecting the company to its customers. Shoolery, who in 1950 had been ranked "the best experimental man that has come to our department" by Yost,[101] provided badly needed expertise. Because the staff of Varian Associates consisted mainly of physicists and engineers, they were eager to incorporate parts of the culture of their most important customers, the chemists. As Martin Packard, a former student of Bloch and top manager of Varian Associates, put it in 1980:

> One of the challenges in running our instrument business was to provide for communications between the customer, who was usually a chemist, and a physicist who understood the concepts of NMR, and the engineer who had to translate the ideas of the physicist and chemist into designs which could be manufactured. I found that it was relatively easy to learn each other's language and jargon, but very difficult to bridge the cultural gap and to recognize that the physicist and chemist approach problems in quite different ways.[102]

---

[99] Varian Associates, *An Early History*, n. d., pp. 9–10, Varian Associates Papers, box 14, folder 4.

[100] James N. Shoolery, interview by Sharon Mercer, April 1990, p. 2, Varian Associates Oral History, SC M 708, box 1.

[101] Yost also appreciated the theoretical knowledge of Shoolery highly, comparing him to the performance of E. Bright Wilson. See Yost to Eugene Rochow, 16 November 1950, Yost Papers, Caltech Archives, folder 11.11.

[102] Martin Packard, "The Varian story. As presented at the 1980 Pittsburgh conference on analytical chemistry," pp. 15–16, Varian Associates Papers, box 4, folder 23.

The extent of the gap between chemists and physicists is highlighted by the statement of Packard that he never really believed the chemical structure of ethanol before he could observe the evidence brought about by the physical technique of NMR.[103] The application laboratory of Varian Associates soon started major marketing efforts. From 1954, a series of advertisements on the back cover of the *Journal of the American Chemical Society* appeared on a monthly schedule. Three years later, Varian Associates began organizing annual workshops, providing training and sales information for a community of academic and industrial researchers. Needless to say, Varian Associates was present at the major instrument fairs in the country. Physicists like Packard saw the diffusion of knowledge as a linear process, directed from the physicists and engineers of Varian Associates to the customers. The interface was the application laboratory:

> The NMR business was, and still is, characterized by close couplings and interchange between the people who make the machines and the people who use the machines. We expected, and had, scientist-to-scientist interaction. Jim Shoolery and other people could talk with the users in the language which they could understand. We would make some little improvements, and then the user would discover what you could do with that improvement.[104]

Packard did not mention the crucial contributions by customers in academic and industrial physics and chemistry. In technical terms, the users primarily wanted three things: Higher magnetic field strength, greater resolution, and better time stability. For dealing with this, Varian Associates relied on both in-house expertise and hints and ideas from the outside academic community. Varian Associates engineer Forrest Nelson designed better pole faces for homogeneous magnetic fields, and physicists James Arnold and Weston Anderson developed Bloch's idea of the spinning sample technique. While this enhanced the resolution of NMR greatly, the flux or "super" stabilizer allowed the measurement of spectra over a much enlarged time period. The idea of the super stabilizer was brought into Varian Associates via one of their "peers," Charles Reilly of Shell Development.[105] The improvement originated in England, and was first applied in the United States by Joseph Lloyd of Shell Development in Houston.[106] Thus, even with regard to the purely technical development of NMR hardware, the company partially relied on outside experience. In effect,

---

[103] *Varian 25 years*, special issue of the Varian Associates Magazine, April 1973, p. 20, Varian Associates Papers, box 14, folder 4.

[104] Martin Packard, interview by Sharon Mercer, December 1989, Varian Associates Oral History, SC M 708, box 1.

[105] Packard, interview by Mercer, December 1989.

[106] Martin Packard, "The Varian story. As presented at the 1980 Pittsburgh conference on analytical chemistry," p. 15, Varian Associates Papers, box 4, folder 23.

innovation and marketing were intertwined features of both communities, the academic "user" and the corporate "innovator."[107]

Varian Associates staff developed quite elaborate relationships with the academic world. In the 1950s and early 1960s, the management allowed employees to publish and to participate in academic affairs, and it also provided the opportunity for academics to join the company for a limited amount of time:

> We had a number of visiting post-docs, and we also had a program of summer professors, where we brought in people from the academic community to work with us for the summer. Now obviously, this kind of program was extremely important for both parties. It allowed us to identify outstanding scientific talent, and it allowed the fellows to decide that maybe Varian was a good place to work, and in addition the contact was stimulating to our scientists.[108]

The logical extension of this step was the support of academic chemists at their own institutions. Seemingly, no active policy of Varian Associates existed for this strategy, but some university scientists who early on used Varian Associates equipment were very important for the breakthrough of the technique in mainstream organic chemistry. These chemists were very interested in novel equipment for the advance of their research projects and careers. It is important to note that this kind of interaction was not formalized. Though some scientists received special treatment, there were no guarantees, and decisions for the development of special instrumentation were made on a case-by-case basis. Moreover, only indirect financial interests were involved: the scientists were treated as customers, and did not receive revenues. It should be kept in mind that because the instruments for lead users were prototypes, specified by the user and designed by the company, the instrument manufacturer did carry a large part of the development costs. Royalties did not play a role, at least in the case of John Roberts. The profits for Roberts were in enhanced scientific productivity, the opening of new and exclusive research fields, and the income he created by his publication activities on the novel uses of NMR.

After delivery of Roberts's instrument in the summer of 1955, Roberts was both excited and dismayed about the fast pace of Varian Associates' innovation of novel equipment. His own instrument had already been upgraded to a resonance frequency of 40 MHz, improving on the 30 MHz of the first generation. At the end of 1955, Varian Associates announced the introduction of a control device to stabilize the magnetic field, the super-stabilizer. The quality of the spectra was improved substantially, and Roberts had to buy the additional equipment to stay in a competitive position. The relatively high price of

[107] See von Hippel, *Sources of Innovation.*

[108] Packard, interview by Mercer, December 1989.

the device (ca. $1,500), and the fact that he was forced to modernize equipment that was less than a half year old, caused him and the Caltech administration to question the sales policy of Varian Associates. Shoolery defended the pricing and innovation procedure of his employer, emphasizing that they had made great efforts to ensure the compatibility of improvements with older instrumentation.[109] Later, Varian Associates referred to this system as the "Living Instrument" policy, enabling users of older instruments to modernize their equipment with prepackaged kits.[110]

Roberts's concern was directed to potential customers of Varian Associates at universities. Department heads would especially resent long-term financial obligations. Roberts reported to Shoolery that the question he most often heard was about maintenance. Though Roberts gave Varian Associates "superb marks" for efficient service, he was worried that the machine did need a lot of attention by qualified personnel:

> You can see why the general reaction of a department head is one of 'How can we afford it?' The other reaction which follows this one immediately is, 'Let's wait for a few years.' . . . My concern is with respect to some of your future and potential customers, many of whom come to me for advice and all of whom want to know how much it's going to cost for maintenance and improvements.[111]

Roberts reassured Shoolery that he by himself would stay in the field of NMR: "I'll grumble and complain, but after all I am using the equipment and must have it in perfect condition to be able to compete with other researchers in the field."[112] Before he received the upgrade, Roberts had the chance to run samples on the prototype that was installed at Varian Associates. In February 1956, after a visit to Palo Alto, he reported that "the spin-spin splitting problem seems more and more interesting and I can hardly wait until we can do high resolution work here."[113] In response, Shoolery let Roberts know that he would obtain the improved capabilities at Caltech soon: "Let me assure you that we

---

[109] Roberts to Shoolery, 3 January, and Shoolery to Roberts, 6 January 1956, Roberts Papers, folder "Varian correspondence through 1969."

[110] Varian Instrument Division, announcement of upgrading possibilities with field/frequency control for the HR-60, DP-60, and HP-100 models, 24 February 1964, Roberts Papers, folder "Varian correspondence through 1969."

[111] Roberts to Shoolery, 11 January 1956, Roberts Papers, folder "Varian correspondence through 1969."

[112] Roberts to Shoolery, 11 January 1956, Roberts Papers, folder "Varian correspondence through 1969."

[113] Roberts to Shoolery, 1 February 1956, Roberts Papers, folder "Varian correspondence through 1969."

are most anxious to see your stability brought to this point in the shortest possible time."[114]

The relative geographical proximity of Caltech to Varian Associates proved to be advantageous for Roberts. The company offered to break the announced chronological order of installations of the super stabilizer with respect to the purchase dates for the benefit of having a test site for the first installation at hand.[115] Roberts was delighted with the prospect of receiving serial number one, though he later offered to wait in case Varian Associates' decision should create ill feelings among other customers:[116]

> Naturally, we were very pleased to learn that you wanted a Super Stabilizer tested by a highly vocal customer in a favorable geographical location. We shall certainly be pleased to give you the benefit of our experiences.[117]

The super stabilizer was indeed installed first at Roberts's laboratory (see Figure 4.4).

In the following years, the relation of Roberts and Varian Associates exceeded the type of contacts normally to be expected between user and manufacturer. For example, Shoolery, in accordance with his participation in the scientific community, received one of the first papers of Roberts on NMR for review. Roberts had already handed him the manuscript for comment, as he was quoting unpublished observations of Shoolery. The Varian Associates scientist favorably commented on the paper of his customer and scientific colleague, and he asked for only minor revisions.[118] But soon the friendly coexistence was disturbed, exactly because of the active scientific role of Shoolery. A friend and colleague of Roberts, Max T. Rogers of Michigan State University, collaborated with Varian Associates on a scientific topic. According to Rogers, the Varian Associates scientists published the results of this investigation without his consent, though he alone had initiated this research. As a result of this ill feeling, Rogers pulled back from further cooperation with Varian Associates in another promising field. Roberts took this example as a warning to Shoolery about the consequences of an aggressive scientific strategy of

---

[114] Shoolery to Roberts, 6 January 1956, Roberts Papers, folder "Varian correspondence through 1969."

[115] W. C. Dersch (Service Manager) to Roberts, 15 February 1956, Roberts Papers, folder "Varian correspondence through 1969."

[116] Roberts to Dersch, 24 February 1956, Roberts Papers, folder "Varian correspondence through 1969."

[117] Roberts to Shoolery, 24 February 1956, Roberts Papers, folder "Varian correspondence through 1969."

[118] Roberts to Shoolery, 21 June and 20 July, Shoolery to Roberts, 6 July 1956, Roberts Papers, folder "Varian correspondence through 1969."

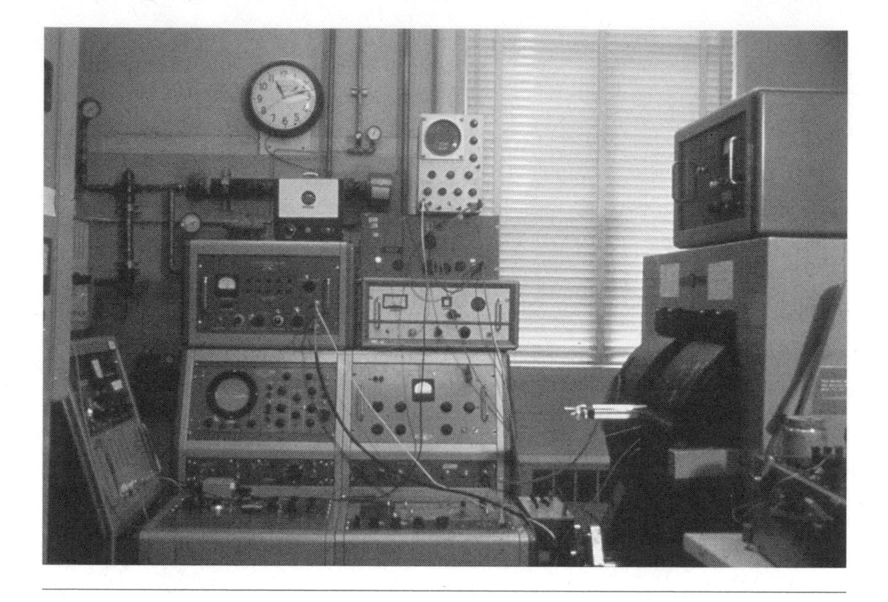

FIGURE 4.4 Roberts's Varian Associates V-4300b NMR spectrometer, ca. 1960. Note all the added peripherals that improved the performance of the spectrometer, the chaotic origins of Varian Associates "Living Instrument" policy of improvements that were compatible to older instrument versions. The super stabilizer is on top of the magnet to the right. Courtesy of John D. Roberts.

Varian Associates, pondering on his own role between the company and his fellow-scientists:

> I am sure that you must realize that unless you wish to do chemistry yourself and compete on all facets of a research problem with the outside world you might get yourself shut out of some very interesting problems if word of this sort gets around. Your function has got to be to help people get started, not to finish their problems, if you want to continue to make sales of your machines and share other people ideas for use of n-m-r. I don't know why I keep getting cast in the role of a hair shirt for Varian. I don't particular like the job and sometimes a nice quiet field like dipole moments where no commercial machines are available sounds attractive.[119]

It is not known how this problem was settled. Obviously, Rogers and Shoolery resumed their cooperation, in 1958 publishing a classic paper on the additivity

[119] Roberts to Shoolery, 2 November 1956 (?), Roberts Papers, folder "Varian correspondence through 1969." Emphasis in the original.

of chemical shifts in steroids.[120] In the meantime, Shoolery seemed to be more cautious with his publishing of results that were close to the interests of his customers. For example, in the summer of 1957 he considered the publishing of effects of molecular asymmetry on the appearance of NMR spectra in the "NMR at work" series of Varian Associates (see Figure 4.5). He planned to use results obtained by his own research but knew about a paper of Roberts on the same subject. Shoolery was anxious to assure him of the good effects his advertisement would have on Roberts's article:

> Although this advertisement may appear about the same time or slightly before your paper, I am sure that it will draw attention to the paper and not detract from it since I have specifically mentioned that you have pointed out the effect in a private communication to be published.[121]

As this incident of intertwined relationships shows, the impact of Varian Associates and possibly other instrument manufacturers was very strong. With its series "This is NMR at work," Varian Associates did more than launch a simple marketing effort. The company set the standards of scientific validity and shifted the values for the future publications of outside scientists. In the opinion of one of the early users of NMR, Edwin Becker, scientists awaited the Varian Associates advertisements "as eagerly as they anticipated research articles."[122] While parts of the advertisements were in the hands of the marketing people, Shoolery insisted that the text and design of the scientific part that showed the spectrum and explained its interpretation was to be under his control only. He wanted it to be "exactly as if it had been submitted to a scientific journal," and was aware of the fact that he enjoyed a privilege "not having to deal with referees."[123] While this was an advantage for rapid publication, advertisements like these also influenced scientific judgments. Varian Associates became involved in many stages of science, not just at the instrument supply side.

The "NMR at Work" series appeared in the 1950s and early 1960s on a monthly schedule on the back cover of the most widely read chemical journal in the United States, the *Journal of the American Chemical Society*.[124] Reprinted in the Varian Associates Technical Information Bulletin, the series had a tremendous impact on chemical research and teaching. While the first issues of the

---

[120] See Bhacca and Williams, *Applications of NMR Spectroscopy in Organic Chemistry*, 14; Becker, Fisk, and Khetrapal, "Development of NMR," 23–24; and Shoolery and Rogers, "Nuclear Magnetic Resonance Spectra of Steroids."

[121] Shoolery to Roberts, 18 June 1957, Roberts Papers, folder "Varian correspondence through 1969."

[122] Becker, Fisk, and Khetrapal, "Development of NMR," 23.

[123] Shoolery, "NMR Spectroscopy in the Beginning," 734 A and 736 A.

[124] Shoolery, interview by Reinhardt, 23 January 1999. For the role of the "NMR at Work" series in the community of organic chemists see Roberts, "Instruments and Domains of Knowledge."

40 of a series EFFECT OF MOLECULAR ASYMMETRY

INTERPRETATION. It has been pointed out by Professor J. D. Roberts[1] that in certain cases the chemical shift between like nuclei attached to the same carbon atom is not necessarily averaged out by rapid rotation. In molecules of the type $R\text{-}C(H)(H)\text{-}C(R_1)(R_3)\text{-}R_2$ in which R_1, R_2, and R_3 are different there may be a chemical shift between the two protons if the residence times of the molecule in the three rotational conformations are not equal. An example of such a molecule[2] is shown below. Its proton spectrum exhibits 12 lines which are assigned in the coupling diagram as arising from 3 non-equivalent protons with coupled spins. The spectrum is independent of temperature over a considerable range which suggests that this model more correctly describes the molecule than one involving a large potential barrier constraining the molecule to one conformation.

(1) PRIVATE COMMUNICATION. TO BE PUBLISHED
(2) THE SAMPLE WAS FURNISHED THROUGH THE KIND COOPERATION OF THE WRIGHT AIR DEVELOPMENT CENTER

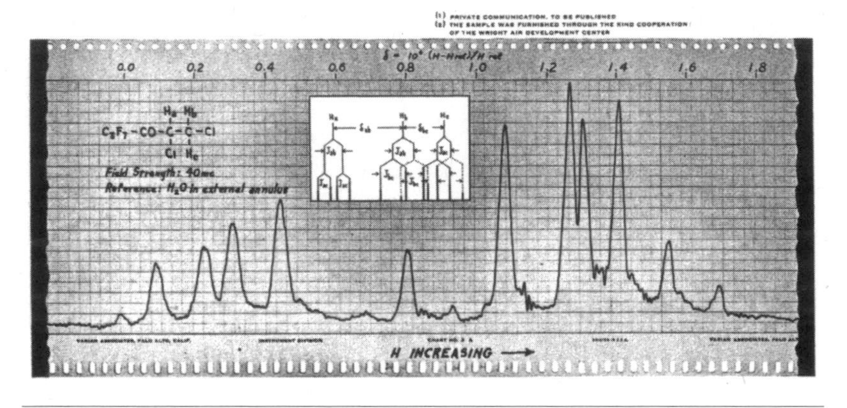

FIGURE 4.5 Varian Associates' advertisement from the "This is NMR at work" series, 1957/58. It shows a new NMR effect, the non-canceling of the chemical shift through rotation of molecular groups. NMR could thus be used to measure the rotational frequency. Besides the didactic and marketing functions of the advertisement, it also acted as a publication of the method. Note that priority was given to John D. Roberts (see text). Reprinted with permission from Varian Associates, *Technical Information Bulletin*, 2, no. 2 (1958), 10. Copy in Roberts Papers, folder "Chemistry 246b."

series were of a largely educational value, real research problems, often tackled in cooperation with academic chemists, began to appear soon. (See Figure 4.5.)

In addition to this informal way of publishing, Shoolery used the normal academic procedure to disseminate his views. Between 1953 and 1959, he published 31 papers in scientific journals, 11 of which he co-authored with university-based chemists, and 12 with scientists at governmental and industrial research institutions. For many of these articles, Shoolery undertook the NMR measurements only. But occasionally he was first, and quite often second author, indicating the importance of his contributions. In addition, Shoolery published three reviews and five instrument-related research articles, most of the latter in cooperation with colleagues at Varian Associates. His network of co-authors was large, and with none of them did he write more than three articles. His journal of choice was the chemical journal with the most general audience, the *Journal of the American Chemical Society* (15 articles).[125]

[125] This paragraph is based on a search done with SciFinder Scholar.

The excellent equipment available to the scientists at Varian Associates provided a unique opportunity to set new records of achievement. Shoolery, though he did so often in the background, was able to influence the quality standards that counted in the small community of NMR scientists.[126] In the 1958 issue of the Varian Associates Technical Bulletin, the green paper mimicked the sports sections in American newspapers. Furthermore, it showed the very best spectrum of acetaldehyde obtained so far. The acetaldehyde spectrum was often used as a standard for the performance of the spectrometer.[127] Consequently, Ray Freeman, then a postdoc in Paris, regarded this as a challenge to the rest of the NMR world. He did not have a chance to follow up, using a self-built 30 MHz spectrometer.[128] Later, Freeman joined the staff of Varian Associates. Varian Associates became a major player in the scientific field. But it was playing by different rules than other participants.

Community and Standardization

In the late 1950s, the community of NMR users became large and coherent enough to foster the need for enhanced and rapid communication. The impetus came from Aksel A. Bothner-By and Bernard L. Shapiro at the Mellon Institute in Pittsburgh, Pennsylvania. In October 1958, they founded the *Monthly Ecumenical Letters from Laboratories of N-M-R* (MELLONMR). MELLONMR was supposed to facilitate the exchange of ideas and information among users of NMR applications in organic chemistry.[129] The newsletter was a very informal undertaking, and Shapiro's and Bothner-By's roles did not involve any editing since original letters were just reproduced and distributed. Moreover, it was strictly forbidden to quote from the newsletter; it had no official place in the public sphere. Only by direct arrangement with the author could the information be cited as private communication. The letters published in the newsletter comprised all aspects of NMR, theoretical, experimental, and instrumental. The subscribers were supposed to contribute regularly to its contents, and the subscription was discontinued if they failed to do so. In October 1958, 32 scientists received the newsletter, mainly organic

---

[126] Compare this with the influence that RCA of Camden, New Jersey, the manufacturer of electron microscopes, had on standards in biological electron microscopy. See Rasmussen, *Picture Control*, chap. 1.

[127] See Chapman and Magnus, *Introduction to Practical High Resolution Nuclear Magnetic Resonance Spectroscopy*, 36–39.

[128] Freeman to Shapiro, 13 August 1990, *TAMU NMR Newsletter* 384 (1990), 15 (Special Section on Shoolery's retirement), copy in Roberts Papers.

[129] Shapiro, "The NMR Newsletter." For a similar newsletter established in the community of Drosophila scientists see Kohler, *Lords of the Fly*, 162–167.

chemists from U.S. universities. After two years of publication, in September 1960, there were 71 subscribers.[130] In March 1964, after Shapiro moved to the Illinois Institute of Technology in Chicago, the newsletter was renamed accordingly, and in January 1967, 260 people received copies, whose production and distribution was funded entirely by the Illinois Institute of Technology.[131] Shapiro's newsletter held together the relatively small, but rapidly growing, community of scientists interested in improving organic chemical applications of NMR. Next to standardization, issues such as new technological and methodical developments played an important role in its pages. For example, methods that enhanced the sensitivity of NMR, such as the computer of average transients (CAT), or entirely novel kinds of applications, such as Zeugmatography, were reported first in the newsletter.[132] No wonder that Shapiro had to remind readers repeatedly that they were not allowed to quote from it. The privacy of the newsletter guaranteed fast and relatively open exchange of information, and it certainly helped the NMR community to define its boundaries.

Among the hottest topics debated in the NMR community of the 1950s and 1960s was the issue of standardization. In order to guarantee comparability of results, both presentation and conditions of measurement of NMR spectra required the agreement on standards. The first to recognize this were Charles A. Reilly of Shell Development and Herbert S. Gutowsky. In March 1955, Gutowsky arranged for an exploratory meeting of interested scientists. The meeting, with Reilly as chairman, was held at Pittsburgh during the conference on analytical chemistry and applied spectroscopy (Pittcon), under the auspices of the American Institute of Petroleum (API) Project No. 44. The API had a long-lasting history in setting standards for spectroscopic data and had done so successfully in mass spectrometry, infrared, and ultraviolet spectroscopy. The participants of this meeting comprised some 14 scientists, with all but two coming from a corporate and military environment, mostly oil and chemical companies.[133] The issues of debate were manifold. At its center were

---

[130] See the mailing lists in *MELLONMR* 1 (October 1958), and *MELLONMR* 24 (September 1960), 11–13, copies in Roberts Papers.

[131] Bernard L. Shapiro, "Policies and Practical Consideration," *IITNMRN* 66 (March 1964); and Shapiro, "Concerning the Future of the [IIT] NMR Newsletter," *IITNMRN* 100 (January 1967), copies in Ernst Papers.

[132] Oleg Jardetzky to Bernard L. Shapiro, 21 August 1962, *MELLONMR* 47 (August 1962), 21–23; Paul Lauterbur, "NMR Zeugmatography," *TAMUNMR Newsletter* 175 (April 1973), 34–35, copies in Ernst Papers. Lauterbur received the Nobel Prize in Medicine (2003) for this technique.

[133] The meeting took place on 2 March 1955 at the William Penn Hotel (the venue of the Pittcon). As it is of interest to know the early users of NMR spectrometers, the names of the participants and their organizations are given here in full: Edward D. Baker (Dow Chemical Company), Harold J. Bernstein (National Research Council, Ottawa, Canada), Norman D. Coggeshall (Gulf Research

the definition of the parameter of the chemical shift and the type of reference compound to be used. Though the group did not reach a unanimous solution to the questions, they did agree that the API Project No. 44 should be informed on the settlement of key points. In a talk at the advisory committee meeting of the API on 8 November 1955, Reilly reported on the state of agreement reached and argued that it was still premature to give more than tentative standards. He named the rapid development of the technique, the still incomplete understanding of the spectra, and the expansion of NMR to other fields of application as reasons for his hesitation. Nevertheless, Reilly was convinced of the necessity of an agreement on a primary reference compound (he preferred water), and for a definition of the chemical shift parameter. He also discussed the type of data that should be given, and the usefulness of different recording systems.[134]

Soon other players appeared on the field, most importantly the editors of MELLONMR, and for a while this newsletter provided the main stage for the exchange of opinions on standardization. In the first issue, and with "considerable trepidation,"[135] Bothner-By and Shapiro discussed the subject of standardization of spectra. In 1958, George Van Dyke Tiers of the Minnesota Mining and Manufacturing (3 M) Company proposed a new system to represent the values of the most important parameter in NMR spectra, the chemical shift, and introduced a new symbol for these values, τ (tau).[136] The new symbol stood in close relation to the reference substance tetramethylsilane (TMS) that Tiers had introduced. Tiers's proposal led to considerable turmoil among the community of NMR spectroscopists during the 1960s, because it threatened the so-called δ-values, established in the early 1950s by Gutowsky.

In the beginning of the debate, Varian Associates carefully avoided taking sides officially, though Varian Associates scientists were not happy about the foothold that Tiers's system had gained. Shoolery expressed the opinion "that as a representative of an instrument company I have to be particularly careful about taking too strong a stand in any direction concerning definitions, symbols, and procedures. After all, it is our job to make instruments which perform

---

and Development Company), Walter Dietz and B. F. Dudenbostel, Jr. (Esso Research and Engineering Company), Harlan Foster (Du Pont Company), Jerome Goldenson (Army Chemical Center, Maryland), Herbert S. Gutowsky (University of Illinois), Paul C. Lauterbur (Army Chemical Center, Maryland), D. E. O'Reilly (Gulf Research and Development Company), Charles A. Reilly (Shell Development Company), Emery Rogers and James N. Shoolery (Varian Associates), R. B. Williams (Humble Oil and Refining Company). Minutes of exploratory meeting on presentation of high resolution NMR, 2 March 1955, compiled by Emery Rogers, Roberts Papers, folder "Varian correspondence through 1969."

[134] Charles A. Reilly, memorandum on presentation of high resolution NMR spectra, 27 January 1956, Roberts Papers, folder "Varian correspondence through 1969."

[135] *MELLONMR* 1 (October 1958), 4, copy in Roberts Papers.

[136] Tiers, "Proton Nuclear Resonance Spectroscopy. I."

in a manner desired by the majority of users."[137] Nevertheless, he added a lengthy statement of his preferred system to the MELLONMR newsletter of April 1959.[138] Tiers's reply was a "pocket guide" for journal referees, establishing his own requirements for papers on NMR. The great advantage of Tiers was his use of TMS, the most convenient internal reference compound available.[139] But in Shoolery's opinion, Tiers's pocket guide would take away all freedom of judgment from journal referees.[140] In April 1960, Varian Associates employees Shoolery, LeRoy Johnson, and Weston Anderson proposed a compromise between their original system and Tiers's, and in May, Varian Associates planned to publish an atlas of spectra, to emphasize it's "past, present and (hopefully) future position in the field."[141] The management pressed forward with a publication date as early as possible, in order to influence the community of NMR users against the Tiers system. But Roberts warned Shoolery about taking a too competitive position. At the same time, he voted for a publication of Varian Associates' spectra atlas with the publishing house of William Benjamin (in which he himself had a business interest):

> I also feel that Benjamin could provide you with much wider distribution and of course reduce the cost to Varian thereby. One thing I would worry about is a corporate image of Varian whereby the company is in scientific competition with its customers. I believe this would not be in the Company's best interests.[142]

Thus, Varian Associates took sides in a conflict inside the community of users. The scientists jealously defended their right to decide on standards on their own, in informally organized circles. Many of them distrusted a dictate by a manufacturer who at that time actually enjoyed a monopoly in the NMR spectrometer market. But the split in the community between followers of the τ-system and defenders of the δ-values enabled Varian Associates to decisively shape the system. Varian Associates published the spectra catalog on their own, in two

---

[137] Shoolery to Roberts, 8 July 1959, Roberts Papers, folder "Varian correspondence through 1969."

[138] Shoolery, 7 April 1959, *MELLONMR* 7 (1959), 6–7, copy in Roberts Papers.

[139] TMS gives a single sharp signal in a region where only a few other kinds of proton resonances appear. Until today, TMS is the reference substance in NMR spectra.

[140] George V. D. Tiers, "The 3M handy pocket guide for journal referees," *MELLONMR* 10 (1959), 2; and James N. Shoolery, 13 August 1959, *MELLONMR* 11 (1959), 8–9, copies in Roberts Papers.

[141] Shoolery to Roberts, 19 May 1961, Roberts Papers, folder "Varian correspondence through 1969."

[142] Roberts to Shoolery, 1 June 1961, Roberts Papers, folder "Varian correspondence through 1969."

volumes in 1962 and 1963. In the introduction, they again carefully explained their decision for the δ-scale.[143] In the end, the δ-values prevailed, but in the meantime often only τ, or both values, were given in the spectra charts. The "victory" of the δ-system was due to the considerable influence that Varian Associates exerted through the introduction of their widely used spectrometer A-60. The A-60 plotted the spectra on calibrated chart paper, using the δ-system. With the A-60, Varian Associates brought NMR within the reach of the average organic chemist, allowing routine use at a reasonable price. The main features of the A-60 were its easiness of operation and the high reproducibility of the spectra. This helped to win the confidence of the chemists, who did not have to consider themselves on "shaky ground," as they had before. The results of the A-60 became trustworthy results, without the reliance on human experts: "A scientific instrument that won't produce the same results twice without an expert tuning it up is no good because you just don't have that many experts around."[144] This allowed NMR experts to use their own instruments more for research, while service work was assigned to the A-60. Consequently, the A-60 became the workhorse of NMR in organic chemistry.[145]

In the 1950s and 1960s, the relationship of Roberts and Varian Associates continued to be fruitful for both sides. For example, Roberts retained first priority in obtaining an upgrade for his machine to the higher frequency of 60 MHz in January of 1958.[146] In 1961, Roberts was among the first to buy the A-60. Varian Associates decided to build on the success of the A-60, and in intervals of five years introduced two successors, the T-60 (with a permanent magnet), and the EM 360, both catering for the low-end side of the market.[147] During this time, Roberts stayed in close contact with the company. In October 1964, he was the only invited speaker from outside Varian Associates at an advanced NMR-EPR workshop. During this workshop, Roberts gave a talk on aspects of application, counterbalancing the heavily instrument-centered contributions of the Varian Associates people.[148]

---

[143] Varian Associates, *High Resolution NMR Spectra Catalog*, iii–iv. See also Shoolery, Johnson, and Anderson, 22 April 1960, *MELLONMR* 19 (1960), 2–3, copy in Roberts Papers; Becker, Fisk, and Khetrapal, "Development of NMR," 36.

[144] Shoolery, interview by Mercer, April 1990, 10.

[145] For the A-60 see Lenoir and Lécuyer, "Instrument Builders and Discipline Makers."

[146] Robert C. Jones of Varian to Roberts, 28 January 1958, Roberts Papers, folder "Varian correspondence through 1969."

[147] Shoolery, interview by Mercer, April 1990, 11.

[148] The other speakers included Harry Weaver on super-conducting solenoids, Forrest Nelson on field-frequency control, Arden Sher on spin-echo techniques, Ray Freeman on double resonance, and Richard Ernst on sensitivity enhancement. Ray Freeman to Roberts, 3 August 1965, and agenda for the conference, Roberts Papers, folder "Varian correspondence through 1969."

Making NMR Teachable

In the preface of his 1959 book on NMR, Roberts apologized for the inclusion of quite many examples of his own research by referring to the fact that "it is always easiest to write about what one knows best."[149] In contrast to other scientists who often regarded the authoring of textbooks as intellectually worthless, and therefore shied away from the writing of books, Roberts had an original interest to make NMR popular by teaching it.[150] This was not a simple matter of advertisement for his own performance. Furthermore, it resembled a complex marketing strategy in a scientists' market, where consumers and providers of scientific information interchanged roles continuously. Those scientists who originated novel lines of inquiry were highly regarded in the scientific community, and the writing of textbooks was one of the possible methods for their proliferation, educating the next generation of users. In NMR, the first books for organic chemists appeared in 1959, a decade after the principal usefulness of the technique had been indicated.

Before 1959, organic chemists had to rely on articles and books written for an audience of physical chemists and physicists. Thus, early users highly valued lectures that made the complex subject approachable. From 1956 on, Roberts was among the most prominent lecturers, at Caltech and at universities throughout the country. He started to give lectures on NMR at Caltech at a time when he was still at the very beginning of his own research. In 1956, only one textbook of NMR existed at all, and clarification of principles and usefulness was badly needed. Thus, Roberts designed his Caltech lectures

> to cover the parts of the basic theory necessary for an organic chemist to use the NMR instrument for practical purposes with reasonable intelligence. No attempt will be made to develop a mathematical basis for resonance absorption—the viewpoint throughout will be qualitative and where desirable much oversimplified (i.e. suitable for electronic morons).[151]

Despite this introduction, Roberts provided his audience with a profound background on the principles of NMR. Beginning with an explanation of the apparatus, the lecture continued with its operation, the origins of the proton signal, the chemical shift, and spin-spin splitting. Applications included the proof of chemical structure (Roberts chose here the example of Feist's acid, one of his first investigations using NMR), electronic effects of functional groups in molecules, and the study of reaction rates. In Roberts's view, the advantages of

---

[149] Roberts, *Nuclear Magnetic Resonance*, v.

[150] Roberts once referred to Woodward and Bartlett, who refused the writing of textbooks for that reason. Roberts, interview by Prud'homme, 79.

[151] Roberts, "Chemistry 246b—Nuclear Magnetic Resonance," manuscript, not dated [1956], p. 1, Roberts Papers.

the technique lay in the speed of analyses that could be made without destroying the sample, in the versatility of information obtained, and in the possibility of using impure samples. The list of disadvantages, however, was long: The equipment was expensive, needed a skilled operator and considerable maintenance; it lacked sensitivity when compared with infrared spectroscopy; and restrictions were quite large, with access to liquid samples and compounds of moderate molecular weights only.

Roberts based his lecture on the available review articles in the field, and he made use of Varian Associates' *Technical Information Bulletin* that had appeared annually since 1953.[152] Moreover, he amply illustrated his manuscript with spectra run on his machine. Practical aspects gained prominence in the description of operation. Roberts described the calibration of spectra, and different possibilities for using a standard reference compound. Rigor was applied when the validity of NMR data was in doubt, and Roberts carefully made clear how pitfalls could be avoided. The greater part of his lecture Roberts reserved for explaining the various applications of NMR in organic chemistry. He used the recent research literature, made clear what could be done with NMR, and very often pointed to doubts, gaps, and unresolved problems. This lecture aimed at the advanced graduate student, with experience in organic chemistry and preferably other spectroscopic techniques.

In the following years, Roberts's goal was to make NMR approachable "for the common man"[153] in organic chemistry. On the basis of the Caltech course, he gave approximately 40 lectures at universities and companies all over the United States, including a sensational success with a lecture given at a conference on reaction mechanisms in New York in the fall of 1958.[154] Roberts put great emphasis on the use of self-drawn color illustrations presented as slides, and considered writing a book for organic chemists about the uses of NMR. At this time, William A. Benjamin, chemistry editor of the publisher McGraw-Hill of New York, saw McGraw-Hill being superseded in organic chemistry by competitors. In order to come back into the field, he set up a series in advanced chemistry that focused on burgeoning subfields. Among other fields, the series covered the two most important novel physical techniques in chemistry, NMR and mass spectrometry, by two books each. One was a fundamental and general treatment, the other concentrated on the largest market: organic chemistry.[155]

---

[152] Roberts cited articles of Pake, "Fundamentals"; Wertz, "Nuclear and Electronic Spin Magnetic Resonance"; and Smith, "Nuclear Magnetic Resonance Absorption."

[153] Roberts, interview by Prud'homme, 77.

[154] Roberts, *The Right Place at the Right Time*, 169–170.

[155] MacDowell, *Mass Spectrometry*; Biemann, *Mass Spectrometry, Organic Chemical Applications*; Pople, Schneider, and Bernstein, *High-resolution Nuclear Magnetic Resonance*; and Roberts's own *Nuclear Magnetic Resonance.*

Planned as a sequel to the more advanced treatises available, Roberts wrote a book along similar lines as he had planned for his Caltech course on NMR. In summary, he intended to persuade the readers of the usefulness of NMR for chemical studies and to provide them with the necessary practical and theoretical background to engage in studies of their own. In the preface, Roberts aligned the success story of NMR in chemistry with that of another technique, gas-liquid chromatography. The latter became ubiquitous in chemistry laboratories in the short time span of a decade, and, according to Roberts, NMR could be the next technique having such a spectacular success. The hook for the organic chemist reading the introduction of Roberts's book consisted—after a very brief demonstration of the principles of NMR—in a proton NMR spectrum of *N*-ethylethylenimine (modern name 1-ethylaziridine), a compound that Roberts had investigated in his own research. Without further ado, Roberts postulated great uses of the NMR spectrum for the unraveling of the compound's structure and the rates of rotation and inversion of groups in the molecule: "Clearly, the NMR spectrum of the compound is a veritable treasure trove of useful information not easily obtainable in any other way."[156] Hoping that the interest of his readership was raised with this direct statement of the potential of NMR, Roberts set out to explain its instrumental and physical principles. Color illustrations had a major role in this endeavor. Roberts's book was the first advanced chemistry book that appeared in four color print. As it is shown in this block diagram (Figure 4.6) of an NMR spectrometer, the colors facilitated the clarification of the interplay of magnetic field, radiofrequency transmitter and receiver. Here, Roberts simplified the instrument, neglecting the details of the electronic set-up. Emphasized and shown in a larger size, out of proportion to the real dimensions, was the test tube containing the sample. The chemical compound was at the core for the chemist approaching NMR.[157]

The shift between chemistry and physics can be seen in the figure explaining the magnetic forces in an NMR experiment. In Figure 4.7, showing the set-up before the start of the experiment, the sample was shown "chemically" in a test tube. In the drawing at the bottom, describing the resonance conditions and the origins of the signal received, the test tube disappeared. What was shown was a representation of the physical forces, here the precession of the magnetic vectors of the nuclei. The transfer of physical explanations to the chemists' world-view was enhanced by replacing chemical symbols by physical ones, given in a simplified manner as representations of complex effects that were not explained.

---

[156] Roberts, *Nuclear Magnetic Resonance*, 3.

[157] The style of Roberts's figure resembles closely the drawing of a simplified NMR apparatus given by Varian Associates in *Technical Information from the Laboratories of Varian Associates*, 1,1 (1953), p. 1. Here, the dimensions of test tube and spectrometer were even more shifted towards emphasizing the test tube. Copy in Roberts Papers.

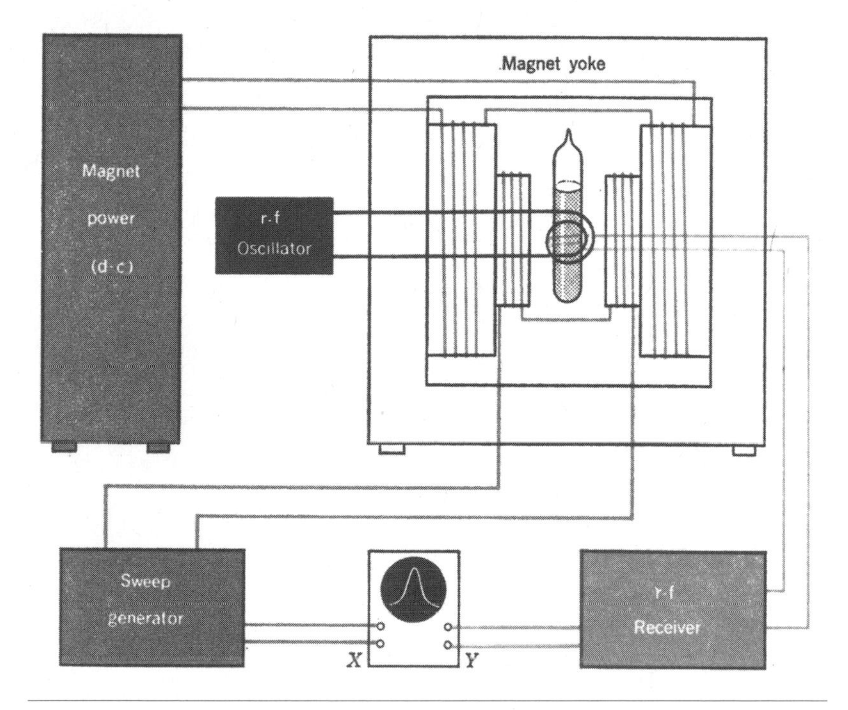

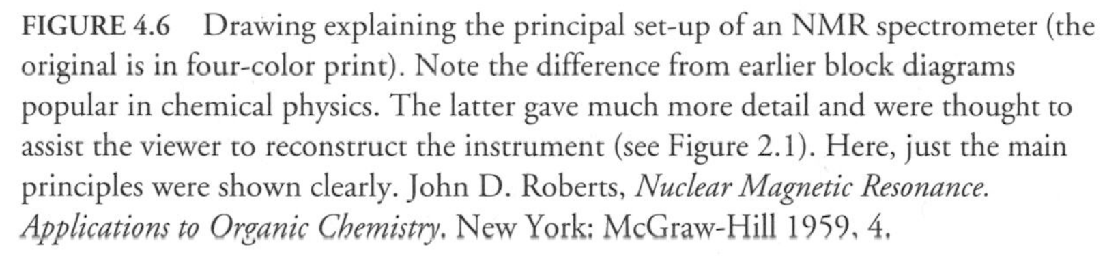
FIGURE 4.6 Drawing explaining the principal set-up of an NMR spectrometer (the original is in four-color print). Note the difference from earlier block diagrams popular in chemical physics. The latter gave much more detail and were thought to assist the viewer to reconstruct the instrument (see Figure 2.1). Here, just the main principles were shown clearly. John D. Roberts, *Nuclear Magnetic Resonance. Applications to Organic Chemistry*, New York: McGraw-Hill 1959, 4.

Roberts himself expressed the opinion that this use of color illustrations was important for the great success of the book, overcoming the first reservations of reviewers that he was not an NMR specialist.[158] The use of color was reserved for the parts explaining the physical principles of NMR and the set-up of the apparatus, the larger part of the 100 pages systematically dealing with the effects that made the use of NMR in chemistry possible: chemical shift, spin-spin splitting, and methods to solve problems in reaction kinetics, exchange processes, and rotation rates. Though Roberts introduced NMR in non-mathematical terms, an appendix described the Bloch equations for the explanation of the line shapes in nuclear resonance. With this, Roberts hoped to explain the

[158] Roberts, interview by Prud'homme, 78.

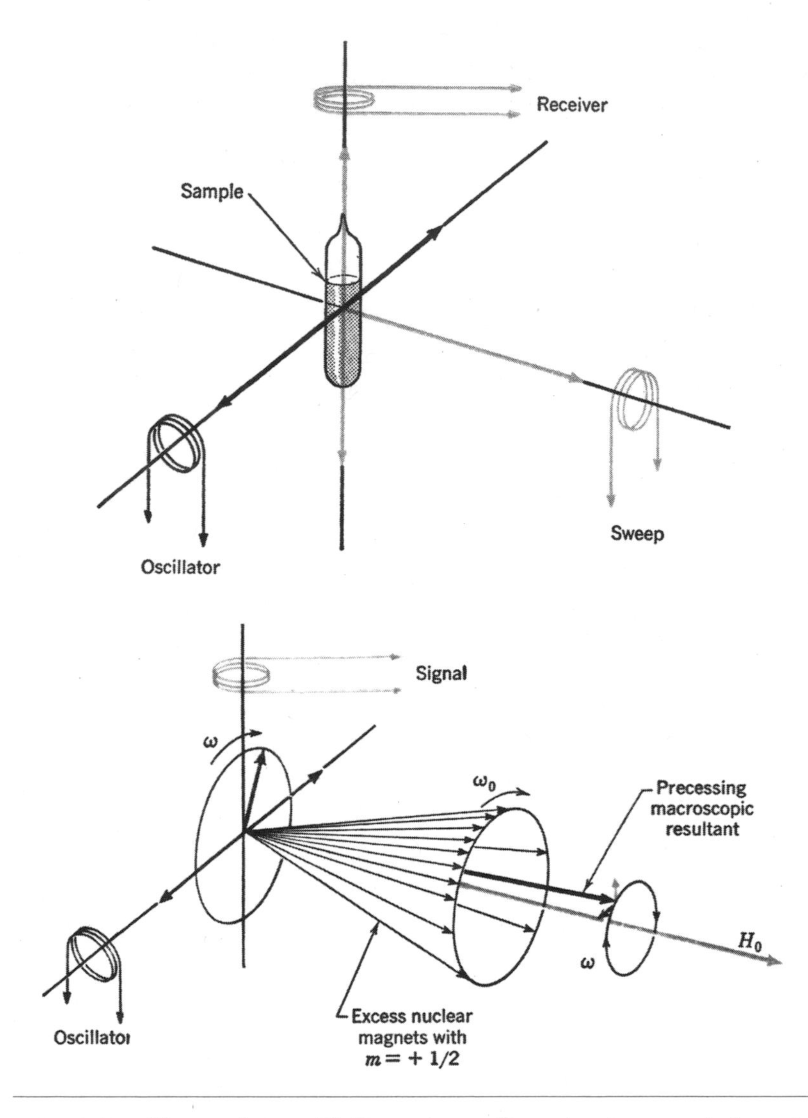

FIGURE 4.7 The sample in an NMR experiment, illustrating the magnetic forces. In the upper figure, the sample was shown "chemically" in a test tube. The lower figure shows the resonance condition "physically" with a vector. John D. Roberts, *Nuclear Magnetic Resonance. Applications to Organic Chemistry*. New York: McGraw-Hill 1959, 14, 16.

origins of the signals obtained, and to clarify the work modes of the spectrometer. A second appendix gave problem sets with empirical formulae and spectra of unknown compounds and was meant to initiate the student into the practice of solving structural problems. Varian Associates used a similar approach in its campaign, explaining the features and modes of operation of their NMR spectrometers in a *Technical Information Bulletin*. For the purpose of convincing potential customers of the usefulness of NMR, the Bulletin included a large selection of spectra and their application to the unraveling of chemical structures.[159] These spectra were assembled from the series *This is NMR at Work*, and reprinted in the Bulletin.

In many respects, Roberts's *Nuclear Magnetic Resonance. Applications to Organic Chemistry* resembled Linus Pauling's famous book *The Nature of the Chemical Bond*.[160] Both treated the relationship of chemistry and physics, and in both chemistry came first and physics second. The authors were aware of potential pitfalls of a complete subordination of chemistry to general physical theory. For example, Roberts emphasized that chemical experience was needed for the correct interpretation of NMR spectra. While Pauling's book concentrated on the principles of chemistry, Roberts's focused on a new experimental technique. Both authors felt the challenge of communicating to an audience of chemists the difficulties of physical theory entrenched in mathematical terms, and both turned to pictorial representations for clarification. As with some early textbooks in quantum chemistry, the early books on NMR shared the approach of creating a certain degree of autonomy of the new field. For that, the consolidation of a peculiar language and a body of theoretical and practical knowledge was mandatory. Here, textbooks of NMR had the same role as textbooks of new scientific disciplines.[161]

Roberts's treatises were among the early textbooks on NMR. They had their unique style, aiming at the middle between a thorough quantum-mechanical treatment and a simple set-up of rules for the interpretation of spectra. Both of these extremes existed. The first textbook on NMR was Edward Raymond Andrew's *Nuclear Magnetic Resonance*, published in 1955.[162] Andrew was introduced to NMR by the physicist Edward Purcell of Harvard, and he later specialized in solid state NMR. His treatise was a general one, and not restricted to uses in organic chemistry. Theory and application of high resolution NMR (which was at that time mainly in organic chemistry) were the focus of the theoretical chemists John Pople, William G. Schneider, and Harold J.

[159] See *Technical Information from the Laboratories of Varian Associates*, 1,1 (1953), copy in Roberts Papers.

[160] For Pauling's book see Nye, "From Student to Teacher."

[161] Gavroglu and Simões, "One Face or Many?"

[162] Andrew, *Nuclear Magnetic Resonance*.

Bernstein in a classic 1959 book.[163] The volume was highly mathematical, and certainly not intended to be used by the average organic chemist. It appeared in the same series and in the same year as Roberts's book. In many ways, Roberts's much shorter treatise complemented Pople's, Schneider's and Bernstein's. Between these two was Lloyd Jackman's *Applications of Nuclear Magnetic Resonance Spectroscopy in Organic Chemistry*, also published in 1959. Though it aimed to be comprehensive and fundamental, it did not demand mathematical knowledge.

A quite extreme example of a pragmatic direction in the teaching of NMR for organic chemists is the book by Roy H. Bible of Searle & Co., published in 1965. Bible intended his book to teach organic chemists how to interpret spectra and to make available the necessary reference materials. Thus, he presented conclusions that were drawn from a theoretical treatment, but he did not present the theory itself. He expressed the opinion that theory was more easily understood after the empirical correlations were mastered:

> Although the early workers in nuclear magnetic resonance placed the subject on a firm theoretical basis, they simultaneously placed a number of stumbling blocks, such as electric quadrupole moments and diamagnetic susceptibilities, in the path of the average organic chemist. Many organic chemists effectively utilize IR and UV spectroscopy without being concerned [about], and in some cases without knowing, the true nature of the particular transition which is involved. It is the purpose of this chapter to show how NMR spectra can be used in the same fashion.[164]

In accordance with this argument, Bible published a second book with exercises on the interpretation of spectra which consisted solely of spectra, formulae of compounds, and questions related to the interpretation of the spectra.[165]

With respect to Roberts's first book, there is no doubt that it was written from the perspective of an organic chemist who wanted to demystify the quantum mechanical foundations of NMR, and to rigorously apply the technique

---

[163] Pople, Schneider, and Bernstein, *High-resolution Nuclear Magnetic Resonance*. In many ways a sequel to this book was Emsley, Feeney, and Sutcliffe, *High Resolution Nuclear Magnetic Resonance Spectroscopy*. In addition, a huge number of articles on NMR appeared in volumes on analysis in organic chemistry. Examples include Foster, "Application of Nuclear Magnetic Resonance Spectroscopy"; Jackman, "Magnetic Resonance Spectroscopy."

[164] Bible, *Interpretation of NMR Spectra*, 7. Another example of this approach is the volume of M. Flett of ICI, England. Flett, who was member of a service laboratory of ICI in Blackwell, in 1962 authored an overview volume of physical techniques in organic chemistry. In his opinion, the chemist should know the kinds of problems that could be tackled by physical methods, and he should be informed about the requirements they posed for the supply of the sample material. A detailed knowledge of theory and the function of the instrument was not asked for. Flett, *Physical Aids to the Organic Chemist*, 2–3.

[165] Bible, *Guide to the NMR Empirical Method*.

to chemical ends. The question of how much mathematics was needed to apply NMR reliably led to his second book about NMR, *An Introduction to the Analysis of Spin-spin Splitting*, published in 1961.[166] Here, Roberts developed a strategy that was similar to that of the quantum chemist George Wheland when he made a plea for mathematical rigor in the teaching of quantum mechanics.[167] In NMR, Roberts chose a middle way between a purely qualitative and the mathematical approach. For him, the late 1950s and early 1960s were ideal for the writing of textbooks, a period before he had learned too much on his own and lost his missionary zeal. Moreover, textbook publishing was a fast-growing industry. Clearly recognizing the market potential of chemistry books, Benjamin decided to leave McGraw-Hill and to start his own company, W. A. Benjamin Inc. of New York. In 1960, Roberts became a member of the board, and—together with Benjamin and a lawyer—a promoter of the firm.[168] In 1963, when he became chairman of the Caltech chemistry division, Roberts resigned from his position on the board of directors of Benjamin Inc., but stayed with the company as consultant.[169]

Roberts's second book on NMR catered to the needs of the newly founded publishing house of Benjamin. Though Roberts and Benjamin succeeded in collecting a substantial list of high-level authors in chemistry and physics (including the famous Caltech physicist Richard Feynman), Benjamin was anxious to get a book out soon after the founding of his company. Thus, his first commissioned manuscript became Roberts's book on spin-spin splitting. Roberts himself considered his work an expansion of ten pages in the highly theoretical book of Pople, Bernstein, and Schneider on high resolution NMR.[170] His interest in the more complicated aspects of spin-spin splitting was caused by observing irregular phenomena that were not interpretable by the simple qualitative rules that were standard for interpretation by organic chemists.[171] But prediction and explanation of more complicated spectra

---

[166] Roberts, *Introduction to the Analysis of Spin-spin Splitting.*

[167] See Gavroglu and Simões, "One Face or Many?" 436. The book is Wheland, *Theory of Resonance and Its Application to Organic Chemistry.*

[168] Roberts, interview by Prud'homme, 79.

[169] Another reason to abandon this position was related to financial matters. Roberts, interview by Prud'homme, 84.

[170] Roberts, interview by Prud'homme, 80–81.

[171] Until the end of the 1950s, most chemists analyzed the splitting pattern by simply following the n+1 rule. For example, the signal of a nucleus with spin ½, coupled to n equivalent nuclei with spin ½, splits into n+1 peaks. The intensities of the lines follow the binomial equation. But this applied to "weak" couplings only. If the couplings were "strong" (the value of the splitting close to that of the chemical shift) a much more complex pattern resulted. From early on, physicists were aware of this problem. In order to facilitate the interpretation of such complex spectra, Pople, Bernstein, and Schneider in their 1959 book introduced a letter notation. Letters close to each other in the alphabet (for example AB) denoted a strongly coupled system; while letters distant in the alphabet (AX)

needed the consideration of quantitative aspects of quantum mechanics. On the basis of his experience in molecular orbital calculations, and with initial explanations of his Caltech colleague, the chemical physicist Harden McConnell, Roberts decided to write a book about these problems.[172] In its preface, Roberts lamented the current lack of knowledge of quantum mechanics among chemistry students. He blamed students and teachers alike: The teachers for their failure to explain step by step the necessary mathematical methods, the students for their reluctance to work through problem sets and equations. He believed that chemists could use NMR with a much greater degree of sophistication after having gained at least a rudimentary knowledge of the quantitative theory of spin-spin splitting.[173] His new book was supposed to achieve this. While treatises by molecular spectroscopists aimed at a community of specialists and were not accessible to the average organic chemist,[174] contributions such as Lloyd Jackman's volume and Roberts's own first book were kept entirely qualitative and descriptive. Needed was a thorough introduction that used accessible mathematics and provided a clear explanation. Roberts was convinced that this could be achieved quite easily, against all odds:

> To learn to apply the methods of wave mechanics in a practical way seems to involve a quantum-like transition with a relatively low transition probability. Chemistry students by the thousands are exposed to the principles and jargon of wave mechanics and are able to talk in a most knowing way about orbitals, overlap, spin, etc. But very few of these students can set about to make any sort of an actual calculation of resonance energies of conjugated systems or the energy levels of nuclear spin systems, and this despite the fact that the mathematics involved, although perhaps tedious, usually does not require more than college algebra.[175]

---

stood for weakly coupled nuclei. Spectra were classified along these lines, and chemists could rely on rules of interpretation which were rooted in the symmetry considerations of quantum mechanics, even if they did not understand the quantum mechanical treatment. Thus, the solution of structural problems with the help of spectroscopic methods resembled in many ways that of the solution of an integral. The researcher followed certain rules, and applied them to his needs. The background of how these rules came into existence and to what reality they referred was important only to the specialist. See Becker, Fisk, and Khetrapal, "Development of NMR," 25–27.

[172] Roberts, *The Right Place at the Right Time*, 172–173. In the preface of his *Introduction to the Analysis of Spin-spin Splitting*, Roberts explicitly acknowledged the help of McConnell and Vernon Schomaker in explaining quantum mechanical principles.

[173] See Roberts, "Applications of Nuclear Magnetic Resonance Spectroscopy" (review of Jackman's book).

[174] Roberts referred here to a publication by scientists at Shell Development Corporation in Emeryville that explains the complex spin-spin coupling patterns: McConnell, MacLean, and Reilly, "Analysis of Spin-spin multiplets."

[175] Roberts, *Introduction to the Analysis of Spin-spin Splitting*, v.

Also NMR spectroscopists Jackman and Sever Sternhell pointed to pitfalls that resulted from an incomplete understanding of the limits of NMR by some organic chemists. But theirs was another style with regard to the importance of fundamental theory. Because of the high resolution that could be achieved with the improved instrumentation of the mid-1960s, users were tempted to rely on small changes of resonance values, not taking into account the complexity of factors leading to the peaks. This had to be avoided. In contrast, "almost universal misconceptions" of the theoretical background of NMR itself were not detrimental, because this seldom led to misinterpretations of spectra.[176] Thus, the chemists only had to learn rules, and they had to be prepared thoroughly to judge the cases in which these rules applied. The general theory of NMR was not so important. The emergence of computer programs that reduced the need for understanding high-level mathematics made this task easier.[177] As an outcome, and contrary to Roberts's inclination, chemists were able to stick to their familiar strategy of trial and error in the elucidation of chemical structures, while they established the correspondence between calculated and experimental spectra. Roberts's second NMR book, though it sold several thousand copies, was not as successful as the first. Roberts assigned this to organic chemists' lack of interest in truly understanding the phenomenon of spin-spin splitting.[178] Perhaps it was the advent of the computer that prevented the explicit incorporation of more complex quantum mechanics into the community of mainstream organic chemists.

Roberts's involvement with Benjamin deeply shaped his later inroads into the publishing of textbooks, most notably his and Marjorie Caserio's book *Basic Principles of Organic Chemistry* (1964). The inclusion of a chapter on spectroscopic methods in a general textbook of organic chemistry was a controversial issue. Some reviewers liked it, some recommended to add it to the appendix, and others thought that it should be excluded from the text. Though Roberts and Caserio did not insist that this chapter had to be included in the teaching of elementary chemistry, theirs was a strong plea for the early use and teaching of spectroscopy: "Despite the qualms of the older generation in this respect, it really is the proper thing to take spectra before determining the melting point of a new compound—vastly more information can be obtained thereby."[179]

---

[176] Jackman and Sternhell, *Applications of Nuclear Magnetic Resonance Spectroscopy*, xii.

[177] Roberts mentioned programs by Wiberg, Bothner-By and Swalen (*The Right Place at the Right Time*, 172). See Wiberg and Nist, *Interpretation of NMR Spectra*; Bothner-By and Castellano, "LAOCN 3"; Swalen and Reilly, "Analysis of Complex NMR Spectra."

[178] Roberts, *The Right Place at the Right Time*, 174.

[179] Roberts and Caserio, *Basic Principles of Organic Chemistry*, ix. An abbreviated version of the book was published in 1967: Roberts and Caserio, Modern Organic Chemistry.

In the years to come, more and more specialized treatises appeared, dealing just with subfields of NMR. Most of these books covered experimental techniques, for example NMR of special nuclei (carbon-13, nitrogen-15). Some concentrated on special compound classes. The work by Norman S. Bhacca and Dudley H. Williams on NMR of steroids is a good example. Bhacca's and William's book aimed to give the basic knowledge necessary to do structural assignments on the basis of NMR, and it complemented books by Williams, Herbert Budzikiewicz and Carl Djerassi on mass spectrometry dealing with the same class of compounds.[180] While Williams worked in Djerassi's group for a couple of years, Bhacca was an employee of Varian Associates. This is another example of the complex interplay of instrument manufacturers and their chemists-customers.

DEVELOPMENT OF CARBON-13 AND NITROGEN-15 NMR

Through his drive to develop special and exclusive instrumentation, Roberts connected the granting agencies with the instrument builders. In the mid-1950s, the National Science Foundation funded a research program for Roberts on the study of small ring compounds and reaction mechanisms. At the same time, a small grant from the Office of Naval Research supported the beginning of Roberts's NMR projects (the instrument had been paid for by Caltech). In the years to come, Roberts's NSF project rapidly moved into the realm of NMR, first emphasizing conformational studies of relatively small organic compounds (1956 to the mid-1960s), then moving on to the development of the special techniques of carbon-13 (1966 to the mid-1970s) and finally working on nitrogen-15 NMR (1974 to the mid-1980s). The National Institutes of Health (NIH) contributed funds in cases where Roberts's research was of value for biological and medical purposes.[181] During these years, Roberts's primary aim was to develop novel kinds of instrumentation. In a report to NIH, he emphasized usability in biological fields:

> A most significant part of the research effort has been to specify and bring into being NMR instrumentation which, at the time of each production, was not only unique but involved state-of-the-art components. The major elements of the instrumentation so developed are now part of all of the current commercial models of ^{13}C and ^{15}N spectrometers. In our view, even if nothing else were accomplished, the conceptualization and commissioning of

---

[180] Bhacca and Williams, *Applications of NMR Spectroscopy in Organic Chemistry.*

[181] See the summary in John D. Roberts, "Final project report," 9 December 1983, NSF CHE81-20508, period 1 May 1979–31 October 1983, Roberts Papers, folder "NSF."

> these instruments has led to a revolution in the use of NMR in biology and biochemistry.[182]

Roberts's goal was to develop instrumentation suitable for later use by a majority of scientists in the field. With the spread of the instrument, Roberts's research was appreciated and valued by members of the scientific community. But when a certain method was widely used, it lost the appeal of novelty for Roberts. Specialists in NMR were always searching for cutting-edge technology to develop new methods. In the mid-1960s, Roberts's studies on proton and fluorine NMR came to an end. Hundreds of chemists used the methods originally developed by Roberts, Gutowsky, Shoolery, and other pioneers. Conformational studies, a specialty of Roberts, had become routine. In search of a new research project, Roberts decided to continue to focus on NMR, but to shift to the use of other isotopes. The elements carbon and nitrogen are, with hydrogen and oxygen, the most widespread elements present in organic compounds. While the abundant isotopes of carbon and nitrogen, carbon-12 and nitrogen-14, do not give a signal in the NMR experiment, the less abundant isotopes, carbon-13 and nitrogen-15 do. The problem was the low ratio of carbon-13 to carbon-12 (and nitrogen-15 to nitrogen-14) which led to a dramatic decrease in sensitivity. Though one could work with carbon-13 and nitrogen-15 enriched compounds, this was not an ideal solution to the problem. To be of real use for mainstream organic chemists and biochemists, carbon-13 and nitrogen-15 NMR had to be run at the natural abundance level.[183]

In the early-to-mid-1960s, several options existed for enhancement of sensitivity. For carbon-13 NMR, Paul Lauterbur of SUNY at Stony Brook developed a technique that involved a fast sweep through the magnetic field. Roberts, to his great disappointment, could not get Lauterbur's technique to work in his own laboratory. Furthermore, it had the disadvantage of a relatively low resolution. At the same time, computers began to appear on the scene, at first used for signal averaging. With the help of a Computer of Average Transients (CAT), data obtained by a huge number of observations could be added to each other. The random noise signals canceled each other out, while the coherent resonance signals did not. CAT required a very stable spectrometer, because the repetition of the sweeps took a lot of time. Therefore, although Roberts obtained funds from NIH to buy such a CAT, this did not solve the

---

[182] John D. Roberts, report of the NIH grant GM 11072-19, "Nitrogen-15 and carbon-13 NMR spectroscopy," period 1 May 1981–30 April 1982, Roberts Papers, folder "NSF."

[183] This was a considerable challenge with regard to the necessary sensitivity. The carbon-12 isotope has a natural abundance of 98.892 percent, carbon-13 only of 1.108 percent. Other values are: Nitrogen-14 (99.635), nitrogen-15 (0.365); oxygen-16 (99.758), oxygen-17 (0.0373), and oxygen-18 (0.2039). For the history of carbon-13 NMR see Becker, Fisk, and Khetrapal, "Development of NMR," 28–29, 49.

problem of obtaining carbon-13 NMR spectra at natural abundance level because the stability of the spectrometer was not good enough.[184]

Browsing through *Scientific American*, the solution came to Roberts in the form of an advertisement for a frequency synthesizer with hitherto unheard-of stability, manufactured by Hewlett-Packard. This meant a shift from the usual sweep of the magnetic field to a sweep of the radiofrequency field and thus implied a major change from a hardware perspective.[185] Together with CAT, and a field-frequency lock for the magnetic field, this combination had the potential to get natural abundance carbon-13 and nitrogen-15 NMR off the ground. Immediately, Roberts bought the quite expensive frequency synthesizer and asked Varian Associates to build a spectrometer around it. After hesitating at first, Varian Associates agreed. Roberts was not eager to construct a novel instrument on his own. In his opinion, scientists who developed instrumentation by themselves often got enamored of its design and improvement. Thus, their scientific productivity was not as high as it could be. Instrument manufacturers, so believed Roberts, were better at instrument construction, and this allowed the chemists to concentrate on the scientific aspects. This cooperation worked satisfactorily as long as the scientists knew exactly what they wanted, and if they were able to communicate their concepts and needs to the manufacturers.[186]

Roberts planned the new instrument as a combined nitrogen-15 and carbon-13 NMR spectrometer, and he used NIH funds for its development.[187] This put certain constraints on both parties of the deal. Roberts had to look for sufficient funds, and he tried to cover the costs with grants extending over several fiscal years, ordering only pieces of equipment in one purchase.[188] NMR data of the isotope nitrogen-15 were even harder to obtain than those of carbon-13. Consequently, both the stakes and the scientific pay-offs were higher. Short of funds, Roberts opted to order the nitrogen-15 probe first. In opposition, Varian Associates opted for an instrument available for the broader market of carbon-13 NMR, and this finally pushed Roberts into the carbon-13 venture. Because the company projected the need for a carbon-13 unit on this

[184] Roberts, in collaboration with his colleague Harden McConnell, also tried another novel technique, ENDOR, demonstrated to him by Rex Richards of Oxford University. Roberts, *The Right Place at the Right Time*, 192–193.

[185] This applies to Varian Associates. The German-Swiss company Bruker-Spectrospin much earlier introduced frequency-sweep capabilities in their NMR spectrometers, though of a limited scope. Personal communication of Thomas Steinhauser.

[186] Roberts, interview by Reinhardt, 28 January 1999.

[187] See Roberts to Forrest Nelson, 19 January 1965, and Roberts to Shoolery, 21 July 1965. Roberts Papers, folder "Varian correspondence through 1969"; Roberts, *The Right Place at the Right Time*, 193–197.

[188] See Roberts to Forrest Nelson, 19 January 1965, Roberts Papers, folder "Varian correspondence through 1969."

basis as being substantially higher than requests for nitrogen-15 NMR, Varian Associates wanted to demonstrate the principle with Roberts's prototype (still in the workshop), and an extra carbon-13 radiofrequency probe. Since the demonstration of the instrument could delay Roberts's research, the company offered him the additional carbon-13 probe on loan for three months free of charge.[189] Roberts agreed to this, knowing that the deadline for the workshop would speed up the engineering of his prototype. But he wanted to make sure that the development of a carbon-13 unit would not delay the engineering of his own nitrogen-15 probe. However, he knew that he could not stem the tides of progress in the field, pointing to the thin line between exclusivity and routine:

> With reference to the ^{15}N spectrometer, I can understand your desire to exhibit it at the Workshop and, although I have pushed for the design and execution of this instrument in hope of starting off at a new tangent to the main stream of n.m.r. research, I guess there is no stopping of progress—hundreds of eager investigators will quickly march in, waving wads of currency to purchase similar instruments for immediate delivery if not sooner.[190]

In the end, Varian Associates policy led Roberts to tackle carbon-13 NMR spectroscopy first. The spectrometer was dubbed DFS-60 (Digital Frequency Sweep spectrometer, the number representing the proton resonance frequency of 60 MHz used for the lock-in of the magnetic field). Vernon Burger of Varian Associates delivered a well-designed and massive piece of equipment, 15 feet long. Though it did not become a production model, Varian Associates, according to Roberts, included many of its features in later spectrometer types.[191] Especially the frequency sweep implied a major break in hardware design, and was adopted in all later spectrometer types. Roberts arranged for photographs of the DFS-60 that showed his secretary at the console, and Roberts used these slides in lectures to generate the impression that carbon-13 NMR by now was routine (see Figure 4.8). In addition, he pointed to the scientific successes achieved with the instrument.

Beginning in spring of 1966,[192] Frank Weigert, a graduate student of Roberts, took care of the measurements and developed the technique. Weigert, whose productivity was praised highly by Roberts, established many essential features of carbon-13 NMR. He analyzed the complex patterns of spin-spin

---

[189] Shoolery to Roberts, 16 July 1965, Roberts Papers, folder "Varian correspondence through 1969."

[190] Roberts to Shoolery, 21 July 1965, Roberts Papers, folder "Varian correspondence through 1969."

[191] Roberts, *The Right Time at the Right Place*, 194–195.

[192] Weigert to Roberts, 15 February 1990, Roberts Papers. In his autobiography, Roberts gives the fall of 1966 as delivery date of the DFS-60. Roberts, *The Right Time at the Right Place*, 194.

FIGURE 4.8 The Digital Frequency Sweep (DFS-60) spectrometer at Caltech, ca. 1967. Varian Associates built this carbon-13 NMR spectrometer along the specifications provided by Roberts. The woman, whose name is unknown, was Roberts's secretary for a short time. To the right of the instrument stands a computer of average transients (CAT). At the left hand of the console, the Hewlett-Packard frequency synthesizer was built in. Behind the secretary is the magnet. Roberts used a slide of this image in his talks, with the aim of demonstrating the feasibility of carbon-13 NMR in a relatively routine way. His use of gender roles resembles advertisements of instrument companies. Courtesy of John D. Roberts.

couplings of the carbon-13 nuclei with protons, and even the couplings of carbon-13 nuclei with each other at the low natural abundance level. In 1968, the use of broad-band proton decoupling made possible the suppression of the complex proton-carbon-13 coupling patterns. This led to a dramatic increase in the intensity of the remaining lines, and made them easier to interpret.[193] Weigert used a second frequency synthesizer system for the decoupling and installed a special combination of units. In May 1968, Weigert in a letter to David Grant described the new set-up allowing "spectacular increase in the decoupling efficiency."[194] Weigert occasionally informed Grant, at the Uni-

[193] Because of the Nuclear Overhauser Effect (NOE), the intensity increases not only twice (as expected from the collapse of a doublet into a singlet line), but sixfold. Roberts, *The Right Time at the Right Place*, 196.

[194] Weigert to Grant, 13 May 1968. See also Weigert to Grant, 12 April 1967 and Roberts and Weigert to Grant, 8 August 1967. The latter described their judgment of Lauterbur's work on the rapid passage technique in carbon-13 NMR. Roberts Papers, folder "Grant, David."

versity of Utah in Salt Lake City, about the progress the Caltech group made in carbon-13 NMR. Grant himself was one of the pioneers in the field of carbon-13 NMR, the third contributor was Paul Lauterbur of the Mellon Institute. Before Weigert left Caltech for employment at the Central Research Laboratory of Du Pont, he achieved with this technique the observation of the carbon-13 resonance lines of cholesterol, a molecule containing 27 carbon atoms.[195] For a while, the spectrum became the star of Roberts's program in carbon-13 NMR. Though Weigert announced that it would take years to make the assignments of the resonance lines, Roberts's co-workers Hans Reich and Manfred Jautelat were able to do this in less than six months.[196] This opened the door for the use of the technique in one of the most competitive fields in bioorganic chemistry and the pharmaceutical industry, steroids. Cooperation with natural product chemists enabled the further study of interesting compounds.

In July 1970, Roberts applied for NSF funds to further modernize the DFS spectrometer, essentially through its conversion to a Fourier Transform (FT) mode.[197] This would enable him to enhance the sensitivity of the carbon-13 signals by a factor of ten or possibly more. The improvement would greatly enlarge the number of compounds accessible, most notably in the field of natural products. Consequently, Roberts described the prospects of carbon-13 NMR for organic structural analysis in glowing terms, referring to its great usefulness in detecting steric effects in relatively complex molecules. In many cases, proton-decoupled carbon-13 NMR could provide much more useful information than proton NMR did. But even with the long-term measurements that could be taken with the DFS-60 (practical limits were about 20–30 hours), most natural products were not available in quantities that would lead to interpretable results. The improvement in sensitivity would be the remedy for this situation, and would allow the use of carbon-13 NMR in biological tracer experiments. In principle, two ways of improvement existed in case the quantity of the sample was limited: higher magnetic fields (with superconducting magnets), or the use of FT NMR. The first solution was rather expensive, because it would require the acquisition of a new magnet for the DFS. The second proved to more feasible.

---

[195] Weigert to LeRoy F. Johnson of Varian Associates, 13 May 1968, Roberts Papers, folder "Varian correspondence through 1969."

[196] Reich, Jautelat, Messe, Weigert, and Roberts, "Nuclear Magnetic Resonance Spectroscopy. Carbon-13 Spectra of Steroids."

[197] John D. Roberts, Supplemental equipment proposal for support of research by the NSF, "Nuclear magnetic resonance spectroscopy, structures and reaction mechanism of organic compounds," 1 July 1970, Roberts Papers, folder "NSF." In FT NMR, the spectrum is excited by using a pulsed radiofrequency. With a mathematical method, first developed by Jean Baptiste Fourier in the early nineteenth century, the time-dependent impulse response (called Free Induction Decay, FID), could be transformed to the frequency-domain spectrum. See Chapter 6.

Richard Ernst and Weston Anderson of Varian Associates invented FT NMR in the mid-1960s (see Chapter 6). Though this company hesitated in going ahead with development, commercial systems became available in 1969/1970, manufactured by the Swiss-German firm of Bruker-Spectrospin and by Varian Associates itself. The main problem remaining concerned the long relaxation times of carbon-13 nuclei. This prevented fast repetition of the radiofrequency pulses necessary to take advantage of the potential of FT NMR. If this problem was not overcome, a gain in sensitivity could be achieved only with even longer measurement time-spans. In order to solve this problem, Roberts proposed to apply a method developed by Thomas Farrar of the National Bureau of Standards. For this, an accessory was available, at a cost of $20,000. Altogether, Roberts asked for $81,125 for the modernization of the DFS,[198] and NSF approved this proposal. But this time, Varian Associates' strategy was different than Roberts's: The company was not willing to add an FT unit to the DFS spectrometer.[199] The reason was that in 1968 Varian Associates had delivered an HR-220 NMR spectrometer to Caltech. This expensive instrument was equipped with a high-field superconducting magnet and constituted the core item of a Southern Californian research facility in NMR.[200] Varian Associates installed a carbon-13 FT unit in this spectrometer, thinking of the larger market for this type of instrument. But Roberts had only one day per week at the HR-220 available, and this was not enough for him in the sense that the progress of his research relied mainly on NMR instrumentation. As a result, Roberts was eager to connect the carbon-13 unit of the HR-220 to the DFS-60.[201] The advantage for him was that the DFS-60 was under his sole control and his group could use it around the clock:

> I am sorry to keep pressuring you and your colleagues on things which are not in your usual line of operation—the trouble is that our survival in this business in a competitive way requires that we are able to function in a different manner than the people who are using nmr as adjunct to their other business.[202]

Still, Varian Associates reacted hesitantly to Roberts's proposal of connecting the FT system to the DFS-60. A change of the DFS-60 would not improve the

---

[198] Roberts, Supplemental equipment proposal for support of research by the NSF, "Nuclear magnetic resonance spectroscopy, structures and reaction mechanism of organic compounds," pp. 2–12, 1 July 1970, Roberts Papers, folder "NSF."
[199] Roberts, *The Right Place at the Right Time*, 197–198.
[200] See memorandum by Sunney I. Chan to the chemistry faculty, research fellows, and graduate students, 21 August 1968, Roberts Papers, folder "Varian correspondence through 1969."
[201] See Roberts to Raymond Ettinger of Varian, 22 April 1970, Roberts Papers, folder "Varian correspondence 1970–1972."
[202] Roberts to Vernon Burger of Varian, 5 February 1971, Roberts Papers, folder "Varian correspondence 1970–1972."

standard instrument line of Varian Associates, as it was a custom-built instrument with minimal documentation. Though they agreed to look into the matter on the spot with a technician,[203] the management finally refused to do the job. In the outcome, Roberts decided to go ahead on his own. His group, the chemists Jean-Yves Lallemand and Bruce Hawkins, together with Chris Tanzer from Bruker-Spectrospin, did manage to connect the FT system to the DFS-60. They were even able to exceed the signal-to-noise ratio that had been achieved at the HR-220. Thus, the negative estimates of Varian Associates were proven wrong, and Roberts saw enough reason to write an angry letter to the people in charge at Varian Associates:

> My general feelings about the infallibility of Varian research and development have been badly shaken . . . by this episode, especially because together these have wasted years of productive research time, along with a couple of man-years of work by our *organic* chemists to learn enough about electronics to do what your people could, and should, have done in short order at considerably reduced cost. At present, my only consolation is that the DFS spectrometer itself was a great job, even if you never thought it could all hold together.[204]

After this incident, Roberts called the DFS-60 the Brukarian-Pulse-Fourier-Transform spectrometer.[205] Soon, Roberts had to shift gears once again. With the routine use of FT NMR, carbon-13 NMR became commonplace among organic chemists and biochemists: "Then everybody started to get C-13 NMR machines and use them, and we had to look for something else."[206] From early on, Roberts had been interested in nitrogen-15 NMR. As we have seen, the DFS-60 was equipped with a nitrogen-15 probe. But its sensitivity was too low to do nitrogen-15 at the natural abundance level. Thus, in the 1960s, Roberts and his group used enriched samples to survey the field, with the constraint that nitrogen-15 labeled compounds were very expensive. They were able to establish the range of the chemical shifts (which is much larger than with protons) and to investigate the spin-spin coupling between carbon-13 and nitrogen-15. In 1969, Frank Weigert was able to take the first natural abundance nitrogen-15 NMR spectrum, using hydrazine (a compound that contains a high percentage of nitrogen, and thus offered good conditions for recording). This, and subsequent work, showed that nitrogen-15 NMR at the natural

[203] Robert S. Codrington to Roberts, 5 April 1971, Roberts Papers, folder "Varian correspondence 1970–1972."

[204] Roberts to Vernon Burger, 10 November 1971, Roberts Papers, folder "Varian correspondence 1970–1972." Emphasis in the original. Roberts refers here to a second episode, a carbon-13 low-temperature probe, where Varian Associates also reacted not favorably and not in a satisfying manner.

[205] Roberts, *The Right Place at the Right Time*, 198.

[206] Roberts, interview by Prud'homme, 109.

abundance level was in principle possible.[207] If sensitivity could be enhanced, and enough instrument time would be available, the use of nitrogen-15 NMR for biochemistry and medicine was within reach.

Consequently, Roberts seriously pushed for the development of a nitrogen-15 spectrometer of the "next generation," using high magnetic field strength and Fourier Transform methods. His first inquiry in 1970 at Varian Associates brought a proposal of $366,000—too much to convince a federal agency to give money for. Also the next proposal of Varian Associates was at a similar level.[208] In the spring of 1972, Roberts knew that less than half of this amount would be available, and that he had to build a new spectrometer incorporating large parts of already existing instrumentation. The problem of sensitivity could be overcome by using very large samples, following a suggestion of David Grant. An increase in sample tube to a diameter of 30 mm would enlarge the quantity and with that the sensitivity by a factor of ten to twenty.[209] Such a large gain was impossible to achieve even with the highest magnetic fields that were accessible around 1970. Moreover, such an instrument could be built much cheaper. But Roberts was skeptical whether Varian Associates would be willing to design a spectrometer following these constrained preconditions:

> The essence of the idea would be to demonstrate feasibility and usefulness of ^{15}N nmr which has so far been marginal at best, rather than take the approach we used with the digital-sweep spectrometer some years ago—that we knew a state-of-the-art instrument would surely be useful and let's build one. If the ultra large sample size approach proved useful and feasible, then one could ask for a large sum to build a more optimum machine.[210]

The problem was that the large sample size called for huge quantities of the compounds under scrutiny—a condition that very often could not be met by biochemists, because biologically active molecules were hard to isolate in large quantities. After some hesitation, new staff at Varian Associates decided to go ahead with Roberts's specifications. In 1972, Varian Associates merged several instrument manufacturers into a single instrument division.[211] For the company, the development of instrumentation meeting the needs of a few users only required careful consideration of the potential market. In doing so, they regarded the "research NMR market" as a special segment:

---

[207] Roberts, *The Right Place at the Right Time*, 200–202.

[208] Raymond Ettinger to Roberts, 16 July 1970, and Richard J. Galetti to Roberts, 10 August 1971, Roberts Papers, folder "Varian correspondence 1970–1972."

[209] Roberts, *The Right Place at the Right Time*, 203.

[210] Roberts to Robert W. Moulton of Varian, 21 April 1972, Roberts Papers, folder "Varian correspondence 1970–1972."

[211] See Robert W. Moulton to Roberts, 2 May 1972, Roberts Papers, folder "Varian correspondence 1970–1972."

> In our analyses, we have concluded that we must be more selective about which special opportunities we support because we cannot spend all of our time handling specials. We also need to select those projects which we expect to relate to future, higher volume products which will be popular and useful for years in the future.[212]

For Roberts, the situation was naturally different. He had an offer from Transform Technology to convert the existing DFS spectrometer, and an offer made by Bruker-Spectrospin within a budget of $150,000. Moreover, he and David Grant had paired up to develop this new kind of instrumentation using similar magnets and probes. Grant concentrated on carbon-13, while Roberts wanted to focus on nitrogen-15.[213] As a result of a meeting of Varian Associates employees with David Grant in late August 1972, the firm offered to develop novel instrumentation according to Grant's and Roberts's specifications and budgets. Varian Associates saw the development of the instrument more as an investment "to broaden the base of NMR in biological problems" than a "product profit model." In doing so, they wanted to make sure that Roberts contributed to the development as consultant during the process.[214] Unfortunately for Varian Associates, at the same time when the negotiations about this contract were at stake, the delivery of a repaired probe to the Caltech HR-220 took much longer than was expected. Varian Associates did not seem to be able to undertake the reorganization deemed necessary for better service.[215] In contrast, Roberts was excited about the prospects of a cooperation with Bruker-Spectrospin, which had a good standing in his research group since Chris Tanzer had helped with the conversion of the DFS-60 to the FT mode. Roberts appreciated this flexibility on the part of Bruker-Spectrospin very much.[216] The German-Swiss company, with its American subsidiary Bruker Scientific Inc. of Elmsford, New York, had begun to cater for the U.S. market. Bruker Magnetics Inc. of Burlington, Massachusetts introduced a line of super conductivity magnets that could be combined with the novel technology of Fourier Transform NMR. Thus, when a decision had to be made on a dedicated nitrogen-15 spectrometer, Bruker-Spectrospin's offer seemed to be more promising for Roberts than the proposal of Varian Associates. The decision of Roberts to give the order to Bruker-Spectrospin was the first break

---

[212] Ken Cruden of Varian to Roberts, 8 August 1972, Roberts Papers, folder "Varian correspondence 1970–1972."
[213] Roberts to Cruden, 16 August 1972, Roberts Papers, folder "Varian correspondence 1970–1972."
[214] Cruden to Roberts, 28 August 1972, Roberts Papers, folder "Varian correspondence 1970–1972."
[215] See Roberts to Cruden, 7 September 1972, Roberts Papers, folder "Varian correspondence 1970–1972."
[216] See Roberts to Tanzer, 31 January 1972, Roberts Papers, folder "Bruker Scientific Inc."

away of Caltech from Varian Associates equipment in NMR, and it proved to be the beginning of a strong position for Bruker-Spectrospin in the United States.[217] Varian Associates, in the 1970s, decided to focus on the low end, "analytical" part of the market, and thus lost the possibility of competing seriously with Bruker-Spectrospin in the high end segment.[218] It can be said that the market became divided between the biochemists, requiring highly sophisticated instrumentation for the study of their more complicated molecules, and organic chemists, with needs for standard, reliable instruments.[219] Especially for the first group the necessity of measuring nuclei other than protons, most importantly carbon and nitrogen, became imperative.

Though Bruker-Spectrospin expressed great interest in having Roberts at Caltech use the firm's latest state-of-the-art technology,[220] as soon as the order was made it "seemed to sink at once to the bottom of their priority list."[221] Due to problems in magnet design, the delivery of the spectrometer was delayed several times. Finally, in April of 1974, Bruker-Spectrospin submitted new—and improved—specifications for the instrument. Enhanced magnet performance allowed the maintenance of a magnetic field of 42 kGauss over a gap of 3.5 inches. The Bruker-Spectrospin engineer Craig Bradley proposed to use the existing design and console of Bruker-Spectrospin's standard instrument WH-90 as the basis for an upgrading. Roberts's instrument was called WH-180, giving the frequency of 180 MHz for the proton-decoupling. Bruker-Spectrospin was appreciative of the fact that the design, made along the lines of Roberts's specifications, had improved the company's understanding of NMR instrumentation. Moreover, they assured Roberts of his priority to use the instrument first:

> Also, as we have previously discussed, no other scientist or research group in North America will be able to purchase a WH-180 until it is fully operational in your laboratory and producing results on a routine basis. We respect your innovative concepts in demanding such a spectrometer and it is indeed a privilege for our company to have been chosen to "put it all together."[222]

---

[217] Roberts to Peter Llewellyn of Varian, 30 November 1972, Roberts Papers, folder "Varian correspondence 1970–1972."

[218] Weston Anderson, interview by Sharon Mercer, February 1990, 23, Varian Associates Oral History, SC M 708, box 1.

[219] Shoolery, interview by Mercer, April 1990, 15.

[220] See Keith Shaw to Roberts, 15 September 1972, where Shaw described in detail the scientific and engineering expertise that was available at Bruker, emphasizing the cooperation with the solid-state NMR specialist John S. Waugh of MIT and the indirect contact to Frank Anet of UCLA through the hiring of Anet's former collaborator Craig Bradley. Roberts Papers, folder "Bruker Scientific Inc."

[221] Roberts, *The Right Place at the Right Time*, 203.

[222] Shaw to Roberts, 5 April 1974, Roberts Papers, folder "Bruker Scientific Inc."

The spirit of this assurance was broken, because Bruker-Spectrospin allowed a European scientist to make measurements with the instrument even before delivery.[223] But in November 1974, the WH-180 (see Figure 4.9) was ready to be installed, and Bruker-Spectrospin's chief engineer Toni Keller asked Roberts to acquit the changes the Bruker-Spectrospin team had made during the development process.[224] The installation at Caltech proved to be complicated, partially because of the instability of the magnet (it had to be changed twice). Originally, the instrument was equipped with a 25 mm diameter probe. Though Roberts congratulated the Bruker-Spectrospin team on the excellent signal-to-noise ratio they had achieved with the magnet, he pushed hard for the construction of a 30 mm probe that would allow the use of much larger samples. This would enhance the sensitivity. Bruker-Spectrospin, eager to achieve a proper design, had laid its emphasis on resolution, while Roberts always argued for better sensitivity: biochemists were glad to just be able to detect relatively broad lines, not caring too much about resolution. In the words of Roberts, sensitivity would be the "sales pitch" for the instrument, and he insisted on the larger probe "because of our hope that this instrument will really test the viability of natural-abundance N-15 spectra in chemical and biochemical research (and indeed this was the pitch on which we made our proposal for support of the purchase of the instrument)." After all, Roberts made clear that he hoped that Bruker-Spectrospin would "benefit from the favorable publicity expected to be generated" by the research that Roberts's group would undertake with the machine.[225]

Indeed, one of the first projects in Roberts's group undertaken by Richard Moon and Devens Gust was the demonstration of utility of nitrogen-15 NMR for medium-size biological compounds such as vitamins, enzymes, and small segments of RNA.[226] Though the quantities needed "caused true biochemists to wince," it showed the potential of the technique in the biomedical sciences.[227] Moreover, they found that the high molecular size of biomolecules acted favorably in the case of fast signal accumulation. In addition, the changes of the Nuclear Overhauser Effect (NOE) with intramolecular mobility were investigated. This established a new technique for the measurement of the mobility of biomolecules. Thereby, Roberts's strategy was a pragmatic one: "Our choice of what we did, I think, quite frankly, was driven by what we

[223] Bruker allowed Ed Randall of Queen Mary College, London, to use the instrument before delivery. See Roberts, *The Right Place at the Right Time*, 204.

[224] Keller to Roberts, 15 November 1974, Roberts Papers, folder "Bruker Scientific Inc."

[225] Citations from Roberts to Shaw, 14 January 1975, Roberts Papers, folder "Bruker Scientific Inc."

[226] Gust, Moon, and Roberts, "Applications of Natural-abundance Nitrogen-15 Nuclear Magnetic Resonance."

[227] Roberts, *The Right Place at the Right Time*, 205.

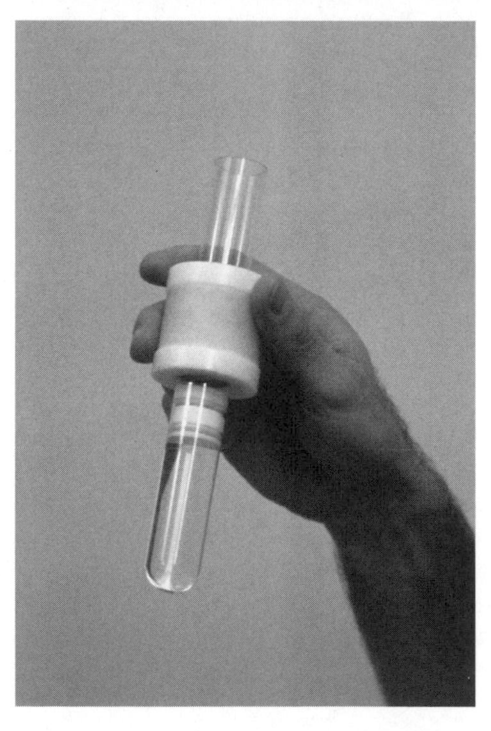

FIGURE 4.9 Console of the Bruker-Spectrospin WH-180 spectrometer, ca. 1981 (upper figure). Below is an example of the sample tubes used for studying nitrogen-15 NMR. Courtesy of John D. Roberts.

could do, and not maybe what the most important things were."[228] Another mechanistic study, the investigation of the cleavage of peptide bonds by enzymes was carried out by William W. Bachovchin in Roberts's laboratory. Bachovchin used a special nitrogen-15 labeled enzyme (a serine protease) and was able to correct the hitherto established mechanism of the action of the enzyme.[229]

In 1976, Roberts's group measured natural abundance nitrogen-15 NMR spectra at rather low concentrations of the compounds investigated. Because the capabilities of his instrument were unique, Roberts collaborated with many outside researchers. He even compared his instrument to a "national facility" in nitrogen-15 NMR—a good argument to obtain more funds.[230] Roberts was eager to develop his instrumentation continuously, with small increments in improvement. For example, in the late 1970s, he applied for funds for a new probe for the WH-180 spectrometer, suitable for a smaller sample size and thus requiring less material. At the same time, the DFS-60 spectrometer was modernized.[231] Roberts's program gained large momentum in the mid-1970s. Of NIH funded equipment alone, in 1976 he had instrumentation that was worth ca. $400,000. The annual cost to keep the instrumentation in operational condition was ca. $25,000. Roberts's exclusive instrumentation was embedded in a large array of instruments at Caltech's chemistry division. In 1978, 14 NMR spectrometers were in service, six of them for routine use (operating at the frequency of 60 MHz), three at 100 MHz, and one high-field instrument, operating at 220 MHz. Altogether, no less than 159 research instruments were reported to be in use, ranging from mass spectrometers to gas chromatographs, and liquid scintillation counters. In the fiscal year 1978 alone, the chemistry division of Caltech made investments in instruments of nearly $540,000.[232]

The high productivity that his group achieved had been made possible by the service support of Caltech and the availability of a large number of fellowships. When in the mid-1970s fellowships became scarce, Roberts asked NIH for an increase in funding to be able to employ enough personnel to use the instrumentation fully. He additionally pondered the possibility of training more graduate students (and through this keep the productivity up). Also, the fast pace of improvements in instrumentation, especially in the case of computers,

[228] Roberts, interview by Reinhardt, 28 January 1999.

[229] Bachovchin and Roberts, "Nitrogen-15 Nuclear Magnetic Resonance Spectroscopy."

[230] Roberts named nine outside collaborations. John D. Roberts, proposal for support of research by the NSF, "Nuclear magnetic resonance spectroscopy, structures and reaction mechanism of organic compounds," 1 April 1976 to 30 May 1981, p. 8, Roberts Papers, folder "NSF."

[231] John D. Roberts, "Application for continuation grant," NIH GM 11072-16, 1 May 1978 to 30 April 1979, pp. 3–3b, Roberts Papers, folder "NSF."

[232] List of major research instruments in the department of chemistry, Caltech, n. d. [1978], Roberts Papers, folder "Nair, P. M."

made his request for "contingency funds" reasonable. This would provide him with a budget that he could spend at short notice, if necessary.[233] The technology of the WH-180 was somewhat dated by the early 1980s. Nevertheless, Roberts's group profited from the exclusivity they enjoyed with an instrument that was available only to them. Government agencies had become reluctant to give scientists an expensive instrument of their own. The commercial development of NMR had reached such an advanced stage that standard spectrometers equipped with extra probes were deemed to be sufficient. This development made the sharing of one instrument between several research groups mandatory. Naturally, sharing meant less measurement time available for each group, and to make it worse, the changing of the probes took extra time.[234] Thus, Roberts appreciated his fully dedicated nitrogen-15 spectrometer:

> Obviously, the continuing availability of our now wholly ^{15}N-devoted, but aging, WH-180 spectrometer is vital to the ^{15}N segment of our research program. The sensitivity of this instrument is hardly state-of-the-art but its full-time availability and usage have permitted studies that few, if any other, laboratories can undertake.[235]

While Roberts was acting as Caltech's provost in the early 1980s, the postdoctoral research fellow Keiko Kanamori directed much of Roberts's program in nitrogen-15 NMR.[236] Beyond the development of instrumentation, Roberts's program defined the properties of compounds that could be investigated by NMR methods, and it provided model solutions for further use by other scientists. At each step, Roberts and his group were able to increase the complexity of the molecules under scrutiny as well as the complexity of the questions asked. Beginning with simple organic compounds, in 25 years Roberts's research moved to include enzymes and their complicated working mechanisms.

CONCLUSION

The standard view rates scientific performance in terms of numbers of citations, amounts of funding, and prestige of prizes. Instruments could help to achieve these scientists' goals, when connected to established research traditions. Lead users of instruments in science came to be, together with the manufacturers, the connectors of high-tech instrumentation and chemical research

---

[233] John D. Roberts, memo attached to letter to John B. Wolff of NIH, 6 August 1976, "Justification for expanded effort on ^{15}N and ^{13}C nmr spectroscopy," Roberts Papers, folder "NSF."
[234] Roberts, interview by Reinhardt, 28 January 1999.
[235] John D. Roberts, report of the NIH grant GM 11072-19, "Nitrogen-15 and carbon-13 NMR spectroscopy," period 1 May 1981–30 April 1982, p. 5, Roberts Papers, folder "NSF."
[236] For a review of this work see Kanamori and Roberts, "^{15}N NMR studies of biological systems."

programs. In doing so, they became first movers and innovative players in a scientific market of citations, grants, and accolades. A lead user paved the way for other scientists, and he did so most efficiently while pursuing and taking advantage of the development and popularization of instruments and methods.

For Varian Associates, Roberts proved to be a "stellar salesman."[237] He combined scientific reputation and high productivity as full professor at one of the leading chemistry departments in the country. Moreover, he became the author of two textbooks on NMR, and co-authored an influential textbook of general organic chemistry that prominently featured the novel spectroscopic techniques. In 1956, he was elected a member of the National Academy of Sciences, one year later he became a member of the chemistry panel of the National Science Foundation. Before he got into the field of NMR, he already had acquired great prestige in the field of physical organic chemistry, and this constituted the basis for his later role in NMR. Chemists with already recognized achievements provided the greatest publicity for the novel instrumental technique, thus ensuring the needs of the instrument manufacturer. The credibility of the user guaranteed the validity of the technique, and high reputation lowered the risk for the scientist to apply unproven instrumentation. Through his substantial influence in a community of users, he was able to connect most easily concepts of instruments with chemical ones. Two major issues played a role in this. First, Roberts's research was open to influences from technology and mathematics, at least when proven applications in chemistry already existed. His field, physical organic chemistry, stood on the borderline to physics and already had a tradition of including mathematical applications. In physical organic chemistry, innovative steps—often involving a transfer from other disciplines—were more easily rewarded with academic success as in more traditional fields, e.g., synthetic organic chemistry. Thus, it is argued, theories and methods of physical organic chemistry prepared the ground for Roberts's later role in NMR. They did so both as stepping stones in his individual career, making him a respected scientist in his field, and in laying the cognitive fundament that made possible the smooth inclusion of many NMR concepts and interpretation schemes. Second, in the social dimension, the success of physical organic chemistry in the United States of the 1940s and 1950s supplied the manpower that was necessary to develop a research methodology that could be used by mainstream organic chemists. Moreover, the increase in science funding during and after World War II allowed and even called for the acquisition of expensive equipment. For the United States, the 1950s were an age of electronics. This cultural impact fostered the emergence and use of high technology, even in the traditionally conservative discipline of organic chemistry.

Early NMR research in chemistry meant to find and to spread research problems that could be tackled with the new method. Structural research had

[237] Roberts, *The Right Place at the Right Time*, 170.

its share in this, and most important became dynamics of molecules. At that time, no other technique in physics and chemistry showed such a promise, and Roberts pushed an area that would not have been accessible without NMR. Thus, NMR participated in creating a new subfield of chemistry, molecular dynamics, and at the same time commanded increasing influence in the old, structural organic chemistry. In the beginning, Roberts worked hard to connect the two sides, as his inclusion of traditional methods and his early focus on structural elucidation shows. But soon, NMR gained enough impetus to stand on its own, and conformational studies became an independent research area. It is this issue that crucially contributed to making NMR a special and very successful chemical technique.

Because of their limited capability for and interest in the development of their own electronic equipment, lead users relied on the services of companies such as Varian Associates. In exchange for uniquely engineered instrumentation, they supplied the manufacturer with critical information on their needs and made proposals for design improvements. Naturally, on one hand, the first users of cutting-edge research instruments wanted to enjoy exclusivity as long as possible. On the other hand, instrument manufacturers wanted to sell as many exemplars of an instrument as possible. Consequently, a delicate balance of the needs of both sides had to be maintained. Arguably, users who enjoyed a high standing in a scientific field could make the most profitable use of cutting-edge research with yet unexplored instrumentation. Their quest for exclusivity required substantial efforts in the development of new instrumentation and collaboration with instrument manufacturers and funding agencies. Their pre-existing reputations would ease the first difficult steps of the establishment of a novel technique in a scientific field. After a successful introduction of a new instrument, it was also in the interest of the scientist to enhance its distribution: his methods spread with the instrument, and his papers were cited. In this sense, Roberts belonged to a group of lead users of NMR in organic chemistry. His function was acknowledged by the instrument manufacturers, for example by Martin Packard of Varian Associates:

> One of the important uses of NMR is now carbon-13, and we first worked on this with people at Cal Tech. Professor Jack Roberts, one of the important chemists at Cal Tech, who later became provost and has had many important government positions, worked with us in the early days. We provided him with special NMR equipment. Another piece of special equipment, I remember, went to a man by the name of Grant, who was or is at Utah, and I think he also did work on carbon-13. The point is, there was money available for pursuing these ideas even when it could not be assured that they were going to be useful concepts. But in practice, many of them turned out to be extremely useful.[238]

[238] Packard, interview by Mercer, December 1989.

Packard emphasized the offices Roberts held, and his standing in the scientific community. Together with David M. Grant of the University of Utah, and Paul Lauterbur of SUNY at Stony Brook, Roberts was instrumental in expanding the routine use of NMR to the measurements of elements other than hydrogen, mainly carbon-13. Carbon-13 NMR greatly widened the utility of NMR, especially in organic chemistry and biochemistry.

With Varian Associates, Roberts found an ally for his endeavors. Their mutual interests in the distribution of the instruments allowed for a relatively smooth relationship until the end of the 1960s. Strains emerged when Varian Associates acted as scientific competitor, and when the company, conscious of its monopoly, exerted too much influence on the community of NMR users. Though these problems could be overcome, the fruitful symbiosis of Varian Associates and Roberts ended when Varian Associates' strategy favored the low-end side of technology and concentrated on a mass market of routine instruments. In the early 1970s, Roberts, in his zeal for exclusive high-end equipment, parted company and cooperated with the major competitor of Varian Associates, Bruker-Spectrospin.

The development of research methods that could be used with soon-to-be commercial equipment became Roberts's specialty. In doing so, he contributed to the fields of organic chemistry, biochemistry, and finally also to medical applications. In his research program, the chemists of his group, the manufacturers of the instruments, and the government agencies that provided the funds were intertwined by common interests. These partners made the breakthrough of NMR in organic chemistry possible. As lead user, Roberts stands for a multitude of scientists who pursued a similar career.

Chapter Five

Data in Process

From the mid-1960s on, the rising demand for mass spectral information in universities, industry, and governmental agencies led to a situation in which the human interpreter increasingly became the bottleneck. Funding agencies, science associations, universities, and instrument manufacturers alike started programs and projects to increase the numbers of trained mass spectrometrists. Moreover, the already existing databases of mass spectra were considerably enlarged, and special matching programs for compound identification were developed. In addition, the sheer mass of detectable compounds made the development of interpretation programs necessary. With the help of these programs, chemists could perform their tasks more efficiently.

At the same time, scientists attempted to emulate the thought process of the mass spectrometrist in a computer. With the advent of artificial intelligence (AI), computer scientists and chemists alike assumed that they had achieved much more than perfecting a simple calculation tool. The object of AI scientists was to understand human intelligence, and to mimic it, mostly with the help of the computer. The efforts that led to a series of AI programs capable of interpreting mass spectra of complex organic compounds belonged to the first AI projects that tackled problems of the "real world." This did not change the main tool box of AI scientists, however, since the methods employed to solve mass spectral problems were essentially the same as those used to play chess. AI scientists' ultimate aim was the self-learning machine, and the crucial advantage of tackling chemical problems with the help of AI was "that the practical utility of what has already been produced should suffice to engage the attention of a considerable number of *human chemists* working on practical problems in a fashion that lends itself to machine observation and emulation of their techniques."[1] But the mimicry of the chemists' reasoning had to fit the self-defined roles of the chemists in their respective fields. In contrast to the rhetoric of AI scientists who thought they had invented an automated process of scientific discovery, Carl Djerassi and other chemists popularized computerized methods in

[1] Lederberg and Feigenbaum, "Mechanization of Inductive Inference," 217–218. My emphasis.

chemistry in a more pragmatic manner. He emphasized the independence of the program from biases, its guarantee to be complete, and the prospects for saving time and amount of sample.

The instrumental basis for the investment in matching and interpretation programs was high resolution mass spectrometry, a technique that from the mid-1960s on supplemented low resolution mass spectrometry (see Chapter 3). The new technique showed higher technical complexity, and demanded different interpretation and representation techniques. Most importantly, high resolution mass spectrometry led to an explosion in the amount of data that could be used for interpretation. In order to make use of these data, chemists immediately invented new modes of representing mass spectra, and in this process they more and more depended on computer applications. This development accelerated the trend of setting up mass spectrometry in special centers that served a whole community of scientists. Funding agencies such as the National Institutes of Health (NIH) and the National Science Foundation (NSF) created special instrument-related programs. Especially NIH provided long-term funding of expensive research equipment with its Special Research Resources Program. Large scientific instruments were set up in regional and national facilities, and made available to the health-related research community on a service basis. Occasionally, instrument and computer manufacturers themselves funded research in mass spectrometry with the supply of instrumentation. Thus, changes in the interpretive and experimental stages of scientific work were locked to new institutions, ranging from centralized facilities to new modes in the interaction of science and industry.

In this chapter, the focus is on specific moments in the career paths of the scientists familiar from Chapter 3, Klaus Biemann, Fred W. McLafferty, and Carl Djerassi. All of them acted and responded to the challenges of fast-moving high-technology instrumentation in their own and peculiar ways. Biemann pioneered the use of high resolution mass spectrometry in organic chemistry, and led a broad research program based on substantial funding from NIH (and, to a lesser extent, NSF and NASA). His focus remained the structural elucidation of complex organic compounds, and in order to achieve this goal, he contributed substantially to the development of new representation techniques. Both Biemann and McLafferty continued to rely on the expertise of the human interpreter, and McLafferty and Djerassi followed Biemann in the technique of high resolution mass spectrometry. McLafferty's specialty became the design of improved instrumentation. More than Biemann, McLafferty invested time and efforts in the up-building of a spectra database, and the development of suitable computer programs for its use. While in the work of both Biemann and McLafferty, mass spectrometry came first, and computer programming second, Djerassi supplied his know-how in mass spectrometry to a large-scale development in artificial intelligence.

Mapping Elemental Compositions

In the mid-1960s, Klaus Biemann's pioneering applications of mass spectrometry to the chemistry of alkaloids found many followers. In order to remain at the forefront of research, he looked for a new technique. As was the case in his beginnings in mass spectrometry, he found it in a method originally developed for the needs of the petroleum industry: high resolution mass spectrometry. With the help of this accurate and precise technique, each peak in a mass spectrum could be assigned one, and only one, elemental composition of the ion causing the peak. By presenting elemental composition as the foundation for subsequent interpretation, Biemann emphasized the chemical connotations of high resolution mass spectrometry. The chemist did not have to deal directly with numerical mass values anymore:

> At this point one realizes that the mass of the ion is irrelevant and it is the elemental composition which represents the set of data that forms the basis of the interpretation of the spectrum. Thus one could say that the term "elemental composition spectrometry" might be more applicable to this approach than mass spectrometry.[2]

While Biemann's term was not adopted, he did develop a style for the presentation of the elemental compositions of a large number of ions that became highly successful. The breakthrough of high resolution mass spectrometry in organic chemistry relied on these presentation techniques as well as on the uses of the computer in undertaking the calculations that led to those "element maps."

The term high resolution mass spectrometry refers to the capability of so-called double focusing mass spectrometers (see Figure 5.1) to measure masses of ions very accurately, and with a precision of several parts in a million. For nuclear physicists and nuclear chemists interested in precision measurements of isotopic masses this was an advantage because, due to the nuclear packing fractions, the isotopic masses differed from whole numbers by different amounts. Thus on the standard scale, carbon-12 had by definition exactly 12 atomic mass units (u), while hydrogen, nitrogen-14, and oxygen-16 had the values of 1.00782465, 14.0030738, and 15.99491415, respectively.[3] But also organic chemists saw a potential use for such advanced instrumentation because, in

---

[2] Biemann, "High Resolution Mass Spectrometry," 196.

[3] Beynon, "Eternal Triangle," 177–178. In 1961, carbon-12 replaced the formerly used oxygen-16 as the standard for the atomic mass scale. This was due to the existence of two different oxygen mass scales, one "physical," and based on the isotope oxygen-16, the other "chemical," and based on natural oxygen (which includes small amounts of oxygen-17 and oxygen-18). The newly adopted carbon-12 scale settled the values approximately between the chemical and the physical scales.

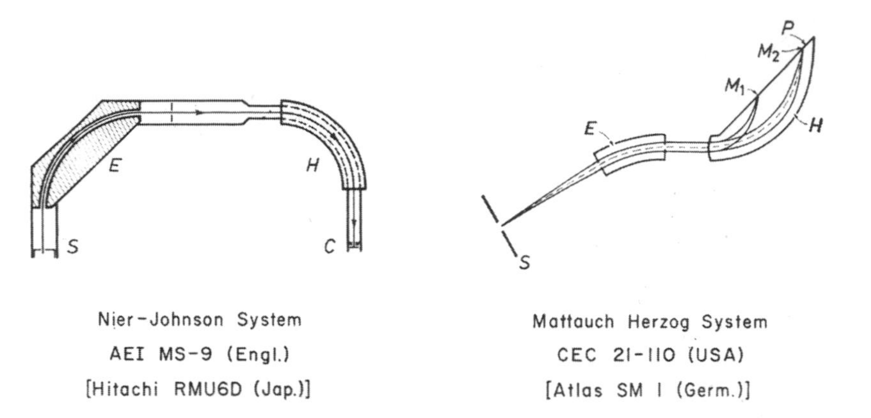

FIGURE 5.1 The Nier-Johnson and the Mattauch-Herzog design of commercial high resolution mass spectrometers. Reproduced with permission from Klaus Biemann, "High Resolution Mass Spectrometry of Natural Products," *Pure and Applied Chemistry* 9 (1964), 95–118, on 98, fig. 2. Copyright 1964, IUPAC.

their measurement range, low resolution mass spectrometers resolved masses to one mass unit only. Consequently, a low resolution mass spectrum could not differentiate between, for example, $C_{24}H_{50}N$ and $C_{25}H_{52}$, since the molecular weights of the two compounds differed only by 0.0126 u. The high resolution spectrum would upon inspection reveal the elementary composition of the ion on the basis of its mass alone.[4] This was a huge advantage and a necessary improvement, because with higher masses the number of possible combinations increased tremendously. If one considered all species of the nominal mass of 135 that contained an unknown number of carbon and hydrogen atoms, and up to five nitrogen and oxygen atoms, 43 possible combinations existed, which could be distinguished by high resolution methods.[5] High resolution mass spectrometers had the potential to revolutionize the art and science of organic mass spectrometry.

[4] Biemann, "High Resolution Mass Spectrometry of Natural Products," 95.
[5] Biemann, "High Resolution Mass Spectrometry," 187–189.

Essentially, a high resolution (or double-focusing) mass spectrometer is a combination of Aston's and Dempster's original designs for single-focusing, low resolution mass spectrometers (see Chapter 3).[6] The impediment of low resolution mass spectrometry was that only in cases where the ions produced in the ion source had all the same energies would a homogeneous magnetic field for ion deflection yield satisfying results. It was shown that an electric field could sort out ions so that their difference in entry point into the magnetic field compensated for their difference in energy. A combination of an electric and a magnetic field became the double-focusing mass spectrometer, with suitable dimensions of both fields to allow for a sufficient intensity of the ion beam.[7] Immediately after World War II, Alfred O. Nier and Edgar G. Johnson designed a version of a double-focusing mass spectrometer. Another often-used configuration was developed by the Austrian-German physicists Josef Mattauch and Richard Herzog in the 1930s, mainly for precision measurements of nuclidic masses.

The advent of high resolution mass spectrometry in organic chemistry had many features in common with the earlier breakthroughs in conventional mass spectrometry. The design of high resolution mass spectrometers was not new in the early 1960s, when such instruments were introduced into organic chemistry. But these highly sophisticated instruments remained in the hands of a few experienced physicists and engineers, and were only applied to the fields of nuclear physics and nuclear chemistry. In the mid-1950s, the physicist John Beynon, working for Imperial Chemical Industries (ICI), the largest chemical enterprise in the United Kingdom, pioneered the use of high resolution mass spectrometry in organic chemistry.[8] The reason that ICI invested in a mass spectrometer was the desire to analyze products of competitors. Through the detection of minor components (impurities, by-products, and remaining starting materials), ICI scientists hoped to unravel the synthetic methods that were used in manufacturing, mostly of synthetic dyestuffs.[9] According to Beynon's needs, the instrument manufacturer Metropolitan Vickers Electrical Co. Ltd. of Manchester (later Associated Electrical Industries Ltd., AEI) constructed a high resolution instrument of Nier-Johnson geometry, the MS 8 (see Figure 5.1).[10] Up to the early 1960s, only Beynon, researchers at Metropolitan-Vickers itself, and, using a prototype of CEC, Milburn J. O'Neal at the Shell Oil

---

[6] For a detailed description of instrument development before 1940, see Jordan and Young, "A Short History of Isotopes"; Beynon and Morgan, "Development of Mass Spectrometry," 25. The first theoretical treatment of a double-focusing design is Bartky and Dempster, "Paths of Charged Particles."

[7] Hintenberger, "Anwendung der Massenspektroskopie."

[8] Beynon, "Eternal Triangle." For a good summary of his work see Beynon, "High Resolution Mass Spectrometry of Organic Materials."

[9] Biemann, interview by Reinhardt, 10 December 1998.

[10] See Craig and Errock, "Design and Performance."

Company[11] were capable of using high resolution mass spectrometers in organic chemistry. An important reason for this delay was that the complicated research instruments of the physicists had to be adapted to the needs of the organic chemists, who did not need the full precision attainable, and instead longed for increased speed and usability. As in low resolution mass spectrometry, the largest potential market was with the petroleum companies. Additionally, some research-oriented chemists felt the need for such accurate mass measurements.

In 1959, the Metropolitan Vickers MS 9, and the CEC 21-110 appeared on the market. Biemann saw both under construction, in Manchester and Pasadena, respectively. For Biemann, who planned to continue work in natural product chemistry with only small amounts of samples available, the CEC 21-110 showed a crucial advantage. Because of the design of its "optical" parts (Mattauch-Herzog geometry, see Figure 5.1), this spectrometer focused the incoming ions in a plane, without the need for scanning the magnetic or electric field. According to Biemann, this made "it perhaps possible to obtain a high resolution spectrum, for qualitative purposes, in a rather short time (30 sec. to 2 min.)."[12] Because the sample pressure had to be kept constant during the recording of the spectrum, a short recording time was mandatory in case only small sample volumes could be used.[13]

In May 1960, CEC had sixteen orders for their novel high resolution instrument. The first was delivered to Esso in the same month, the second to U.S. Steel. For Biemann, this relatively large number of orders was another advantage that the CEC instrument enjoyed over its main competitor, the MS 9 of Metropolitan Vickers:

> The fact that there are quite a number of customers will make sure that the instrument is continuously developed further and that future changes will be made in such a way that the same changes can be made on the earlier models also, as we have seen on the CEC 21-103.[14]

Indeed, this form of continuous, and retrospective, instrument development by further engineering of CEC allowed an increase in the resolution from one part in 2,500 to one part in 30,000 over a period of four years.[15] These improvements were made both by users and at CEC itself. For an efficient exchange of information, Biemann proposed to CEC that it publish a newsletter for the spectrometer "with suggestions for changes or improvements based on the

---

[11] See Voorhies, "Theoretical and Experimental Study," 54.

[12] Biemann to Stoll, 6 May 1960, Biemann Papers, folder "Firmenich et Cie."

[13] Biemann to Clifford C. Berry, 18 December 1961, Biemann Papers, folder "CEC."

[14] Biemann to Stoll, 6 May 1960, Biemann Papers, folder "Firmenich et Cie."

[15] *CEC 21-110 Mass Spectrometer Newsletter*, vol. 3, no. 1, 13 April 1965, Biemann Papers, folder "CEC current 1963–64."

experience of your people in Pasadena."[16] Already three weeks later CEC published the first issue, and provided for a "users clinic" for the 21-110 customers.[17] Biemann, in a style that resembles the cooperation of John D. Roberts with Varian Associates in the field of NMR (see Chapter 4), reported his needs back to CEC. In addition, he made suggestions on how to solve his specific problems and offered to contribute with self-built solutions.[18] To ease the exchange of special experiences of 21-110 users who supposedly worked in organic chemistry, Biemann also asked CEC for the names of these customers. In early 1963, ten of these instruments were in operation, six in U.S. chemical and oil companies, one at the U.S. Army Chemical Center, one at the Max-Planck-Institut für Kohlenforschung, Mülheim, Germany, one at Euratom in Ispra, Italy, and one at the University of Zurich, Switzerland.[19] Thus, Biemann had the only such instrument available in an American university, which he ordered in December of 1961, after the finances had been approved (see Figure 5.2).[20] The instrument was paid for by combining grants from NSF and NIH.[21]

Biemann enjoyed the harvest of academic exclusivity for a short period only in the mid-1960s. In 1964, he acknowledged this fact, and at the same time pointed to its transitoriness:

> For a long time Beynon possessed the only such instrument accessible to organic chemists, and even at the present time practically all the data published were obtained in three laboratories, namely, at ICI (Beynon), at AEI (the manufacturer of one of these instruments), and our own. This situation will perhaps change very rapidly in the near future as high resolution mass spectrometers are now being produced commercially in great numbers.[22]

---

[16] Biemann to S. W. Downer, 31 December 1962, Biemann Papers, folder "CEC."

[17] *CEC 21-110 Mass Spectrometer Newsletter*, vol. 1, no. 1, 10 January 1963, Biemann Papers, folder "CEC."

[18] In the case of the 21-110, problems were related especially to the lack of a suitable heating system for the ion source. See Biemann to Berry, 31 December 1962, and Biemann to Charles G. Blanchard, 18 April 1963, Biemann Papers, folder "CEC."

[19] Philip Wadsworth of Shell Development Company in Emeryville, CA; Norman Coggeshall of Gulf Oil Corporation, Research Center, in Hamarville, PA; Nelson Trenner of Merck and Company, Research Laboratories in Rahway, NJ; Robert Silas of Phillips Petroleum Company, Research Center, in Bartlesville, OK; L. C. Buckles [?] of the Army Chemical Center of Edgewood, MD; Dieter Henneberg of the MPI für Kohlenforschung in Mülheim, Germany; R. A. Brown of Esso Research & Engineering, analytical research division, Linden, NJ; James Guthrie of Sandia Corporation in Albuquerque in New Mexico; scientists at Euratom in Ispra, Italy, and Dr. Ernst Schumacher of the University of Zurich, Switzerland. See S. W. Downer to Biemann, 11 February 1963, Biemann Papers, folder "CEC."

[20] Biemann to Clifford C. Berry, 18 December 1961, Biemann Papers, folder "CEC."

[21] Biemann, NIH grant application "High resolution mass spectra of labeled compounds," 27 January 1964, p. 6, Biemann Papers, folder "Early NIH grant."

[22] Biemann, "High Resolution Mass Spectrometry of Natural Products," 97.

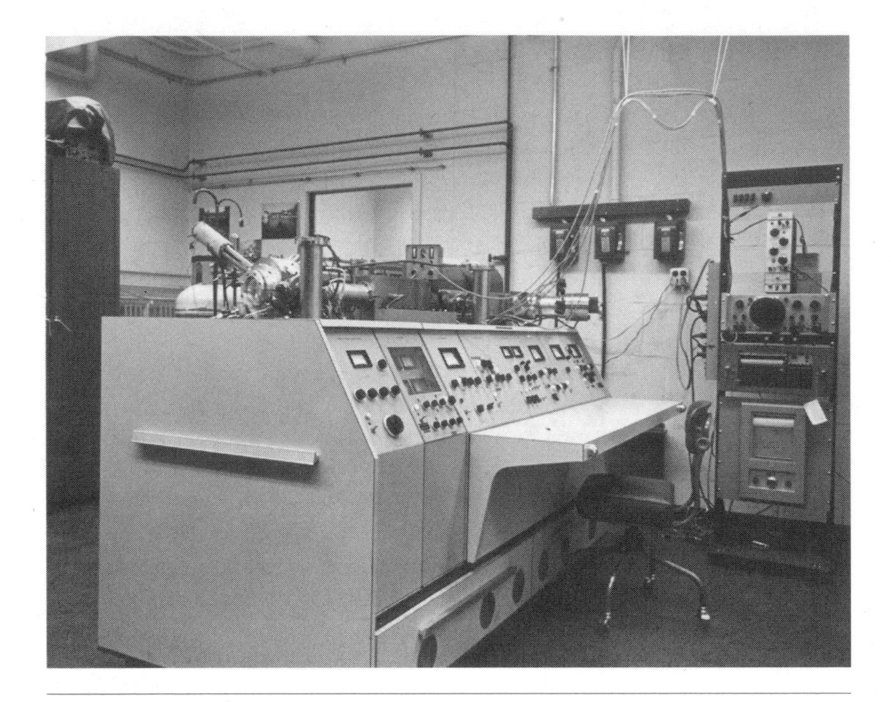

FIGURE 5.2 The CEC 21-110 high resolution mass spectrometer, installed in Biemann's laboratory at MIT. To the far left, a gas chromatography column is attached to the ion source. At the right, an ink pen recorder (bottom) and a galvanometer recorder (middle); on top an oscilloscope to aid in adjusting the spectrum. At the right end of the console, the photoplate holder can be seen. Courtesy of Klaus Biemann.

Biemann was right, his head start did not last long. In 1963, AEI already had 26 orders for the American market, the first one delivered to Humble Oil. In 1962, also Atlas Werke of Bremen brought a high resolution mass spectrometer to the market, the SM 1.[23] Desperate to be in the high resolution business rapidly, the academic competitors of Biemann, notably Carl Djerassi, but also Alma Burlingame, one of Biemann's first graduate students and then at the University of California at Berkeley, based their decisions about which instrument to order on the delivery times.[24]

Biemann's approach differed completely from Beynon's. The rather tedious procedure necessary for high resolution measurements with the AEI

[23] See prospectus of February/May 1962 in Biemann Papers, folder "Burlingame, Dr. A."
[24] Burlingame to Biemann, 27 February and 8 March 1963, Biemann Papers, folder "Burlingame, Dr. A."

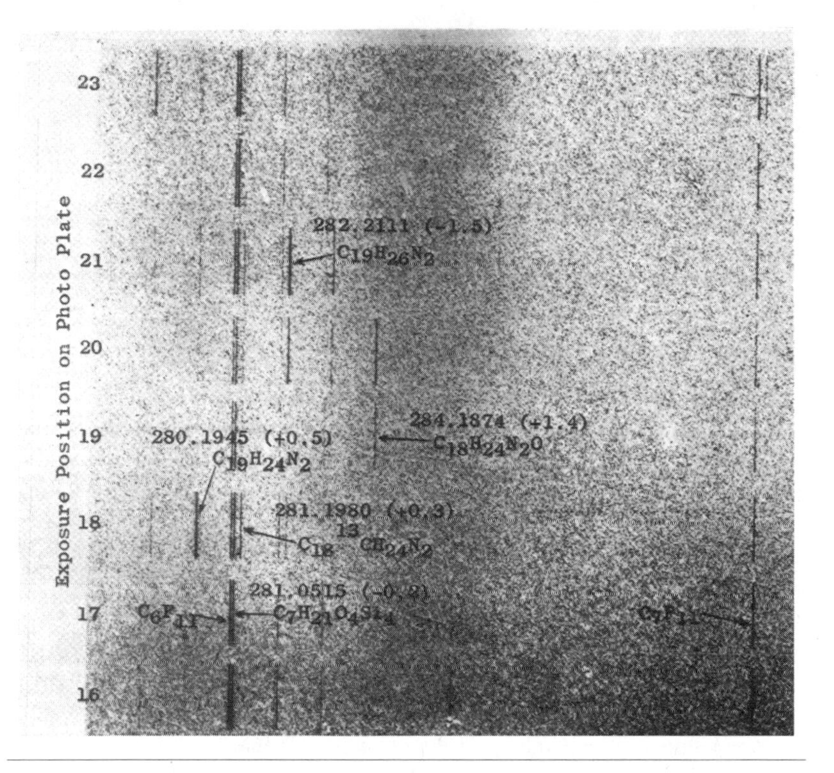

FIGURE 5.3 Photographic plate of high resolution mass spectrum of an alkaloid. Reprinted with permission from J. T. Watson, Klaus Biemann, "Direct Recording of High Resolution Mass Spectra of Gas Chromatographic Effluents," *Analytical Chemistry* 37 (1965), 844–851, on 848. Copyright, American Chemical Society.

spectrometers prevented the scanning of the whole spectrum at high resolution. Thus, Beynon scanned the whole spectrum with low resolution and measured only a few interesting peaks in high resolution mode. In contrast, and using the "seemingly outmoded techniques of allowing the ions to impinge on a photographic emulsion placed in this focal plane,"[25] Biemann practically disconnected the measurement process from the spectrometer. The mass spectrum was recorded on a photographic plate as a series of lines with different densities (see Figure 5.3). The positions of these lines depended on the square root of the ion masses. Thus, determination of mass was reduced to an accurate measurement of line distances.

[25] Biemann, "High Resolution Mass Spectrometry," 193.

FIGURE 5.4 One of the earlier versions of what finally became a fully automated, computer controlled photoplate reader. Here the optical table is driven by a variable speed motor (far left) while a digital encoder on the other end of the precision screw reads its position. The box at left on the microscope contains a photomultiplier, the output of which goes through the electronics of a millivoltmeter to a strip-chart recorder (behind the head of the microscope). On its y-axis drive is mounted another digital position encoder that measures the output of the photomultiplier, i.e. the optical density of the line passing under the microscope. The output of both encoders is fed to an automated IBM-card punch (far right). The cards were then taken to the IBM 7094 computer for off-line processing to print the complete mass spectrum in terms of the elemental composition of each ion and its relative abundance (Klaus Biemann, personal communication, 6 October 2005). Courtesy of Klaus Biemann.

In the beginning, this was done manually, but soon the procedure was converted to function automatically:

> The main limitation to this approach was not so much the problem of making all the distance measurements, which is quite easy once the plate is aligned on a suitable measuring microscope (comparator), but the work involved in noting the measured distances and computing their masses, a tedious process which is subject to human error due to fatigue.[26]

[26] Biemann, "High Resolution Mass Spectrometry," 194.

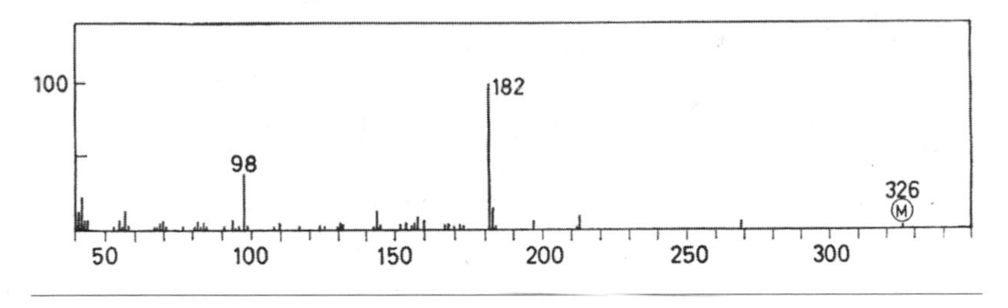

FIGURE 5.5 Low resolution mass spectrum of Deoxydihydro-N_b-methylajmaline. Reprinted with permission from Klaus Biemann "High Resolution Mass Spectrometry of Natural Products," *Pure and Applied Chemistry* 9 (1964), 95–118, on 103, fig. 4. Copyright 1964, IUPAC.

Originally, the photo plate technique had been developed by physicists for exact atomic mass measurements. For them, an entirely manual measuring technique was sufficient. For the large amount of data produced by organic mass spectrometry, the automation of the technique became mandatory. From the beginnings in 1964, these techniques were constantly improved.[27] In Biemann's first set-up, an automated card punch recorded the positions along the photo plate (see Figure 5.4). Manually operated, it yielded five boxes of cards for every one plate. A computer program calculated the masses. Soon, he introduced a semi-automatic version, with a plate bed driven by a stepper motor.

While at first Biemann sent the cards to a central computer facility at MIT, he soon acquired his own computer.[28] The card punch became fully automated, and the density profile of each line was recorded. In 1966, the card system was replaced by magnetic tape and one year later the computer was connected on-line to the comparator. As a result, this enabled Biemann's group "to measure an entire mass spectrum consisting of more than 1000 lines, including the computation of all accurate masses within a total of 6 min."[29]

These refinements of mass measuring techniques made it possible to determine without any doubt the composition of each ion that hit the photographic plate. Human interpretation, for example along the lines of fragmentation mechanisms, played no part in this. The data of high resolution spectra and their direct conversion to elemental compositions required the use of the computer to assist the human interpreter. In a conventional spectrum, the masses of the ions were presented in bar graphs, and interpretation was based on a few peaks only. For example, the interpretation of the low resolution mass spectrum of the specific alkaloid shown in Figure 5.5 was difficult, because it contained only a few large, characteristic peaks.

[27] Biemann, Bommer, and Desiderio, "Element-mapping."
[28] Biemann, interview by Reinhardt, 10 December 1998.
[29] Biemann, "High Resolution Mass Spectrometry," 194.

FIGURE 5.6 Klaus Biemann in front of the CEC 21-110, holding a photoplate, and discussing it with postdocs Walter McMurray and Peter Bommer (seated), 1963. Courtesy of Klaus Biemann.

Because in the conventional interpretation of low resolution mass spectra small peaks had to be neglected, Biemann and his group members "always considered that it is a pity to discard all the information contained in this part of the spectrum and [it was] felt that their elemental composition might be the needed additional criterion to make them useful."[30] In high resolution mass spectrometry, the elemental compositions of all fragment ions could contribute to spectra interpretation, and Biemann decided to make them the centerpiece of graphic presentation of the high resolution spectrum. He coined the name "element map" for his new presentation scheme.

Together with his associates Peter Bommer, Dominic Desiderio, and Walter McMurray (see Figure 5.6), Biemann in the mid-1960s developed high res-

[30] Biemann, "High Resolution Mass Spectrometry of Natural Products," 103.

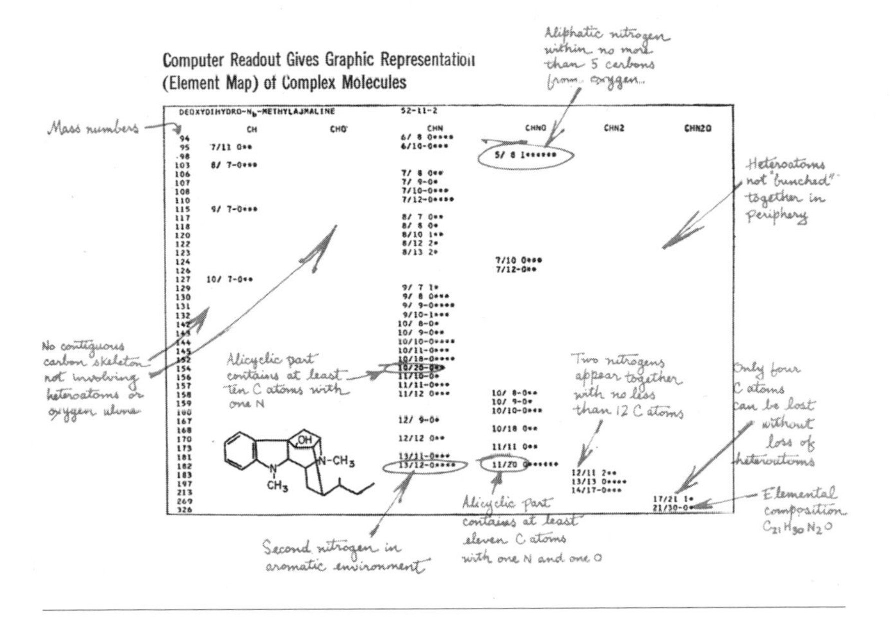

FIGURE 5.7 Element Map of the high resolution mass spectrum of Deoxydihydro-N$_b$-methylajmaline, showing Biemann's reasoning in interpretation. Reprinted from Anonymous, "Element Maps Depict Complete Mass Spectra. MIT Scientists Use all Lines in High-resolution Mass Spectra to Determine Element Distribution in Large Molecules," *Chemical and Engineering News*, 6 July 1964, 42–44, on 43. Published 1964, American Chemical Society.

olution mass spectrometry into a highly efficient tool for the structural elucidation of complex organic molecules. The distribution of the elements in each fragment was given as well as the relative intensity. From left to right of the element map, fragments with increasing heteroatom (nitrogen and oxygen) content were listed. In the example shown in Figure 5.7, fragments containing only carbon (C) and hydrogen (H) were listed at the extreme left of the map, and ions with oxygen (O), and nitrogen (N) in the following columns. The numbers referred to the number of carbon and hydrogen atoms in each species; the last digit showed the deviation in millimass units of the mass found from the calculated mass of the species shown. The asterisks were symbols for the relative intensities. Obviously, such element maps required a skilled interpreter. A 1964 article in *Chemical & Engineering News* gave pedagogical insights into Biemann's methods with the example of the spectrum of an alkaloid of known structure (for its low resolution mass spectrum see Figure 5.5).[31]

[31] Anon., "Element Maps Depict Complete Mass Spectra."

After Biemann and his group members had developed the main features of element mapping, the new interpretive technique had to be checked with a number of mass spectra of different compound classes. This would ensure the connection of high resolution to conventional mass spectra. More importantly, a lot still had to be learned even about mass spectra of simple compounds. Only with a thorough knowledge of the effects of molecular groups on the mass spectra of different compounds could element mapping gain the reliability that was necessary for routine use. With this, more complex compounds such as alkaloids and steroids could be tackled. Already in late 1964, Biemann planned to investigate natural products that had hitherto not been studied by mass spectrometry:

> This would be a test for the prediction that the element map reveals sufficient information about a compound to arrive at once at a reasonable working hypothesis for a possible structure which enables one to conceive a few specific experiments which prove the complete and correct structure. If this can be achieved, mass spectrometry would have gone an important step further toward a less empirical technique.[32]

In sum, Biemann was convinced that element mapping was a "new and entirely different way of interpreting mass spectra." With the help of this presentation scheme, the full amount of information contained in a high resolution mass spectrum was taken into account. Biemann, whose hobby was photography, pointed to a familiar analogy:

> The element map . . . adds a new dimension to mass spectrometric information which may be likened to the comparison of a black-and-white picture with colour photography. In the former one we used to associate various tones of grey with colour, which may or may not have been the correct one. The colour picture relays this information directly. Analogously, we were used to associating mass numbers with various combinations of elements but as was illustrated above, the element map shows this directly and we have hardly even used mass numbers . . . nor was it necessary to draw detailed mechanisms to support conclusions.[33]

The great advantage for Biemann was based on the fact that many conclusions about the structure of the compound could be drawn on the basis of elemental compositions alone. This diminished the need to use fragmentation mechanisms of sometimes doubtful validity, and efforts to prove their correctness with time-consuming labeling studies. As the latter was the main task of the expert mass spectrometrist in the mid-1960s, high resolution mass spectrome-

---

[32] Biemann, application for research grant NSF G-21037, "High resolution mass spectra of complex organic molecules," 1964, p. 6, Biemann Papers, folder "Early NSF grant."
[33] Biemann, "High Resolution Mass Spectrometry of Natural Products," 117.

try threatened to destroy a research field that had come into existence only a few years earlier. Fortunately, for the traditional mass spectrometrist, high resolution mass spectrometry was very expensive, and needed highly sophisticated operation procedures that could not be installed at each university. Still, many problems in structural organic chemistry could be solved with low resolution mass spectrometry. Thus, high resolution mass spectrometry complemented but did not replace its low resolution counterpart. It did so with the massive use of the computer, and with the foundation of special centers where the necessary human skills for operation and interpretation could be accumulated. Biemann did not see this as a danger that threatened the chemist, one that removed him from his tasks. On the contrary, the chemist was to be set free for more creative work:

> If provided with the necessary instructions, it [the computer] may also handle those steps requiring experience and thus arrive at a pre-interpretation which may represent the solution of the problem in simple cases, or the basis for the final interpretation by the chemist who, being relieved by the computer from the bookkeeping portion of the work, may use his intuition more effectively and thus be able to solve more formidable problems.[34]

Still, the complex relationships of the various data in an element map, and the information that could be gained from inspection of all of them made a thorough training in interpretation mandatory. It was important that the mass spectrometrist "reached a stage at which one can visually inspect the vast amount of information contained in a high resolution mass spectrum and can interpret it sensibly."[35] The availability of the computer, and the numerical nature of the data, consequently led to a replacement of the calculation efforts by human interpretation through the computer. But even this approach was restricted through the vast amount of chemical knowledge that had to be included in the program:

> For a given set of data there are too many possibilities which have to be considered and evaluated; in short, we know far too much about chemistry and mass spectrometry to be able to incorporate all this into a single program capable of interpreting the spectra of a wide variety of compounds.[36]

Only in cases where the mass spectral behavior of a specific compound class was highly predictable could the computer take over all parts of the interpretation. The amino acid sequencing of peptides was a suitable task in this regard. The uniform character of peptide chains and the constitutive amino acids enabled

---

[34] Biemann and McMurray, "Computer-aided Interpretation," 647–648.
[35] Biemann and Fennessey, "Problems and Progress," 232.
[36] Biemann and Fennessey, "Problems and Progress," 232.

algorithms that could be easily expressed in computer programs.[37] For other, more general problems, Biemann found ways to use the computer in an interactive procedure. This could

> combine the computer's capacity of quickly and exhaustively performing many numerical comparisons and evaluations, with the human mind's enormous capacity of recognition and decision-making based on knowledge and experience.[38]

Biemann proposed using a time-shared computer, in this case the project MAC at MIT. An IBM 7094 central processing unit was connected via telephone lines to electrical typewriters throughout the institute, some of them to other places in the United States. In direct dialogue with the central computer, the chemist could interrogate the stored spectra, and establish, for example, which elements were present in the compound, and what the most intense peaks were.[39] Later, the group around Biemann developed methods for the on-line handling of mass spectral data. Dominic Desiderio wrote the first program for calculating masses from line distances, in FORTRAN. In 1965, Norman Mancuso joined the group as a postdoctoral associate. While trained as an organic chemist, he took charge of the programming. Two years later, James E. Biller entered as graduate student. During his undergraduate years at Occidental College, California, he had become experienced in computer programming and his Ph.D. thesis with Biemann was concerned with the acquisition and processing of GC/MS data. He led the computational efforts for the next two decades.[40] With the new technique of high resolution mass spectrometry, new funding measures and new organizational modes came within reach of the high-budget technology of mass spectrometry.

CENTERS AND FACILITIES

Apart from its large research laboratories in Bethesda, Maryland, the National Institutes of Health (NIH) supported a multitude of health-related research projects all over the United States. Because many fields in chemistry, and even physics, could be related to the biomedical area, NIH also funded investigations in the physical sciences. In the beginning, the "extramural program" of NIH comprised a project-related funding system only. But in the early 1960s, NIH officials fully recognized the importance of expensive instrumentation,

---

[37] Biemann, Cone, and Webster, "Computer-aided Interpretation."

[38] Biemann and Fennessey, "Problems and Progress," 232.

[39] Biemann and Fennessey, "Problems and Progress," 233–234.

[40] Personal communication of Klaus Biemann, 10 April 2003. See Biemann, "Massachusetts Institute of Technology Mass Spectrometry School," 337.

and set up a program to finance special instruments, organized in centers. In NIH terminology, these centers were called facilities, or Special Research Resources. They resided in universities and other non-profit institutions, and provided service to the academic community in their respective fields. The scope of such a Special Research Resource could range from the entire United States to a group of scientists belonging to several departments within a single institution. In most cases, an NIH research resource covered a certain geographical area. In addition to service work, the personnel of the facility were expected to pursue "core research" in their respective fields, mainly to enable the scientists to continuously upgrade the facility. Part of this work was done in collaboration with outside users. The third important task of a facility was to train researchers. Special Research Resources in mass spectrometry were equipped to the extent that their capacity was large enough to provide mass spectral data to a substantial number of users working on a wide variety of research problems. Behind this development stood the fact that instrumentation was often so expensive that it could not be financed by a single research group. Thus, the sharing of the instrument became essential. NIH officials regarded sharing especially useful in fields where the state of research was still in flux:

> First, knowledge relevant to resource techniques and associated instrumentation should be in a dynamically evolving status and, therefore, an important subject for intensive research in its own right. When such is true, the Special Research Resource becomes a potentially near ideal administrative form for ensuring that the results of scientific and technical innovation find application in support of continuing biomedical research and, conversely, for ensuring that further innovative efforts to upgrade resource capabilities will be initiated, at least in part, in response to the recognized needs of the associated biomedical community.[41]

Facilities of such a kind thus became centers for innovation of scientific methods, and the institutional locus of scientists specializing in advanced uses and improvements. For a mass spectrometrist pursuing cutting-edge research in the field of instrumentation, a grant of this kind supplied crucial equipment and gave access to a large community of clients that relied on his work. For NIH, the Special Research Resources became a means to influence the development of instrumentation along the lines of greatest interest to NIH. At the same time, many biomedical scientists were granted access to specialized instruments and scientific expertise that would be too time- and resource-consuming for them to do on their own. Needless to say, instrument manufacturers saw an

---

[41] Memorandum, "Characteristics of a special research resource," n. d., Biemann Papers, folder "NIH SRR Grant Renewal of RR00317-05, August 1, 1971–July 31, 1976."

opportunity to make their instruments better known, and to provide contacts to potential buyers. Thus, instrument facilities became meeting grounds for funding agencies, instrument makers, and user-scientists, with the instrument specialists such as Biemann and McLafferty in the middle. In 1977, 52 such centers existed (18 Computer Resources, 3 Biomedical Engineering Centers, 8 Mass Spectrometry Resources, 12 NMR Resources, 6 Electron Microscopy Centers, and 5 others).[42]

In 1965, Biemann applied for such a research resource grant entitled "Mass Spectrometry Facility for Biomedical Research." This was the result of his earlier application for an NIH training grant, thought to support the education of graduate students in specialized fields of importance for the biomedical sciences. Because Biemann needed a second high resolution mass spectrometer if the training grant were approved, he was given advice by NIH officials to try to get the funds through a research resource grant.[43] Biemann's resource was the first in mass spectrometry financed by NIH. For the period from 1966 to 1971, he received a total of more than $800,000. The mass spectrometer, a CEC 21-110B, came at a price of $135,000, and Biemann needed nearly $50,000 for the renovation of the space to house the laboratory. The rental of an IBM 1800-2D computer required around $50,000 a year. Altogether, NIH spent 75% of the grant on equipment and supplies, the rest for personnel. Biemann was assisted by an assistant director, one chemical technician (to operate the spectrometer), one electrical technician (to keep the system in operating condition), a programmer, and a secretary. The training grant, funded at $510,000 over five years, supported graduate and postdoctoral students who made use of the equipment.

In his application, Biemann emphasized the aims of a Special Research Resource, and the need to have available the necessary trained manpower. His research group provided this expertise, and Biemann regarded high resolution mass spectrometry as ideally suited to a centralized facility:

> The simplicity of the basic principle, namely the determination of elemental compositions by accurate mass measurement is so convincing that it has obtained the kind of publicity which stimulates the desire of having such an instrument in each university or laboratory.[44]

Notwithstanding such wishes, most problems encountered in a typical chemical laboratory could be solved by low resolution mass spectrometry. Moreover,

---

[42] Division of Research Resources, NIH, *Biotechnology Resources. A Research Resources Directory*, Bethesda: U.S. Department of Health, Education, and Welfare 1977 (DHEW Publication No. (NIH) 77-1430). Copy in Jardetzky Papers, folder "SMRL research resources info."

[43] Biemann, "Massachusetts Institute of Technology Mass Spectrometry School," 334.

[44] Biemann, application for research grant at NIH, FR 00317-01, 30 September 1965, p. 3, Biemann Papers, folder "SRR progress report Nov. 1967, FR 00317."

the operation of a high resolution instrument required a large investment in equipment and personnel:

> Thus, to make use of such an instrument and to economically digest all the data which it can produce, a relatively large, experienced and efficient group of people is required as well as a quite complex data acquisition system and a considerable amount of time on a large and fast computer. Once this is accomplished a large body of data can be obtained routinely, much more than any given laboratory can require or use.[45]

The automation of high resolution mass spectrometry thus caused the change in its social organization: High resolution mass spectrometry became a centralized venture. This, in turn, made necessary the efficient and error-free communication of research results to the clients. The members of the facility had a huge responsibility for achieving this, as the users required substantial training in interpretation of the results. Biemann thought that the computer would annihilate the risks of miscommunication, reducing the danger of error on the side of technicians, and thus make his facility more reliable than others that had to rely on the inconstant output of human labor.[46] In summary, the facility served the scientific community with spectra not including any interpretation, interpreted spectra, and collaborative projects on a larger scale. In addition, the efficient data handling system of the automated densitometer and the IBM 1800 computer allowed acceptance of photo plates of other laboratories equipped with a similar mass spectrometer, but lacking the accessories.[47]

Based on his experience with requests for mass spectra, Biemann did not fear that the facility would be under-used. On the contrary, he thought that he would have to exclude problems that were "submitted to merely collect another set of data on trivial compounds or to impress others by the fanciness of the methods." In order to set up a working mechanism for such judgments, he proposed an advisory board, consisting of an "impartial group of scientists of high reputation, either making this selection directly or overseeing the decisions made by the person in charge of the mass spectrometry laboratory." These scientists should live in relative geographical proximity, and Biemann proposed two faculty members of MIT, one from Harvard University, and one from Tufts University. Three of them were biochemists, two, in addition, members of Medical Schools. Thus, Biemann chose them out of the circle of the largest

---

[45] Biemann, application for research grant at NIH, FR 00317-01, 30 September 1965, p. 4, Biemann Papers, folder "SRR progress report Nov. 1967, FR 00317."

[46] Biemann, application for research grant at NIH, FR 00317-01, 30 September 1965, p. 5, Biemann Papers, folder "SRR progress report Nov. 1967, FR 00317."

[47] Biemann, description of facility for high resolution mass spectrometry, n. d., Biemann Papers, folder "Facility."

group of intended future users.[48] With the approval of the NIH Special Research Resources grant, Biemann considerably expanded his laboratory. Complete with facilities for data collection and handling, he acquired a second new CEC 21-110B. In addition, his old high resolution spectrometer was rebuilt with funds from the NIH training grant. Thus, two up-to-date instruments were available to Biemann's laboratory. One was installed in a specially shielded room to reduce the effects of nearby power transformers and cables, and to permit higher stability of the ion beam for longer exposures.[49] In addition, a Hitachi RMU6-D single focusing mass spectrometer had been acquired in May 1966, funded by a NASA grant for preparing for detection and identification of organic matter in the lunar crust. With the old CEC 21-103C, four mass spectrometers were assembled in Biemann's laboratories, funded by grants from NIH, NASA, and NSF.[50] For the years from 1965 to 1970, Biemann had been awarded around $2 million, including the NIH Special Research Resource and training grants, worth around $1.3 million.[51] His budget provided by public agencies for 1967 alone comprised more than $400,000, excluding a large grant to develop mass spectrometry at the NASA Lunar Receiving Laboratory, in Houston, Texas.[52] Research in mass spectrometry was an expensive business. In one grant application alone, Biemann had to ask for $5,000 a year for maintenance. This included money for the "chart paper, photographic supplies, liquid nitrogen, filaments, spare parts, [and the] service engineer." This was approaching the annual salary of a research assistant, around $6,000.[53] Many of the items needed for the continuous operation of a

[48] John M. Buchanan, head of the division of biochemistry, MIT; George Büchi, MIT; Alton Meister, chairman of the department of biochemistry, Tufts University School of Medicine; and Bert Vallee, professor of biological chemistry, Harvard University Medical School. Biemann, application for research grant at NIH, FR 00317-01, 30 September 1965, p. 9–10, Biemann Papers, folder "SRR progress report Nov. 1967, FR 00317."

[49] Klaus Biemann, informal preliminary progress report on NIH grant FR 00317-01, enclosed with Biemann to George N. Eaves, 31 August 1967, Biemann Papers, folder "SRR progress report Nov. 1967, FR 00317."

[50] Resource equipment list, 1 June 1966 to 30 September 1969, Biemann Papers, folder "SRR progress report 10/15/69."

[51] See Biemann, application for research grant at NIH, FR 00317-01, 30 September 1965, p. 26, Biemann Papers, folder "SRR progress report Nov. 1967, FR 00317."

[52] Biemann, grant application "MS methods in protein and nucleic acid chemistry," 28 December 1967, p. 3, Biemann Papers, folder "Renewal GM 05472 (11-14), NIH 5 yrs. 9.01.68." The total of the grant covering the lunar receiving program was $283,328 from July 1967 to June 1968. Another NASA grant is included with $60,000 in the above mentioned sum. All together, in 1967, Biemann was the recipient of three NASA grants, one NSF grant, and four NIH grants.

[53] Biemann, grant application "MS methods in protein and nucleic acid chemistry," 28 December 1967, p. 4, Biemann Papers, folder "Renewal GM 05472 (11-14), NIH 5 yrs. 9.01.68."

mass spectrometer were minor pieces of equipment that had to be replaced frequently. Biemann considered these devices "just as much as 'supplies' as an organic chemist or biochemist would call a piece of glassware slightly more sophisticated than a test tube."[54]

In 1971, Biemann applied for a continuation of the research resource grant, keeping the size of it constant. To extend the scope, he added gas chromatography/mass spectrometry, data processing techniques, and chemical ionization methods to the research program of the facility.[55] The combination of gas chromatography and mass spectrometry (GC/MS) of Biemann's design[56] was developed mainly under the support of the NASA grants, and was greatly enhanced by the availability of the IBM 1800 computer. It became possible to scan the effluent from the gas chromatograph continuously with the mass spectrometer, and to record the data on-line with the help of the computer.[57] The enormous amount of data had to be presented in a form that made it suitable for interpretation, without losing the information as to which peak in the gas chromatogram each mass spectrum was related. Biemann and his graduate student Ronald A. Hites chose to plot a characteristic mass peak during the entire gas chromatogram, a presentation dubbed "mass chromatogram."[58] This made the interpretation of complex mixtures easier, and added the enormous detection capabilities of the mass spectrometer to the efficient separation achieved with a gas chromatograph. To enhance the comfort of interpretation, Biemann and his group chose to record these spectra on 16 mm film. This seemingly old-fashioned feature made the whole inspection process independent of the computer. The films also made it possible to mail voluminous data sets to outside collaborators. The interpretation of the GC/MS data was improved and automated with a matching program using a file of known spectra,[59] compiled in cooperation with the Jet Propulsion Laboratory and a British group, and later taken over by NIH and the Environmental Protection Agency.[60] The output of

[54] Biemann to McClure, 4 March 1968, Biemann Papers, folder "Facility."

[55] Biemann, grant application "Mass spectrometry facility for biomedical research," 18 December 1970, pp. 14–15, Biemann Papers, folder "NIH SRR grant renewal of RR00317-05, August 1, 1971–July 31, 1976."

[56] Watson and Biemann, "Direct Recording."

[57] Hites and Biemann, "Mass Spectrometer-computer System."

[58] Hites and Biemann, "Computer Evaluation."

[59] Hites and Biemann, "Computer Evaluation"; and Hertz, Evans, and Biemann, "A User-oriented Computer-searchable Library." For the whole paragraph see Biemann, grant application "Mass spectrometry facility for biomedical research," 18 December 1970, pp. 15–17, Biemann Papers, folder "NIH SRR grant renewal of RR00317-05, August 1, 1971–July 31, 1976."

[60] Biemann, interview by Reinhardt, 10 December 1998.

Biemann's facility grew steadily from 1967 to 1970.[61] With the knowledge that this had not been the case with all of the other mass spectrometry facilities of NIH, he indicated the preconditions that favored the concept of the MIT facility. First, the start of the facility was eased by Biemann's existing research group that provided experienced personnel. This proved to be a constant feature, because Biemann always had additional research support, mainly from NASA and other NIH grants. The research associates of Biemann could tackle non-routine problems sent in by outside scientists, building on their experience in a variety of topics. This support, through his graduate students and postdoctoral fellows, according to Biemann, provided the major part of the expertise necessary to satisfy the needs of external users. In this process, they individually benefited from the research questions that were raised by the scientists. It was Biemann's custom to assign such research problems to the member of his group who had the greatest experience with the problem in question, either with regard to the chemistry or the mass spectrometric part. Thus, members of the whole group came into contact with outside researchers and their problems, a task that took time, but extended their experience:

> The great training potential of activities related to the mass spectrometry facilities are quite obvious: The steady influx of problems stemming from chemical, biochemical and biomedical research, covering the entire range from simple to very complex ones exposes and involves all members of this research group, pre- and postdoctoral students alike, and even the technical personnel, to a variety of 'real-life' research problems hardly covered by any single research laboratory.[62]

Biemann liked his students and associates to be involved with most of the research problems in his laboratory. Thus, many of his associates mastered the

[61] From 1967 to 1970, the outside use was as follows, in quantitative terms:

Year	HR MS	LR MS	GC-MS, runs (no. of spectra)	Plates, no. of outside spectra measured
1967	60	85	—	—
1968	84	242	6 (4,000)	—
1969	160	247	145 (57,000)	8
1970	374	468	230 (92,000)	20

HR MS: High resolution mass spectra. LR MS: Low resolution mass spectra. GC-MS: Gas chromatography-mass spectrometry. Source: Biemann, grant application "Mass spectrometry facility for biomedical research," 18 December 1970, p. 22, Biemann Papers, folder "NIH SRR grant renewal of RR00317-05, August 1, 1971–July 31, 1976."

[62] Biemann, grant application "Mass spectrometry facility for biomedical research," 18 December 1970, pp. 20–21, Biemann Papers, folder "NIH SRR grant renewal of RR00317-05, August 1, 1971–July 31, 1976."

techniques fully, and many of his students who had been trained in chemistry learned to write programs for the IBM 1800 computer, for example.[63] Biemann's approach enabled the training of experts for instruments that in the 1970s became so vital to industrial, governmental, and academic science and technology. Most of his students came from the organic and the analytical side. While the latter were mainly interested in instrument development, the organic chemists were inclined more towards biochemistry and solved mechanistic and structural problems. Because both groups worked closely together they knew about the potential of each other side.[64]

The combination of a Special Research Resource and a training grant proved to be fortuitous for Biemann and his group of students and research associates. Between 1963 and 1975, the end of the training grant, Biemann had 22 graduate students. In these times of expanding federal support for scientific research, and the boom of mass spectrometry in chemistry departments, more than half of them obtained senior faculty positions. In addition, a substantial number of postdoctoral associates, in the beginning many from Austria, learned mass spectrometry in Biemann's laboratories. Until 1993, Biemann claims, his research school comprised 54 graduate students, and 83 postdoctoral associates, visiting scientists, and technical assistants having permanent careers in mass spectrometry.[65]

Already in the mid-1960s, mass spectrometry had gained enough academic visibility that university departments hired mass spectrometry experts from industry. Examples include John Beynon from ICI, and Fred W. McLafferty from Dow. The training of mass spectrometrists was regarded as essential to further progress in the field, mainly in industry. Consequently, substantial efforts went into the teaching of mass spectrometry, not only in programs at the universities, but also on the national level with short courses arranged by chemical societies and instrument manufacturers. The fact that mass spectrometry was not only an analytical technique, but also a tool to study chemistry of a novel kind, certainly helped to overcome the prejudices in academic departments that analytical chemistry was a boring, fact-gathering technology, not a science, and in decline. The reasoning behind the transfer of knowledge from industry to academia, and back, was the training of graduate students and postdocs, who after their studies could strengthen the analytical capabilities in industry. A precondition for this was that universities, industry, and research funding institutions should supply the necessary research infrastructure (instruments and facilities), and provide opportunities for special training courses.

[63] Biemann, grant application "Mass spectrometry facility for biomedical research," 18 December 1970, p. 21, Biemann Papers, folder "NIH SRR grant renewal of RR00317-05, August 1, 1971–July 31, 1976."

[64] Biemann, interview by Reinhardt, 10 December 1998.

[65] Biemann, "The Massachusetts Institute of Technology Mass Spectrometry School," 334–337.

McLafferty began a series of courses and lectures at Purdue University, and under the auspices of the American Chemical Society (ACS) also on a national level. Two hundred and fifty students and staff members attended his first course at Purdue in spring 1965, an "Introduction to the Interpretation of Mass Spectra." (See Figure 5.8.) A two-day course at the ACS meeting in Detroit from 3 to 4 April 1965 found 100 participants, and was repeated in the following year. The latter was part of the ACS program of continuing education. Moreover, a year after he had moved to Purdue, McLafferty already had requests from six organizations to send their research people there to get acquainted with high resolution mass spectrometry. This enormous interest

FIGURE 5.8 The News Gazette, Champaign-Urbana, April 28, 1965, offered the following account of the occasion shown on the photograph: "Experts on analysis technique. Authorities on mass spectrometry, scientific analysis technique, inspect one of Purdue University's mass spectrometers during a break in a Purdue short course on using the instruments to determine molecular structure. From left are guest lecturers Prof. K. L. Rinehart, University of Illinois, and Prof. A. L. Burlingame, University of California, and Prof. F. W. McLafferty, Purdue, course chairman. Representatives of industrial and government laboratories and universities from 18 states and two Canadian provinces are taking the course, sponsored by the Purdue chemistry department. The course is the first in the nation offered on the subject." The instrument shown is the CEC 21-110B. Courtesy of Fred W. McLafferty.

demonstrates the demand for the new technology in the chemical sciences and industries.[66]

In 1963, Fred McLafferty received an offer from Purdue University of a full professorship in analytical chemistry. There, in Lafayette, Indiana, Lockhart B. "Buck" Rogers (who came from MIT) tried to build up a strong program, and Purdue officials resisted a trend in U.S. chemistry departments that reduced analytical chemistry to a bare minimum. Though McLafferty at first rejected the job, in February 1964 he finally decided to leave Dow Chemical for Purdue, keeping track of Dow research as a consultant. In a letter that announced his decision to his colleagues and friends, McLafferty emphasized the facilities made available to him at Purdue, and found "the prospect of teaching a most intriguing attraction."[67] Most crucially, Purdue University invested considerably in McLafferty's research. A CEC 21-110B high resolution mass spectrometer at a cost of $125,000 (see Figure 5.9), a special vibration-free and air-conditioned laboratory ($20,000), and a Bendix model 12 time-of-flight spectrometer were acquired for McLafferty. In addition, a single-focusing Hitachi RMU 6A spectrometer had been bought with NSF funds. To make full use of the high resolution mass spectrometer, McLafferty applied for a Datex automatic microdensitometer-comparator at a cost of $50,000. At that time, the automation of data processing permitted a considerable advantage in the productivity of McLafferty's group with respect to the main competitors in this field: Biemann at MIT, and Alma Burlingame at Berkeley.[68] Compared with Dow's Framingham laboratory, at Purdue University McLafferty could boast of having much better equipment.

McLafferty originally planned to cover nearly all aspects of instrumentation, fundamental research of principles, and application. In a 1964 grant application submitted to NIH, he emphasized the potential that mass spectrometry had especially for biomedical research. His correlation studies of spectra and structure, supported by newly available high-resolution data, should be expanded to compounds of biological interest. A precondition for this was the development of introduction systems that allowed the study of low-volatility samples. Additionally, other ionization techniques than electron impact needed to be followed up, most notably field ionization. A specific field was the devel-

---

[66] McLafferty, Application for research grant at NIH, GM 12755-02S1, 29 June 1965, p. 5, McLafferty Papers, folder "NIH FR 00354."

[67] McLafferty, form letter, copy sent to Carl Djerassi, 20 March 1964, Djerassi Papers, McLafferty correspondence.

[68] Biemann, who had pioneered this technique, still had to convert the photoplate data semi-automatically to a punched card device; Burlingame's automatic system was regarded as not precise enough. McLafferty, addendum to NIH research grant application "Analytical mass spectrometry," 28 August 1964 (grant no. GM 12755-01, application of 9 May 1964). McLafferty Papers, folder "NIH FR 00354."

FIGURE 5.9 Purdue University, ca. 1965. High resolution mass spectrometer CEC 21-110B with Fred McLafferty (middle), Ph.D. student Edward Chait (above), and William Haddon. On the lower end of the spectrometer, the ion source is seen, followed by the electric and the magnetic sectors. In the upper part, the detection system is visible. Courtesy of Fred McLafferty.

opment of ultramicroanalysis, taking advantage of the high sensitivity and specificity of the mass spectrometer. All these research directions could be made available to the research community in the biomedical sciences on a service and cooperation basis. The CEC 21-110B high resolution instrument was a vital part in McLafferty's research plan, and would enable him to embark on a research program at a time when few high resolution correlation studies had appeared in the literature. For the analysis of these spectra, a collection emulating the Dow low resolution collection seemed a necessary condition, and McLafferty planned to use spare instrument time for the building-up of such a database.[69] Moreover, he was very well aware that the significance of correlation studies had two purposes, a better understanding of the fragmentation processes, necessary for structural elucidation, and the coining of mass spectrometry in chemical terms to serve a marketing function among scientists:

> On the one hand, these [correlations] make possible the application of the mass spectrometer for structure determination of compounds in new important fields. On the other hand, publication of these also helps to make scientists in a wide variety of disciplines more aware of the unique and versatile capabilities of mass spectrometry.[70]

Together with a Bendix time-of-flight spectrometer, McLafferty had three mass spectrometers in working condition within a few months after his arrival at Purdue. Partially through university funds, he was able to employ the qualified research personnel necessary to operate the instruments and to undertake the investigations envisaged: Maurice M. Bursey from Johns Hopkins University was McLafferty's first postdoc, and started with research on the effects of substituents.[71] Thomas Shannon came in May 1965 from the University of Toronto, already with research experience in mass spectrometry through his work with Alex Harrison. He planned to work on field ionization techniques. Martin Senn, who obtained his Ph.D. at the University of Innsbruck, arrived in July 1965 after a postdoctoral fellowship at the University of California at Berkeley with Burlingame. His work was to be the development of techniques to unravel the amino acid sequences of polypeptides. Included in the group was a synthetic organic chemist, Grant Warner, who had considerable experience with instrumentation at the University of Vermont, taking care of the department service laboratory there. Accordingly, the work assigned to him comprised fundamental studies of metastable ions, design parameters of the Mattauch-Herzog geometry of double focusing instruments, and a special retarding device for the Bendix time-of-flight instrument. Rengachari Venkataraghavan, from the National Research Council of Ottawa, Canada, took over as computer specialist. In addition to these postdoctoral fellows, five graduate students and two MS students worked with McLafferty. In summer of 1965, a group of twelve people, equipped with four large instruments (including the Grant-Datex reader) occupied a series of seven adjacent labora-

[69] McLafferty, application for research grant at NIH, GM 12755-01, 9 May 1964, pp. 6–7, McLafferty Papers, folder "NIH FR 00354."
[70] McLafferty, application for research grant at NIH, GM 12755-01, 9 May 1964, p. 7, McLafferty Papers, folder "NIH FR 00354."
[71] See Bursey, "Appreciation of Fred W. McLafferty."

tories at a main ground-floor laboratory of Purdue University.[72] Thus, the effect McLafferty had on the training of mass spectrometrists was already considerable soon after his commencement of an academic career.[73]

The immense investment that was necessary in equipment and trained personnel explained the relative lack of facilities of this type. In the United States, only Biemann's mass spectrometry center offered a full-range service in high resolution mass spectrometry. It was a consequent move by McLafferty and the Purdue officials that in May of 1966 they applied to NIH for a $2 million grant for a mass spectrometry center for biomedical research. The plan was to install a national facility with programs in core research (theory, methodology, instrumentation, and data handling); cooperative research with other investigators on problems in the biomedical sciences; routine service analysis, both in low resolution and high resolution mass spectrometry; and training of Ph.D. students, postdocs, and visiting scientists.[74] Emphasizing the service work of the planned center, and pondering on his experience at an analytical laboratory at Dow Chemical, McLafferty asked for a continuous and sufficient funding for the center:

> The size of the investment required makes it imperative that the total productivity of the organization is not seriously hampered by a shortage in either personnel or equipment . . . . Thus the operation of this mass spectrometry center will resemble the operation of large high-speed digital computer centers or high energy particle accelerators.[75]

The supposed core of the new center was a new "super" high resolution mass spectrometer that McLafferty considered essential. In the planning, he was reminded of his refusal ten years earlier to enter the venture of high resolution mass spectrometry, as he wrote to John Waldron of Metropolitan Vickers, the Manchester-based instrument manufacturer that had built Beynon's first high resolution mass spectrometers:

> In 1954 or 1955 while I was at Dow we requested a quotation on a high resolution mass spectrometer from you. You offered to build a companion instrument to the one you were planning for John Beynon. As anyone's hindsight could tell now, I have of course regretted many times my decision not

[72] McLafferty, application for research grant at NIH, GM 12755-02S1, 29 June 1965, p. 11, McLafferty Papers, folder "NIH FR 00354."
[73] McLafferty, application for research grant at NIH, GM 12755-02S1, 29 June 1965, p. 4, McLafferty Papers, folder "NIH FR 00354."
[74] McLafferty, application for research grant at NIH, FR 00354-01, 31 May 1966, p. 2, McLafferty Papers, folder "NIH FR 00354."
[75] McLafferty, application for research grant at NIH, FR 00354-01, 31 May 1966, pp. 11–12, McLafferty Papers, folder "NIH FR 00354."

to go ahead with that instrument. The following is an attempt to prevent falling behind in the same way again.[76]

In McLafferty's opinion, the new spectrometer should represent a "quantum leap" compared to the then state-of-the-art spectrometer, the CEC 21-110B. He asked for a substantial increase in resolution and put much weight on an improved capability to measure mass accurately. Automatic transfer of the results to the computer should be possible, and McLafferty included specific details in the inlet and ionization systems. He was optimistic that funds would be provided by NIH for the construction of such a spectrometer, having "had considerable encouragement on a sizable expansion of our research program in mass spectrometry from one of the government agencies that support our work."[77]

Despite these promises, McLafferty's inquiries about the potential for meeting these demands from instrument manufacturers were not widely taken up.[78] Finally, he was able to receive a quotation from the Japanese company Hitachi (in the United States represented by Perkin-Elmer). Their guaranteed specifications represented an improvement in resolution by a factor of three, and in mass measuring accuracy of a factor of 200, compared to the CEC 21-110B. The higher input of data asked for an efficient data handling system. The complex nature of the research projects planned, and the high potential of the new high resolution mass spectrometer, led to the development of a system for man-machine communication for control of the instrument itself. This, however, would place great strain on Purdue's central computer, and McLafferty asked for direct computer input and output systems in his laboratories.[79]

It was projected that from the third year of operation 28 people would work in the center, and that no outside support, for example through other grants, would exist.[80] A very important member of the center's team was Jonathan W. Amy, associate professor of chemistry, who directed the instrumentation part of the analytical facilities of the department of chemistry. Started as a one-man operation in 1953, Amy's group had sixteen employees in 1968, developing, modifying and operating analytical methods and instructing

---

[76] McLafferty to John Waldron, 6 August 1965, McLafferty Papers, folder "AEI/Kratos."
[77] McLafferty to Waldron, 6 August 1965, McLafferty Papers, folder "AEI/Kratos."
[78] Letters of inquiry were sent to AEI, CEC, Varian Associates, Perkin-Elmer, and Nuclide Corporation. See McLafferty to C. E. Johannsen of CEC, 6 August 1965, McLafferty Papers, folder "RMH-2."
[79] McLafferty, application for research grant at NIH, FR 00354-01, 31 May 1966, pp. 6–8, McLafferty Papers, folder "NIH FR 00354."
[80] See organizational chart of Purdue Mass Spectrometry Center, from McLafferty, application for research grant at NIH, FR 00354-01, 31 May 1966, p. 15, McLafferty Papers, folder "NIH FR 00354."

students and faculty members in their use.[81] McLafferty proposed that NIH would pay for 25 percent of his time to support instrumentation research at the center. Moreover, he asked for two assistants, one for each high resolution spectrometer. While for the CEC 21-110B spectrometer he could already name Jack R. Barnes, he suggested hiring a specialist from Japan when the Hitachi instrument arrived: "Our experience with the CEC 21-110 indicates that the full time availability of such a man is a real necessity for the success of our rather ambitious research and service plans for this 'super' mass spectrometer."[82] Barnes had obtained an MS in chemistry at Purdue after a two-year stay as instrument development engineer at Fisher Scientific Company, of Pittsburgh, Pennsylvania. There, Barnes had focused on the design of automatic means for chemical analysis.[83] In addition, the half-time service of an electronics engineer for the data handling and computational facilities was deemed a necessity, and McLafferty proposed to hire Edward D. Schmidlin, who had five years of experience in the Nuclear Chemistry Division of Purdue University.

Rengachari Venkataraghavan was supposed to head the computation part of the center with the title of co-investigator. At the National Research Council at Ottawa, he had worked with a pioneer in infrared spectroscopy, R. Norman Jones, on mathematical techniques and computer applications to correct distortions in infrared spectra.[84] He developed the connection of the automated densitometer, the results of which were stored on magnetic tape, to the central IBM computer. The next step was to bypass the photoplate and magnetic tape transport to the central computer by acquiring the data on-line, directly from the high resolution mass spectrometer. For this work, Venkataraghavan used a Digital Equipment Corporation (DEC) PDP-8 computer. For years, this was the computer of choice for doing on-line calculation. McLafferty had good contacts to DEC, and through this connection managed to acquire this computer that had 4 k memory, and a 32 K storage. The resulting system was made available commercially by Perkin-Elmer.[85] For the new center, a Sigma 5 from Scientific Data Systems was planned, representing a medium-size computer.

---

[81] See the biographical sketch in application for research grant, NIH FR 00354-02S1, p. 12, McLafferty Papers, folder "NIH FR 00354."

[82] McLafferty, application for research grant at NIH, FR 00354-01, 31 May 1966, p. 16, McLafferty Papers, folder "NIH FR 00354."

[83] Biographical sketch of Jack R. Barnes, in application for research grant at NIH, FR 00354-01, 31 May 1966, p. 25, McLafferty Papers, folder "NIH FR 00354."

[84] See the biographical sketch in application for Research Grant, NIH FR 00354-02S1, p. 9, McLafferty Papers, folder "NIH FR 00354"; and personal communciation from McLafferty, 16 April 2003.

[85] McLafferty, interview by Reinhardt, 16 and 17 December 1998. See Venkataraghavan, Klimowski, and McLafferty, "On-line Computers in Research."

The cooperative research, and parts of the core research efforts, were to be overseen by an assistant director, unnamed at the time of application in May 1966. He was supposed to have experience in general chemical research as well as in mass spectrometry. In charge of the day-to-day operations of the center, he would need the assistance of two chemists and three instrument operators. McLafferty planned to run the facility 16 to 24 hours per day, according to demand. For research in mass spectrometry, plans were made to accommodate eight graduate students, five postdoctoral fellows, and up to three visiting scientists.

As a consequence of this elaborate and expensive organization, and in the rhetoric of McLafferty, the existence of such an efficient institution capable of handling requests from all over the United States made the need for local high resolution mass spectrometers less urgent. This should justify the substantial investment of more than $2 million (including indirect costs) over a period of five years. Finally the cost of the Sigma 5 computer system alone was more than $300,000, while the Hitachi RMH-2 came at a price of more than $182,000.[86] In 1968, the Hitachi instrument had not yet been delivered. Thus, all of the service and research work-load in high resolution mass spectrometry was carried by the CEC 21-110B spectrometer. In addition, although the facility was a national one, for the period between March and September 1967 the large majority of the spectra were commissioned by staff of Purdue University (especially McLafferty himself).[87]

McLafferty's and Biemann's organizations of their respective NIH mass spectrometry centers shows that organic mass spectrometry had come close to the scale and scope of Big Science. Their facilities centered around a specific technology, and provided service work and a communication platform for a community of scientists. The characteristic features, core research, service, and training, made it possible to provide the innovation and diffusion of novel methods. Ideally, such centers were two-way stations for research, the clients providing the problems, cooperating with instrument experts who developed solutions. The latter could request help from instrument manufacturers, in cases where new method development required substantial changes in the hardware.

Although he received the NIH grant for the mass spectrometry center at Purdue, in 1968 McLafferty decided to accept an offer from the chemistry department of Cornell University in Ithaca, New York. Because McLafferty

---

[86] Special research resource report, NIH grant FR 00354-03, from 1 January 1969 to 31 December 1969, pp. 18, 21, Biemann Papers, folder "NIH site visit Purdue Univ. Dec. 15–16, 1969."

[87] Around 90 percent of the 476 high resolution spectra were used inside Purdue (400 by McLafferty's group alone), and 2,000 of the 2,073 low resolution spectra (ca. 1,000 each by McLafferty and Purdue staff, respectively). McLafferty, progress report, special research resource project grant, 26 October 1967, section II-A. McLafferty Papers, folder "NIH FR 00354."

moved to Cornell before the Purdue instrumentation was delivered, he wanted to transfer a substantial part of the grant to Cornell. This would include the Hitachi mass spectrometer and the Sigma 5 computer. In addition, all graduate students and Venkataraghavan decided to move to Ithaca. Therefore, McLafferty proposed to transfer about 50 percent of the personnel funds of the grant to Cornell, too.[88] At that time, Purdue University had already announced the founding of this mass spectrometry center, the second nationwide facility after Biemann's at MIT. The response of NIH officials to McLafferty's move was to urge Purdue University to keep the mass spectrometry center fully intact:

> While we realize that Dr. McLafferty has relatively rare skills, it is by no means inconceivable that Purdue would be able to find a replacement to provide leadership for this resource, especially when one considers the enormous attraction of this very expensive equipment.[89]

Referring to the earlier encouragement from NIH that he should drop all other research support, McLafferty argued that the Hitachi instrument, and additional equipment and funds, represented the research, and not the service, activities of the center, and he was of the opinion that this research program should move with him to Cornell. Because Cornell, since 1967, also had a facility in mass spectrometry funded by NIH, but lacked the large research component of McLafferty's at Purdue, he actually requested to strengthen this research at Cornell, and to reduce those efforts correspondingly at Purdue.[90] Both NIH and Purdue denied this, and decided to keep the full grant at Lafayette, with the understanding that Purdue University would find a suitable successor for McLafferty within the next 12 to 18 months.[91] As an intermediate successor to McLafferty, Purdue hired John Beynon for one year, beginning January 1969. The hiring of this acknowledged specialist in high resolution mass spectrometry allowed Purdue to run the mass spectrometry facility at a high level of scientific quality, but under different circumstances than before, as Beynon's interests were more instrumental and physical than McLafferty's.[92]

At Cornell, McLafferty joined a department where mass spectrometry played a crucial role in many fields, and with considerable differences in out-

[88] McLafferty to Bruce Waxman of NIH, 15 March 1968, copy in Biemann Papers, folder "NIH site visit Purdue Univ. Dec. 15–16, 1969."

[89] Waxman to Frederick L. Hovde, president of Purdue University, 21 March 1968, copy in Biemann Papers, folder "NIH site visit Purdue Univ. Dec. 15–16, 1969."

[90] McLafferty to the advisory board of the Purdue mass spectrometry center, 25 March 1968, copy in Biemann Papers, folder "NIH site visit Purdue Univ. Dec. 15–16, 1969."

[91] Michael A. Oxman of NIH to Biemann, 2 December 1969, Biemann Papers, folder "NIH site visit Purdue Univ. Dec. 15–16, 1969."

[92] McLafferty, interview by Reinhardt, 16 and 17 December 1998.

look and purpose. Jerrold Meinwald ran an NIH mass spectrometry facility, equipped with an AEI-MS 9 high resolution instrument. After McLafferty's group arrived, a new PDP-9 computer was obtained as a chemistry department facility, and Venkataraghavan provided for the direct linkage of the PDP-9 with the mass spectrometer.[93] In addition to Meinwald, Franklin A. Long, who had a long-standing interest in theoretical aspects of mass spectrometry, had succeeded Peter Debye as department chairman. Simon Harvey Bauer used time-of-flight instruments in his shock tube fast reaction kinetic experiments, and Richard Francis Porter had constructed high temperature mass spectrometers for his research in boron and carbon hydrides. George Morrison specialized in neutron diffraction with the nuclear reactor on campus, spark source mass spectroscopy, and ion microscopy. Other faculty members used quadrupole mass spectrometry in molecular beam and catalyst surface investigations. On a routine basis, two CEC instruments were used for isotope and gas analysis under Gordon Wood.[94]

After he had left Purdue, McLafferty's funding situation completely changed. His application for a new NIH grant to replace the funds left behind at Purdue did not go through. Though he applied for NSF funds to develop Ion Cyclotron Resonance Spectroscopy (more than $200,000 for January 1969 to December 1972), and was named principal investigator of a departmental grant application to NSF for a "departmental computer facility with advanced on-line and graphics capabilities,"[95] his main research interests were endangered. In the outcome, McLafferty's research in his early years at Cornell was supported by the Army Research Office at Durham, North Carolina (AROD).[96] AROD funded McLafferty from 1969 to 1975 with a budget of ca. $175,000. For McLafferty, this army support rescued his research program when "things went wrong" after his move to Cornell, and allowed him to continue his research program, though at a considerably limited scale.[97]

---

[93] See McLafferty to Charles C. Sweeley, 23 December 1970, McLafferty Papers, folder "NIH RR 00649."

[94] McLafferty, interview by Reinhardt, 16 and 17 December 1998, and McLafferty, grant application to NIH, RR 00649-01, 25 January 1971, pp. 22–23, McLafferty Papers, folder "NIH RR 00649."

[95] See McLafferty, grant application to U.S. Army Research Office, 10 October 1968, p. 5, McLafferty Papers, folder "Army Res. Proposal."

[96] In general, AROD funded unclassified basic research in mathematics, the physical, engineering, and environmental sciences. They differentiated between "Oriented Basic Research" on fifteen military topics, and "Exploratory Basic Research," the latter comprising ca. 60 percent of the available funds. Chemistry, as the largest division, accounted for about 25 percent of AROD's research funds. Peter A. Curtiss, Office of Sponsored Research, Cornell University, "Information memo," 25 July 1969, McLafferty Papers, folder "Army Res. Proposal."

[97] See McLafferty to George M. Wyman, director of the chemistry division of AROD, 5 July 1972, McLafferty Papers, folder "Army Res. Proposal."

In his need for a modern instrument, Hitachi crucially supported McLafferty, offering to give him an RMH-2, the instrument he had helped to develop for Purdue, for free (see Figure 5.10). Behind this offer was the fact that McLafferty was a consultant of the company. Moreover, in the mid to late 1960s, Hitachi instruments enjoyed high sales, and this caused substantial interest and investment in McLafferty's research. But in 1968, Finnigan Instrument Corporation appeared on the market with a much cheaper technology, the quadrupole, computerized and easily connectable to a gas chromatograph. As a result, Hitachi sales (and with them Perkin-Elmer's) began to decline. Though Perkin-Elmer developed a high resolution mass spectrometer that could be connected to a gas chromatograph, this instrument was too expensive to compete with the quadrupole. Finally, Perkin-Elmer stopped their mass spectrometry program, and tried to convince Hitachi to withdraw the offer to donate the instrument to Cornell. Naturally, McLafferty tried hard to change their mind. In the end, they made a deal: Cornell bought for $40,000 one of Perkin-Elmer's GC/MS instruments that could not be otherwise sold, and Hitachi was allowed to give McLafferty the $150,000 double focusing instrument for free.[98]

When McLafferty started at Cornell, his plans were to set up a state-of-the-art computer-based high resolution mass spectrometer, emulating his experiences at Purdue, but free of the service work that came with a national facility. Moreover, he wanted to connect the high resolution instrument to a gas chromatograph, thus for the first time expanding the peculiar performance of GC/MS to high resolution work. But biomedical scientists doubted whether the results attainable with this highly expensive combination of instruments were worth the investment. McLafferty found himself in a dead-end situation. To obtain the funds, he had to convince the experts of the utility of his novel instrumentation. To convince them, he had to show an excellent research record, possible only with the instrument. Henry M. Fales, chief of the chemistry laboratory of the National Heart and Lung Institute of NIH, remarked in January 1971 with regard to extra high resolution mass spectrometry:

> I am afraid it is up to you fellows who have the necessary equipment to show the biochemists how valuable it can be by solving some problems of current biochemical importance which could not have been solved using radioactivity, or regular mass spec methods. I think that once they see something like this in print, they will rapidly join the band wagon.[99]

While other responses, for example by Karl Folkers of the University of Texas at Austin, were much more positive,[100] in the end NIH did not see itself able

---

[98] McLafferty, interview by Reinhardt, 16 and 17 December 1998.

[99] H. M. Fales to McLafferty, 15 January 1971, McLafferty Papers, folder "NIH RR 00649."

[100] Folkers to McLafferty, 15 February 1971, McLafferty Papers, folder "NIH RR 00649."

FIGURE 5.10 The Hitachi RMH-2 double focusing mass spectrometer installed at Cornell University in 1969/1970. John Michnowicz, the postdoctoral fellow shown in this photograph, later went to Hewlett-Packard of Palo Alto. Courtesy of Fred McLafferty.

to fund the necessary computer for the set-up of the system, DEC's PDP-11/45. Though McLafferty gained high priority scores in the review process of NIH, the agency finally refused the funding of the instrument in the fall of 1973, because of the general shortage of funds available.[101]

The PDP-11/45 system had a price tag of more than $125,000, roughly the same amount as the high resolution mass spectrometer itself. In 1972, while the application for NIH funding was still pending, McLafferty asked the marketing representatives of DEC to deliver a PDP-11/45 without being certain that funding could be obtained. DEC, though hesitant in the beginning, agreed under one condition: They wanted to use McLafferty's mass spectrometry database as a sales incentive for their instruments. If prospective customers would like to include it in their order, McLafferty would receive a payment. In

[101] McLafferty, grant application to NIH, RR 00649-01, 25 January 1971, and William F. Raub of NIH Biotechnology Resources Branch, 5 September 1973, McLafferty Papers, folder "NIH RR 00649."

exchange for this willingness, DEC loaned a computer to Cornell.[102] This agreement secured McLafferty the use of a high performance computer system although the governmental agencies had refused to fund his application. In addition to the mass spectra database, he used his efforts in software development as a marketing ploy on behalf of DEC. Especially through his contacts to the Environmental Protection Agency (EPA), he was "convinced that there is a very large market for a next-generation small computer in the GC/MS business."[103] In a few cases during 1973 and 1974, McLafferty wrote letters on behalf of DEC, describing the high potential of the system to prospective customers. He also used his acquaintances in chemistry departments and industrial laboratories, especially when he shared software with them, to underline the utility of the PDP-11 series.[104] McLafferty's programs provided the software basis for sales of the computer. When in May 1974, McLafferty could finally offer DEC a partial payment for the computer with either NSF or NIH funds, he argued for a wished-for substantial price reduction in terms of market publicity:

> This [would] be on the basis of how our research and publicity can help spread the gospel among mass spectrometrists and analytical chemists; we are firmly convinced that the tremendous additional capability of this computer will make it indispensable in a wide variety of laboratories once the possible applications are better known.[105]

The representatives of Digital Equipment Corporation were not too impressed. In the fall of 1974, when the company had a large backlog in deliveries of PDP-11 computers, they increased the pressure on McLafferty to pay for the computer. McLafferty even considered selling the mass spectral data files he had collected, and he asked NSF to come up with $20,000 to $25,000 to contribute to the payment for the computer. At that time, five of McLafferty's research people relied on the use of the PDP-11/45.[106] This was a clear sign of how heavily McLafferty's research centered on the collection of mass spectral data, their efficient handling, and use for interpretation. At the same time, this incident showed how much McLafferty's research depended on agreements with various instrument and computer manufacturers. With the help of his

---

[102] Dell D. Glover from DEC to McLafferty, 4 May 1972, and McLafferty to Glover, 25 May 1972, McLafferty Papers, folder "Dig. Equip. Corp."

[103] McLafferty to Glover, 14 December 1972, McLafferty Papers, folder "Dig. Equip. Corp."

[104] See McLafferty to Glover, 20 September and 26 November 1973, McLafferty Papers, folder "Dig. Equip. Corp."

[105] McLafferty to Glover, 17 May 1974, and the attached list of contacts in academe and industry, McLafferty Papers, folder "Dig. Equip. Corp."

[106] McLafferty to Arthur F. Findeis of NSF, 9 October 1974, McLafferty Papers, folder "PBM/STIRS."

methods, and his data collections, McLafferty collected research funds from various sources, with instrument and computer manufacturers supplying equipment, and governmental agencies providing funds for personnel. Naturally, McLafferty had to offer something in return. Besides his talents in improving instrumentation, his development of programs for the matching and interpretation of mass spectra proved to be the most successful incentive for instrument manufacturers to invest in his research.

Matching and Interpreting

Beginning in the mid-1970s, a new environmental regulation policy triggered the demand for new and sensitive analytical techniques. This provided a huge push for mass spectrometric techniques. McLafferty's group contributed through the development of computer programs for identification of small amounts of substances, using databases originally developed in the chemical industry (partially by McLafferty himself at Dow). Computer applications for mass spectrometry and for combined systems of mass spectrometry and gas chromatography became one of McLafferty's research and development specialties. His cooperation with the publisher John Wiley & Sons made possible the administration and publication of a large database of mass spectra. McLafferty's and Wiley's private enterprise very successfully competed with similar joint efforts of the Environmental Protection Agency (EPA) and the National Bureau of Standards (NBS). The inclusion of a high-performance computer program with an up-to-date file of mass spectral data became a crucial sales argument for medium-size mass spectrometers fitted for routine applications in industry, academia, and the government. Data banks were the central factor in this development.

An early, and not computerized, example is McLafferty's book *Mass Spectral Correlations*, using data from a collection of more than 4,000 mass spectra. Though the initial response to the book by Biemann was lukewarm, McLafferty was optimistic about the potential of mass spectrometry and the role of his book in bringing this potential to reality:

> One of the biggest hopes that I have for this book is that it might help to get more organic chemists into mass spectrometry. I don't think of this only as the kind of expert, full time mass spectrometry that you practice, but also as a handy tool in the same way the organic chemist has learned to use IR, GC, and NMR in the last decade. I look for the day when the organic chemist will have his GC and TLC hooked up with the mass spectrometer for quick identification of all sorts of samples of interest to him. Besides the pure new compounds which he has synthesized, there would be identification of reaction mixtures, study of the change of product composition by taking samples at

> different changes during the reaction, checking starting material purities, solvents, etc., etc.[107]

In one important respect, mass spectrometry differed from the spectroscopic techniques that were familiar to chemists of the 1960s: The amount of data present in a mass spectrum far outweighed that of an NMR or infrared spectrum. Contrary to the uses of the latter techniques, mass spectrometrists only took a small percentage of the available spectral information for the elucidation of an unknown molecule into account. The sheer amount of data blocked progress and dissemination of the technique. The solution lay in a more rational organization of mass spectral data. Already in the late 1940s, and still limited to quantitative analysis of mixtures of hydrocarbons, the American Petroleum Institute (API) had set up a database for reference spectra. Supplementing the efforts of the API, CEC itself published reference spectra on McBee keysort cards.[108] Together, the two databases included roughly 1,500 reference spectra in 1956. To be useful in quantitative analysis, the reference spectra and the spectra of the unknown compounds had to be run under identical conditions, even with the same type of instrument. With the rising interest in using mass spectrometry for qualitative analysis, the need for an expansion of the datafiles increased, while at the same time the necessity to include only spectra that were recorded under the same conditions disappeared. Thus, the American Society for Testing Materials (ASTM) E-14 committee on mass spectrometry set up a subcommittee dealing with the issue. Its collection became known as the file of "uncertified" spectra, distinguishing it from the certified spectra of the API and CEC.[109] In July of 1958, the task group of the ASTM E-14 committee reported that they had obtained offers of 700 mass spectra from eleven laboratories. Seventy companies showed interest in receiving copies of these spectra.[110]

In 1960, McLafferty informed the Swedish mass spectrometrist Einar Stenhagen of Gothenborg about his reasoning in collecting uncertified mass spectra:

> Our largest difficulty has been to get holders of large mass spectral files, such as Dow Chemical, to take the time and effort to put them in condition and

---

[107] McLafferty to Biemann, 12 November 1963, McLafferty Papers, folder "22." McLafferty seemingly compared the role of his book to the importance of the Colthup chart in infrared spectroscopy (see Chapter 2). For Biemann's criticism see an early draft of his review held with the same papers. TLC is thin layer chromatography.

[108] The McBee keysort cards were a collection of cards with edge-notched punched holes. Peaks in the spectrum were punched at the edge. Sticks were inserted, and the card that matched the spectrum fell out of the box. See Zemany, "Punched-card Catalog."

[109] McLafferty, Stauffer, Loh, and Wesdemiotis, "Unknown Identification," 1230. See Chapter 3.

[110] McLafferty, "Minutes of meeting of ASTM committee E-14 on mass spectrometry, 11 July 1958," McLafferty Papers, box "Meeting notes."

> make them available for such publication. Most such laboratories are quite willing to donate their spectra, but do not find the time to do the necessary work involved. The 'uncertified' file was our answer to the problem of how to make it as little work as possible.[111]

In 1965, the uncertified spectra file of the ASTM subcommittee contained more than 2,000 spectra, with 2,400 at the processing stage. This included the collection of ca. 2,000 spectra in a volume "Uncertified Mass Spectral Data," edited by Roland S. Gohlke.[112] Distribution and the supply of back copies was informally organized. In order to put this on a more stable basis, McLafferty discussed the further procedure at the Dallas meeting of the ASTM E14 committee in May 1966. In June, he distributed a proposal containing the main topics for the improvement of data handling made necessary by the increase in data and members:

> There are more and more laboratories interested in taking part; new laboratories want the old data; data is needed in new forms, especially those suitable for computer processing; and the mechanics of assembling and distribution are becoming too big a job for a strictly volunteer effort.[113]

McLafferty had in mind to set up a non-profit organization, along the lines of the group behind the publication of *Organic Synthesis*. The plan was to cooperate with the newly established Mass Spectral Data Centre of the Atomic Weapons Research Establishment (AWRE) in Aldermaston, UK,[114] and the group around Einar Stenhagen and Sixten Abrahamsson at the University of Gothenborg in Sweden. It was planned that this worldwide system of mass spectral data would be distributed through the Aldermaston group. But the ways of the two projects parted. In 1969, the collaboration of McLafferty with Einar Stenhagen and Sixten Abrahamsson led to the publication of 6,800 spectra in bound volumes and on magnetic tape by Wiley-Interscience.[115] Essentially, the Wiley *Registry* was the continuation of the ASTM efforts. In their joint endeavor, Abrahamsson was the computer specialist, and he had insisted on publishing the data both in book form and on magnetic tape from the

---

[111] McLafferty to Stenhagen, 27 July 1960, McLafferty Papers, folder "Book chap. Nachod, Phillips, 19."

[112] H. R. Harless, "Annual report subcommittee IV, June 1964–June 1965," McLafferty Papers, folder "Biennial review Anal. Chem. 32"; and personal communciation from McLafferty, 16 April 2003. See Chapter 3.

[113] Fred W. McLafferty, proposal "Mass spectral data committee," 3 June 1966, p. 2, McLafferty Papers, folder "Atlas of mass spectral data, 76."

[114] For a description see Ridley, "Mass Spectrometry Data Center." See also McLafferty to A. T. Morse, Manager of the Isotopic Products Department of Merck Sharp & Dohme of Canada, 24 March 1967; circular authored by N. R. Daley and R. G. Ridley of the UK Atomic Energy Authority, 9 March 1966, McLafferty Papers, folder "Atlas of mass spectral data, 76."

[115] Abrahamsson, Stenhagen, and McLafferty, *Atlas of Mass Spectral Data.*

beginning on. In 1978, the Wiley *Registry of Mass Spectral Data* comprised 41,000 spectra. In the same year, a competing collection, managed by the National Bureau of Standards, appeared with 26,000 spectra.[116] Stenhagen died in 1974, and Abrahamsson in 1979, after which the Wiley database was run only by McLafferty. The Wiley funds supported his research group by travel grants and other measures, and the access to this largest file of mass spectra was a large impetus for McLafferty's research projects in general.[117] At the end of the 1990s, McLafferty's database published with Wiley contained 338,000 reference spectra.

In the 1970s, the future, it was recognized, belonged to computer programs for the searching processes in mass spectra data files. Not only the availability of computers, but also the better performance of mass spectrometers drove this development. Especially high resolution mass spectrometry led to an enormous increase in output of data, and the combination of gas chromatography and mass spectrometry (GC/MS) could yield hundreds of spectra from only one mixture of compounds. Interpretation became the bottleneck, as complicated spectra could take hours for the assignment of a probable structure. Thus, one modern GC/MS could keep several interpreters busy. Automation of interpretation was needed, "computerizing the spectra puzzle" was a called-for strategy. Only with the help of on-line computing, it was thought, "pollutants in water supplies, illicit drugs, insect pheromones, and metabolites of new medicinals"[118] could be identified in an efficient way.

Matching techniques relied on the completeness of the underlying data files, while interpretation programs had to mimic chemical understanding. Both possible ways were tried, and many attempts combined the two features. Klaus Biemann of MIT, for example, developed his own method of matching. Another matching process, based on a "conversational" technique (CYPHER-NET) was used in a system developed by NIH, AWRE at Aldermaston, and EPA.[119] Other systems applied information on structural sub-entities of the molecules. With the help of these systems, compounds could be identified even if they were not included in the data file: the closest neighbors in the matching process gave important hints as to the nature of the compound. One of these systems, based on "learning machine theory," could answer questions

---

[116] McLafferty, Stauffer, Loh, and Wesdemiotis, "Unknown Identification," 1230. See also Stenhagen, Abrahamsson, and McLafferty, *Registry of Mass Spectral Data*; Hilne and Milne, *EPA/NIH Mass Spectral Data Base*. The NBS (today the National Institute of Standards and Technology, NIST) cooperated with NIH and EPA in this project.

[117] McLafferty, interview by Reinhardt, 16 and 17 December 1998.

[118] McLafferty, Dayringer, and Venkataraghavan, "Computerizing the Spectra Puzzle," 78.

[119] For a review of the early algorithms see Pesyna and McLafferty, "Computerized Structure Retrieval."

such as: "Does this compound contain oxygen?"[120] McLafferty chose a midway between matching and interpreting, elaborating on the interpretive aspects of the retrieval systems when they found closely related matches. On the basis of structural features of the spectrum, the compound class was determined. Each of the classes was optimized to be selective for a specific type of substructure.[121] Thus, McLafferty combined the matching process with the interpretation, using a large reference file. McLafferty's program was called *Self-Training, Interpretive, and Retrieval System* (STIRS), and found spectra in the file that were of a similar kind than the sample spectrum. Essentially, this followed the same line of reasoning as McLafferty used in his book on interpretation of mass spectra. Though it was a difficult goal to achieve, McLafferty thought that the emulation of the interpretation procedure of a mass spectrometrist by the computer showed great promise:

> We find that the computer often does a better job, not only because it remembers the behavior of particular compound types more accurately than the human interpreter but also because it can take the time necessary to do all of the detailed steps which the interpreter often short-cuts.[122]

This approach led to an inclusion of the computer into the experimental system comprising the experiment, read-out, computer, and researcher. The researcher would make "high-level decisions" only and would exert them through the computer interface. McLafferty was convinced that the experimenter was relieved from the routine work in "observing, calculating, and recording the results," and that the use of the computer could improve considerably in terms of accuracy and speed. Most importantly, it made possible the inclusion of the large amount of data that previously had not been utilized for the interpretation process.[123] At the same time, McLafferty defended his style of hands-on experience with the novel instrumentation:

> Indeed, we feel that for many experiments the chemist should design, assemble, and/or modify his own hardware (equipment) and software (program instructions); the incentives are similar to those which have induced chemists

[120] Jurs, Kowalski, and Isenhour, "Computerized Learning Machines"; Jurs, Kowalski, Isenhour, and Reilley, "Computerized Learning Machines"; Isenhour and Jurs, "Chemical Applications of Machine Intelligence"; Wangen, Woodward, and Isenhour, "Small Computer."

[121] Kwok, Venkataraghavan, and McLafferty, "Computer-aided Interpretation of Mass Spectra. III," 4186. See McLafferty to A. F. Findeis of NSF, 23 December 1975, McLafferty Papers, folder "NSF application 1975."

[122] McLafferty, grant application, not dated [1968], GM 16609-01, p. 10, McLafferty Papers, folder "1968 NIH MS application file."

[123] McLafferty, addendum to application GM 16609-01, 26 July 1968, McLafferty Papers, folder "1968 NIH MS application file."

> to become familiar with the operation and maintenance of specialized instruments and the interpretation of various types of spectra.[124]

The concept of STIRS was based on the selection of different data classes supposed to have high structural significance.[125] Originally, STIRS used a database of 13,000 spectra. In a second step, alluding to the capabilities of the program to recognize new compound classes simply through the addition of more spectra to the datafile, this was expanded to include 25,000 spectra, in 1973. These spectra were chosen from a variety of sources, including the *Atlas of Mass Spectral Data*, the database of the AWRE at Aldermaston, and published reference books and journals. The majority was checked for errors before including them in the file.[126]

There were doubts, however, as to whether McLafferty's self-training interpretive and retrieval system really was "self-training." Because humans did the organization of the data files and the final steps of interpretation, the concept of a self-training machine did not seem applicable. Moreover, in 1973, there was no feedback of successful or unsuccessful interpretations that could enhance further performance, "an essential part of training" as one referee remarked.[127] Nevertheless, in the opinion of McLafferty and his co-workers, the system was self-training with regard to the fact that it "can interpret an unknown spectrum using only the related reference spectra; no human correlation of spectral behavior is necessary."[128] Thus, in effect it was more a "self-finding" system. The main advantage of STIRS with respect to other programs was the fact that it did not have to be specially programmed for new product classes by adding the characteristics of their spectra-structure correlations. Only the respective reference files had to be added to the database. Here, McLafferty and his group relied on the computation power of the DEC PDP-11/45.[129] McLafferty's approach seemed most promising especially with the needs of the many GC/MS users in mind. They were more interested in identifying a compound whose structure was known already, and not in structural elucidation of

---

[124] Venkataraghavan, Klimowski, and McLafferty, "On-line Computers in Research," 158–159.

[125] See McLafferty, "Self-training Interpretive and Retrieval System," 49; Wiswesser, "107 Years of Line-formula Notations."

[126] McLafferty, "Self-training Interpretive and Retrieval System."

[127] Journal of the American Chemical Society, comments of referee on Kwok, Venkataraghavan, and McLafferty, "Computer-aided Interpretation of Mass Spectra III," McLafferty Papers, folder "STIRS, 113."

[128] Kwok, Venkataraghavan, and McLafferty, "Computer-aided Interpretation of Mass Spectra. III," 4186, n. 19.

[129] McLafferty, grant application to NIH, GM 16609-06S1, January 1974, p. 18, McLafferty Papers, folder "GM 16609, 1971."

molecules from scratch. Consequently, McLafferty chose to cater to this large market.[130]

The overall goal of McLafferty's project was to improve STIRS to such an extent that it could be used routinely in a condensed version on a small GC/MS/computer system. In 1972, when he applied for an EPA grant, McLafferty suggested that this small system would be operable at the end of the second grant year. For this, STIRS had to be reduced in size, because it was developed on a large computer, the PDP-11/45.[131] In 1975, STIRS was made available through Cornell University's IBM-370/168 computer over the TYMNET international network.[132]

In developing STIRS further, McLafferty and his collaborators proposed to first try to match the spectrum of the unknown with a database, and to start the interpretation algorithm only in case this was unsuccessful. He developed a special matching method, called Probability Based Matching system (PBM). PBM served as a prefilter for STIRS, and had been initiated because most online users of the latter program clearly were in need of a matching program, and not an interpretive one.[133] The beginning of PBM was McLafferty's idea of reverse searching: "That's where I came up with this reverse search idea. You are going to be hunting for your targeted compound, but . . . a mass spectrum is a mixture of all compounds coming in."[134] Because of the great demand in identification of compounds in mixtures, it was preferable to check the spectrum for the presence of the data of each reference compound, and not vice versa, as it had previously been the case. Consequently, more than one right answer for each sample spectrum could be found, and PBM was best at identifying components of mixtures.[135] McLafferty had this idea when he consulted for a small GC/MS company, Universal Monitor Corporation of Pasadena, California. In the early 1970s, Universal Monitor employed Duane Littlejohn, whom McLafferty knew from his consulting for Varian Associates. In 1969, and together with Peter Llewellyn of Varian, Littlejohn invented a new

---

[130] McLafferty, memorandum "Computer matching and interpretation of mass spectra, meeting of August 12th," 16 August 1971, McLafferty Papers, folder "STIRS, 113."

[131] McLafferty, application for research, development, and demonstration grant, "Computer interpretation of pollutant mass spectra," 1972, pp. 9a and 9f, McLafferty Papers, folder "EPA 1976."

[132] McLafferty, research proposal to NSF, "Computer interpretation of mass spectra," 13 May 1975, p. 11, McLafferty Papers, folder "NSF application 1975."

[133] McLafferty, report of grant "Computer interpretation of mass spectra," EPA R-801106, March 1976, p. 1; McLafferty to John M. McGuire of EPA, 4 June 1976, McLafferty Papers, folder "EPA 1976."

[134] McLafferty, interview by Reinhardt, 16 and 17 December 1998.

[135] See McLafferty to John McGuire of EPA, 15 October 1976, McLafferty Papers, folder "EPA 1976."

GC/MS interface.[136] The plan of the company was to use GC/MS to monitor the environment and plant process streams. McLafferty published his first paper on PBM in 1974 with Universal Monitor employees Robert H. Hertel, a physicist, and Robert D. Villwock, a computer and electronics expert.[137] Though Universal Monitor patented PBM, the company was on shaky grounds and later was sold to McDonnell-Douglas.

PBM used a quite sophisticated algorithm, based on the experiences of Gerard Salton, Cornell's "guru" of document retrieval from libraries.[138] While the documentation of books and other printed materials needed quite a lot of data fields (such as author, subject, year of publication, etc.), in mass spectrometry only mass and abundance had to be compared. The algorithm used the probabilities of occurrence of individual peaks in the spectrum to weight the matching. For example, the matching of peaks of high mass values was less probable than with peaks of low mass.[139] If a peak was matched that had a low probability of appearing in a mass spectrum, it was weighted higher in the assignment of a structure than a more probable one. In addition, the use of the molecular ion, discarding of peaks, and other interpretive techniques used by experts in mass spectrometry were included. Thus, a lot of mass spectrometry know-how was needed to retrieve the full information of a mass spectrum.[140]

To make PBM a matching program compatible with a variety of instruments, the program retained only those peaks with the highest confidence values. Thus, the statistical weightying of PBM proved to be a crucial part for its broad usefulness. Until 1976, PBM was developed to become an efficient search program using the largest and most diverse reference file on the market. In cases where no match could be found by PBM, STIRS was used to provide information on the structures of the molecules under scrutiny. Nevertheless, a completely computerized interpretation of a molecule not contained in the spectra file was an exception, not the rule, "because STIRS is meant to be an aid to, not a replacement for, the trained mass spectrometrist, the best answer in difficult cases will be achieved through human interpretation of computer results."[141]

The greatest part of the development of PBM and STIRS was supported by EPA. For EPA, the detection of pollutants was a crucial step in control of

[136] Peter M. Llewellyn, Duane P. Littlejohn (Varian Associates), "Gas Analyzer System Employing a Gas Chromatograph and a Mass Spectrometer With a Gas Switch Therebetween," U.S. patent application (1969), no. 19,670,327.

[137] McLafferty, Hertel, and Villwock, "Probability Based Matching."

[138] Salton, *Automatic Information Organization and Retrieval.*

[139] Pesyna, McLafferty, Venkataraghavan, and Dayringer, "Statistical Occurrence"; id., "Probability Based Matching System."

[140] McLafferty, interview by Reinhardt, 16 and 17 December 1998.

[141] McLafferty, report of grant "Computer interpretation of mass spectra," EPA R-801106, March 1976, p. 7; McLafferty to McGuire, 4 June 1976, McLafferty Papers, folder "EPA 1976."

the environment, and in the early to mid 1970s, GC/MS became the method of choice for such measurements. During a single GC run, many mass spectra could be made, meaning that hundreds of spectra had to be interpreted in a single day. Consequently, EPA had a major interest in the development of efficient computer programs for interpretation and matching of mass spectra.[142] As a result, GC/MS showed its great advantages over other analytical methods. John McGuire, McLafferty's contact official at EPA, emphasized in 1975 that GC/MS was much more efficient than GC alone and yielded results of high confidence at a part-per-billion level.[143]

At the end of the first EPA grant period, a conflict of interests arose involving the stakes that Universal Monitor Corporation had in PBM, and the goals and wishes of EPA for the collaboration with McLafferty. In order not to infringe the company's patent, McLafferty was forced to refrain from renewing the EPA grant in the summer of 1975, with serious financial consequences for the continuation of his research. To prevent such a conflict happening again, in April of 1976, when McLafferty applied a second time for an EPA grant, he asked representatives of EPA and Universal Monitor to clarify the situation and to find out the status of the company's "patent rights on any research work which I might do at Cornell under EPA auspices."[144] If Universal Monitor felt that the existing consulting agreement with McLafferty would not harmonize with his work for EPA, McLafferty announced that he would not renew the consultancy. This turned out to be unnecessary, though, since the company did not see "any finite chance of patent interference."[145]

In general, McLafferty's work in this direction was not unanimously greeted with approval by the scientific community, in contrast to EPA. When, in 1976, McLafferty planned to develop STIRS further for high resolution mass spectrometry, the reviews were contradictory, and NSF did not fund the application. This threatened to be the end of McLafferty's efforts in computer applications in mass spectrometry, and he announced to EPA that their grant would probably be used to finish the PBM/STIRS work.[146] One reviewer's criticism had been that McLafferty's proposal included two major challenges, one of them not being scientific: "McLafferty is now faced with refinements of the LRMS [Low Resolution Mass Spectrometry] - STIRS approach to make it economically *feasible*, and exportable on a network, and the inclusion of HR

---

[142] McLafferty, draft for report to EPA, "Improvements to probability based matching for unknown mass spectra," grant R-804509, January 1979, p. 1, McLafferty Papers, folder "EPA 1976."

[143] Heller, McGuire, and Budde, "Trace Organics by GC/MS."

[144] McLafferty to Philip P. Sharples of UMC, 20 April 1976, McLafferty Papers, folder "EPA 1976."

[145] McLafferty to McGuire, 4 June 1976, McLafferty Papers, folder "EPA 1976."

[146] McLafferty to McGuire, 4 June 1976, McLafferty Papers, folder "EPA 1976."

[High Resolution] mass spectra."[147] McLafferty's next grant application to NSF reflected these remarks. In 1979, he applied for an industry/university cooperative research activity in the NSF program. The goal of the project was the extension of both PBM and STIRS to the GC/MS systems of Hewlett-Packard. This cooperation continued the established connection of McLafferty to this company: One of the first Purdue Ph.D. students of McLafferty, Robert Board,[148] had gone to Hewlett-Packard and built the first quadrupole spectrometer there. A PBM algorithm was in 1979 installed in the HP 5985 GC/MS system, and McLafferty planned to concentrate on the more scientific questions as his side of the project. The loss of Rengachari Venkataraghavan, who had gone to American Cyanamid's Lederle Laboratories in 1977, made the collaboration with Hewlett-Packard necessary. Venkataraghavan had done much of the computer science in McLafferty's laboratory.

The NSF grant application brought funding that between May 1980 and October 1983 enabled improvement in both the performance of PBM and STIRS, and enlargement of the database up to 80,000 spectra.[149] McLafferty wrote to member of Congress, Matthew F. McHugh, that Hewlett-Packard was "a major factor in the approximately $150,000,000 per year GC/MS market, but they have strong competition from manufacturers in Japan, Britain, France, and Germany."[150] The performance of PBM in the automatic and fast identification of components of mixtures certainly provided a competitive edge for the U.S. company. An earlier sign of success had been that the Water Research Laboratory of EPA in Athens, Georgia, installed PBM to process a "backlog of a million unknown mass spectra from 20,000 GC/MS runs."[151] This was a consequence of the increase in stringent legislation regarding environmental standards in the United States during the late 1970s. Moreover, when instrument manufacturers such as Finnigan Instruments Corporation demonstrated the cost-effectiveness and availability of computerized GC/MS systems, this analytical technique became the method mandated by EPA for analyses in industry.[152]

The story of how McLafferty and his collaborators developed PBM showed the importance of his connection to industry. This proved to be true also in other fields of McLafferty's research. For example, his consultancy for

---

[147] Comments of reviewer 1, enclosed with Findeis's letter to McLafferty, 8 June 1976, McLafferty Papers, folder "NSF application 1975." Emphasis in the original.
[148] Robert D. Board, "Increased mass measurement accuracy for photoplate recorded high-resolution mass spectra," Ph.D. thesis, Purdue University, Lafayette, Indiana, USA (1969).
[149] McLafferty, project report, 24 May 1984, McLafferty Papers, folder "PBM/STIRS."
[150] McLafferty to McHugh, 12 March 1982, McLafferty Papers, folder "PBM/STIRS."
[151] McLafferty, grant application to NSF, "Computer identification of unknown mass spectra," 4 January 1979, McLafferty Papers, folder "PBM/STIRS."
[152] Finnigan, "Quadrupole Mass Spectrometers," 973A–974A.

Dow Chemical kept him fully aware of the needs of the chemical industry. An additional benefit of such industrial relationships came with his close contacts to instrument manufacturers, the first in 1965 being Varian Associates with its new venture in mass spectrometry.[153] Through this, McLafferty obtained first-hand access to the new technology of Ion Cyclotron Resonance. His consulting for the Waters Corporation allowed him to be among the first users of the Waters liquid chromatography (LC) instrument.[154] The prototypes in the laboratory provided McLafferty with a competitive edge over other academic mass spectrometrists. Often, as was shown with DEC's PDP-11/45 and Hitachi's RMH-2, computers and instruments made possible the very existence of the research projects, and formed the basis for further grant applications. Clearly, funding agencies and instrument builders found their point of interaction in university laboratories such as McLafferty's at Purdue and Cornell. In this interaction, access to mass spectral data files proved to be an indispensable feature of the research process, and—as McLafferty's example makes clear—in times of scarce governmental funds constituted an important asset in convincing instrument manufacturers to support work in academe.

Artificial Intelligence

In contrast to McLafferty's matching and interpreting approach, Carl Djerassi, in close cooperation with the geneticist Joshua Lederberg and the computer scientist Edward A. Feigenbaum, chose to mimic human understanding of mass spectrometry with the help of methods that originated in Artificial Intelligence research (AI).

AI scientists strived to understand and to mimic human intelligence. Projects of the late 1950s and mid-1960s included programs designed for playing chess, solving problems in logic and algebra, and explaining the rules of baseball (in increasing difficulty).[155] Mass spectrometry was one of the first AI efforts that tackled problems in the natural sciences, an especially tempting aim, as AI pioneers were convinced that they understood the principles of sci-

---

[153] McLafferty Papers, folder "Varian."

[154] The Waters Corporation in Framingham engaged in scientific activities at the same time as McLafferty was director of the Dow Eastern Research Laboratory. James L. Waters had developed a differential refractometer. McLafferty, through common relatives and friends, had a close connection to Waters. When Dow looked for a detector to use in gel permeation chromatography (invented by Dow employee John Brown of the Texas division), Waters's differential refractometer was the tool of choice. Dow invested in Waters, and pushed it further into liquid chromatography. Waters later became the major manufacturer of High Performance Liquid Chromatography (HPLC). During 1973–75 McLafferty was a consultant for Waters. McLafferty, interview by Reinhardt, 16 and 17 December 1998. See *Today's Chemist* 3,5 (Oct. 1990), 29–30.

[155] Cartwright, *Applications of Artificial Intelligence in Chemistry*, 1.

entific reasoning. In their view, a specific method of scientific discovery was embodied in the programs: induction, the generating of hypotheses from experience and evidence, first successfully popularized by Francis Bacon in the early seventeenth century. According to this view, experiment and observation could—in principle—exclude the invalid hypotheses, leaving the valid or "true" fact. In the 1960s, most philosophers of science agreed that this was not the method of scientific discovery and put forward a hypothetico-deductive scheme. One argument of adversaries of inductivism was that it would never be possible to generate all the possible hypotheses that could explain a certain set of experimental data. Some AI scientists thought they could do so with the help of specific heuristics that would discard on rational grounds large parts of theoretically possible hypotheses. Thus, the AI method of scientific discovery would be an exhaustive exclusion of hypotheses that were not in accordance with experimental data:

> The method advanced in our work is in some sense a revival of those old ideas on induction by elimination, but with machine methods of generation and search substituted for exhaustive enumeration. Instead of enumerating all sentences in the language of science and trying each one in turn, a computer program can use heuristics enabling it to discard large classes of hypotheses and search only a small number of remaining possibilities.[156]

Needless to say, the mechanization of scientific induction met with opposition, among others by social constructivists who were convinced that human induction would never converge with processes done within a machine. Exactly this approach was quite common among AI scientists, as BACON, a program that was supposed to rediscover fundamental scientific laws such as Kepler's laws and Boyle's law, shows.[157] This program was put forward by one of the pioneers of AI, Herbert A. Simon of the Carnegie-Mellon Institute. On his side, Lederberg was convinced that the advent of high performance computers would make possible the unification of the practical (Baconian) and aesthetic (Newtonian) ideals of science. While the mathematization of the sciences mainly had strengthened the aesthetic ideal, in the late 1960s computers afforded a new kind of mathematical rationalization that allowed the entry into new fields of practical importance:

> If we could give biology sufficient formal structure, it might be possible to mechanize some of the processes of scientific thinking itself. Many of the most striking advances in modern biology have come about through the formulation of some spectacularly simple models of important processes, for example, virus growth, genetic replication, and protein synthesis. Could not

---

[156] Lindsay, Buchanan, Feigenbaum, and Lederberg, *Applications of Artificial Intelligence for Organic Chemistry*, 161.

[157] See Collins, *Artificial Experts*, 128–132.

> the computer be of great assistance in the elaboration of novel and valid theories? We can dream of machines that would not only execute experiments in physical and chemical biology but also help design them, subject to the managerial control and ultimate wisdom of their human programmer.[158]

Mainly because of its systematic structure, elementary organic chemistry seemed to be a much better candidate for the uses of AI in science than genetics, Lederberg's original field. From early on, Lederberg was interested in mass spectrometry, and in 1964 he devised a method of calculation for the empirical formula of a compound based on its high resolution mass spectrum.[159] This problem was behind Lederberg's idea of using the AI approach in establishing plausible solutions to the interpretative problems encountered in mass spectrometry. In 1965–66, and in cooperation with Feigenbaum, one of Simon's students and then director of the computation center of Stanford University and faculty member of the computer science department, Lederberg began to explore the potential of AI to find the most plausible structure of a molecule among those possible. This approach resembled in many ways a chess-playing program that tried to find the best moves among the possible legal moves. Djerassi and his group were close at hand, and together they applied mass spectrometry as one of the first real-world test cases of AI.[160] Obviously, Lederberg's enthusiasm for the use of AI in organic chemistry convinced Feigenbaum that this scientific problem would be an ideal test case for a novel approach to AI. With Feigenbaum's input, the program became one of the first expert systems in AI. Expert systems opposed the then prevalent paradigm of AI, based on general problem-solving methods.[161]

In chemistry, as in most other scientific disciplines of the late 1960s and early 1970s, the use of computers for numerical transformation of data was routine. Without computerized calculations, neither X-ray crystallography nor Fourier Transform NMR would have been thinkable. These methods were quickly accepted by chemists who realized that manual operations could be replaced reliably and with a large gain in productivity. In contrast to *numerical* calculations, the foundations of AI in chemistry rested on *symbolic* manipulations. AI scientists converted the familiar structural formulae of organic

---

[158] Lederberg, "Topology of Molecules," 38.

[159] Lederberg's computational scheme assisted the chemist before high performance computers would become available and was thought to make large volumes of tabular data superfluous. See Lederberg, *Computation of Molecular Formulas for Mass Spectrometry*; and the appendix in Budzikiewicz, Djerassi, and Williams, *Structure Elucidation of Natural Products by Mass Spectrometry*, vol. 2, 279–297.

[160] Djerassi, NIH grant application, "Resource related research—computers and chemistry," 20 December 1973, p. 41, Djerassi Papers, ACCN 1999-02 Dje, box 12, folder "Resource related research—computers and chemistry."

[161] See Crevier, *AI. The Tumultuous History*, 146–150.

compounds into topological graphs, accessible to calculation by algorithms. This made possible the planning of chemical syntheses as well as assisting in the structural elucidation of unknown compounds. Thus, the computer moved to the more creative parts of human reasoning in chemistry, with the inevitable outcome that AI methods suffered from a strong recalcitrance in the chemical community, especially in organic chemical synthesis: "The synthetic chemist wishes to be both architect and building contractor—the former function being the intellectually and aesthetically more pleasing one—and it is precisely this architectural role that the computer is perceived partially to usurp."[162] In contrast, chemists interested in structural elucidation of natural products were always prone to accept any help they could get, including physical instruments. Thus, they were better prepared to incorporate AI methods into their repertoire. An argument related to the norms of scientific behavior assisted the acceptance of AI methods in structural natural products chemistry:

> No synthetic organic chemist claims, or needs to claim, that he or she has thought of all possible synthetic paths to a given molecule. The structural chemist, on the other hand, must be able to claim that every possible structure compatible with the available chemical and physical evidence has been considered.[163]

According to Feigenbaum and his fellow AI scientists, organic chemistry was a well-suited field for the computerization of inductive inference. The range of topics could be chosen in such a way that the treatment by AI methods could be simplified, but remained useful. Step by step, more data domains could be added to secure generality. The data on which the inductive inferences rested could be presented in uniform, tabular format. And, most importantly, hypotheses in organic chemistry were essentially structural formulae. Treated as topological models, these structure diagrams were accessible to algebraic algorithms that could be used in few other scientific fields, if any. The algorithm that formed the basis for AI methods in organic chemistry was named DENDRAL, from DENDRitic ALgorithm. DENDRAL was supposed to create an exhaustive and not redundant list of possible structural formulae based on the empirical formula of an unknown compound. This was a straightforward task for simple molecules, but with the numerical increase of atoms present in the molecule the possible number of isomers literally exploded. For example, in the homologous series of alkanes with the formula C_nH_{2n+2}, the number of isomers for $n = 7$ is nine, for $n = 10$ it is 75, and for $n = 30$ more than 4×10^9 possible isomers exist.[164] No human would have the patience, and the capability, to generate so many possible solutions. However, if the computer would apply

[162] Djerassi, "Foreword," ix.
[163] Djerassi, "Foreword," ix–x.
[164] Cartwright, *Applications of Artificial Intelligence in Chemistry*, 6.

the method of exhaustive calculation only, the computation time necessary would soon exclude practical application. Thus, heuristic programming was used in such a way that unproductive areas defined by experimental data were excluded very early on, and experimental information was applied to generate a ranking list of possible hypotheses that were examined in this order: "The program thus simulates a systematic idealization more than it does the haphazard evocation of new concepts in human intelligence."[165]

DENDRAL started in 1964 with Lederberg's conceptualization of an algorithm that framed molecular structure in terms of topological graph theory and thus made them amenable for calculation in digital format. In DENDRAL, molecular formulae were represented by nodes (representing atoms) and edges (representing bonds). With the help of DENDRAL, all possible topological combinations of a set of atoms could be generated. In its first version, DENDRAL could tackle only acyclic structures, and it took ten years before a workable structure-generating algorithm for cyclic structures was devised.[166] The topological mapping of organic molecules was a statement of the connections of each atom to the others in a compound, based on chemical valence only and not taking into account the bond character or stereochemical considerations. For acyclic molecules, Lederberg used a tree structure to standardize them, starting with the unique center (either a node or an edge) of the graph. Lederberg developed a notational system that was based on the "canonical mapping of the tree following simple rules of precedence."[167] This linear notation allowed the generation of acyclic structures, in a complete and irredundant fashion, with knowledge of the valence values of the elements only. The atoms were numbered according to their positions in the linear string representing the molecule. Lederberg chose strictly trivalent graphs, represented by their polygonal and polyhedral forms. The translation of these graphs into a linear notation followed the Hamilton circuit in each graph which was a "round trip" that traversed each node only once. This circuit was transferred into a systematic numbering code, and on its basis Lederberg could compute all possible Hamilton circuits, and thus the structures. With this method, it became possible to transfer chemical structures to electronic computers, and handle them digitally. Since Lederberg was a prominent figure in exobiology and the early NASA programs for seeking life on the moon and Mars, equipping the landing systems with intelligent instruments seemed to be a promising goal.[168] A more mundane object was the computerization of mass spectrometry on earth, and here Lederberg joined forces with Djerassi.

---

[165] Lederberg and Feigenbaum, "Mechanization of Inductive Inference," 188.
[166] For a description see Lindsay, Buchanan, Feigenbaum, and Lederberg, *Applications of Artificial Intelligence for Organic Chemistry*, chap. four.
[167] Lederberg, "Topological Mapping," 134.
[168] Lederberg, "Topology of Molecules," 39, 49.

The first years of the DENDRAL project were funded by the Advanced Research Projects Agency (ARPA) of the U.S. Department of Defense. ARPA funding was restricted to "frontier computer science research." Thus, the agency was reluctant to finance the more chemistry-oriented applications of DENDRAL in the later periods of its development. From 1968 on, the DENDRAL project was closely coupled with the NIH-funded resource Advanced Computer for Medical Research (ACME) at Stanford University that in 1973 was extended to the Stanford University Medical Experimental Computer Resource (SUMEX). DENDRAL was only one of many scientific applications of AI that were undertaken at SUMEX, but certainly one of the most popular and successful ones. Furthermore, with the exception of one program in Japan, in the early 1970s DENDRAL was the only AI-based project for structure elucidation of organic chemicals. In addition to ARPA, NASA funded a large part of the mass spectrometry facility, especially the direct link of the computer to the mass spectrometer. This was achieved in the Instrumentation Research Laboratory of the department of genetics, and linked the AEI MS-9 high resolution and the Atlas CH-4 low resolution mass spectrometer of Djerassi's group to the ACME computer facilities.[169]

In the late 1960s, heuristic DENDRAL had reached a quite sophisticated level. The prefix heuristic referred to the fact that the program searched only in a small subset of all possibilities (structures), according to rules set up by chemists. The program was fed with data from a mass spectrometer and produced scientific hypotheses in the form of structural formulae that were consistent with the information given. In order to be able to do this, the algorithm of DENDRAL tried to emulate some of the decision-making procedures of chemists. This was based on Lederberg's original algorithm that generated all possible topological structures of a given empirical formula. In 1968, DENDRAL was structured in five sub-programs: the preliminary inference maker, data adjustor, structure generator, predictor, and an evaluation function (see Figure 5.11).[170]

The preliminary inference maker looked for patterns in the mass spectrum that favored specific subgroups and excluded others. On the basis of this information, this sub-program constrained the structure generator in its function. The preliminary inference maker was fed with some rules of mass spectrometry, for example the one that identified the conditions for the presence of a ketone group (C=O) in a molecule. This group had the combined atomic masses of carbon (12), and oxygen (16), thus a molecular weight of 28. The

[169] Carl Djerassi, NIH grant application, "Resource related research—computers and chemistry," 20 December 1973, pp. 36–41, Djerassi Papers, ACCN 1999-02 Dje, box 12, folder "Resource related research—computers and chemistry."
[170] Buchanan, Sutherland, and Feigenbaum, "Heuristic Dendral," 211.

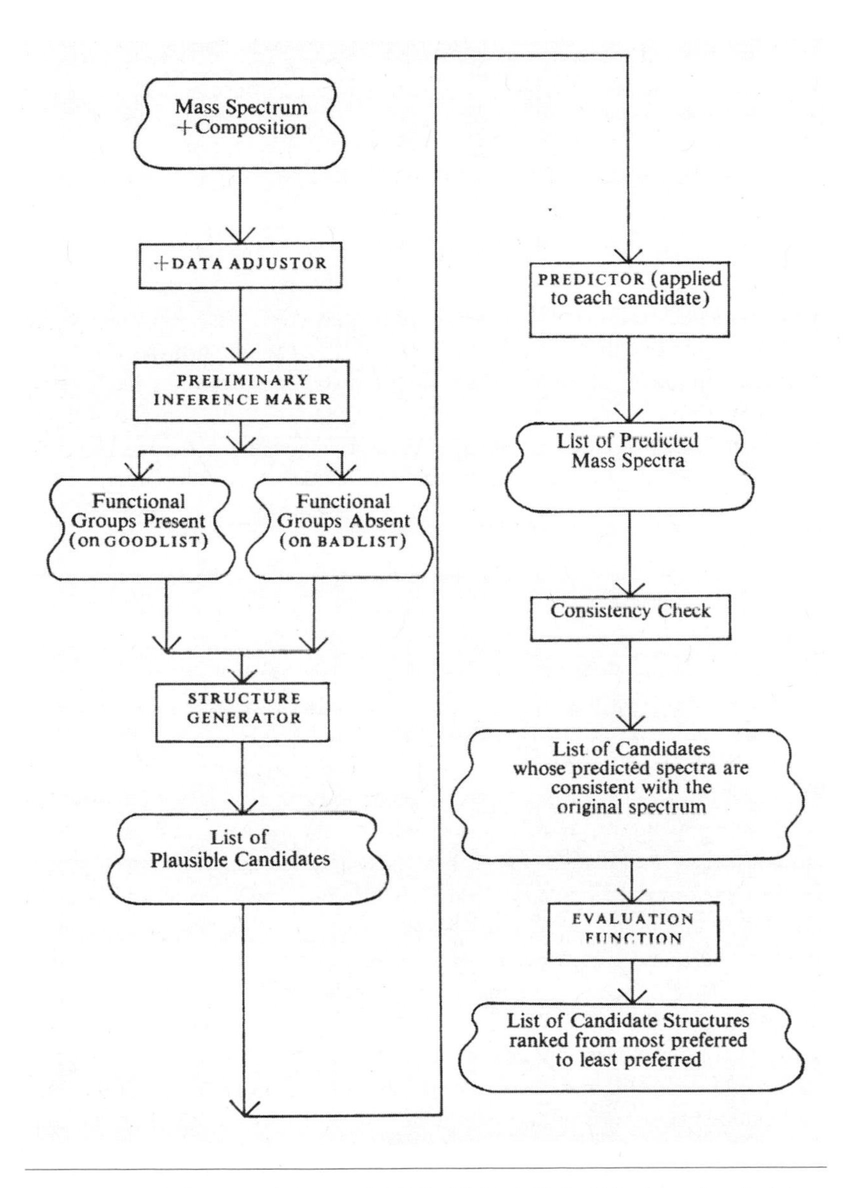

FIGURE 5.11 The general design of HEURISTIC DENDRAL. Reproduced with permission from Bruce Buchanan, Georgia Sutherland, Edward A. Feigenbaum, "HEURISTIC DENDRAL. A Program for Generating Explanatory Hypotheses in Organic Chemistry," *Machine Intelligence* 4 (1969), 209–254, on 212. Copyright 1969, Edinburgh University Press (www.eup.ed.ac.uk).

program assumed that a ketone group was present in the molecule if the following conditions were fulfilled:

1. There are 2 peaks at mass units x1 & x2 such that
   a. x1 + x2 = M [molecular weight of the compound] + 28
   b. x1 – 28 is a high peak
   c. x2 – 28 is a high peak
   d. At least one of x1 or x2 is high.[171]

This is just one example of the kind of rules applied in mass spectrometry that were incorporated in DENDRAL, enhanced by the possibility of a dialog between computer and human expert. The program thus could filter a relatively small number of probable structures out of a multitude of possibilities. For example, taking into account the constraints generated by the mass spectrum, DENDRAL proposed three probable structures for a compound with an empirical formula of $C_8H_{16}O$, out of 698 isomers that were chemically possible.[172]

The data adjustor looked for significant peaks of a mass spectrum, given rules that resembled those used by mass spectrometrists, while the structure generator proposed the structural formulae that were consistent with the empirical formula, the spectrum, and a list of likely substructures and impossible ones. The predictor constituted the first step towards a computer theory of mass spectrometry, in accordance with Lederberg's and Feigenbaum's plans for the mechanization of induction. This sub-program later became the beginning of META DENDRAL, a program that predicted and established hypothetical rules of fragmentation in a mass spectrometer on the basis of a given set of empirical data. In HEURISTIC DENDRAL, the predictor was necessary to evaluate the proposed structures. For each structure, a mass spectrum was created (by computation only, without the real compound or access to a mass spectrometer), and compared to the mass spectrum of the real compound in question. The predictor calculated the mass of the molecule, broke step by step one bond after the other, and determined the extent of rearrangements and eliminations of each fragment ion. In addition, the intensities of the peaks were calculated. This was possible only with a huge programming effort that incorporated large parts of mass spectral theory in the computer. For example, the bonds next to a ketone group were prone to break quite easily, and, in addition, ketones with a hydrogen atom in a specific position underwent the McLafferty rearrangement. Thus, the presence of a ketone group led to the prediction of a number of peaks with certain intensities in the spectrum. The evaluation function compared the artificially produced spectra with the real ones. Its output was a ranking list of possible structures of the compound.

---

[171] Buchanan, Sutherland, and Feigenbaum, "Heuristic Dendral," 214.
[172] Buchanan, Sutherland, and Feigenbaum, "Heuristic Dendral," 219.

In the late 1960s, HEURISTIC DENDRAL's capability to handle ring structures was still very poor. Thus, it had no real application value for chemists, who were mostly interested in more complicated structures than simple aliphatic ones. Moreover, the separation of the graph-manipulating processes from the chemical input was of crucial importance if the program was not to get stuck because of the sheer amount of data. In addition, this separation would make changes and updates in mass spectrometric theory easier.[173]

In mid-1969, DENDRAL was subjected to a test by Djerassi when he submitted the mass spectra of three unknown acyclic compounds both to the program and to the participants of his graduate seminar on mass spectrometry. The program successfully determined two of the three structures. The results of the graduate students and one postdoctoral fellow were of a comparable performance. The "human chemists" spent from 15 to 40 minutes on each problem, while the computer needed two to five minutes. In sum, the mass spectrometrists (who belonged to a group that was close to the development group of DENDRAL) said that the program performed as well as they themselves in its limited domain.[174] Most of the users of DENDRAL came from the local Stanford community, especially from Stanford University's medical school and the chemistry department. Outside users were admitted on a case-by-case basis, reflecting Djerassi's conviction that "rather than provide routine service, experience has shown that discretionary selection of problems results in better utilization of our people and instrumentation resources."[175] A close connection to SUMEX guaranteed access to high performance computer facilities, while the contact to the Stanford Genetics Research Center provided an important user of the program, especially in the analysis of body fluids with combined gas chromatography and mass spectrometry. Thus, the DENDRAL project was vitally linked to Stanford's biomedical community, financed mainly by NIH.

One of the greatest problems during the development of DENDRAL was the slow transfer rate of information from the chemists to the computer via the programmers. The performance of the program could be improved only when it was "criticized" by experienced mass spectrometrists. "Eliciting a theory from an expert" was crucial for the endeavor, a difficult task because the theory of mass spectrometry itself was not fully developed, in the eyes of the programmers at least:

> The theory of mass spectrometry is in very much the same state as the theory of good chess play: there exists a collection of principles and empirical gener-

[173] Buchanan, Sutherland, and Feigenbaum, "Heuristic Dendral," 251.

[174] Buchanan, Sutherland, and Feigenbaum, "Rediscovering Some Problems," 255.

[175] Djerassi, NIH grant application, "Resource related research—computers and chemistry," 20 December 1973, p. 45, Djerassi Papers, ACCN 1999-02 Dje, box 12, folder "Resource related research—computers and chemistry."

> alizations laced throughout with seemingly ad hoc rules to take care of exceptions. No one has quantified these rules and only a few attempts have been made to systematize them.[176]

The precarious state of mass spectrometric theory contributed to the lack of performance of the program. In order to improve DENDRAL, joint sessions of a programmer and a chemist at the console of the teletype tied to the computer were necessary. In a 1969 publication, such a dialog was condensed and transcribed, showing the basic difficulties of knowledge transfer between members of different disciplines. In the following quote, A is the chemist, B the programmer and text in parentheses refers to metacomments.

> A: Since El Supremo [Carl Djerassi] and the rest want us to work on ketones, I guess we should get started.
> B: OK. Incidentally, why are ketones important?
> A: Besides being very common in organic chemistry we also know something of their mass spectrometry because they've been studied a lot.
> B: What subgraph exactly will cause a molecule to be classed as a ketone?
> A: The keto, or carbonyl, radical, That is —C=O (noticing B's puzzled look).
> B: Then all of these are ketones?
> CH_3—CH_2—C=O—R
> CH_3—C=O—R
> H—C=O—R
> A: Wait a minute. The first two are ketones, but the last one is a special case which we should distinguish in the program. It defines the class of aldehydes.
> B: So we can formulate the general rule that a ketone is any molecule containing C—C=O—C (thinking of the LISP list '(C(2O)(1C)(1C))').
> A: That's it.
> B: Now what mass spectrometry rules do you have for ketones?
> A: Three processes will dominate: alpha-cleavage, elimination of carbon monoxide from the alpha-cleavage fragments, and the McLafferty rearrangements.
> B: OK. I wrote those down—now tell me exactly what each one means. Start with alpha-cleavage—do you mean the bond next to the heteroatom?
> A: (Digression on notation—often alpha-cleavage would mean this bond, but not for mass spectrometry.) . . . Here alpha-cleavage is cleavage of the C—C=O bond, i.e., cleavage next to the carbonyl radical—on both sides don't forget.
> B: All right. That's an easy rule to put in . . . .[177]

---

[176] Buchanan, Sutherland, and Feigenbaum, "Rediscovering Some Problems," 257.
[177] Buchanan, Sutherland, and Feigenbaum, "Rediscovering Some Problems," 257–258. LISP (from List Processing) is the programming language the group used.

This certainly is an idealized description of the many encounters between organic chemists and computer scientists, and presents a stage where both sides already knew exactly what they wanted to do. As they reported, the interdisciplinary team ran into three main difficulties. First, the mass spectrometric experts sometimes wanted to keep the explanations simple, because they were directed to a layman. Thus, essential information was overlooked. Second, there were so many gaps and special cases in mass spectrometry that theory and experiment were often in disagreement. Sometimes, the repeating of real experiments became necessary. A positive result of this was that the theory could be improved in the process. The third problem was "false starts," when it was discovered that different classes of chemical compounds needed different configurations of the hierarchy of subgraphs: "Typically it has taken weeks of interaction with a chemist at a console to proceed past the first two difficulties, never knowing whether we were making a false start."[178] In all, around fifteen person-years were used to develop DENDRAL to such a stage that it could be used profitably.[179] No wonder that the computer scientists looked for means to automate the teaching process, mainly with interface programs that would displace the programmer in the dialog.

DENDRAL was the first of the expert systems that pushed aside the older paradigm in AI, the general problem-solving systems.[180] Feigenbaum became the founder of this direction in AI, which he preferred to call "knowledge engineering." The new paradigm experienced a breakthrough with the program MYCIN, a diagnosis program for blood infections and meningitis that gave advice on the treatment with antibiotics. Feigenbaum, Bruce G. Buchanan, and their co-workers developed MYCIN in the mid-1970s. It worked on an interactive dialog basis, and allowed questions on how it reached certain conclusions. The capturing of human expertise, much along the lines described in the dialog between chemist and programmer quoted above, was a typical style for developing the early expert systems in AI.[181] It was important that both the knowledge base, the "facts," and the implicit heuristic procedures be incorporated. Feigenbaum defined heuristics as "experiential, judgmental knowledge," and as "rules of thumb."[182] The combination of both assured workability. Moreover, Feigenbaum, Buchanan, and Georgia Sutherland were quite confident about their success in providing organic mass spectrometry with a better theoretical framework:

---

[178] Buchanan, Sutherland, and Feigenbaum, "Rediscovering Some Problems," 263.

[179] See Harmon and King, *Expert Systems*, 134–135.

[180] For an intermediate stage in Feigenbaum's development of expert systems see Feigenbaum, Buchanan, and Lederberg, "On Generality and Problem Solving."

[181] See Feigenbaum and McCorduck, *Fifth Generation*, 83–91; and Harmon and King, *Expert Systems*, chapters two and six.

[182] Feigenbaum and McCorduck, *Fifth Generation*, 351.

> Eliciting a theory from an expert . . . has been our key to the wealth of knowledge not yet accessible in textbook packages. And it has benefited the scientist since it provides a means of codifying a loose collection of empirical generalizations into a theory.[183]

In Djerassi's opinion, the interdisciplinary character of the DENDRAL project was its most important asset. People with skills in computer science, organic chemistry, molecular biology, and instrumentation engineering worked closely together. Buchanan, a mathematician and philosopher, took charge of large parts of the computing part of the project. Altogether, in 1973 fifteen people were assigned to work for the chemical applications of the project. On the side of mass spectrometry, Alan Duffield and Dennis H. Smith contributed to the DENDRAL project. Smith was in charge of the extension of the program to compounds of biological importance, notably steroids,[184] while Duffield had a dual appointment in both the department of genetics and the department of chemistry and concentrated on the identification of trace metabolites in urine and organic impurities in other fluids.[185] These projects were thought to be a test case for DENDRAL and naturally were of special interest for the funding agency NIH. Thus, the direction was changed from the systematic inclusion of more functional groups to the tackling of the more complex, but medically important, steroids. The group started with estrogenic steroids, with the object to screen body fluids for metabolites.[186]

At this time, other spectroscopic data, mostly from carbon-13 NMR, were also included in the program, thereafter called CONGEN (CONstrained structure GENerator).[187] The complexity of the program grew considerably in the 1970s, especially with its extension to stereochemical problems.[188] Djerassi was eager to show CONGEN's performance both in the classroom and to his academic colleagues. He used CONGEN to test the unambiguity of structure elucidations published in the chemical literature, based on chemical and spectral data. With the help of the program, graduate students found out that in no single instance was the published structure the only one consistent with the data given. Since additional possible structural assignments existed, the pub-

---

183 Buchanan, Sutherland, and Feigenbaum, "Rediscovering Some Problems," 265–266.

184 Djerassi, NIH grant application, "Resource related research—computers and chemistry," 20 December 1973, pp. 6–8, Djerassi Papers, ACCN 1999-02 Dje, box 12, folder "Resource related research—computers and chemistry."

185 Feigenbaum, supplement to grant application of 24 April 1970, dated 7 September 1970, Djerassi Papers, ACCN 1999-021, Dje, box 12, folder "Resource related research: Computers and chemistry."

186 Feigenbaum, NIH grant application, "Resource related research: Computers and chemistry," 24 April 1970, p. 39, Djerassi Papers, ACCN 1999-021, Dje, box 12, folder "Resource related research: Computers and chemistry." See also Smith et al., "Applications of Artificial Intelligence."

187 See Carhart et al., "Applications of Artificial Intelligence for Chemical Inference. XVII."

188 Djerassi, *Steroids Made it Possible*, 109–114.

lished structures were not as certain as the authors of the papers had indicated, or if they were they had a basis in data not mentioned in the paper. Thus Djerassi, only partly tongue-in-cheek, as he recalled, proposed to establish the program as an automated journal referee.[189]

With this development, DENDRAL changed from being an interesting task environment for AI to an intelligent assistant of structure elucidation in general and mass spectrometry in particular. CONGEN especially was also used outside the Stanford community. This service was made available on-line. Interestingly, this caused reluctance on the side of industrial companies. They feared that the connection was not secure enough and that their use could reveal the topics they were working on. Consequently, Eli Lilly Research Laboratories, for example, preferred an exportable version of the program.[190] Djerassi used the program intensively for his own research in marine sterols. Additionally, his industrial connections, Syntex and Zoecon, certainly were in positions to make use of the expert systems generated by Djerassi and his collaborators.[191] In the early 1980s, the program was available commercially through Molecular Design, Ltd. At the same time, Djerassi became board member of Teknowledge, a company founded by Feigenbaum to exploit the industrial applications of AI.[192]

The acceptance of DENDRAL and CONGEN in the community of chemists had a central place in Djerassi's thinking. Thus, he made clear that the programs were written to assist and not to displace the chemist. This was ensured through a strong interaction between the user and the program, and the ease with which even inexperienced organic chemists learned to make use of it. The maturity of the programs was described in a characteristic vein: "Chemists now use them as 'black boxes' just as most physical methods are used."[193] Nevertheless, in the long run, the AI approach to organic mass spectrometry was the least successful, compared with McLafferty's and others' matching-and-interpreting style. The main problem with AI was that these expert systems had to be fed with an enormous amount of data and rules. Djerassi and his group did that for steroids only, their own area of interest. As a result, AI programs never were of much use in general organic chemistry. In the end, human experts with their acquired competence prevailed.

---

[189] Djerassi, *Steroids Made it Possible*, 110–111; and Djerassi, Smith, and Varkony, "A Novel Role of Computers," 14.

[190] Douglas E. Dorman of Lilly Research laboratories to Djerassi, 6 May 1976, attached to Carl Djerassi, NIH grant application "Resource related research: Computers and chemistry, May 1976, p. 2A, Djerassi Papers, ACCN 95-168.

[191] See for example the case of warburganal, an insect antifeedant, reported in Djerassi et al., "Dendral Project."

[192] Carl Djerassi, interview by Jeffrey L. Sturchio and Arnold Thackray, 31 July 1985, p. 36, Chemical Heritage Foundation, Philadelphia, PA.

[193] Djerassi, Smith, and Varkony, "A Novel Role of Computers," 10.

Demand-pull on the side of the biomedical and environmental sciences led to far-reaching changes in the methods of mass spectrometry. For the elucidation of large and complex molecules, high resolution mass spectrometry became the method of choice, while GC/MS dominated the field in identification of small amounts of compounds in mixtures and the environment. High resolution mass spectrometers became as close to instruments of Big Science as instruments in chemistry could. Characteristically, the new organizational modes combined different features of Big Science: Relatively large centers, built around a variety of high performance instrumentation, catered for a geographically and disciplinary defined community of scientists. Thus, the network around an instrument became the crucial issue of "big" mass spectrometry: It enabled mass spectrometrists to employ instrumentation that otherwise would have been out of reach, and to concentrate on methods development. The price they paid was the service work they had to offer, and the expected willingness to take the scientific problems of the biomedical community seriously. In this way, instrument centers and facilities became the nodes in the network of an emerging scientific specialty: organic and biomedical mass spectrometry.

In the form of GC/MS, Little Science had its share, too. These relatively small-scale instruments with their task to routinely search and reliably find traces of substances in the environment merged governmental agencies—most notably EPA, instrument manufacturers, and mass spectrometrists in an alliance of common interests. Not localized in centers, this style of mass spectrometry depended on loose configurations of the actors, trading instrument access, grant money, scientific data and know-how.

At the core of both Little Science and Big Science in mass spectrometry stood data, and their processing. Access to data and the capability to make sense of them meant the first, and most important, step to success. In the 1960s and 1970s, when computer technology made huge leaps, new opportunities emerged for the mass spectrometrist: Could he happily delegate the calculation part of interpretation to the new device, or should he in addition allow the machine to take over the interpretative parts of his tasks? Here, different styles fought for domination: The first, matching, was a straightforward undertaking of "search and find," and limited to known compounds. The second, interpretation, called for the imitation of human thought processes, and was theoretically capable of dealing with any new compound that could be imagined. In the end, a combination of both methods as the most pragmatic was also the most successful. Moreover, it allowed mass spectrometrists to keep their key positions as experts, continuously working on the improvement of instrumental technology and scientific methods.

Chapter Six

A Spin Doctor in Resonance

> The world of the nuclear spins is a true paradise for theoretical and experimental physicists. It supplies, for example, most simple test systems for demonstrating the basic concepts of quantum mechanics and quantum statistics, and numerous textbook-like examples have emerged. On the other hand, the ease of handling nuclear spin systems predestinates them for testing novel experimental concepts. Indeed, the universal procedures of coherent spectroscopy have been developed predominantly within nuclear magnetic resonance (NMR) and have found widespread application in a variety of other fields.[1]

With these words, Richard R. Ernst (b. 1933) began his Nobel lecture on 9 December 1991. It is the affectionate description of his life-long research field as well as the self-confident statement of the importance of NMR for scientists and engineers alike. During the longest part of his career, Ernst was an academic scientist, a physical chemist at the Eidgenössische Technische Hochschule (ETH) in Zurich, Switzerland. During a crucial part of his career, from 1963 to 1967, he was employed as research scientist by Varian Associates of Palo Alto, California, manufacturer of NMR spectrometers. During all this time, Ernst developed novel measurement methods in NMR spectroscopy. His work was based on the understanding of the interaction patterns of nuclear spins that govern the appearance of NMR spectra. With the help of theoretical insight, a manipulation of these interactions came within reach. Ernst belonged to a small group of NMR spectroscopists that commanded the theoretical and practical know-how, having an interest in the manipulation of spin interactions as a goal of research in itself. Their second, and equally important, aim was the enhancement of sensitivity and resolution of NMR, i.e., the improvement of the measurement technique. These NMR scientists produced knowledge for

[1] Ernst, "Nuclear Magnetic Resonance Fourier Transform Spectroscopy," 12.

the scientific community as well as the instrument market, mostly working at universities, but also employed by companies, and bridging the boundaries between the commercial and the academic spheres. Thus, their work is close to, but different from, that of research-technologists.

The notion "research-technology" has its origins in the work of sociologist Terry Shinn in the 1990s. Research-technology comprised instrument-related work that enabled both scientific research and the production of other goods than knowledge. It crossed disciplinary and geographical boundaries, while it was dis-embedded from the local context of its origins and re-embedded in novel application niches. Originally, Shinn restricted investigations of research-technology to prosopographical studies of communities of practitioners, "research-technologists." In going beyond the purely sociological sphere, the term research-technology entails three main features: Interstitial communities, generic devices, and metrology. Shinn's concept thus approaches not only social issues. It allows one to tackle the design and uses of instruments as well as the topics of standardization, representation, and paradigmatic shifts in scientific theory. Interstitial communities are meant to throw light on the transverse careers of research-technologists, their back-and-forth movements between university, industry, and governmental agencies and institutions. Their identities "cannot be mapped in terms of an organizational or professional referent." The notion of generic devices, in turn, refers to the in-built general functionality of the instruments designed by research-technologists. Often, the interest of research-technologists focused on instrument-science itself, and they took care of the innovation of "systems of generic detection, measurement, and control devices that focus on particular parameters which are potentially of interest to scientists, laboratory technicians, test personnel, production engineers, and planners." This far-reaching audience is held together by "systems of notation, modeling and representation," its metrology. Metrology is an important issue, ensuring the crossing of boundaries while guaranteeing research-technologists an independent realm.[2]

In contrast to research-technologists, the institutional basis of methods-oriented scientists was the university, and typical career paths did not involve frequent crossings between academe, industry, and government, as those of many research-technologists did. Even when employed by an industrial instrument manufacturer, a method maker saw himself as member of an academic scientific community. Work at a university included the possibility for close contacts with instrument manufacturers, which were an important precondition for successful methods development. Thus, a method maker networked his science through cooperation and collaboration, and not so much with actual personal movements, or transverse orientations, between the professional

[2] Joerges and Shinn, "Fresh Look at Instrumentation," 8–9.

spheres. Characteristically, a method maker worked for the distinctive application of methods in special scientific fields, and not for generic uses. Interested in a thorough grounding of methods in theory and experiment, a method maker worked from the inside of a scientific community, concentrating on the development of methods, and not instruments. Ernst regarded methods even as the basis of instruments, not vice versa. Thus, in their own opinion, method makers laid the foundations for the innovation of instruments. As we will see, both sides depended on each other. The only subject in which a direct relation of research-technology to methods-oriented science seems to be possible is that of metrology. Here, participants in science, industry, and government had their own interests at stake, and they took part equally in its evolution. In the 1960s, NMR became a specific, and autonomous, scientific domain. Its autonomy rested on the wide and diversified range of clients and users, including many—if not most—scientific disciplines, medicine, and technology, and involving scientists from many different professional fields.

In this chapter, Ernst's career will serve as an example of a method maker, and we will carefully draw the line between his working style and that of a typical research-technologist. Ernst only moved once from university to industry, and back, and during the longest period in his career he worked at a university. But he kept close contacts to instrument manufacturers during his academic tenure. Through Ernst, we will sketch the community of NMR spectroscopists in the 1960s and 1970s, and draw a picture of the development of a technique-based scientific specialty. In these two decades, Ernst's contributions deeply changed the design of NMR spectrometers and their general working mode, as well as important parts of its metrological system. At first sight, Ernst's two most important innovations—Fourier Transform (FT) NMR spectroscopy and two-dimensional (2D) NMR spectroscopy—certainly are examples of generic devices in Shinn's sense. Especially FT spectroscopy was a method that allowed a dramatic increase in sensitivity not only in NMR, but also microwave spectroscopy and electron spin resonance spectroscopy. 2D NMR enhanced the interpretation of complex overlapping spectra, and soon led to a tremendous enlargement of the scale and scope of NMR. For Ernst, these methods were at the same time measures for realizing quantum mechanical states and facilities for improving the performance of NMR in terms of sensitivity and resolution. He worked hard for a general theoretical foundation of both techniques, and looked for applications in certain cases. While FT NMR required alterations of the hardware components of NMR spectrometers, and left the notational and representational system largely untouched, 2D NMR had a deep impact on how NMR spectra were presented. Moreover, 2D NMR was largely responsible for the breakthrough of NMR in the structural elucidation of large biomolecules, notably proteins. FT NMR and 2D NMR together reunited the formerly methodologically separated NMR communities of solid-state physicists and liquid-phase chemists, as both fields of application could be tackled then with the same methods.[3] Despite all this generality, Ernst continued to be an NMR specialist, and he continuously developed, in cooperation with other academic chemists and molecular biologists, methods for use in the respective fields. Especially the local environment, in the case of Ernst the company Varian Associates and ETH Zurich, provided the framework for applications. While fostering many new uses, the novel NMR methods were firmly rooted in the community of NMR spectroscopists, and their network of interests. For these method makers, a thrilling technique was the first goal, and its application their second.

INSTRUMENTATION FOR ITS OWN SAKE

From 1952 to 1956, Richard R. Ernst studied chemistry at ETH Zurich. Though he had to submit as many as four theses for his diploma (one in each of the subdisciplines of chemistry that were taught at ETH, analytical, inorganic, organic, and physical chemistry), the subject of NMR was not part of the curriculum at that time. Analytical chemistry, taught by William Dupré Treadwell, dealt just with classical, "wet" chemical methods, and physical chemistry under Gottfried Trümpler comprised thermodynamics and kinetics, but no quantum mechanics at all. After military service (Ernst later used his scientific expertise as a lieutenant specializing in chemical and nuclear arms during his frequent call-ups in the Swiss army), he began his Ph.D. studies under Hans Heinrich Günthard in 1958. This was the crucial turning point in Ernst's career in the sense that Günthard steered him into NMR. Günthard, a physical chemist, was in charge of spectroscopic methods, first in the division of organic chemistry, then as head of the division of physical chemistry. Though being in close contact with organic chemists—among them the famous chemists Leopold Ruzicka and Vladimir Prelog, for whom he introduced infrared spectroscopy—Günthard only occasionally was engaged in service work. His interests were the theoretical and experimental foundations of optical spectroscopic methods, later supplemented by microwave spectroscopy and electron spin resonance spectroscopy. NMR fitted well in this array of physical methods. Already in 1953, Günthard assigned a co-worker, Hans Primas, to develop NMR technology.[4] Later, Primas became the supervisor of Ernst's Ph.D. thesis.

Primas followed a remarkable career path. Originally educated as a laboratory technician (Chemielaborant), he studied chemistry at the Technikum

---

[3] See Ernst, "Nuclear Magnetic Resonance Fourier Transform Spectroscopy," 18.

[4] Ernst, "Success Story of Fourier Transformation in NMR," 293–294; and Richard R. Ernst, interview by Carsten Reinhardt, 25 February 2002.

Winterthur, near Zurich, and from 1951 to 1954 mathematics and theoretical physics at ETH and the University of Zurich. Although he never graduated, he was made associate professor of physical chemistry at ETH in 1961, and full professor of theoretical chemistry in 1966.[5] During the mid-1960s, his interests switched from NMR to theoretical chemistry, later including philosophical aspects of quantum mechanics.[6]

In 1953, Primas got acquainted with practical aspects of NMR during a brief cooperation with the solid-state physicist Hans Staub, a former co-worker of Felix Bloch at Stanford, then at the University of Zurich. At ETH, Primas designed and built an NMR spectrometer from scratch, including the magnet. In 1956, he convinced the Swiss company Trüb-Täuber, a manufacturer of general measurement devices and electron microscopes, to build and market this instrument, which had a proton resonance frequency of 25 MHz, and was equipped with a field stabilizer.[7] Although the performance of the spectrometer could not be compared with the 40 MHz instruments that Varian Associates built at that time, Trüb-Täuber sold several NMR spectrometers in Europe. In the early 1960s, Primas constructed an improved spectrometer, using an electromagnet. The pole caps of this electromagnet had a special design to avoid saturation effects and afforded a better homogeneity of the magnetic field. This was crucial in obtaining high resolution with moderate field strengths. With a resonance frequency for protons of 90 MHz, this spectrometer was also marketed by Trüb-Täuber. More or less, the company copied Primas's design, including the details, and did not engage in its own R&D work.[8] In 1965, Trüb-Täuber was dissolved, and its NMR division became the kernel of Spectrospin AG in Fällanden, Switzerland. About the same time, Spectrospin formed an association with the German company Bruker Analytische Meßtechnik GmbH, a manufacturer of magnetic resonance instrumentation and laboratory magnets.[9]

Primas's research program was a curious one. He focused on the quantum mechanical theory of NMR, and the practical aspects of instrumentation.

---

[5] See Primas's curriculum vitae in Ernst Papers, box 4, folder "Ehrensymposium Hans Primas, 8. Nov. 1995."

[6] Primas, *Chemistry, Quantum Mechanics and Reductionism.*

[7] The performance of NMR spectrometers was usually given by the resonance frequency of protons. This figure depended on the magnetic field strength. 25 MHz corresponded to a magnetic field strength of 5.875 kGauss (0.5875 Tesla). The relationship between resonance frequency and magnetic field strength is expressed by the Larmor equation: $\nu_0 = (\gamma B_0)/2\pi$. With ν_0: resonance frequency; γ: magnetogyric ratio; B_0: magnetic field strength. The higher the magnetic field, the higher the resolution and sensitivity of the spectrometer.

[8] A major contribution of Trüb-Täuber engineers was the introduction of frequency-sweep capabilities, if only in a limited frequency range. Thomas Steinhauser, personal communication.

[9] Ernst, "Success Story of Fourier Transformation in NMR," 293–294; and Ernst and Primas, "Gegenwärtiger Stand," 262–263.

Largely absent were experiments, both in chemistry and solid-state physics, although Primas was naturally aware of the potential applications of NMR in these fields.[10] As Richard Ernst put it succinctly in retrospect: "We never did experiments . . . . What we really did were calculations. One had an application in mind, but actually did not believe in its realization."[11] This style of Primas had a deep impact on the later work of Ernst, who described science as being

> partially simply a playful occupation, partially you don't know what else to do, you don't have a better idea. Then, you are occupied with doing something and you hope that finally something useful will result. But you don't really believe in it. I mean, this was very often the case in my later work. I have done things because I did not know better, and not because I was convinced that this would be the great breakthrough. I never believed that of my work.[12]

In 1958, Ernst became Primas's second Ph.D. student, and in the beginning engaged mainly in practical electronics.[13] In addition, he heard lectures in mathematics, experimental and theoretical physics (the lecture course in wave mechanics was given, in the beginning at least, by the famous theoretical physicist Wolfgang Pauli who died in December of 1958). The practical part of Ernst's Ph.D. thesis, submitted in 1962, was the calculation and construction of a probe head for the 90 MHz NMR spectrometer designed by Primas. By a careful choice of the dimensions of the probe, Ernst could improve considerably on the sensitivity of the spectrometer.[14] In Günthard's division, each Ph.D. candidate had to construct an instrument, or a part thereof: "This was, so to speak, my duty that I have fulfilled, the construction of something. That

---

[10] See, for chemistry, the review article by Primas, "Anwendungen der magnetischen Kernresonanz."

[11] "Experimente haben wir nie gemacht . . . . Es waren also wirklich nur Rechnungen, die durchgeführt wurden. Man hatte schon eine Anwendung im Kopf, aber man hatte eigentlich nicht an die Realisierung geglaubt." Ernst, interview by Reinhardt, 25 February 2002. All translations from the German are my own, if not otherwise noted.

[12] "Zum Teil ist es einfach eine Spielerei, zum Teil weiß man nichts anderes zu tun, hat keine bessere Idee, dann beschäftigt man sich halt damit und hofft, daß schlußendlich vielleicht doch etwas dabei herauskommt; aber so richtig daran glauben tut man dann eigentlich doch nicht. Ich meine, das war auch später sehr oft in meinen Arbeiten so. Ich habe Sachen gemacht, weil ich nichts Besseres zu tun wußte und nicht, weil ich davon überzeugt war, das gibt jetzt den großen Durchbruch. Das habe ich nie geglaubt mit meinen Arbeiten." Ernst, interview by Reinhardt, 25 February 2002.

[13] He constructed, among other devices, a "high stability oscillator, frequency multipliers, phase synchronized oscillator, thermostats, stabilized power supply." Ernst, "Biographical and professional summary," n.d. [1962], Ernst Papers, box 7, folder "Diverse papers, 1962."

[14] Ernst, "Summary of PhD dissertation," n.d. [1962], Ernst Papers, box 7, folder "Diverse papers, 1962"; and Ernst and Primas, "Gegenwärtiger Stand," 264–265.

is the *Gesellenstück* that you had to perform before you were accepted."[15] Thus, the application of the instruments was the last point only in a long agenda. As a consequence, the Zurich division could point to original contributions in instrument design, but not to novel applications, and the connection to chemistry was completely lost. For example, although Ernst ran some spectra for ETH organic chemist Prelog, he did so unsuccessfully and finally both sides, disappointed, gave up.[16]

Next to the instrumental aspects, Ernst's thesis was a theoretical treatment of the NMR measurement process. In following up some ideas of Primas that were based on Norbert Wiener's work on non-linear systems, Ernst investigated the stochastic excitation of nuclei in the NMR experiment. Ernst's and Primas's objectives were purely theoretical, and they did not think about an experiment: "Our aim was more a theoretical excursion to explore the nonlinear stochastic response of a quantum mechanical system by means of an example."[17]

For Ernst, the neglect of experimental applications was a huge disappointment in his work. Moreover, he missed a clear direction in his research at ETH, and therefore decided not to start an academic career. In order to have the experience of living in another country, he applied at companies in the United States. He received some offers, including one of Atlantic Refining Co., but did not want to engage in routine analytical work in an oil company. Fortunately, scientists of Varian Associates were aware of the performance of Primas's group in instrument development. After they heard that Ernst was looking for a job, they made him an offer. In the fall of 1963, Ernst and his wife crossed a stormy Atlantic, by ship.[18]

Sensitivity for the Market

In the mid-1960s, Varian Associates still dominated the NMR spectrometer market, but competitive manufacturers from Great Britain (Perkin-Elmer of Beaconsfield), Japan (JEOL and Hitachi), Germany (Bruker of Karlsruhe), and Switzerland (Trüb-Täuber) already tried to obtain a share in this growing segment of analytical instruments.[19] Around the arrival date of Ernst, Varian Associates brought two new instruments to the market, two spectrometers that were

---

[15] "Das war sozusagen meine Pflicht, die ich erfüllt habe, Bauen von irgend etwas. Das ist das Gesellenstück, das man zu leisten hatte, bevor man akzeptiert wurde." Ernst, interview by Reinhardt, 25 February 2002.

[16] Ernst, interview by Reinhardt, 25 February 2002.

[17] Ernst, "Success Story of Fourier Transformation in NMR," 294.

[18] Ernst, interview by Reinhardt, 25 February 2002; and Ernst, "Richard R. Ernst," 8.

[19] See the list of NMR manufacturers and their instruments in Becker, Fisk, and Khetrapal, "Development of NMR," 52–54.

completely different in design and potential application. The A-60 was a low-cost instrument, designed for the average organic chemist. It had a remarkable stability and reliability, and its compact construction enhanced a user-friendliness previously unknown. It was as close to a bench-top instrument that an NMR spectrometer could come, and soon one exemplar stood in almost every chemistry department of the world. It was first marketed in 1961, and in the subsequent six to seven years, 1,000–1,500 pieces were sold. But with a proton resonance frequency of 60 MHz, the A-60 was far below the limit that was technically within reach. A stronger magnetic field could bring about higher sensitivity and resolution. In addition, interpretation of the spectra would become easier, because the values of chemical shift and spin-spin coupling, the two major parameters of interest for the organic chemist, could be distinguished better. Moreover, with higher magnetic fields, larger molecules came within reach of NMR methods. Thus, in 1964, Varian Associates launched an instrument at the high-end side of the market. The HR-220 (proton resonance frequency of 220 MHz) already had a superconducting magnet. The development of this instrument was commissioned by William Phillips of Du Pont, who applied it mainly to work in the field of biopolymers.[20] Thus, in the mid-1960s, with the A-60 and the HR-220, Varian Associates covered both the "analytical" and the "research" side of the NMR business. Still, they were at the forefront of NMR instrumentation.

NMR was the most profitable part of Varian Associates's instrument division. In the 1960s and 1970s, the firm had developed a nearly complete array of analytical methods of interest to chemists, some of them through their own research, others through the acquisition of competitive companies (see Chapter 4). But the tube division remained by far the largest part of the company, building and marketing microwave tubes (klystrons) that were used for missile navigation. The instrument division was never more than an annex to this part of the company, and it profited from the services that such a large high-tech company could give.[21] In contrast to the military-funded tube enterprise, the R&D work in analytical instrumentation was largely financed by the company itself.[22] Weston A. Anderson, a former Ph.D. student of Felix Bloch, co-inventor of NMR, led the NMR research division. Bloch's students dominated the technical ranks of the instrument division, as Bloch in the 1950s continued to cooperate with Varian Associates. Anderson had joined Bloch on his one-year venture as director of the newly founded European center for nuclear research (CERN) in Geneva, Switzerland. In 1955 he entered the instrument division of Varian Associates, soon concentrating on the improvement of NMR spec-

[20] Ferguson, "William D. Phillips," 312–313.

[21] James Shoolery, interview by Sharon Mercer, April 1990, p. 8, Varian Associates Inc. Oral History Project, Stanford University Archives, SC M 708, box 1.

[22] Lenoir and Lécuyer, "Instrument Makers and Discipline Builders."

trometers for chemical applications. In this process, high importance was assigned to the needs of the customers, articulated through the application laboratory, run by James Shoolery. But somehow disregardful to all this chemistry, Anderson continued to be a physicist:

> It wasn't a blind 'Let's try something and see if it goes.' We had a good idea of the importance of certain areas and where to put the emphasis. The emphasis was on higher magnetic fields, greater sensitivity, better resolution, field/frequency stability and double resonance.[23]

In late February of 1963, Warren Proctor of the Zurich branch of Varian Associates approached Ernst and asked him to think about accepting a position in Palo Alto. Although Ernst had become an accomplished expert in NMR instrumentation during his four-year Ph.D. studies with Primas, he did not insist on working in NMR at Varian Associates. Most importantly, Ernst was tired of the purely academic style of instrument development at ETH and was on the search for ways to combine theoretical and practical aspects:

> My intention is not to work now entirely on electronic problems . . . nor to treat theoretical problems of only academic interest, but I would like to mark out new methods and to treat more practical applications in strong connection with the theory.[24]

Answering Ernst's questions regarding the conditions and style of research as well as the area in which Varian Associates wished to employ him, Anderson wrote in a telex in March 1963 that they wanted him to concentrate on basic aspects in NMR, but that

> work can be changed to a different field within the instrument division provided the chances for a significant contribution are high. In addition the company is involved in other fields of science which offer opportunity for a wide range of experience.[25]

In Palo Alto, Ernst joined a small group, led by Anderson, and comprising the physicist Ray Freeman and the technician William Siebert. Next to this group, the development department under Robert Codrington, the application laboratory under James Shoolery, and engineering under Forrest Nelson were important parts of the NMR R&D organization. Ernst described the organization as rather fluid, a "continuum from research to development."[26] The research physicists and engineers at Varian Associates focused on instrument-

---

[23] Anderson, "Early NMR Experiences," 171.
[24] Ernst to Anderson, draft 28 February 1963, Ernst Papers, box 5, folder "Varian Consult."
[25] Anderson to Proctor, telex, n.d. [early March 1963]. See also Ernst to Proctor, 5 March 1963, Ernst Papers, box 5, folder "Varian Consult."
[26] Ernst, interview by Reinhardt, 25 February 2002.

science, but with an applied bent. The Varian style to develop methods for the market suited the Swiss scientist. Quantum mechanical calculations of nuclear spin systems and advanced knowledge of electronic devices made it possible to point to bottlenecks in the quest for improvements, and to possible solutions. In his four-year-period at Varian Associates, Ernst worked on techniques that completely changed the practice of NMR: FT NMR, broad-band decoupling, and computerization. With that he contributed in a major way to the most urgently felt disadvantage of NMR, its lack of sensitivity, as well as to methods that enabled chemists to decrease the complexity of spectra. But only the importance of decoupling with broad-band noise was realized immediately inside the company. The full impact of computer applications, and especially FT NMR, was felt only years later, and—in addition to Ernst's and Varian Associates' contributions—through the work of other companies and scientists.

Ernst's beginning at Varian Associates connected to his research done at ETH with Primas. At Varian Associates, Anderson had spent a long time improving so-called shim coils that served to enhance the homogeneity of the magnetic field and thus to bring up resolution.[27] Primas at ETH Zurich had managed to greatly improve homogeneity by a better design of the magnet pole caps. With this design, he hoped to avoid the decrease of homogeneity while going to higher magnetic fields, which was caused by saturation effects in the iron alloy of the caps. Naturally, Ernst was very much aware of Primas's work in this area, and the mathematical treatment of the design of the pole caps was the subject of his first talk after he arrived in Palo Alto in November of 1963. Varian engineers indeed built some caps following this design, but because of problems in manufacturing, and the advent of superconducting magnets (they did not show the same saturation effects as iron magnets), this project never bore fruit.[28]

In addition, at first, Ernst had to continue working with the "cat," the "dog," and even the "mouse," all methods that had been invented by NMR scientists in order to improve sensitivity. CAT was probably the first acronym in NMR. It referred to Computer of Average Transients, a device storing, adding, and averaging the signals of repeated scans. As a result, the signal-to-noise ratio was substantially improved because the incoherent noise canceled out when the data of many scans were added to each other. The main precondition for the applicability of this method was a very stable spectrometer, in order to add the scans coherently. The A-60, with its field/frequency lock, guaranteed such a high stability, and CAT became the method of choice to upgrade its sensitivity. The use of CAT in NMR was developed at several independent sites in the middle of 1962.[29] One of them was the laboratory of biochemist Oleg Jardet-

---

[27] Anderson, "Early NMR Experiences," 171.

[28] William Siebert, "The Post Doc," manuscript, 1991, Ernst Papers, folder "Correspondence Wes Anderson."

[29] Becker, Fisk, and Khetrapal "Development of NMR," 37–38.

zky at Harvard Medical School. He coined the acronym from his Mnemotron Model 400 Computer of Average Transients, and reported an increase of sensitivity by a factor of 50 in studies on a coenzyme. Jardetzky used an HR-60 spectrometer of Varian Associates, a model that was not yet equipped with the stabilization system so successfully employed in the A-60. Thus, though the stability of the HR-60 was good enough for measuring periods of up to six hours, slow drifts of the magnetic field had to be corrected with the traditional procedure, "the graduate student-in-attendance method."[30] Ernst and Primas discussed CAT with colleagues at a conference of the Faraday Society in September 1962. They were convinced that the relatively expensive[31] and complex technique using a computer was not always necessary. In their response to Jardetzky's announcement of CAT, and answering remarks by James Shoolery of Varian Associates at the Faraday conference, Ernst and Primas proposed to apply a single scan, but with a very long sweep time. Based on calculations, they showed the equivalence of the two methods, provided a very stable spectrometer was available. They aptly called their technique DOG (this time no abbreviation, but the natural enemy of the CAT), in a letter "The story of the CAT and the DOG" to Bernard L. Shapiro, the editor of the MELLONMR newsletter.[32] It was in this newsletter that CAT, DOG, and also MOUSE were reported first. The latter method was an inexpensive timing device for the A-60 that could enhance the sensitivity by a factor of ten.[33] After Ernst had accepted the position at Varian Associates, and while still in Zurich, Anderson asked Ernst to find out the limits of the CAT technique theoretically, given a certain resolution of the spectrometer.[34]

But the break-through did not come with step-by-step improvements of existing technology. An entirely novel approach of how to obtain high-resolution NMR spectra came up during the interaction of Anderson and Ernst: Fourier Transform (FT) spectroscopy.

Fourier Transform NMR

The basic idea behind FT NMR is the excitation of all resonance frequencies in a spectrum at the same time, and not sequentially as in the "normal" con-

---

[30] Jardetzky to Barry Shapiro, 21 August 1962, *MELLONMR* 47 (mailed 29 August 1962), 21–23, on 23, copy in Ernst Papers.

[31] In 1964, the computer was available for $11,500. Alexander V. Robertson, "A-60 sensitivity: MOUSE by CAT out of DOG," letter to Bernard L. Shapiro, 19 March 1964, *IITNN* 67 (April 1964), 16–19, copy in Ernst Papers.

[32] Ernst and Primas to Shapiro, 20 September 1962, *MELLONMR* 48 (mailed 28 September 1962), 1–4, copy in Ernst Papers. For the newsletter, see Chapter 4.

[33] Robertson, "A-60 sensitivity: MOUSE by CAT out of DOG."

[34] Anderson to Ernst, 15 April 1963, Ernst Papers, box 5, folder "Varian Consult."

tinuous wave techniques, where either the magnetic field or the radiofrequency field was swept through. The traditional mode led to a reduced sensitivity per time unit, because only a narrow frequency band brought about the signal. In FT NMR, the whole spectrum is excited at once by using a pulsed radiofrequency. Using a mathematical method, first developed by Jean Baptiste Fourier in the early nineteenth century, the time-dependent impulse response (called Free Induction Decay, FID), is transformed to the frequency-domain spectrum. In the same time that you needed to run a conventional, continuous-wave spectrum, you could perform several pulsed runs, add the data in a digital computer, form the average, and finally transform them to an NMR spectrum suitable for interpretation. The gain was increased sensitivity, up to a factor of 100 (see Figure 6.1).[35]

Coincidentally, Fourier Transform methods made their start in infrared spectroscopy,[36] but the two parties seemed to have not influenced each other. The reason was that the two methods are conceptually different: FT infrared spectroscopy is an experiment in space, employing an interferometer. FT NMR works in the time domain. A spatial FT NMR instrument would need an interferometer with a path length of 3×10^8 m, while the masses of data that would have to be handled in a time-domain FT infrared experiment would exceed the capabilities of even modern computers. Thus, we have here the case of two separated and specialized communities of method makers, in the beginning not taking too much notice of each others' most recent developments.[37]

In general, for simultaneous excitation of an NMR spectrum, several possibilities did exist: radiation with a wide band of frequencies by a noise generator (stochastic resonance), the construction of a multi-channel device for both excitation and detection of resonances, and the use of a series of short radiofrequency pulses of relatively high power level.[38] All approaches were checked. Ernst, especially in the late 1960s and early 1970s, intensively worked on stochastic resonance (see below). Anderson, on his side, between 1962 and 1964 thought about constructing a cylinder or wheel that would transmit a light beam at different oscillating frequencies. This light source modulated the radiofrequencies that excited the spectrum, and the same cylinder demodulated and detected the spectrum. In principle, the device would have worked in real time, but mechanical difficulties prevented its practical use, and the "prayer wheel" or the "wheel of fortune" as it was called, was never finished. Never-

[35] Forsén, "Nobel Prize in Chemistry," 3–5; and Anderson, "Fourier Transform Spectroscopy," 2126–2127.

[36] See Johnston, "In Search of Space."

[37] Ernst, interview by Reinhardt, 25 February 2002; Ernst, "Nuclear Magnetic Resonance Fourier Transform Spectroscopy," 14.

[38] Freeman, "Fourier Transform Revolution," 744A–745A.

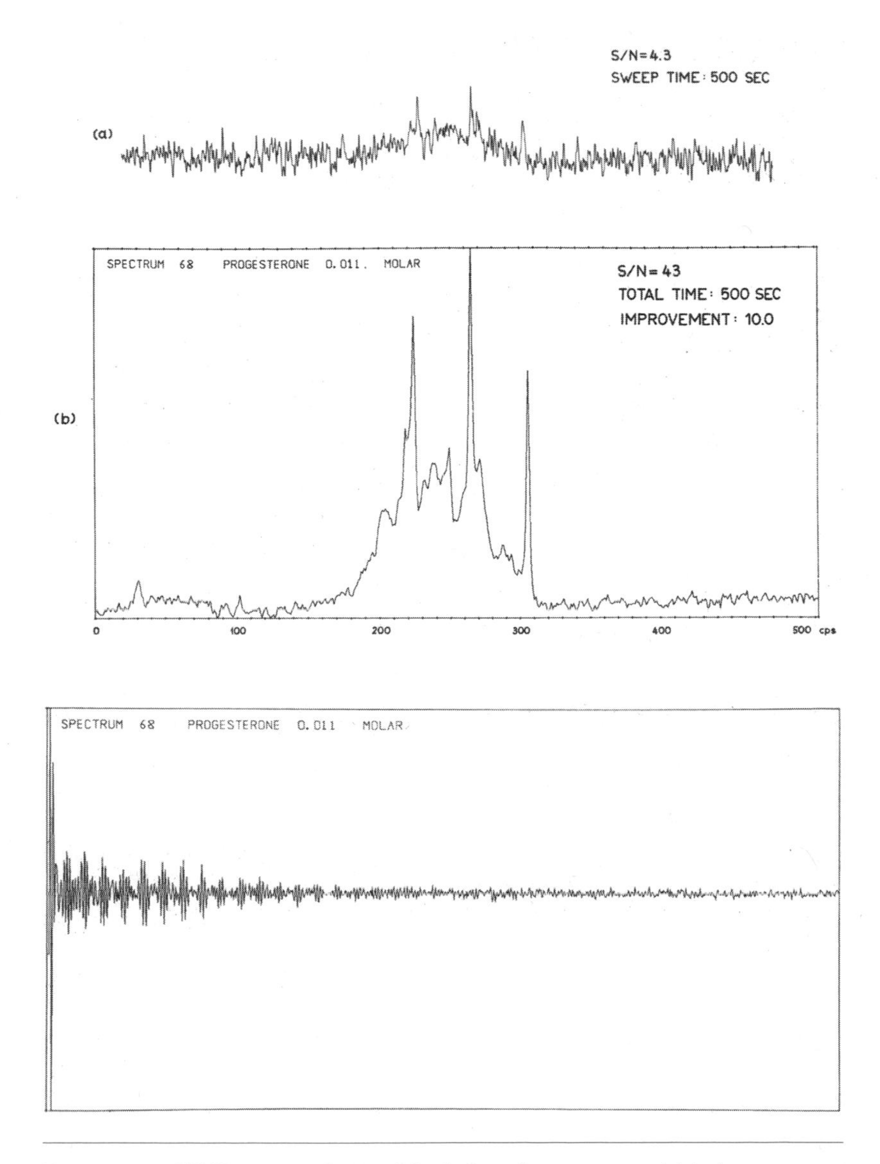

FIGURE 6.1 NMR spectra of 0.011 M solution of progesterone. (a) is the spectrum in continuous wave mode, measured with a single sweep of 500 s. (b) shows the Fourier transformed spectrum of the same compound, obtained by 500 pulses, one second apart, and added to each other. The gain of sensitivity here is tenfold. Below is the Free Induction Decay (FID), the basis of the spectrum shown in (b) before its Fourier transformation. Reprinted with permission from Richard R. Ernst, Weston A. Anderson, "Application of Fourier Transform Spectroscopy to Magnetic Resonance," *Review of Scientific Instruments* 37 (1966), 93–102, figs. 7 and 8 on 101–02. Copyright 1966, American Institute of Physics.

theless, Anderson donated the prototype years later to the Smithsonian Institution in order to illustrate the thought process that led to FT NMR.[39]

The basic idea to use radiofrequency pulses for NMR was as old as the experiment of NMR in bulk matter itself. Already Felix Bloch mentioned this possibility in one of his early publications.[40] Among the physicists that early on explored pulsed NMR techniques was Erwin Hahn of the University of Illinois, and later on also Herbert S. Gutowsky made use of the technique for broadline NMR studies of solids (see Chapter 2). Nevertheless, their methods did not relate to the special techniques employed in high resolution NMR, and did not involve Fourier Transformation. However, the fact that the free induction decay and the continuous-wave spectrum could be transformed to each other by Fourier Transform methods was well known theoretically, and proven explicitly for NMR by Irving Lowe and Richard Norberg in 1957 during investigations in solid-state NMR.[41] But the potential gain of sensitivity in NMR of liquid samples was not obvious at that time. Therefore, the method kept being an interesting, but technically useless academic toy, and no experiments were performed until Ernst and Anderson took up the issue, building on a 1956 patent application of Russell Varian. Russell Varian had proposed to use a wide-band radiofrequency source, and to apply a Fourier transformation of the resulting data with the help of an analog device. The disadvantages consisted mainly in the incoherences brought into the system by the use of a wide band of frequencies, and the crudeness of the Fourier analyzer. Both points could be circumvented with the application of a train of radiofrequency pulses, and the use of a digital computer for Fourier transformation.[42]

Following Anderson's suggestion, accompanied by a detailed block diagram of the apparatus needed,[43] Ernst set out to work. He asked William Siebert, the technician of the group, to build a high power radiofrequency amplifier (unusual for NMR experiments, where normally only very low radiofrequency power was needed), and disappeared for a few weeks to work in

---

[39] Freeman, "Fourier Transform Revolution," 744A–745A; Anderson, "Early NMR Experiences," 172–175.

[40] Bloch, "Nuclear Induction," 461.

[41] Lowe and Norberg, "Free-induction Decays in Solids." See also Lowe, "My Life in the Rotating Frame"; and Norberg, "NMR in Urbana and St. Louis."

[42] Anderson, "Early NMR Experiences," 172–175. In contrast to many other countries, U.S. patents were not disclosed to the public before they were issued, and it was possible to amend the patent. In the case of this patent (US patent no. 3,287,629, issued 22 November 1966), Varian Associates added the use of pulsed methods after Anderson and Ernst had developed their method on FT NMR. In contrast, the concept of applying FT methods by wide-band excitation was contained in the original patent file. See Anderson to Ernst, 28 and 29 January 1993, Ernst to Anderson, 29 January and 11 February 1993, Ernst Papers, folder "Correspondence Wes Anderson."

[43] See Anderson, "Early NMR Experiences," 175.

his laboratory.[44] There, he established the experimental parameters of FT NMR, and explored some potential uses of pulsed NMR in general. Next to the enhancement of sensitivity, the possibility of designing new decoupling experiments seemed most promising.[45] Through his studies at Zurich, he was accustomed with the application of Fourier transform equations, and aware of their potential pitfalls: if applied in an uncritical manner, the solutions of the time-dependent and the frequency-dependent equations could diverge.[46] A lot of the early work concerned details of the experimental set-up. Ernst's main concern here was the necessary stability of the spectrometer for the time averaging methods. In addition, he established the conditions for a suitable appearance of the spectrum after the transformation and the phase adjustments. Because of the high cost of computer time, and the waste of time in transferring punched cards from one place to the other, Ernst considered other possi bilities for Fourier transformation, including an analog Fourier transformer. In addition, he pondered about optical methods, but practical problems prevented their use in commercial spectrometers.[47]

Difficulties with FT NMR remained. In 1964, the computer time needed for the calculation of the spectra was an insurmountable obstacle. The promise of FT NMR, the gain of sensitivity per time unit, seemed to be a bad joke:

> When one considers the cumbersome treatment of the data acquired in a time averaging CAT 400 computer, into paper tape, being converted into punched cards at IBM San Jose, then converted to magnetic tape at the Service Bureau Corporation, Palo Alto, Fourier-transformed on an IBM 7090, and plotted on a Calcom plotter, nobody could have been convinced by us of a time-saving advantage![48]

Thus, the technical impediments were great. In addition, from its very principle, it was clear that FT NMR would appeal to chemists, and not to physicists. The large potential gains were mainly in the field of high resolution NMR spectra of liquid samples, and those were of no great concern to physicists. The chemists, on their side, were still busy in exploring the wonderful world of traditional continuous-wave spectra.[49] But in Ernst's opinion, the time had come for novel solutions of sensitivity enhancement, and FT NMR was one of them.

---

[44] William Siebert, "The Post Doc," manuscript, 1991, Ernst Papers, folder "Correspondence Wes Anderson." See also Ernst, "Success Story of Fourier Transformation in NMR," 295.

[45] See Richard R. Ernst, "Some thoughts about the Fourier transform spectrometer," engineer's notebook no. 934, 4 to 6 July 1964, copy in Ernst Papers.

[46] Ernst, "Zur Fourier-Transformation der Bloch'schen Gleichungen," laboratory notebook "Theorie VIII," 21 May 1963, pp. 34–36, Ernst Papers.

[47] See Ernst, laboratory notebook "Experiments I," started ca. 10 June 1964, Ernst Papers; Ernst, interview by Reinhardt, 25 February 2002.

[48] Ernst, "Success Story of Fourier Transformation in NMR," 295.

[49] Freeman, "Fourier Transform Revolution," 744A.

One has suddenly recognized that there were possibilities to enhance sensitivity, and to make NMR more attractive. It became fashionable to discuss this problem. I mean, this was expressed in the work of Primas and Günthard already. They have tackled this problem, and through them I was introduced to linear response theory for example, and all these possibilities of data acquisition and the problem of signal-to-noise ratio. That was something, if you have built a spectrometer that did not work, you finally could do . . . calculations and keep yourself busy. This was done very often, and I was accustomed to it. I knew the whole set of problems, and other people had similar ideas during this time. It was just a topic that was ripe to talk about.[50]

Already in 1965, Ernst was convinced that FT NMR was a reasonable method to gain sensitivity. In principle, simple time averaging could enhance sensitivity endlessly, but only at the cost of instrument time. Thus, improvements not increasing the measurement time would meet considerable needs, especially in the biomedical community. Addressing his colleagues at the 6th meeting of the Experimental NMR Conference (ENC) in Pittsburgh in February 1965, Ernst compared the potential of the new method to the limitations of older approaches:

Those of you which have thought more seriously about further possibilities to improve the still rather poor sensitivity of nmr, have certainly recognized how difficult it is to advance significantly towards the needs of biochemists and physiologists.[51]

Ernst knew well the still existing disadvantages of the FT NMR method. The instrumentation was relatively complicated, and some inherent limitations of the technique were overcome only later. This concerned lack of resolution, the question of the reliability of intensity values, and difficulties in FT carbon-13 NMR. Especially the latter method showed the greatest potential among

[50] "Man hat dort plötzlich gesehen, es gibt Möglichkeiten wie man die Empfindlichkeit erhöhen könnte um die Kernresonanz noch attraktiver zu machen und es wurde dann einfach aktuell, über dieses Problem zu diskutieren. Ich meine, das kam eben schon in den Arbeiten von Primas und von Günthard zum Ausdruck. Die haben mit diesem Problem eben auch schon gekämpft und dort wurde ich eigentlich schon irgendwie vorbelastet, vorgespannt, mit eben linearer Responsetheorie zum Beispiel und allen diesen Möglichkeiten der Datengewinnung und das Problem vom Signal zum Rauschverhältnis, das war etwas, wenn man ein Spektrometer gebaut hat, das nicht funktioniert hat, dann konnte man schlußendlich immerhin noch . . . Rechnungen durchführen und auf diese Art und Weise sich beschäftigen. Das wurde dann eben sehr oft gemacht und das war bei mir irgendwie intus und daher kannte ich die ganze Problemstellung und andere Leute in jener Zeit sind auf ähnliche Ideen gekommen und das war einfach ein Thema, das spruchreif war." Ernst, interview by Reinhardt, 25 February 2002.

[51] Ernst, talk manuscript, "Sensitivity enhancement by Fourier transform techniques," 6th ENC, Mellon Institute, Pittsburgh, PA, 25–27 February 1965, 1, Ernst Papers, box 11, folder "Frühere Konferenzen."

organic chemists and biochemists, and was in greatest need for enhanced sensitivity (see Chapter 4). Thus, Ernst had "the feeling that it is a promising new method and only the future can show whether it will ever become useful for the organic chemist."[52]

Initially, the technique met considerable skepticism. The referees and editors of the *Journal of Chemical Physics* rejected the manuscript for publication twice. In their opinion, FT NMR did not constitute anything new, and they considered the paper as too technical.[53] The article finally appeared in the *Review of Scientific Instruments*.[54] To make matters worse, the engineers of Varian Associates refused to include FT accessories in the production lines of NMR spectrometers. Technical considerations were the reason for this, as traditional field modulation—a crucial feature of spectrometer design—was not compatible with the pulsed FT technique. Also the response by expert NMR spectroscopists was lukewarm; only Jardetzky, with his experiences in time averaging using the CAT, and a few others showed interest.[55] For a few years, the method was almost forgotten. Ironically, a company other than Varian Associates profited most from the new technique. Varian Associates already was too big to react quickly, and the investments at stake for the change of a whole production line were too high. In contrast, the German competitor, Bruker Analytische Meßtechnik of Karlsruhe, a much smaller but more versatile company, took the lead in 1968, when a first prototype instrument for proton resonance FT NMR was installed at Seymour R. Lipsky's laboratory at Yale University.[56] In 1969, the race for the true goal, carbon-13 FT NMR, started, and Varian Associates was still behind Bruker, as Ernst's former colleague Ray Freeman wrote in October of that year:

> You were four or five years ahead of your time with the Fourier transform work. It has now come back to haunt us. We shipped our first proton machine a few weeks ago, but now of course there is a big panic to do carbon-13 since Brucker [sic] recently showed some carbon spectra . . . . My ex-Postdoctoral Fellow and I are working on a very naive and oversimplified description of Fourier transform spectroscopy to help "educate" the average chemist-spectroscopist in this field. I feel convinced that the boom in Fourier transform NMR has started at long last and it is suddenly the fashionable thing to do.[57]

The breakthrough of FT NMR was possible only with the advent of dedicated computers, situated in the laboratory itself. Together with the invention of a

---

[52] Ernst, "Sensitivity enhancement by Fourier transform techniques," 8.
[53] Ernst, "Success Story of Fourier Transformation in NMR," 296.
[54] Ernst and Anderson, "Application of Fourier Transform Spectroscopy to Magnetic Resonance."
[55] Ernst, "Success Story of Fourier Transformation in NMR," 296.
[56] Ernst to Anderson, 15 and 16 May 1968, Ernst Papers, "Correspondence Wes Anderson."
[57] Freeman to Ernst, 13 October 1969, Ernst Papers, folder "Corr. Freeman."

less time-demanding algorithm for Fourier transformation in 1965,[58] the computer ensured the broad use of FT NMR. With the advent of FT NMR, Bruker obtained the leadership in sophisticated NMR instrumentation during the 1970s. Varian Associates decided to cater for the mass market of routine analytical instruments, a decision that was reversed only much later. In the meantime, Varian lost its practical monopoly in the field, and had to put up with a much larger Bruker group. The patents that Varian Associates held were no real obstacle. A rather friendly licensing policy existed between the firms, at least before the gigantic market of medical diagnosis came within reach of NMR methods in the late 1970s.[59] At that time, Fourier Transform methods already had changed the scale and scope of NMR in chemistry. But in 1965, Ernst at Varian Associates looked for another research topic, one still in line with his interest in the improvement of sensitivity and decreasing complexity of spectra. He found it in broad-band noise decoupling, and the uses of the computer in NMR.

Broad-band Noise Decoupling and Busy Boxes

In retrospect, Ernst described his initial move into broad-band noise decoupling as "being out of better ideas for the time being."[60] Thus, he continued the work of his Ph.D. thesis on stochastic resonance, but this time with a clear application in mind: heteronuclear decoupling. This method could make the nascent carbon-13 NMR methods much more sensitive, and the spectra easier to interpret. Actually, carbon-13 NMR was just one of many potential applications of Ernst's work in this field. Moreover, this time his paper was accepted by the *Journal of Chemical Physics*. Thus, broad-band noise decoupling did not share the fate of early FT NMR. But the two techniques had one aspect in common: In both cases, a relatively large range of nuclear spin resonance frequencies was excited at once. While in FT NMR, the pulse served for the excitation of the measured resonances, broad-band noise decoupling used a wide frequency band to saturate resonances. This led to the removing of some spin-spin coupling patterns, and consequently to the simplification of complex spectra. In contrast to a normal double resonance experiment, where a single frequency was used for decoupling, a much broader range of spin-spin couplings could be decoupled with Ernst's technique. Double resonance methods with single frequencies were known since the mid-1950s, and Freeman and

[58] The Cooley-Tukey algorithm. Becker, Fisk, and Khetrapal, "Development of NMR," 58.
[59] Ernst, interview by Reinhardt, 25 February 2002, and Weston Anderson, interview by Sharon Mercer, February 1990, p. 18, Varian Associates Inc. Oral History Project, Stanford University Archives, SC M 708, box 1.
[60] Ernst, "Success Story of Fourier Transformation in NMR," 296.

Anderson intensely studied them in the early 1960s.[61] Ernst chose a fluorine-proton system for his early investigations, decoupling either the fluorine nuclei or the protons, and produced the necessary binary random noise with the help of the even/odd numbers of the last digits in the Palo Alto telephone directory. Consequently, the method was dubbed "phone-book resonance." Later on, a more efficient technique was used.[62]

The idea went back to Primas's proposal for Ernst's Ph.D. thesis. In Zurich, Ernst found out that broad-band noise decoupling could not be used to decouple strongly coupled spins of a homonuclear species, but worked for weak couplings only. As the latter was not of much practical interest, he did not pursue it further. In Palo Alto it became clear that the broad-band technique was of great use in decoupling different kinds of atoms, i.e., heteronuclear systems that were weakly coupled. In a 1966 paper, Ernst adapted the Zurich theoretical solutions for the heteronuclear case, and performed experiments that showed the feasibility of his method. His treatment was closely connected to established theories of NMR. He demonstrated that the use of random radiation frequencies caused resonance line patterns that were related to relaxation effects and chemical exchange phenomena.[63] Also the theoretical treatments of the effects were equivalent. Thus, he could artificially emulate certain "natural" phenomena of molecular systems.

> The artificial introduction of a random time dependence into a spin system bears a strong similarity to the effects of relaxation and chemical exchange in NMR. Experiments with a random rf [radiofrequency] magnetic field can establish a continuous transition from the phenomena of conventional double resonance with the characteristic coherence effects (narrow lines, sidebands) to the incoherent phenomena of relaxation and chemical exchange (line broadening).[64]

Ernst worked in the tradition of the specialists who invented experiments to manipulate the response of spin systems to radiation with radiofrequency power. A good example was some double resonance techniques that did not cause spin-spin splittings to disappear fully, but perturbed them systematically. These methods were called "spin-tickling" and allowed the study of complex spin systems.[65] Interestingly, Ernst emphasized the "artificial introduction" of double resonance effects, thus implying the naturalness of relaxation and chem-

[61] Becker, Fisk, and Khetrapal, "Development of NMR," 45–47.

[62] Ernst, "Success Story of Fourier Transformation in NMR," 296.

[63] Relaxation is the process of how nuclei return to the lower energy levels, after excitation by radiofrequency in the NMR experiment; chemical exchange means actual movements of molecular groups. Both effects shape the appearance of lines in an NMR spectrum.

[64] Ernst, "Nuclear Magnetic Double Resonance," 3846.

[65] See the literature given in Ernst, "Nuclear Magnetic Double Resonance," 3845.

ical exchange. Certainly, all these effects were caused by and could be measured with complex instrumentation only. But in his wording, Ernst brought the manipulative aspects of the double resonance technique to the fore.

Ernst carefully connected the theoretical part to the experimental and instrumental sides of the novel method, and showed the huge advantages that broad-band noise decoupling had in important cases over the older frequency-fixed decoupling methods. Moreover, he explicitly mentioned carbon-13 NMR spectroscopy as a potential application, aware of the fact that this was the largest market in organic chemistry with the greatest need for improvements. His method was a suitable one for the average organic chemist who was always reluctant to put up with the newest complicated (and expensive) gadgets of the physicists:

> One of the main advantages of this method is the ease of operation. It does not require a field-frequency lock system nor a means of sweeping the frequency. It can well be used in field-sweep experiments on instruments with a relatively poor stability. This favorably distinguishes the method from most other double-resonance techniques.[66]

Ernst's method was used, largely unchanged, until the early 1980s, when Malcolm Levitt, Ray Freeman, and John Waugh proposed better-performing decoupling techniques based on pulsed methods.[67]

The last project Ernst tackled during his stay at Varian Associates refers back to the first task that Anderson formally assigned Ernst after his arrival at Palo Alto. It concerned the optimization of the performance conditions of the spectrometer, a complex interplay of the parameters that had to be chosen for running an NMR experiment. Ernst's and Anderson's goal was nothing less than the automation of NMR spectroscopy with the help of the computer:

> I have done many many calculations on the computer of the Service Bureau Corporation, an IBM computer. I punched thousands of cards, wrote programs and calculated data, and made huge surveillance calculations in order to get hold of the variability of possible parameters and to find out where the optimum is, actually in a purely empirical manner. Through calculations, I found certain rules, how to sweep a spectrum, how fast, what the repetition rate is, how much the frequency, just to get the best signal-to-noise ratio for a normal sweep-experiment, without losing too much resolution.[68]

---

[66] Ernst, "Nuclear Magnetic Double Resonance," 3860.

[67] Becker, Fisk, and Khetrapal, "Development of NMR," 47.

[68] "Ich habe dann viele, viele Rechnungen gemacht auf dem Computer der Service Bureau Cooperation, ein IBM Computer, und dann tausende von Lochkarten gestanzt und Programme geschrieben und Daten berechnet, große Übersichtsrechnungen gemacht, um die ganze Mannigfaltigkeit der möglichen Parameter in den Griff zu kriegen, um zu sehen, wo das Optimum liegt; eigentlich rein empirisch. Durch Rechnungen gewisse Regeln herausgefunden, wie man eben

In 1963–64, when Ernst was doing these calculations, Varian Associates did not have an in-house computer, and Ernst had to buy external computer time. During his stay, this situation changed for the better, and this was a critical point in the development of the techniques that Ernst and Anderson had in mind. Partially based on treatises of random radiofrequency noise, filtering and basic information theory, Ernst calculated the potential of time-averaging methods, and the impact that instrumental parameters such as the sweep rate had on instrument performance.[69]

These were modest beginnings, and with one exception Ernst even did not dare to publish them. But this start sparked a deep transformation of NMR, as it did with other spectroscopic methods. When, in 1991, he finally published some of his early work on computers and NMR, Ernst was convinced that "computers indeed have revolutionized NMR in a more dramatic manner than most other fields . . . . They allow radically new approaches, they lead to a new style of life for NMR spectroscopists, and they have changed the philosophy of experimenting."[70] Twenty-five years earlier, when this "revolution" began, a low performance PDP 8 computer (4096 words of 12 bit memory) was attached to the low-cost NMR spectrometer of the time, the A-60. Ernst's immediate goal was the computerization of the A-60, focusing on the automation of the measurement process. In a remarkably short span of time, from the fall of 1966, when the computer was delivered, to the 8th ENC in Pittsburgh in March 1967, Ernst was indeed able to tackle most of the issues that later formed the basic understanding of computer applications in NMR: control of the spectrometer, and data accumulation, reduction, and evaluation.[71] Again, as in the case of FT NMR, this met with resistance on the part of traditionalists:

> However, at that time, this was ridiculed by 'true' scientists as child's play in the same manner as computer scientists were not properly accepted by the 'true' mathematicians. For this reason, we never published this piece of work in the open literature.[72]

At the very same time, in mathematics, the different work style with the computer led the way to an enhancement of number calculations, something that was resisted by many mathematicians who were used to thinking in abstract

---

durch ein Spektrum durchsweepen mußte, wie rasch, wie die Repetitionsrate, wieviel war die Frequenz, einfach um für ein normales Sweepexperiment das beste Signal-zum-Rauschverhältnis zu erhalten, ohne all zu viel an Auflösung zu verlieren." Ernst, interview by Reinhardt, 25 February 2002.

[69] Ernst, "Sensitivity Enhancement in Magnetic Resonance. I"; and Ernst, "Sensitivity Enhancement in Magnetic Resonance. II."

[70] Ernst, "Without Computers—No Modern NMR," 1.

[71] Ernst, "Without Computers—No Modern NMR," 9, fig. 7.

[72] Ernst, "Without Computers—No Modern NMR," 9–10.

terms.[73] In the case of NMR spectroscopy, the designation of a child's toy, "busy box," was chosen to ridicule the computer. In this case, "true spectroscopists" distrusted the black box of the computer:

> Oh yes, that was named 'busy box.' I don't know if you have heard about the term busy box. A busy box is a device for children that was sold then in the USA. That is simply a plate, with many knobs on top, and you can turn these knobs and something mechanical happens. The knobs are coupled on the other side of the plate in such a way that you don't know what happens. There are little windows where something appears if you turn a knob. That is the busy box. And such a busy box was the computer, where you could do such playful games. But we did not take it really serious. Naturally, and with the same effect, you could realize a filter electronically to avoid noise or to enhance resolution.[74]

One example of such child's play was the automatic shimming of the NMR spectrometer.[75] Shimming referred to methods to improve the homogeneity of the magnetic field and thus to enhance the resolution. This had to be done before each experiment. The A-60 was equipped with shim coils applying small electric currents and through this homogenizing the magnetic field. In the very beginning of NMR, and before the advent of the A-60, a suitable spot in the magnet gap had to be found by moving the sample. With electric shimming, the magnetic field was changed until the operator achieved sufficient resolution. Still, the electric shimming took a lot of time: "Normally, an NMR spectroscopist is spending an appreciable part of his lifetime in front of the spectrometer desperately turning shim knobs, more or less at random, in the hope of improved resolution."[76] The computer could do that automatically and much faster, focusing on a single sharp line in the spectrum. In addition to automatic shimming, Ernst applied data acquisition with a correction of the

---

[73] On this issue in mathematics see MacKenzie, "Negotiating Arithmetic, Constructing Proof." I thank Kristen Haring for guiding me to this reference.

[74] "Ja, ja, das war dann Busy Box. Ich weiß nicht, ob Sie den Begriff Busy Box kennen. Eine Busy Box ist so ein Kindergerät, das wurde damals in den USA verkauft. Das ist einfach so eine Platte, da hat es viele Knöpfe drauf und diese Knöpfe kann man drehen und da passiert irgend etwas mechanisch; die einzelnen sind miteinander verkoppelt hinten, so daß man nicht genau weiß, was passiert und dann hat es irgendwie Fensterchen, wo irgend jemand dann erscheint, wenn man da an einem Knopf dreht. Das ist die Busy Box. Und so eine Busy Box war der Computer, wo man solche Spielereien machen konnte. Aber eben so ganz ernsthaft haben wir das eigentlich nicht genommen. Man konnte auch elektronisch natürlich gerade so gut eben einen Filter realisieren, um Rauschen auszuschalten oder um Auflösungen zu verbessern." Ernst, interview by Reinhardt, 25 February 2002.

[75] This was the only of Ernst's inroads into the computer field that was published immediately: Ernst, "Measurement and Control."

[76] Ernst, "Without Computers—No Modern NMR," 10.

sweep range, and signal-to-noise filtering to improve the resolution. He was well aware that filtering did not add more information to the spectrum. Nevertheless, a filtered spectrum could lead to appreciable assistance in interpretation, as some lines became clearer. Data reduction, the automatic identification of multiplets, the identification of sample compounds by a library search, and phase adjustment made the computer tasks complete. Ernst also made preliminary attempts to apply computer techniques in double resonance, but interrupted this project because of his move back to Zurich in early 1968.[77]

Back in Service at ETH Zurich

In late 1967, Ernst received the offer of ETH Zurich to come back and take over the NMR group of Primas in the division of physical chemistry. Primas had moved more and more into theoretical chemistry, and Günthard was looking for somebody to fill this void. In many ways, it was an inadequate offer. The salary was considerably lower than in Palo Alto, and Ernst would be on "soft money," as funds were supposed to come from the Swiss "Nationalfonds," the national science funding agency. Moreover, the equipment was inferior to that at Varian Associates, consisting essentially of two self-built spectrometers that did not function properly, and a Varian DP-60. To compensate for this, Günthard planned to apply for funds for a Swiss high-field NMR laboratory, equipped with the most modern of the Varian spectrometers, the HR-220. Ernst should take care of this instrument also. Overall, in the beginning, the return to Switzerland was disadvantageous for Ernst, in scientific terms at least. But he was Swiss, and wanted to return to his native town of Winterthur, near Zurich.[78] Only reluctantly, Varian Associates, and especially Weston Anderson, let him go. Repeatedly, Anderson asked him to return, but Ernst refused.[79] In the end, this was a wise decision. At Varian Associates, the research-intensive situation changed and the structure of the research department was rather disrupted with the moving of Weston Anderson to a management position.[80]

On 1 March of 1968, Ernst and his family left Palo Alto, and after a month's trip through Asia returned home to Winterthur. Among the countries they saw, the Ernsts liked Nepal most, because it was "unique in its pure character and not yet much influenced by the Western materialistic civilization."[81]

---

[77] Ernst, "Without Computers—No Modern NMR," 16–21.

[78] Ernst, "Success Story of Fourier Transformation in NMR," 296; and Ernst, interview by Reinhardt, 25 February 2002.

[79] See Ernst to Anderson, 27 July 1969, and Anderson to Ernst, 17 February 1970, Ernst Papers, folder "Correspondence Wes Anderson."

[80] Hanspeter Benz to Ernst, 25 October 1969, Ernst Papers, folder "Corr. Benz."

[81] Ernst to Anderson, 28 April 1968, Ernst Papers, folder "Correspondence Wes Anderson."

After his return to the Western style of life, he faced the challenge of building up an NMR group from almost nothing, because Primas had neglected the more practical parts of his projects at ETH, and only one Ph.D. student was working in the field. Immediately after his arrival in Zurich, Ernst fulfilled the procedure that was necessary in the Swiss system to start the career of a university professor, the "Habilitation." As Primas and Günthard guaranteed success, he submitted a rather short version of a thesis, concentrating on the control of NMR spectrometers with the help of a computer.[82] Although he successfully surmounted this hurdle in the spring of 1968, he remained on grant money for two years and only after the end of this period was hired as assistant professor at ETH. Moreover, during the first months of his stay in Zurich, he did not have co-workers, and he could employ a technician only from October 1968 on.[83] Thus, Ernst faced the task "to do some down-to-earth electronics" by himself again. He tried to continue to focus on the development of multipurpose methods, and consequently chose a broad range of topics. He mentioned relaxation phenomena in liquids, NMR of the solid state, and optical pumping[84] as a novel method that would greatly increase the sensitivity of NMR. For the time being, Ernst decided to split his research time between relaxation phenomena in liquids and NMR of solids. These fields, in Ernst's opinion, had the potential for novel applications in chemistry, but so far had been neglected by chemists. Moreover, he proposed to work predominantly on the development of new methods, not so much on the application side itself, and he decided not to invest much energy in instrument development. This was reminiscent of the style of work he got accustomed to at Varian Associates, described in a letter to Primas:

> As you will see from the draft [for the research proposal], my actual interest is more in a methodological direction, and not in very specialized applications. That is partially caused by the atmosphere here at Varian, where the development of instruments dominates, and novel methods are the basis for new instruments. However, I do want to keep apart as far as possible from the development of instruments, but understand that this will not be completely possible . . . . Beginning with this methodological motivation, I want to go deeper into physical-chemical problems, especially with respect to the occupation of Ph.D. students.[85]

---

[82] Ernst to Primas, 18 November 1967, Ernst Papers, box 5, folder "Early Varian."

[83] Ernst to Hanspeter Benz, 2 September 1968, Ernst Papers, folder "Corr. Benz."

[84] Optical pumping could lead to an increase in the polarization of nuclear spins. Consequently, a larger number of nuclei would "participate" in the measurement. In the normal NMR experiment, only a fraction of all nuclei of a sample is oriented in the direction of the magnetic field, due to random movements. Although Ernst invested a considerable amount of time in this project, optical pumping did not lead to a breakthrough in NMR.

[85] "Aus dem Entwurf ist ersichtlich, dass meine gegenwärtigen Interessen mehr in methodischer Richtung gehen als zu sehr spezialisierten Anwendungen. Dies ist zum Teil bedingt durch die

When he started in Zurich, Ernst's projects were financed by the Swiss Nationalfonds, through a grant applied for by Hans Primas. While Ernst wanted to concentrate more on the methodological side of NMR, Primas, obviously for tactical reasons, decided to include an important application field, the structural elucidation of large molecules. Primas divided their joint project into two parts, a theoretical part on the quantum mechanical characterization of molecular systems, and an experimental part. Primas was, in the first two years, the supervisor of both parts, but Ernst could work independently on the experimental side.[86] This grant application, and even more so the acquisition of a high-field NMR spectrometer in 1969, in the long run guided Ernst more towards the application of NMR for structural problems than he had expected originally.

Describing his situation at ETH, Ernst wrote to Anderson at Varian that "the start will be rather slow. Money and people are scarce. One feels pretty much isolated and does not know where to start, it is completely different from the start at Varian."[87] But the prospects of finding money for an HR-220 Varian spectrometer seemed promising, as quite a lot of interest had accumulated at the chemistry divisions of ETH and the Swiss chemical industry. Soon, most of Ernst's time was used up by the installation of the new HR-220 spectrometer for the purpose of service work at ETH. The crucial innovation of the HR-220 was its superconducting magnet (5.17 Tesla) that more than doubled the magnetic field compared to the field strength in reach with electromagnets. This led to an increase in resolution and sensitivity, and made larger molecules, especially biomolecules, accessible for effective studies with NMR. But it came at a price: the spectrometer, especially its magnet, was unstable, and the technology not ripe for continuous use. Nevertheless, in the following years, the HR-220 became the prestige object of many universities and research centers around the world. Because of its complexity, its price and high operating costs, the HR-220 became the first NMR spectrometer whose sharing was necessary to allow its use. In the second half of the 1960s, among others, Varian Associates engineers installed HR-220 spectrometers at Bell Laboratories (Robert Shulman), at Caltech (Sunney Chan), and at the University of Pennsylvania in Philadelphia under Mildred Cohn. The two latter instruments were set up in

---

Atmosphäre hier bei der Varian, wo die Entwicklung von Instrumenten im Vordergrund steht, und neue Methoden sind die Grundlage zu neuen Instrumenten. Doch möchte ich mich so weit als möglich von der Instrumentenentwicklung fern halten, ganz wird es natürlich nicht gehen . . . . Von der ursprünglich methodischen Motivierung möchte ich gerne tiefer in echt physikalisch-chemische Probleme eindringen, besonders auch in Hinblick auf die Beschäftigung von Doktoranden." Ernst to Primas, 18 November 1967, Ernst Papers, box 5, folder "Early Varian."

[86] Primas to Ernst, 2 February 1968, and draft "Forschungsgesuch an den Schweizerischen Nationalfonds," n.d. [January 1968], Ernst Papers, box 6, folder "Nationalfonds."

[87] Ernst to Anderson, 28 April 1968, Ernst Papers, folder "Correspondence Wes Anderson."

regional facilities, financed by NIH and NSF grants. In addition, the chemical industry was among the earliest customers. In Europe in 1968, altogether four instruments were installed: three in chemical companies (Bayer of Leverkusen, Germany, BASF of Ludwigshafen, Germany, and ICI in Runcorn, UK), and one in a research center for macromolecular chemistry at the University of Freiburg in Southern Germany.

At the latter place, the Swiss chemical industry got to know the advantages of this high-field spectrometer. Thus, chemists in Switzerland felt the need for an HR-220 available on a nation-wide basis. Zurich, with the experience of Ernst and the support of ETH, provided a suitable home. A consortium of the ETH laboratories of organic chemistry (Oskar Jeger and Wilhelm Simon), molecular biology (Robert Schwyzer), physical chemistry (Hans H. Günthard), and technical chemistry (Piero Pino) provided the necessary finances. Costs for the HR-220 were relatively high. The spectrometer itself had a price of $160,000 (Swiss francs 696,000), and an extra carbon-13 unit would cost nearly $40,000. The capital investment of nearly 900,000 Swiss francs was huge, even when compared with the projected annual costs of Swiss francs 36,000 for a chemist and a technician that were needed for the operation. The price for the annual needs of liquid helium was at the same level. Thus, the annual costs were less than 10 percent of the initial investment, but still substantial. Ernst considered both a closed shop (operation by technician) and an open shop (operation by trained user) operation of the spectrometer on a 12 to 16 hours-a-day basis. In late 1968, the laboratory of organic chemistry of the nearby University of Zurich joined the consortium as an equal partner with an investment of 20 percent of the costs. At the University of Zurich, Wolfgang von Philipsborn had considerable experience with the Varian spectrometers A-60 and HR-100, and wanted to expand his research with the novel instrument.[88]

Although the five members of the consortium were equal with respect to decision-making processes, the operation and maintenance of the spectrometer was in the hands of Ernst and the laboratory of physical chemistry.[89] The spectrometer was delivered in April of 1969, and it took Ernst and the Varian engineers half a year to finish the installation process and to reach the given specifications of resolution, sensitivity, and stability.[90] One of the major issues

---

[88] "Projekt zur Anschaffung und zum Betrieb eines 220 MHz Kernresonanzspektrographen," n.d. [1968]; "Absprache zwischen den Laboratorien fuer organische Chemie, Molekularbiologie, technische Chemie und physikalische Chemie der ETH einerseits und dem organisch-chemischen Institut der Universitaet Zuerich andererseits," n.d. [May 1969]. For the experiences of Wolfgang von Philipsborn see "Forschungsgesuch Schweizerischer Nationalfonds," 12 December 1968. All in Ernst Papers, folder "HR 220 Projekt."

[89] See "Reglement ueber den Betrieb eines HR-220 Kernresonanz Spektrometers," Ernst Papers, folder "HR 220 Projekt."

[90] Richard R. Ernst, "Betriebsbericht des HR-220 Kernresonanz-Spektrometers," 28 October 1969, Ernst Papers, box 3, folder "Betriebsberichte."

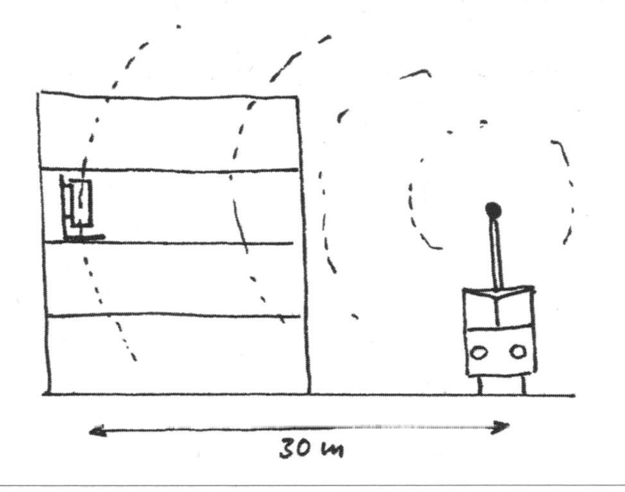

FIGURE 6.2 Cartoon of the Zurich street-car disturbing the magnetic field of the Varian HR-220 NMR spectrometer. From Ernst to Anderson, 27 July 1969, Ernst Papers, folder "Correspondence Wes Anderson."

that kept Ernst busy with the operation of the HR-220 NMR spectrometer was the disturbing effects of a nearby street-car line (see Figure 6.2). The stray field from the 500 dc Volt line caused considerable shifts of the resonance lines. Even worse, in the whole area of ETH in the center of Zurich, there was no place that guaranteed an undisturbed operation of the instrument. The only solution was to correct the influences of the street-car line with an internal field-frequency lock, or to work between midnight and 4 A.M. Naturally, Ernst chose to build an NMR stabilizer, which took a lot of effort before it worked.[91]

This was made possible following the decision of the consortium members not to buy a carbon-13 unit, but instead to equip the spectrometer with a frequency sweep unit and an internal reference stabilizator of the magnetic field. With that, the distortions that were caused by the neighboring street-car line could be avoided and much improved spectra could be recorded.[92] In late 1969, the stability of the spectrometer was improved to such a degree that Günthard could invite Swiss universities and chemical companies to send samples for investigation.[93] But in general, the spectrometer kept being used by the

[91] Ernst to Anderson, 27 July 1969, Ernst Papers, folder "Correspondence Wes Anderson."

[92] "Protokoll der Diskussion über den Betrieb des HR 220 Spektrometers," 5 May 1969, and consortium members Günthard, Pino, Schwyzer to J. Burckhardt, 14 July 1969, Ernst Papers, folder "HR 220 Projekt."

[93] Günthard to H. R. Denzler, 16 December 1969 and letter of Günthard, 22 January 1970, Ernst Papers, folder "HR 220 Projekt."

members of the consortium only, who also covered the operative costs on a time-dependent basis.[94] Here, a lack of balance in use can be noted: The laboratories of molecular biology and of physical chemistry together used more than 80 percent of instrument time between 1972 and 1974, while the use rate of organic chemistry (both at ETH and at the University of Zurich) was under 5 percent. Technical chemistry (investigations of polymers) used the rest of the time.[95] Obviously, the relatively simple molecules that were investigated by organic chemists did not make necessary the use of such high fields. In contrast, the methods development of physical chemists, and the study of large molecules by molecular biologists made the application of high fields either useful or mandatory. As an outcome, the capital investment of the laboratories of organic chemistry was in vain. More and more, molecular biology under Kurt Wüthrich became the dominant user.

Already in the early 1970s, the consortium members made the decision not to modernize the HR-220, but to save funds to acquire a succeeding instrument. In the mid-1970s, Bruker-Spectrospin of Fällanden offered the use of spectrometers with higher fields (270 and 360 MHz) on a service basis. Consequently, the user rate of the HR-220 decreased considerably in 1974. In addition, the number of non-ETH users became practically zero.[96] In order to compete with the better-performing instruments of the new generation, the consortium in 1975 bought a 360 MHz spectrometer of Bruker-Spectrospin.[97] The new instrument came under the management of the molecular biologist Kurt Wüthrich.[98] His group was the largest user, and at his laboratory outside of the inner city of Zurich no disturbing fields had to be expected. The old HR-220 spectrometer remained in the care of Ernst and was modified for solid-state studies. This meant the end of the service responsibilities of Ernst's laboratory, and he was freed to concentrate entirely on his own work.

While he was developing new ways to increase the usability of NMR at Varian Associates, Ernst had become acquainted with the needs of the instrument market. Back in Zurich, he was looking for academic freedom, and the possibility of embarking on more basic features. But in the beginning years, he was entangled in a net of expectations. On one hand, there was the old style of concentrating on instrument development. On the other hand, and somewhat

---

[94] Only the chemical company Geigy of Basel used the HR-220 on a regular basis in 1970. See Richard R. Ernst, "Betriebsbericht des HR-220-NMR-Labors," 1 July 1970, Ernst Papers, box 3, folder "Betriebsberichte."

[95] Kurt Wüthrich to H. Bühlmann, 23 May 1975, Ernst Papers, folder "HR 220 Projekt."

[96] Richard R. Ernst and A. Frey, "Betriebsbericht 1974 des HR-220-NMR-Labors," 24 February 1975, Ernst Papers, box 3, folder "Betriebsberichte."

[97] Kurt Wüthrich to H. Bühlmann, 23 May 1975, and Wüthrich to Ernst, 31 July 1975, Ernst Papers, folder "HR 220 Projekt."

[98] "Betriebsordnung für das HXS-360 Hochfeld-NMR-Spektrometer," 5 January 1976, Ernst Papers, folder "HR 220 Projekt."

in contradiction, the needs of Swiss organic chemists led to a focus on service work. In retrospect, Ernst described the period between 1968 and 1974 as the "dark middle ages" of his career, after the "classical" years at Varian Associates. Indeed, it was a transition period, but important in the respect that it initiated his cooperation with biochemists, especially Kurt Wüthrich. Though Ernst's work suffered under technical difficulties with the equipment, he was able to tackle NMR of solid-state materials with the help of a self-built spectrometer, and he included the study of relaxation effects in his program. Using the new Varian HR-220 spectrometer, Ernst studied the structures of small organic compounds. In addition, he and his co-workers continued to investigate Fourier Transform methods and explored the application of pulse techniques to other NMR methods. Thus, Ernst's relatively small group of about seven Ph.D. students and one postdoctoral fellow in the first half of the 1970s worked on nearly every aspect of modern NMR.[99] It was clear to him that his work would be of interest to other spectroscopic techniques also. NMR was the ideal technique to test principles, because of its straightforward concepts in theoretical and experimental perspectives.[100] Ernst's function as consultant for Varian Associates helped in meeting these demands, and, especially, in securing opportunities and instruments to tackle topics that were close to his own scientific interests. Thus, in moving to academe, he remained in touch with an industrial environment, struggling for the independence to concentrate on basic features of NMR.

The Consultant

After he left Varian Associates in 1968, Ernst became a consultant for the company, and stayed in this position for a decade. In return for a monthly payment, his professional service agreement called for eight hours per month of consulting "in nuclear and electronic magnetic resonance, and the application of computing techniques." Moreover, Ernst was expected to report the inventions he made in connection with his consulting activities.[101] In practice this meant that

[99] The graduate students, in chronological order until 1973, were: Thomas W. Baumann (solid state NMR), Alexander Frey (molecules in the nematic phase), Dieter Welti (relaxation effects), Enrico Bartholdi (Fourier spectroscopy), Stefan Schäublin (chemically induced nuclear polarization), Anil Kumar, and Luciano Müller. The postdoc was Robert E. Morgan. See Ernst, "Success Story of Fourier Transformation in NMR," 296, and the tabular summary of Ernst's co-workers in Ernst Papers, folder "Mitarbeiter."

[100] Ernst, "Forschungsgesuch Schweizerischer Nationalfonds, Fortsetzung des Projektes Nr. 5185.2," 10 March 1970, Ernst Papers, box 6, folder "Nationalfonds."

[101] Professional service agreement between Ernst and Varian AG, Switzerland, 20 July 1970, Ernst Papers, box 5, folder "Varian Consult." Ernst made his contract with the Swiss branch of Varian Associates but reported to the headquarter in Palo Alto. The agreement continued until April of 1978.

Ernst assigned the patent rights to some of his inventions to Varian Associates for a very small extra payment, but without receiving any royalties. Ernst decided which of his inventions fell under the agreement, and he assigned the important patents covering stochastic resonance, 2D NMR, and Fourier imaging to Varian Associates. Especially the patent on Fourier imaging later proved to be very valuable for the company, because of licensing arrangements that Varian Associates made with manufacturers of medical equipment.

The idea of starting a consultancy began after Ernst had emphasized his needs for an advanced computer system to Anderson when he was back in Zurich. It was clear that he was not able to pay for it, as the price of the Varian computer was four times as high as his entire budget for two years.[102] Already in late 1967, before he left Varian Associates, Ernst had written a proposal for a computer system at the laboratory of physical chemistry at ETH. He emphasized that the system would be relatively small-scale, and that it should be versatile enough to be used with the different spectrometers present in the laboratory. Above all, he wanted to avoid doing research on computer methods themselves: "Research in relation with the computer shall not be an aim in itself. To the contrary, the computer should stay a practical tool in the sense of an oscilloscope or a counter."[103]

The system was constructed in such a way that the data acquisition unit could be used independently from the calculation unit. Furthermore, it included two digital-to-analog converters and one analog-to-digital converter, whose speed was crucial for the performance of the system. Ernst considered two computer systems in his proposal, the PDP 8I of Digital Equipment Corporation, and Varian Associates' 620 I. Both were similarly priced. While the PDP 8I had a better design, and was easy to program, the 620 I showed better performance, and Ernst thought that Varian Associates would do more on specialized software for applications in spectroscopy than Digital Equipment Corporation.[104] Finally, Varian Associates did lend a 620 I computer system to ETH, for one year without charge from late 1968 on.[105] Varian managers

---

[102] Ernst to Anderson, 28 April 1968, Ernst Papers, folder "Correspondence Wes Anderson."

[103] "Forschung im Zusammenhang mit dem Computer soll nicht Selbstzweck sein, sondern der Computer soll praktisches Hilfsmittel bleiben im gleichen Sinn wie ein Oszilloskop oder ein Zähler." Richard R. Ernst, "Vorschlag für ein Daten Aquisitions System für das Laboratorium für Physikalische Chemie," 1 December 1967, 2, Ernst Papers, box 6, folder "Nationalfonds."

[104] Together with teletype, converters, paper-tape reader and magnetic-tape recorder, the Varian 620 I system cost $25,900, and the PDP 8I $26,680. Richard R. Ernst, "Vorschlag für ein Daten Aquisitions System für das Laboratorium für Physikalische Chemie," 1 December 1967, 6, Ernst Papers, box 6, folder "Nationalfonds." A later offer by Varian Associates of a more elaborated system came to $48,000. David Cunningham to Ernst, 26 February 1968, Ernst Papers, box 6, folder "Nationalfonds."

[105] See "Auftragsbestätigung" of the Varian Associates branch at Zug, Switzerland, 11 and 16 August 1968, Ernst Papers, box 5, folder "Varian Consult."

"decided that it would be in the best interests of Varian" to lend Ernst a Varian computer system in order that he might continue the investigations of computer-aided spectroscopy on a Varian product.[106] Writing to Ernst, his former colleague Ray Freeman expressed his conviction that Varian Associates did "badly need your help in programming useful features" into the Varian computer systems.[107] Ernst was glad to receive this crucial tool for his research. Both Ernst and Günthard used the computer for their work, with Ernst focusing on double resonance and pulse experiments. In a letter to Primas, Ernst mentioned that this might bind him as consultant for Varian Associates and inquired if something at ETH would speak against this.[108] It did not, and a relationship began that lasted to 1978.

Already in the beginning, Anderson was concerned about the progress of their working arrangement. As Ernst had predicted, the start of his research at the ETH was rather slow, mainly for lack of graduate students. Anderson thought of launching a special project, and Ernst considered doing some work on computerized pulse experiments on the 620 I. Most of all, he was convinced "that much of what I intend to work on might be of interest to you."[109] Soon, a research project in cooperation with Varian Associates began to take shape: stochastic resonance. Stochastic resonance had a quite long history. Its main layout was included in a patent that Russell Varian filed in 1956.[110] Independently from Russell Varian, Ernst and Primas in 1959 had the same idea, but did not proceed in the direction of practical application. In 1966, while at Varian, Ernst thought about methods to correlate the output with the input noise, which is a critical practical feature of the technique. Back in Zurich in June 1968, an inquiry by Anderson finally sparked off the thought process of Ernst which led to the concept of practical realization.[111] Following up a question of Anderson on how to cover a wider spectrum with FT NMR, Ernst considered using random radiofrequency noise for the excitation of the nuclei. In principle, this was a consequent move, as it was in accordance with his interest in broad-band decoupling. This time, the stochastic radiofrequency power would not be used for decoupling, but for the measurement process itself.[112] At the end of June, Ernst could report progress on the side of FT NMR with random

---

[106] Dick Rabin to David Cunningham, 17 July 1968, Ernst Papers, box 5, folder "Varian Consult."
[107] Freeman to Ernst, 14 June 1968, Ernst Papers, folder "Corr. Freeman."
[108] Ernst to Primas, 18 November 1967, Ernst Papers, box 5, folder "Early Varian."
[109] Ernst to Anderson, 10 November 1968, Ernst Papers, folder "Correspondence Wes Anderson."
[110] This is the patent that covered FT methods, U.S. patent no. 3,287,629, issued 22 November 1966.
[111] This follows Ernst to Anderson, 23 March 1970, Ernst Papers, folder "Correspondence Wes Anderson."
[112] Anderson to Ernst, 10 May 1968, Ernst to Anderson, 15 May 1968, Ernst Papers, folder "Correspondence Wes Anderson."

noise excitation.[113] Stochastic resonance had several advantages over pulsed methods: the necessary radiofrequency peak power was lower, and the operator could adjust sensitivity and resolution independently from each other. Independently from Ernst, Reinhold Kaiser from the University of New Brunswick also developed stochastic resonance methods. This led to a cooperation with this scientist in the early 1970s. Stochastic resonance was one of the main research projects of Ernst during this time, and showed a high potential to replace pulsed FT NMR methods, but this did not materialize.

Next to his research work, Ernst criticized Varian Associates' products from the standpoint of a customer. Through his contacts to other NMR scientists, he was able to collect their criticism, and reported this opinion. In July 1968, after a considerable number of talks on computer applications in spectroscopy, Ernst gave Anderson a briefing on what spectroscopists thought of Varian Associates' computer product policy. Varian Associates had decided to deliver pre-packaged and completely programmed computers only. To make matters worse, the technical description of the 620 I in the company's leaflets was wholly inadequate, and potential customers did not learn how to interface the computer to their already existing spectroscopy systems, which were of considerable diversity:

> I did not find anybody who simply has an A60 and would like to have an exclusive, completely programmed computer for it and thus could rely on Varian's interfacing experts. Most people have instruments of various kinds for which they would like to use the computer from time to time . . . . A second point, which is related to the first one, is the tendency to exclusively deliver complete and fully programmed systems. I mentioned my opinion about busy-boxes several times in Palo Alto and I am more than ever convinced that this policy is wrong . . . . The customer has the right to know what he orders.[114]

Thus, Ernst sought for ways to circumvent the "busy-box" appeal of computers that had been strengthened by Varian Associates' policy. Although Varian Associates' approach ensured the performance of the computer, it would limit the market to "computer morons" only. In the late 1960s, and this was the argument of Ernst, most companies and universities interested in computer applications had considerable knowledge in-house. They had to be served with more technical information, and not just with black boxes. Also with his teaching, Ernst tried to spread the gospel of the uses of modern instrumentation. In the winter term of 1968 to 1969, he gave an introductory course in magnetic resonance, and a course in data handling in spectroscopy: "There is a lot of

---

[113] Ernst to Anderson, 25 June 1968, Ernst Papers, folder "Correspondence Wes Anderson."
[114] Ernst to Anderson, 9 July 1968, Ernst Papers, folder "Correspondence Wes Anderson."

interest at the ETH in these problems. People are just about . . . to discover the use of big and small computers."[115]

In matters FT NMR, only weeks after his arrival in Zurich, Ernst was invited by Toni Keller from Spectrospin in Fällanden to visit the company, which belonged to the Bruker group. Spectrospin was the successor of the NMR branch of Trüb-Täuber, the company that had built the NMR spectrometers of Primas's design. Keller, a self-taught electronic engineer, considered taking up FT NMR methods. The new spectrometer of Spectrospin already had a suitable design to implement FT methods quite easily, and with Lipsky of Yale University, Keller had a potential customer at hand who was able to do the computer interfacing on his own. Thus, as early as in the spring of 1968, Spectrospin prepared to launch a beta-test site for an FT NMR spectrometer, at a time when Varian Associates did not offer such a product, even though they possessed the patent rights. In a letter to Anderson, Ernst described in detail the specifications of Bruker's standard 90 MHz spectrometer, which he considered to be "certainly the best presently available instrument on the market." Thus, through Ernst, Anderson acquired important information about technical details as well as about the marketing policy of a close competitor.[116] Keller of Spectrospin naturally knew that Ernst had worked for Varian Associates. On the side of Varian Associates, the conversations of Ernst with Bruker and Spectrospin people caused gossip that he would change sides and become a consultant for Bruker.[117] Indeed, this happened, but ten years later.

The role of Varian Associates' consultants was poorly defined by the management. Wanting to change this unsatisfying situation, Ray Freeman in late 1972 proposed that Ernst should analyze specific problems, with the end result of "position papers" for the use of "Varian scientists, engineers, and product management." In addition, Freeman invited Ernst to spend a week in Palo Alto for discussions.[118] The meeting took place in September of 1973, and covered research and engineering aspects as well as the image of Varian among customers in Europe.[119] Among the results was Ernst's advice to enter the field of solid-state high resolution NMR, a topic about which he and scientists at Caltech, MIT, and Sheffield made considerable methodological progress at this time. In September 1974, Varian Associates' staff arranged a second consultant meeting in Zurich. The agenda covered the improvement of instrument perfor-

---

[115] Ernst to Anderson, 10 November 1968, Ernst Papers, folder "Correspondence Wes Anderson."

[116] Ernst to Anderson, 15 and 16 May 1968, Ernst Papers, folder "Correspondence Wes Anderson."

[117] Freeman to Ernst, 13 October 1969, Ernst to Freeman, 21 October 1969, Ernst Papers, folder "Corr. Freeman."

[118] Freeman to Ernst, 30 November 1972, Ernst Papers, box 5, folder "Varian Consult."

[119] Codrington to Ernst, 26 July 1973, Ernst to Codrington, 19 August 1973, tentative schedule of the meeting, 10 to 14 September 1973, Ernst Papers, box 5, folder "Varian."

mance, the finding of novel NMR applications, and the evaluation of the firm's competitive position. Already at this time, Ernst advised them to begin with research in the novel field of Fourier imaging, a medical application of NMR.[120]

In connection with Ernst's activities in Fourier imaging, in December of 1975 General Electric offered Ernst a grant to support this research project.[121] General Electric had a product line in X-ray tomography and ultrasonic techniques, and the management was interested in novel technologies for medical imaging. The giant U.S. firm thought of supporting Ernst's activities in studying the potential of Fourier imaging for medical purposes with $12,000 a year, for a period of up to three years. In exchange, the company expected to review article manuscripts before publication, to acquire the patent rights in case that ETH Zurich might not file its own patents, and to receive progress reports by Ernst. Both companies, Varian Associates and General Electric, foresaw a potential conflict of interests if Ernst worked for both companies at the same time. After Ernst informed General Electric that he had assigned his Fourier imaging patent to Varian Associates, General Electric withdrew its offer. For Ernst, this was a considerable loss of research money, and he was concerned about the interests of ETH Zurich. The university funded most of his research, but did not get any return from Varian Associates in the form of patent license fees. Therefore, Ernst proposed that Varian Associates should sponsor his research in form of an annual grant channeled through an account at ETH. Varian Associates indeed supplied a grant at the same level as the General Electric offer had been, and Ernst could use the money for fellowships and salaries of technical personnel. In return, Ernst reported on all of his NMR work, and he agreed to assign the patents to Varian Associates that would eventually come out of the research funded.[122]

Ernst's consulting was important for Varian Associates' research activities. During the negotiations with Ernst in 1976, Robert S. Codrington, head of NMR research in Varian's instrument division, expressed his satisfaction with the organization of the consulting:

> The main thing we are interested in is maintaining a fairly close contact with your activities since we rely on this contact to keep us abreast of developments in NMR in those special areas in which you are concentrating your efforts.[123]

---

[120] Ernst to Freeman, 18 December 1972, Ernst to Peter Llewellyn, 20 October 1973, Agenda for consultants' meeting in September 1974, 3 July 1974, Ernst Papers, box 5, folder "Varian Consult."

[121] See the notes of Ernst, "Developments in zeugmatographic imaging," 17 December 1975, and letter of John K. Wolfe of General Electric to E. Freitag of ETH Zurich, 18 December 1975, Ernst Papers, box 5, folder "General Electric."

[122] See Ernst to Codrington, 17 December 1975, 24 January, 18 March, 7 June 1976, Codrington to Ernst, 8 January, 4 February, 26 March, 6 May 1976, D. F. Minnery of Varian AG to Ernst, 11 June 1976, Ernst Papers, box 5, folder "Varian Consult."

[123] Codrington to Ernst, 26 March 1976, Ernst Papers, box 5, folder "Varian Consult."

Nevertheless, in March of 1978, Ernst terminated his consultancy with Varian Associates and accepted an offer by Bruker-Spectrospin. His reasoning was that he felt obliged to support a Swiss company because almost all of his research funding came from Swiss sources. Possibly more important was the geographical proximity of Bruker-Spectrospin scientists and engineers, and their being a smaller and more dynamic company than Varian Associates. Above all, the present and future research interests of the Swiss company and Ernst were in better agreement, although Ernst explicitly expressed his gratitude for fifteen years of collaboration with Varian Associates:

> Spectrospin-Bruker is mostly active in the field of sophisticated instrumentation both in solid state and liquid state NMR. This harmonizes better with my own interests than the general tendency of Varian towards the routine-type instruments . . . . Leaving Varian, I have at least the satisfaction that we have achieved something together in the past and I am sure that Varian has also profited from our collaboration . . . . I do not know any person or institution from which I have profited more for my scientific career.[124]

For Ernst it was not so much the financial benefit (the payment of Varian was in the order of one fifth of his ETH salary, and the royalties for patents were in the order of $100 to $200 each) that led him to consult for instrument manufacturers. Furthermore, there were some issues that spoke against the filing of patents originating in publicly funded research, and to assign them to a private enterprise. Among those issues were the interests of his co-workers, and of ETH itself. But Ernst felt obliged to contribute to the technical development of the company he was consulting for. He and his research group would benefit scientifically from the consulting, as it was the case first with Varian Associates and then with Bruker-Spectrospin:

> All of our instruments, actually from beginning on, were Bruker spectrometers. [It is important] that you are in contact, that you know the people, that you can exchange parts if something goes wrong, and that you receive newly developed instruments for testing. You keep at the technical frontier. These are advantages, and the contacts make sense of it in the way that you have a partner who is in a decisive position and responsible for the instruments. Thus you can influence the course of development of the instruments you are finally working with. This kind of interaction makes the whole so valuable.[125]

---

[124] Ernst to Codrington, 8 March 1978, Ernst Papers, box 5, folder "Varian Consult."

[125] "Alle unsere Spektrometer waren eigentlich von Anfang an Bruker Spektrometer und [es ist wichtig] daß man den Kontakt hat und dann eben die Leute kennt, Teile austauschen kann, wenn es einmal nicht funktioniert und neue Geräte erhält zum Ausprobieren, die gerade entwickelt worden sind, daß man wirklich immer an der Spitze ist. Das hat schon Vorteile und es sind eigentlich diese Kontakte, die das Ganze sinnvoll machen, daß man eben einen Partner hat in der Industrie, der an der entscheidenden Stelle sitzt und für die Geräte verantwortlich ist, daß man auch die Geräte beeinflussen kann, die man schlußendlich kriegt und diese Interaktion ist eigentlich das, was das Ganze so wertvoll macht." Ernst, interview by Reinhardt, 25 February 2002.

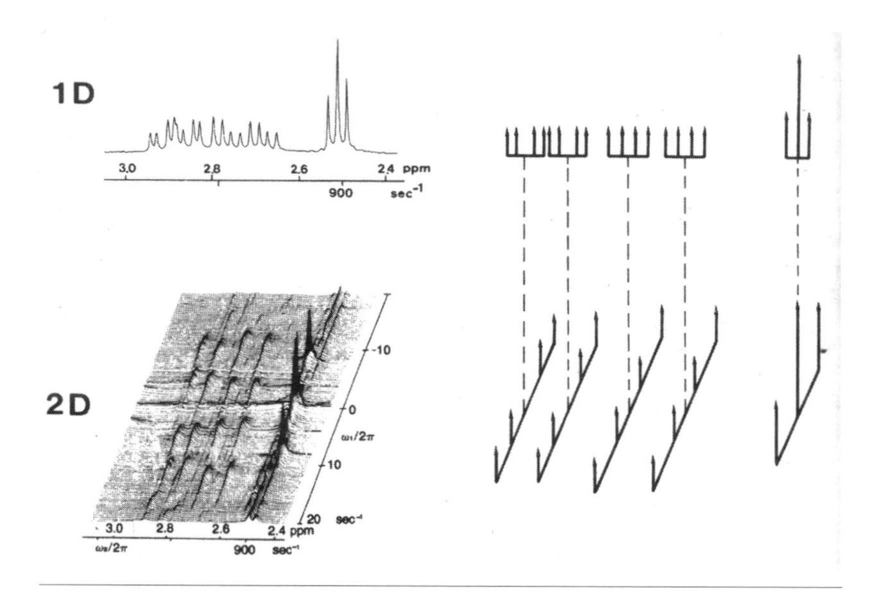

FIGURE 6.3 Principles of 2D NMR spectroscopy. Reprinted from Kurt Wüthrich, Kuniaki Nagayama, Richard R. Ernst, "Two-dimensional NMR spectroscopy," *Trends in Biochemical Sciences* 4 (1979), N178–N181, fig. 1 on N178. Copyright 1979, with permission from Elsevier.

Thus, through close interactions with instrument manufacturers, Ernst and his group could stay at the forefront of research in NMR. Building on this, in the mid-1970s, he made an innovation that substantially expanded the outreach of NMR methods, as well as changing its representational modes: NMR in two dimensions, 2D NMR.

Two Dimensions of NMR

Following up an idea of the Belgian scientist Jean Jeener, Ernst and his co-workers developed 2D NMR experimentally from 1974 onwards. 2D NMR is probably best understood with a brief description of the method directly based on Jeener's original idea. In Figure 6.3, a traditional NMR spectrum with 19 lines is shown. The appearance of such a spectrum depended on the chemical shifts and the spin-spin coupling values. Spin-spin couplings split single resonance lines whose position depended on the chemical shift further into multiplets. The interpretation of the resulting complex spectral patterns was difficult. In the 2D spectrum reproduced below, the multiplets were rotated around their central, chemical shift, frequencies. Thus, the spectrum was spread in a second dimension, and the number and type of spin-spin multiplets

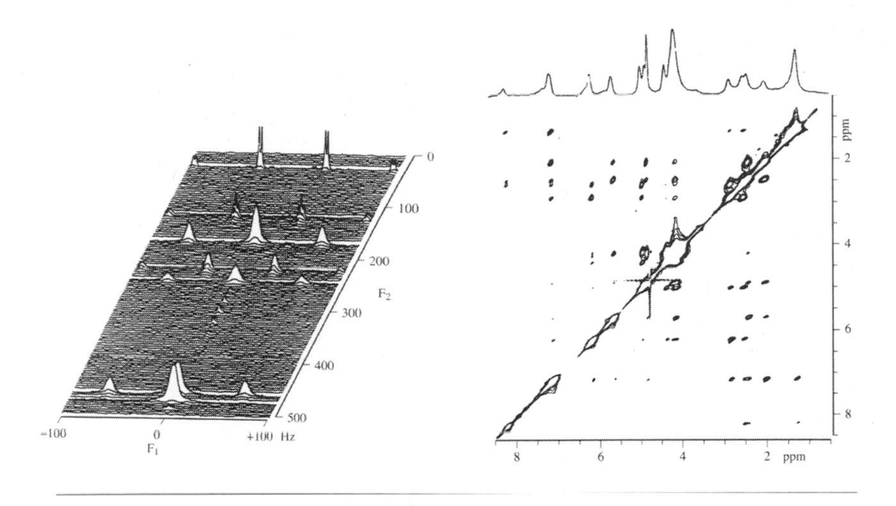

FIGURE 6.4 Stacked (left) versus contour (right) plot of 2D NMR spectra. From Edwin Becker, Cherie L. Fisk und C. L. Khetrapal, "The development of NMR," in David M. Grant, Robin K. Harris, eds., *Encyclopedia of Nuclear Magnetic Resonance*, vol. 1, *Historical Perspectives*, Chichester: John Wiley & Sons 1996, 1–158, on 94, fig. 76. 1996. Copyright, John Wiley & Sons Limited. Reproduced with permission.

was revealed upon inspection of the 2D spectrum. This was also the case if multiplets would overlap in the traditional spectrum.

The spectra in Figures 6.3 and 6.4 (left) show a stacked plot presentation. To spectroscopists, stacked plots were more familiar, but time-consuming to produce and difficult to interpret. Another possibility of displaying 2D spectra was the contour plot, encircling high-intensity peaks in a two-dimensional presentation (see Figure 6.4, right). This followed the earliest approaches of Ernst and co-workers, who used teletype notation to print the peaks of a 2D spectrum (see Figure 6.7). Although "old-time spectroscopists," according to a historiographical survey, "had difficulty visualizing the results without the familiar spectral curves," the contour plot gained acceptance.[126]

2D NMR was not simply a different graphic representation. In contrast to traditional methods, the molecular spin system was excited with two pulse sequences, called evolution and detection periods. The length of the evolution period was altered sequentially, and the information content of both periods entered the observed signals, which were Fourier transformed in two dimensions (see Figure 6.5). Through changing the pulse types and sequences, a great variety of 2D NMR experiments could be designed, measuring all kinds of

[126] Becker, Fisk, and Khetrapal, "Development of NMR," 94.

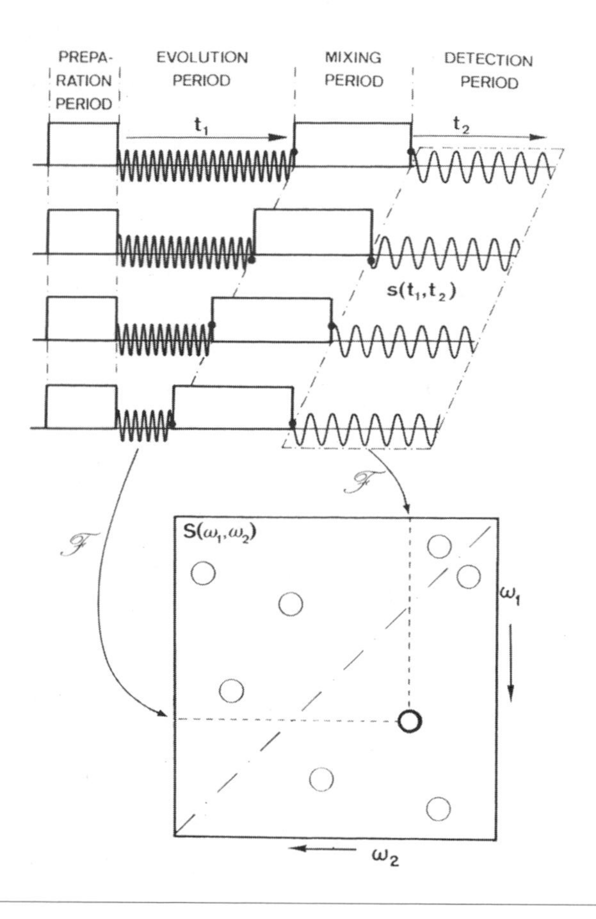

FIGURE 6.5 General 2D NMR experiment showing the preparation, mixing, and detection periods as well as the altered evolution periods. From Richard R. Ernst, "Nuclear Magnetic Resonance Fourier Transform Spectroscopy," in Bo G. Malmström, ed., *Nobel Lectures Chemistry 1991–1995*, Singapore: World Scientific 1997, 12–57, fig. 7 on 22. Copyright, the Nobel Foundation, 1991.

spectral properties. In the mid and late 1970s, Ernst and his group enjoyed the benefits of relatively little competition in 2D NMR. They transferred many of the already existing double resonance techniques of conventional NMR to 2D NMR,[127] and together with some other NMR spectroscopists developed three main 2D NMR methods up to the late 1970s: J-resolved 2D NMR (called 2D correlation spectroscopy (COSY)), nuclear Overhauser effect spectroscopy

[127] See the description in Ernst, "Success Story of Fourier Transformation in NMR," 298.

based on internuclear cross relaxation (NOESY), and 2D exchange spectroscopy, based on chemical exchange (EXSY). These methods measured different transfer processes of nuclear spins. An important aim of the 2D NMR spectroscopist was to increase the information content while decreasing the complexity of the spectra. In the late 1980s, with the advent of higher performing computer systems, even multidimensional NMR methods came into vogue.[128]

In 1976, with the first successes of two-dimensional (2D) NMR, Ernst felt that the "dark middle ages" in his career were over. At the 17th ENC in Pittsburgh, he began and concluded his lecture as follows:

> In the next 30 minutes, I would like to present to you just a few possibilities as examples of a very rich multitude of new exciting tricks that can be played on an NMR spectrometer for your own fun and perhaps also for profit . . . . I hope to have animated you to try to play this exciting game in two dimensions. The playground is almost unlimited and it is well possible that you may find many more treasures. So I would like to invite you to contribute to the liberation of the one-dimensional spectroscopist.[129]

Later, Ernst mentioned that they "felt at this moment more like graphic artists and playboys than scientists."[130] Indeed, for Ernst, art and science were inseparable entities of human culture. In both science and art, creativity and intuition were necessary ingredients:

> It is well-known that also in the sciences beauty, simplicity, and symmetry often distinguish the superior theories from the less successful ones. We must 'love' a theory before we are ready to accept it. Aesthetics clearly has great importance in the sciences.[131]

Since his Asian journey in 1968, Ernst had developed an "insatiable interest" in Tibetan art, especially thangkas, scroll paintings in the Buddhistic tradition. Thangkas depicted the large pantheon of deities in Mahayana Buddhism, characterized by a precise iconography of numbers and positions of heads, limbs, and certain attributes (see Figure 6.6). As in spectroscopy, "verbal descriptions . . . alone remain largely inadequate." Probably also appealing to Ernst was the fact that Buddhists did not assign an objective reality to their deities, but thought of them as "symbolic representations of abstract spiritual concepts

---

[128] For a non-mathematical brief introduction to 2D NMR see Freeman, *Handbook of Nuclear Magnetic Resonance*, 318–327. One of the standard textbooks is Ernst, Bodenhausen, and Wokaun, *Principles of Nuclear Magnetic Resonance*.
[129] Ernst, "ENC, Motor of Progress in NMR," 210–211.
[130] Ernst, "Success Story of Fourier Transformation in NMR," 298.
[131] Ernst, "Arts and Sciences," 901.

FIGURE 6.6 Thangka, painted between 1540–1560, showing the lineage of Jigten Wangchug (1454–1532). From Richard R. Ernst, "Arts and Sciences. A Personal Perspective of Tibetan Painting," *Chimia* 55 (2001), 900–914, fig. 4, on p. 906. Copyright, Swiss Chemical Society.

which are essential for mastering our life."[132] In some thangkas, according to Ernst, the deity was not shown but only indicated by symbols and the onlooker had to create the deity in his mind during meditation. The concept of abstraction, the formalism, and in some cases even the graphical design of Tibetan

[132] Ernst, "Arts and Sciences," 902.

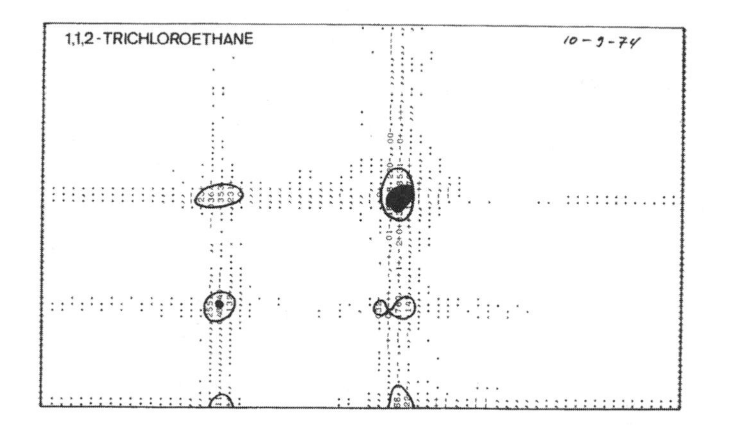

FIGURE 6.7 2D NMR spectrum of 1,1,2-trichloroethane, recorded on 10 September 1974 in Ernst's laboratory. From Ernst, "The Success Story of Fourier Transformation in NMR," in David M. Grant, Robin K. Harris, eds., *Encyclopedia of Nuclear Magnetic Resonance*, vol. 1, *Historical Perspectives*, Chichester: John Wiley & Sons 1996, on 298, fig. 9. 1996. Copyright, John Wiley & Sons Limited. Reproduced with permission.

scroll paintings obviously supported Ernst's ventures into 2D NMR. Clearly, Ernst liked the artistic appeal of 2D NMR, but did not think that interpretation of 2D NMR spectra required the feeling of an artist. Nevertheless, the NMR spectroscopist needed a good eye, and he or she had to recognize patterns, in many cases performing better than computer filtering. To find one's way through the jungle of data points and to see the regularities was part and parcel of the tacit knowledge of the successful NMR spectroscopist (see Figure 6.7). Ernst never included this ability in his teaching: his students and co-workers "just did it."[133]

Ernst had plenty of reasons to embrace 2D NMR as a novel tool. When in 1971, Jean Jeener proposed the outlines of an experiment that included the basic idea of 2D NMR, Ernst's graduate student Thomas Baumann brought back the news of this proposal from a conference in Baško Polje, Croatia. In Ernst's laboratory, Jeener's method fell on prepared ground. Ernst had a rich experience in double resonance experiments, and already in 1970 had pondered on computerized double resonance spectroscopy in two dimensions. His aim then was the resolution of over-lapping resonance lines and the complete measurement of energy-level diagrams.[134] He did not follow up this idea, aware of

[133] Ernst, interview by Reinhardt, 25 February 2002.

[134] Ernst, "Forschungsgesuch Schweizerischer Nationalfonds, Fortsetzung des Projektes Nr. 5185.2," 10 March 1970, 10.4B, Ernst Papers, box 6, folder "Nationalfonds."

the complexity of two-dimensional spectra of this type if modeled after the plots of Anderson and Freeman, known since a 1962 publication.[135] Although this idea had nothing in common with the method of Jeener, Ernst on this basis clearly saw that 2D NMR could do to a much greater extent what conventional double resonance experiments were supposed to achieve: to disentangle the complexity of NMR spectra. Thus, Ernst included the study of this method in his research program, side by side with systematic double resonance and stochastic resonance. Moreover, the high appeal of 2D NMR to him obviously was rooted in its applicability to the study of the non-linear responses of molecular systems, a goal that previously largely eluded the efforts of NMR spectroscopists.[136]

Another graduate student of Ernst, Enrico Bartholdi, calculated the possibilities of performing two-pulse experiments, keeping in mind the potentially disturbing factor of relaxation (something that Jeener had not included in his talk). This showed the feasibility of the concept. But Ernst, Bartholdi, and Bachmann did not plan experiments, because they regarded 2D NMR as Jeener's "property," and because the existing data storage facilities seemed to be insufficient.[137] On his side, Jeener had difficulties with the experimental set-up and, until much later, did not publish the concept.[138] Obviously, the 2D experiment was not a trivial one, calling for considerable experience in high resolution NMR. In 1973 and 1975, Jeener, Ernst, and Bartholdi had some exchange of information. In a letter from November 1973, Bartholdi and Ernst indicated their hopes for and troubles with the new method:

> We think that the pulse pair technique is certainly a very clean and simple method to obtain 'double resonance information.' But the information which one obtains is somewhat different . . . . We feel that the pulse pair technique is particularly suited to treat situations with close-by or even overlapping resonance lines. Here, the conventional double resonance techniques usually fail as the utilized mathematical approximations generally applied are invalid in this case . . . . The treatment of overlapping lines with the pulse pair technique obviously requires very high resolution and therefore causes an enormous amount of required experimental data. If one could apply this technique to a limited portion of the total spectrum, it would be much easier to realize it. We will continue to think about possible modifications of the general technique to reduce the experimental requirements.[139]

---

[135] Ernst, "Success Story of Fourier Transformation in NMR," 297. See Freeman and Anderson, "Use of Weak Perturbing Radio-frequency Field."

[136] Ernst, "Forschungsgesuch Schweizerischer Nationalfonds, Fortsetzung des Projektes Nr. 2.288.70," 25 March 1972, 10.4, Ernst Papers, box 6, folder "Nationalfonds."

[137] Ernst, "Success Story of Fourier Transformation in NMR," 297.

[138] See Jeener, "Reminiscences About the Early Days of 2D NMR."

[139] Enrico Bartholdi and Ernst to Jean Jeener, 17 November 1973, Ernst Papers, box 18, folder "2D Spektroskopie RIER."

Ernst's hesitation to publish on the 2D technique was overcome by two events. In April 1974, at the 15th Experimental NMR Conference, he heard a lecture by Paul Lauterbur that described the basic outline of Zeugmatography, a method of producing images of three-dimensional objects by NMR. Lauterbur used a conventional continuous wave technique. Ernst realized that a method in analogy to the 2D NMR experiment would be a much better option. This concept of Fourier imaging (later called Magnetic Resonance Imaging, MRI) became the basis of applications in medical body-visualizing techniques. Knowing of the potential importance and being aware of the originality of his method, Ernst filed a patent and assigned it to Varian Associates. In addition, Ernst planned to speak on his work on Fourier imaging at the 6th International Conference on Magnetic Resonance in Biological Systems in Kandersteg, Switzerland, in mid-September 1974. Knowing that his experimental work was still premature, he included the first 2D spectra of chemical molecules to add to the imaging part. In his opinion, two novel methods not at an advanced state were just enough material to be presented:

> In my eyes, both methods were bagatelles. One was perhaps not enough, but with two trifles you can better enter the public sphere. Perhaps, in necessity, if I would have had nothing else, I would have made a show with only one.[140]

In fact, Ernst knew that each of the two techniques had a tremendous potential. In August 1974, and with his co-workers Anil Kumar and Dieter Welti, Ernst submitted a manuscript on Fourier imaging to the *Journal of Magnetic Resonance*. At this early stage, he was able to report the first experimental results, NMR "pictures" of two glass tubes filled with water. The authors developed suitable instrumentation and showed the feasibility of the technique. Moreover, they pointed to the analogous 2D technique of Jeener. Although the latter had completely different aims and was a principally different method, it could use the same experimental set-up and the same computer programs as Fourier imaging.[141] It was a natural step to follow up 2D NMR in parallel to Fourier imaging. Still, in early 1975, Ernst was concerned about the priority rights of Jeener and asked Jeener directly if he would allow him to talk on the subject at a forthcoming Gordon conference on magnetic resonance.[142] Seemingly, Jeener had nothing against it. The first publication of Ernst that included

[140] "Ich meine, es waren eigentlich beides Bagatellen in meinen Augen und wiederum eigentlich Spielereien und eine Spielerei war vielleicht etwas wenig. Zwei Spielereien, damit kann man eher an die Öffentlichkeit treten, aber vielleicht in der Not, wenn ich gar nichts anderes gehabt hätte, hätte ich auch mit einer Spielerei eine Show aufgezogen." Ernst, interview by Reinhardt, 25 February 2002.

[141] Kumar, Welti, and Ernst, "NMR Fourier Zeugmatography," 82.

[142] Ernst to Jeener, 13 January 1975, Ernst Papers, box 18, folder "2D Spektroskopie RIER."

2D NMR spectra appeared in 1975.[143] In producing reliable and useful data, Ernst and his co-workers had to overcome severe experimental difficulties. With the availability of larger disk memory, the matrix of the spectra could be considerably improved, and filtering allowed obtaining partial spectra.

Among the earliest scientists taking up 2D NMR was Ray Freeman, who had moved from Varian to Oxford University. Together with Geoffrey Bodenhausen (a former student of Ernst, but in Zurich not involved in 2D work), Freeman developed many of the heteronuclear methods. In conversation with Ernst, Freeman revealed the picturesque appeal of the 2D technique, at this time not believing in its practical usefulness: "Oh, this would be a wonderful method to produce the title page of a book!"[144] In addition, John S. Waugh of MIT early on developed 2D methods in solid-state NMR. In the very beginning, Ernst undertook 2D NMR experiments with protons only, although it was clear that carbon-13 NMR would also benefit greatly. In a letter to Jeener, Ernst expressed his opinion that 2D carbon-13 NMR "could become one of the more interesting applications of 2D spectroscopy, at least from the user's standpoint."[145] Freeman shared this opinion, and regarded carbon-13 2D NMR "the landmark where two-dimensional high resolution NMR really got started."[146] Consequently, he underlined Ernst's article on the method in an early review.[147] Soon, 2D NMR became a competitive undertaking. This is shown for example with Ernst's method of detecting carbon resonance spectra indirectly via the proton resonances, first presented at an ENC meeting in Asilomar, California, in April of 1977. In following up Ernst's idea, Freeman and Bodenhausen detected the proton resonances with a carbon-13 NMR spectrometer (that is, vice versa to Ernst's method), and submitted an article manuscript already in July of 1977. This caused some ill-feeling on the side of Ernst, who thought that his priority was endangered.[148] Also Kuniaki Nagayama, a member of the Zurich group, in a brief biographical account on the beginning of 2D NMR of proteins remembered an intense competition

---

[143] Ernst, "Two-dimensional Spectroscopy."

[144] "Ich habe einmal mit Ray Freeman darüber gesprochen. Der hat gesagt: 'Oh, das wäre eine wunderschöne Methode, um ein Titelbild für ein Buch zu produzieren.' Das hat ihn eigentlich sofort angesprochen, das Bildhafte von der zweidimensionalen Spektroskopie; aber geglaubt an die praktische Anwendung hat er eigentlich auch nicht." Ernst, interview by Reinhardt, 25 February 2002.

[145] Ernst to Jeener, 10 November 1975, Ernst Papers, box 18, folder "2D Spektroskopie RIER."

[146] Freeman to Ernst, 3 August 1978, Ernst Papers, folder "Corr. Freeman."

[147] Müller, Kumar and Ernst, "Two-dimensional Carbon-13 NMR Spectroscopy"; and Freeman and Morris, "Two-dimensional Fourier Transformation in NMR."

[148] See Freeman to Ernst, 13 and 25 July, 4 August 1977, Ernst to Freeman, 18 July, 1 August 1977, Ernst Papers, folder "Corr. Freeman." The papers were Maudsley and Ernst, "Indirect Detection of Magnetic Resonance"; and Bodenhausen and Freeman, "Correlation of Proton and Carbon-13 NMR Spectra."

between the Oxford and Zurich groups.[149] While this issue was settled immediately between the two close colleagues, the case shows the fast exchange of ideas, and the rapidity with which experiments could be made. Indeed, actually no changes in the hardware of Fourier pulse spectrometers had to be made. The scientists only had to adapt the software that arranged the pulse sequences and the data handling.

For the practical breakthrough of 2D NMR, Ernst's cooperation with the molecular biologist Kurt Wüthrich became crucial from late 1976 on, when both sides agreed to a mutual development of the novel method for structural elucidation of biomolecules. After having taken over Ernst's former responsibility of directing the Swiss high-field NMR facility, Wüthrich was eager to keep Ernst's group involved, knowing that his group would benefit from the ideas of the physical chemists. Ernst suggested the development of 2D NMR, and Wüthrich agreed. In hindsight, Ernst was aware of the importance of finding suitable applications: "In my hands it probably would have stayed an exotic thing, but he [Wüthrich] has seen the practical application and forced it. The whole development benefited a lot from that."[150] Wüthrich had considerable experience in NMR of biomolecules through a postdoctoral stay at the biophysics department of Bell Laboratories in Murray Hill, New Jersey.[151] There, from 1967 to 1969, Wüthrich worked in the group of Robert G. Shulman, one of the pioneers of biological NMR,[152] and specialized on hemoproteins. Back in Switzerland, at ETH Zurich, and establishing his own research group, Wüthrich decided to avoid the main lines of research due to heavy competition by leading groups, mostly in the United States. Wüthrich and co-workers achieved their first success in 1974 with NMR studies of a small globular protein, basic pancreatic trypsin inhibitor.[153] Incidentally, Wüthrich reported on this work at the very same conference on magnetic resonance in biological systems in Kandersteg where Ernst announced his first 2D spectra. At this time, they did not know yet that they soon would start a collaborative project on this subject.

In the mid-1970s, NMR methods could not solve the problem molecular biologists were most interested in: the structural elucidation of large biomolecules, especially proteins. Without prior knowledge of the molecular structure, often acquired by X-ray crystallography, the interpretation of NMR data of

---

[149] Nagayama, "The First Protein Two-dimensional (2D) NMR."

[150] "In meinen Händen wäre es wahrscheinlich irgend so ein Exotikum geblieben, aber er hat dann eben die praktische Anwendung gesehen und forciert. Das hat dem Ganzen sehr viel genützt." Ernst, interview by Reinhardt, 25 February 2002.

[151] Wüthrich, "NMR Structures of Biological Macromolecules."

[152] On biological NMR see Markley and Opella, *Biological NMR Spectroscopy*, 3–25; Becker, Fisk, and Khetrapal, "Development of NMR," 80–92; and Shulman, "My Years in NMR."

[153] Wüthrich, "NMR Structures of Biological Macromolecules," 712; and Wüthrich, "Pancreatic Trypsin Inhibitor."

biomolecules was a fruitless endeavor. Wüthrich knew that he had to deal with this bottleneck, should NMR have a future in molecular biology:

> I realized that we could only attain broader acceptance by the biochemistry community if we worked with a technique capable of de novo determination of molecular structures, other spectroscopic data alone being too difficult to explain or visualize.[154]

Wüthrich decided to tackle the complex topic of the Nuclear Overhauser effect (NOE) in NMR, causing the enhancement of signal intensities. Work along these lines, still in the one-dimensional domain of NMR, enabled one to obtain proton-proton distances in proteins, a crucial information to establish three-dimensional structures. In the beginning of the joint project on 2D NMR, the new technique in Wüthrich's group was applied at a relatively low level. In Ernst's laboratory, Peter Bachmann did a lot of computer programming, and Kuniaki Nagayama worked on the experimental side in Wüthrich's group. Only in 1980, when the research fellow Anil Kumar developed the 2D version of Wüthrich's one-dimensional NOE experiments, NOESY, the whole group of Wüthrich began to apply 2D NMR in their daily work. Around the same time, Nagayama proved the practical usefulness of 2D correlation spectroscopy (COSY) in the protein field.[155] Thus, in the early 1980s, Wüthrich's group possessed the necessary methods to tackle three-dimensional protein structures in solution.[156]

The work with the Bruker 360 MHz spectrometer at ETH Hönggerberg had some important advantages for the project. The spectrometer was very stable, and Wüthrich, in a consequent and critical manner, applied the methods to "real" problems in biochemistry. In a 1998 nomination letter for the Robert A. Welch Award in chemistry, Ernst praised the performance of his colleague, especially for the practical implementation of 2D NMR: "What was originally a very cumbersome procedure became through his contributions a truly routine operation. A great number of protein structures have already been determined by using the protocol invented by Professor Kurt Wüthrich."[157] In the eyes of Ernst, his cooperation with Wüthrich was not always easy, but scientifically successful. Both partners had ill feelings about their joint work, even before Ernst received the 1991 Nobel prize in chemistry (Wüthrich received this honor in 2002):

> Very often it was a problem for me, because I had the feeling that ideas developed by us moved too fast to the Hönggerberg [Wüthrich's laboratory]. On

---

[154] Wüthrich, "NMR Structures of Biological Macromolecules," 713.

[155] Nagayama, "The First Protein Two-dimensional (2D) NMR."

[156] Wüthrich, "NMR Structures of Biological Macromolecules," 717; and Ernst, "Success Story of Fourier Transformation in NMR," 299–301.

[157] Ernst to Norman Hackerman, 23 January 1998, Ernst Papers, folder "Corr. Wüthrich."

> the other side, Kurt Wüthrich also had an ill feeling that all the beautiful spectra of biomolecules produced in his laboratory were misused by me.[158]

In 1987, Ernst and Wüthrich terminated their decade-long cooperation. The initial reason for this was probably a controversy about a planned cooperation of Ernst's laboratory with a group in Munich in the field of 3D NMR. Wüthrich aimed for an exclusive relationship with Ernst in the biological NMR field, and he feared the uncontrolled flow of information to the Munich scientists. In his eyes, this endangered the leading position of the Zurich group. Ernst, in contrast, argued for a free exchange of scientific know-how, also with regard to the careers of their younger co-workers. Though many characteristics of their cooperation, their geographical and institutional proximity, their high standing in their respective communities, and the fact that both scientists consulted for the same instrument company, Bruker (in 1978, Ernst terminated his consultancy with Varian and accepted a similar position with Bruker), spoke in favor of a continuation, the cooperation finally ended in a struggle over power and access to know-how.[159]

Nevertheless, the cooperation with Wüthrich enabled Ernst to build up the most important application field of 2D NMR, biological NMR spectroscopy. At the same time, his ongoing consultancies first for Varian Associates and then for Bruker-Spectrospin provided the necessary technical capabilities. The flow of knowledge between Ernst's laboratory and the instrument manufacturers was always in two directions. With respect to the importance of his and Wüthrich's contributions to 2D NMR for the instrument manufacturer Bruker, Ernst remarked:

> The development of two-dimensional spectroscopy by the two Zürich NMR groups has much stimulated European technology. In particular Bruker Analytische Messtechnik through Spectrospin AG depends at this moment very heavily on this innovation. This company has gained worldwide leadership through their access to the knowledge of the Wüthrich group and the undersigned. These two groups have developed the necessary computer software for performing two-dimensional experiments and the required two-dimensional Fourier transformation. In addition also software was developed for the automatic analysis of two-dimensional spectra.[160]

---

[158] "Für mich war es schon sehr oft ein Problem, daß ich das Gefühl hatte, es seien Ideen, die bei uns realisiert worden wären, dann allzu rasch auf den Hönggerberg geflossen. Auf der anderen Seite, Kurt Wüthrich hatte auch ein schlechtes Gefühl, daß all die schönen biomolekularen Spektren, die bei ihm produziert worden sind, von mir dann mißbraucht würden." Ernst, interview by Reinhardt, 25 February 2002.

[159] See the notes of Ernst, May 1987, Ernst Papers, folder "Corr. Wüthrich."

[160] Ernst, "Appreciation of the scientific contributions of Prof. Dr. K. Wüthrich," 10 May 1989, Ernst Papers, folder "Corr. Wüthrich."

Indeed, Bruker, at that time still independent from Ernst, in 1977 marched quickly into developing 2D methods for their instrument line. In contrast, Varian Associates did not show the same enthusiasm, although Ernst had made the company aware of the method's potential[161] and assigned a patent to Varian. Probably, Varian Associates' decision to concentrate on the mass market of NMR instruments negatively influenced their position with regard to 2D NMR. Moreover, the development was surprisingly fast, exemplified by a remark of Varian Associates' employee Robert Codrington on the talks on 2D NMR at the Experimental NMR Conference (ENC) in the spring of 1976:

> I enjoyed your talk at the ENC but I was a little startled at how quickly people have picked up two-dimensional spectroscopy as a new tool for NMR research studies. Although we had advance information of your work, we haven't yet been able to take advantage of it in our commercial instruments.[162]

Ernst, for his part, kept Codrington informed on the progress he made in 2D NMR during 1976, especially with regard to the improvement of resolution and data storage on disks. In addition, he offered to send Varian the computer programs his group had written.[163] 2D NMR did not require changes in the hardware of NMR spectrometers, and access to software was the only, but crucial, condition for a fast implementation of the technique. As Varian Associates in 1976 changed the operating systems of their computer lines, they wanted to wait with the introduction of 2D programs until they finished this task.[164] In July of 1977, Ernst pushed Codrington again to take up the development of 2D NMR, a more urgent task after Bruker hired Enrico Bartholdi, a former member of Ernst's research group. Bartholdi was the theoretician who first calculated the feasibility of 2D NMR. Furthermore, Bruker got into touch with Wüthrich, the latter using Ernst's programs, to find out about details of 2D programming. Although Ernst did not support these ventures, a flow of information to Bruker could not entirely be prevented. This became critical when Wüthrich considered accepting a formal cooperation with Bruker's subsidiary Spectrospin. In exchange for 2D NMR programs, Spectrospin offered to modernize Wüthrich's computer equipment. As Ernst had a close cooperation with Wüthrich, knowledge of his group would be transferred to Varian's competitor.[165] Part of Varian Associates' difficulties in taking up 2D NMR was the lack of available manpower, as in 1977 members of the NMR research group were

---

[161] Already in 1972, when his venture into 2D NMR was still in its very beginning, Ernst informed Wes Anderson about this project of his, in accordance with the terms of his consultancy with Varian. Ernst to Anderson, 15 August 1972, Ernst Papers, folder "Correspondence Wes Anderson."

[162] Codrington to Ernst, 6 May 1976, Ernst Papers, box 5, folder "Varian Consult."

[163] Ernst to Codrington, 9 September 1976, Ernst Papers, box 5, folder "Varian Consult."

[164] Codrington to Ernst, 22 September 1976, Ernst Papers, box 5, folder "Varian Consult."

[165] Ernst to Codrington, 14 and 15 July 1977, Ernst to F. Wehrli, 10 October 1977, Ernst Papers, box 5, folder "Varian Consult."

busy with problems in engineering. Ernst offered to send one of his students, Luciano Müller, as a postdoc to Palo Alto to expedite the technology transfer.[166] These events took place near the end of Ernst's consultancy with Varian Associates.

Richard R. Ernst and the Calling of Methods

Sensitivity was NMR's ruling topic of the 1960s, and Ernst made inroads into many of the possibilities that were available to increase it. Although NMR spectroscopy had made great progress in the early 1960s, Ernst was convinced that there was still room for improvement: "A major demand remains—the demand for the higher sensitivity necessary for many applications in biochemistry and in other fields."[167] In the 1970s, the breakthrough achieved by FT NMR made sensitivity a less urgent theme. Suddenly, complexity constituted the most important problem of instrument manufacturers, method makers, and their users. As a result, 2D NMR methods dominated the research program of Ernst's group from the mid-1970s on. For Ernst, though, the real benefit of 2D NMR was the "hunting for hidden treasures" in the realm of nuclear spins and their interactions. Thus, the greatest feature of 2D methods was the possibility to observe phenomena that were principally not observable with one-dimensional NMR.[168] In competition with Ray Freeman's group at Oxford University, Ernst developed many of the basic techniques, pulse schemes, and programs in 2D NMR that served this purpose. 2D NMR came to be the major "playground" for NMR spectroscopists in the 1980s. In 1991, the scientific community honored Ernst's achievements with the most prestigious award in science, the Nobel Prize.

Ernst lived the career of a method maker. He was member of a community of experts, employed by companies and universities alike. His aim was to develop methods, and although he concentrated on NMR, his contributions proved to be fruitful in many other spectroscopic techniques. Moreover, he made use of general techniques of data processing and signal theory and connected NMR in this regard to broader developments in electronics. His laboratory at ETH, the ENC meetings and his contacts to the instrument manufacturers Varian Associates and Bruker-Spectrospin provided the necessary context and support for his research projects. Through ETH, and the Swiss community of chemists, Ernst took up service work in high resolution

---

[166] Codrington to Ernst, 22 July 1977, Ernst to F. Wehrli, 10 October 1977, Ernst Papers, box 5, folder "Varian Consult."

[167] Ernst, "Sensitivity Enhancement in Magnetic Resonance. I," 1689.

[168] Ernst, "ENC, a Motor of Progress in NMR," 213.

NMR from 1969 to 1975. Together with the molecular biologist Kurt Wüthrich, he found important applications in the structural elucidation of biomolecules. Here, Ernst's NMR technology became embedded in a new application, protein science.

For the scientific community at large, NMR spectroscopists were problem solvers. In their outreach, NMR specialists could build on already established techniques, be it in NMR itself, or other spectroscopic methods. In doing so, they used their consultancies for instrument manufacturers, spreading the novel physical gadgetry among chemists, biologists, and physicians. In order for potential buyers to become convinced of new needs and new opportunities, the value of a new method had to be shown and advertised. Here, the cooperation of NMR specialists with scientists in chemistry, biology, and medicine was indispensable, as the example of Ernst's collaboration with the molecular biologist Kurt Wüthrich has shown. Wüthrich, by himself a considerable expert in NMR, commanded know-how and standing in protein chemistry. An important task of NMR scientists was the actual solving of problems in a form of service work for a wider scientific community, and Ernst took over such a function in operating the Swiss high-field NMR facility. Thus, consultancy for instrument manufacturers, collaboration with user-scientists, and service to the scientific community were the crucial features of the method maker Ernst in NMR. The central core of his research, however, remained the theoretical and experimental dealing with nuclear spins.

At the 33rd Experimental NMR Conference in 1992 in Asilomar, California, Ernst chose to present a potpourri of his talks at former meetings. This was in reflection of his receiving the Nobel Prize in Chemistry for the year 1991. The ENC guided and structured Ernst's professional life more than anything else:

> Indeed from 1964 to very recently, I was working and living truly from ENC to ENC, having only one thought in mind: to be properly prepared with a new experiment or a new revealing insight to be presented at the next conference. I think it is important in the unstructured life of a scientist to have these regular caesuras to account for one's achievements.[169]

These conferences not only provided much of the structure in an NMR spectroscopist's career; the organizing committees also largely decided about who belonged to the community and who did not. More than any other conference, the ENC was the meeting ground for the specialized NMR spectroscopists. ENC was founded as an informal meeting in 1960 by William Ritchey, the NMR spectroscopist of the research laboratories of Standard Oil of Ohio.[170] In

[169] Ernst, "ENC, a Motor of Progress in NMR," 201.
[170] Becker, Fisk, and Khetrapal, "Development of NMR," 39.

the beginning, a few dozens of users and manufacturers exchanged know-how and announced novel methods. From 1961 to 1970, the Mellon Institute in Pittsburgh was the host for the gatherings; since then the ENC took place at the West coast in Asilomar, and changing venues in the East. The topic of the ENC was simply NMR, in its instrumental, experimental, and theoretical variations. Routine applications were shunned as being of no importance:

> I mean that applications did not count at all, for example to record a nice spectrum of a new molecule and to interpret it. This was not a good performance. Called for were new methods, methods were entirely in the foreground. This suited me, naturally, and additionally strengthened my basic tendencies.[171]

ENC became the meeting ground for the NMR specialist, the same way the MELLONMR newsletter (see Chapter 4) was the communication medium of this community. While to become a recipient of the newsletter one simply had to contribute to its contents, speakers at ENC meetings were selected by the organizing committee. Ernst entered the circle through Varian Associates, whose scientists from the beginning on played important roles in the ENC administration. This procedure guaranteed a high homogeneity of the group, as contributors were selected mostly because they were known by the core group:

> The NMR community was a relatively closed association, and was actually not accepted by other spectroscopies and by analytical chemistry. One felt to be an outsider in the scientific community, and consequently this led to unification. All the lame people, looked at with contempt by the others, have built an association. Naturally, these people were not lame, to the contrary . . . . I have the feeling . . . that especially in the beginning the most creative people went into NMR, and contributed crucially to it. These were intellectuals, and the level was actually astonishingly high . . . . This club organized these conferences, and existed through these conferences. One lived, so to speak, from conference to conference, actually worked in order to be able to present something at an ENC meeting . . . . I do not know how many people shared this feeling; I think quite a lot. At Varian, and later here in Zurich, we worked very much under this influence.[172]

---

[171] "Ich meine, Anwendungen zählten überhaupt nicht—also irgendwie ein neues Molekül zu spektroskopieren, ein schönes Spektrum, und das zu interpretieren, das war keine Leistung, sondern Leistungen waren neue Methoden. Es stand ganz das Methodische im Vordergrund. Das lag mir natürlich, das hat mich zusätzlich bestärkt in meinen Grundtendenzen." Ernst, interview by Reinhardt, 25 February 2002.

[172] "Die Kernresonanz-Community war ein relativ geschlossener Verein, auch eigentlich von anderen Spektroskopien oder von der Analytischen Chemie nicht akzeptiert. Man empfand sich irgendwie als Außenseiter in der ganzen Wissenschaftsgemeinde und hat sich entsprechend zusammengeschlossen. Es waren alle die Lahmen, die von den übrigen verachtet worden sind, die dann

The exclusion that Ernst experienced with respect to the other spectroscopic methods was a result, in his opinion, of the relatively high complexity of NMR. One needed to know quantum mechanics, electronics, and had to learn the terminology of NMR. In contrast, in optical spectroscopy, the relevant aspects of the molecules under scrutiny, such as rotation and vibration, were the basis of the effects. While this could be grasped relatively easily, the basic theory of NMR that rested in nuclear spin remained quite unfamiliar to the average chemist. For most analytical chemists of the 1950s and 1960s, NMR stayed an esoteric technique undertaken mostly by physicists. Beginning in the 1960s, this picture changed. The instruments became easier to handle, and NMR showed its greatest strength: its versatility. Most other physical methods, even mass spectrometry and X-ray crystallography, were more limited in the scope of potential applications. While the instrument manufacturers pushed the instrumental development of NMR to higher magnetic fields and with it to better sensitivity and resolution, novel methods originated with scientists at universities. Inside the small, but rapidly growing community of NMR spectroscopists, the exchange of information, know-how, and personnel became a crucial feature for the success of the "club." For Ernst's group, next to the cooperation with Kurt Wüthrich, the contacts to Ray Freeman at Oxford, Alexander Pines at Caltech, Robert Griffin at MIT, and Horst Kessler in Frankfort and Munich, were the most important ones.[173] This showed the breadth of Ernst's research interests. The molecular biologist Wüthrich was joined by the full-blooded NMR specialist Freeman, the experts on solid-state NMR Pines and Griffin, and the organic chemist Kessler. One might expect that as members of a relatively closed and specialized community, the NMR experts would have had difficulties connecting to a broad range of scientific disciplines. The contrary proved to be the case. In Ernst's view, the traditional disciplines of medicine, physics, chemistry, and biology partially overlapped and were united by NMR:

> I benefited from my acceptance in medicine, in physics, in chemistry anyway, and in biology. You can present your talks everywhere, and they find reso-

einen Verein gebildet haben. Natürlich waren das gar keine Lahmen, im Gegenteil . . . . Ich habe das Gefühl gehabt . . . die kreativsten Leute sind eigentlich in die Kernresonanz gegangen und haben dort ganz Wesentliches beigetragen und das sind Intellektuelle. Das Niveau war eigentlich erstaunlich hoch . . . . Dieser Club hatte eben diese Konferenzen organisiert und von diesen Konferenzen auch gelebt. Man hat sozusagen von Konferenz zu Konferenz gelebt, hat eigentlich auf diese Konferenzen hin gearbeitet, daß man wieder etwas einer ENC vorzutragen hatte . . . . Ich weiß nicht, wie viele Leute sonst auch dieses Gefühl hatten; ich glaube schon eine erhebliche Anzahl, aber wir bei Varian damals und später hier in Zürich, wir haben sehr stark unter diesem Einfluß gearbeitet." Ernst, interview by Reinhardt, 25 February 2002.

[173] Ernst, interview by Reinhardt, 25 February 2002.

> nance. You do not have to modify them strongly. Slight changes in the wording and motivation suffice, and you can enter these circles.[174]

Organic chemists would literally close their laboratories if their NMR spectrometers would not function. Molecular biologists relied on NMR as much as they did formerly on X-ray crystallography. And Magnetic Resonance Imaging became an expensive, but standard, technology in medical diagnosis. More than other spectroscopic techniques, NMR proved to be extraordinarily versatile, especially in chemistry. Although complex in its own right, its application was made to be relatively straightforward. Once the technology had gained a foothold in the chemist's laboratory, it displaced to a certain extent other technologies.

With the impact of NMR, questions of a novel kind appeared on the chemists' agenda: Dynamic molecular structures; the analysis of minuscule amounts of samples; and the study of short-lived reaction intermediates are some examples. The states of molecules became new foci of chemical research, and NMR showed a characteristic feature that distinguished it from the older, structural approach in chemistry:

> You can draw a structure on a piece of paper, make a photography of it and use it as a proof. A dynamic process of a molecule is a diffuse thing, complicated to portray, and perhaps less convincing. A structure is more like an oil painting, in contrast to a mobile that moves and is hard to describe.[175]

Ernst's story sheds light on the establishment of a community of NMR spectroscopists. He worked in the context of specialists who designed methods for the manipulation of spin systems, the conceptual basis of NMR, and emphasized the "artificial introduction" of such effects. Certainly, all NMR phenomena were caused by and could be measured only with complex instrumentation. In his wording, Ernst brought the manipulative aspects of modern NMR techniques to the fore. In his work, he connected the theoretical part to the experimental and instrumental sides of NMR, and in addition took care of the needs of users. Ernst, and his colleagues, made technoscience and brought it to other scientists in the form of methods.

---

[174] "Von dem habe ich eigentlich gelebt, daß man damit eben in der Medizin akzeptiert ist und in der Physik ist man akzeptiert, in der Chemie sowieso und in der Biologie, kann überall Vorträge geben und die finden Anklang, muß die Vorträge gar nicht allzu stark modifizieren, muß nur die Wortwahl etwas ändern, die Motivierung ein bißchen ändern, dann kann man in diese Kreise eben eindringen." Ernst, interview by Reinhardt, 25 February 2002.

[175] "Eine Struktur kann man auf ein Blatt Papier zeichnen, man kann ein Foto davon machen und kann das als Beweismaterial benützen. Ein Prozeß, eine Dynamik von einem Molekül, das ist das Diffuse und das ist schwieriger darzustellen und wirkt vielleicht weniger überzeugend. Eine Struktur ist mehr Kunstwerk, ein Ölgemälde, im Gegensatz zu einem Mobile, das sich bewegt und das diffus ist, das man nicht fixieren kann [und das] schwierig zu beschreiben ist." Ernst, interview by Reinhardt, 25 February 2002.

Chapter Seven

The Spectrum of Methods

> Quantum mechanics cannot give a conceptually consistent *and* chemically relevant description of molecular structure. But molecular structure is too important a concept to be thrown overboard only because it is foreign to the edifice of ideas belonging to some fundamental theory.[1]

Theoretical chemist Hans Primas opined that chemistry could not be construed to be a branch of spectroscopy.[2] He was convinced that chemistry could not be reduced to physics, and that the chemical theory of molecular structure had a purpose in its own right. Indeed, with the introduction of physical instrumentation in chemistry, chemistry was not reduced to physics, but adapted the physical methods to its own paradigms. This book analyzes some of the transitions that enabled spectroscopic evidence to acquire great importance in the experimental practice of chemistry. Physical instruments empowered chemists to observe, measure, and construct novel phenomena at an unprecedented pace. Chemical physics, analytical chemistry, and, especially, physical organic chemistry served as junctures between physical theory, instrumental techniques, and chemical methods. Chemists adapted physical instrumentation by building up theories and rules that related physical data to chemical structures. They created the chemistry of the instruments, as lead users paving the ways for the diffusion of instruments in the chemical community, as method makers engaged in building technoscientific specialties, and as communities of experts in their own right. At the organizational level, chemistry changed from a labor-intensive scientific discipline to an undertaking that, equipped with high-technology instrumentation, was one of the more expensive scientific enterprises. This process gained momentum at such a rate that even by 1962, the chairman of the National Science Foundation chemistry advisory panel called the instru-

---

[1] Primas, *Chemistry, Quantum Mechanics and Reductionism*, 345. Emphasis in the original.

[2] Primas, "Foundations of Theoretical Chemistry," 105. See Nye, *From Chemical Philosophy to Theoretical Chemistry*, 281.

ments "pieces of equipment without which no department could pretend to be supporting modern chemical research."[3] The needs and opportunities of chemistry rapidly changed through the incorporation of physical instrumentation.

How did physical phenomena became chemical instruments?[4] In the process, entities that had been studied in their own right in physics switched into the means for achieving an end in chemistry. During the twentieth century, the instrumentalization of physics for the needs of chemistry concerned a multitude of techniques. Arguably, it reached its peak in the three decades after World War II, when scientific and industrial demand, availability of funds, and progress in electronics greatly accelerated the change-over and broadened its scope. High technology constituted the core of the material side of the "instrumental revolution" in chemistry. Soon, the instrument industry came to be regarded both as a measure of technological achievement and of scientific ranking of a nation. The availability of high-performance instrumentation was indeed a necessary, though not sufficient, condition for the actual use in the chemical and biomedical sciences. In order to secure usability and credibility in communities whose members were at first indifferent, and sometimes even hostile, to the inroads of unfamiliar research methods, the instrumentation had to be adapted not only to the needs, but also to the concepts and research styles, of chemistry.

From the outset, we assumed that chemists did not remain passive users in the physicalization and technicization of their laboratories. But what does passivity mean in this respect? What did the majority of chemists do, when facing the inroads of a large array of novel methods? Among the possible activities were, first of all, defense and rebuttal. Indeed, this did happen, though not very often; one example was the initial rejection of organic mass spectrometry by the editors of the *Journal of the American Chemical Society*.[5] Second, there was neglect, but this soon became a minority position. Third, careful weighing of opportunities, cost, and needs, and that was the most widely seen activity of chemists. If the result was positive, they included the novel methods in their interpretation schemes. They did so in the form of standardized instrumentation and methods, available either in isolated instrument centers or by black-boxed apparatus brought into the laboratory. This was the contribution of the majority of chemists. But in order to respond to an influx of instrumentation, this influx had to be stimulated. Most chemists, for their part, did not have the knowledge to judge the potential of, e.g., infrared spectroscopy in the 1940s, NMR in the 1950s, and mass spectrometry in the 1960s for answering their own questions. For that, expert knowledge, and the time to pursue and expand

---

[3] William von Eggers Doering to Geoffrey Keller, 1 August 1962, NARA, RG 307, office of the director, general records, 1949–63, 1957–59, box 76, folder "M, P, & E chemistry program."

[4] See Rabkin, "Uses and Images of Instruments in Chemistry," 31.

[5] See Chapter 3.

it, was necessary. This expert knowledge, accumulated in disciplines and fields such as chemical physics, physical organic chemistry, and the chemical industry, played the crucial role, and its impact on general chemistry was assisted by those user-communities that generated the most urgent demand, e.g., analytical and natural product chemistry.

In the mid-twentieth century, both chemical theory and research mode changed. Chemistry's characteristic feature before the advent of quantum chemistry in the 1930s consisted in the fact that its empirical material was underexplained. Its enormous amount of data was ordered and classified, but not reduced to fundamental theory. Quantum chemistry considerably eased this problem, but did not make it disappear. The strategy chosen by the period's most influential theoretical chemist, Linus Pauling, consisted in a plea for working side by side. The aim was to explain as much as possible with the new theory, but not to the detriment of chemistry's empirical knowledge base. Though Pauling put the discussion of the properties, structures and reactions of substances in the framework of electrons and calculations, he allocated "descriptive chemistry" as much space as "theoretical chemistry" in his influential 1947 textbook *General Chemistry*. Thus, chemists of his generation approached quantum chemical and physical concepts via their immediate use in the memorizing and the conceptualization of chemical facts. As a consequence, chemists were willing to accept a similar challenge if novel methods for the acquisition of facts required understanding of concepts that originated in physics. The heuristic function of theory joined forces with its explanatory success. E. Bright Wilson, chemical physicist and Pauling's co-author of a textbook on quantum chemistry, once commented that the introduction of NMR methods into chemistry had contributed more to the acceptance of quantum chemistry than the teaching of quantum chemistry itself. When Herbert S. Gutowsky complimented him on his educational achievements, Wilson replied, laughing:

> 'You have done more to teach quantum mechanics to chemistry students than anybody.' I said: 'I have?' He said: 'Yes. Organic chemists have had to learn second order quantum mechanics to solve the AB [NMR] spectrum. That taught more chemists more quantum mechanics than anyone realized.'[6]

Thus, the influx of quantum theory and physical methods into chemistry was a dialectical process, decisively influenced by the chemists' pragmatism. Chemists wanted to know about the benefits of methods for their own research. This pragmatism created a void for a new group of scientists, connecting physics, high technology, and chemistry through the making of methods. Seen in simplified perspective, such method makers came from two

---

[6] Herbert S. Gutowsky, interview by Carsten Reinhardt, 1 and 2 December 1998.

different spheres. On the one hand, they represented a community of chemists close to physics and technology, supplying general chemists with instruments and methods. On the other hand, they were members of the adopting sub-disciplines, adapting the techniques to their interests. Soon, the momentum of method-directed research let these communities merge in new hybrid fields that connected users and suppliers of instrumentation. Each time that an apparatus passed from physics to chemistry, it had to be—metaphorically—opened, partially emptied, and filled again with chemical concepts. Methods had to be developed, and proofs had to be presented that the results were meaningful, accurate, and reliably to obtain. Increasingly, the opening and re-closing of instrumentation, encompassing the development of useful methods, became a techno-scientific field in its own right. Competence in methods was the word of the day, and carried forward by a new type of scientist. Neither just experimentalist nor theoretician, an expert in methods worked across many fields, ranging from physics to chemistry and the biomedical sciences.

Chemical physicists asked if the phenomena they found and followed up with the new instrumentation represented not only physical entities but could also be attached to chemistry. Do the effects have an explanation in accepted physical theory? How can we manipulate them? Do they relate to chemical concepts? Can we use them to measure data of interest to chemists? In this book, Gutowsky and Richard R. Ernst are shown to have pursued questions of this kind in the field of NMR. The analytical chemist Fred W. McLafferty, and the organic chemists Klaus Biemann and Carl Djerassi, went one step further in making mass spectrometry an original chemical method. Through close analogy with reactions in the test tube, they argued, mass spectrometry functioned by the micro-degradation of molecules. In doing so, they strongly affirmed the simple response that if mass spectrometry was organic chemistry then it was a technique of value to the organic chemist. They proceeded even further, in using mass spectrometry to expand the outreach of organic chemistry. Next to structural elucidation of complex natural products, they contributed to the installment of data systems designed to decrease the need for time-consuming interpretation. The "computerization of the spectra puzzle" even dared, though unsuccessfully, to replace the chemist by methods based on artificial intelligence. In the end, the human expert prevailed. Once the chemistry of the instrument had been achieved, one of the most urgent questions with respect to modern instrumentation was how it could be used. This question, posed by many organic chemists, decided on the career path and research style of physical organic chemist and NMR specialist John D. Roberts. Always at the cutting edge in the design of methods for use in the communities of chemists, he had to stay in front of his fellow chemists, that is, his clients. In order to achieve this, he continuously developed new methods, building on the latest instruments. These one-of-a-kind instruments served as path-breakers and show-pieces of NMR: they were embodiments of novel methods, and they served as models for subsequent, routine-use instruments built in larger numbers. The development of methods dictated the research pathways of the six chemists described in this book, and possibly of many others. Methods connected NMR and mass spectrometry to various fields in chemistry and the biomedical sciences.

The hierarchy of methods ranges from laboratory practice to scientific paradigms and knowledge domains. Examples of the first kind include measurement procedures such as NOESY or other protocols (see Chapter 6), and element mapping (see Chapter 5). Both 2D NMR and high resolution mass spectrometry served as frameworks for further research, and as such as paradigm. According to this view, NMR and mass spectrometry would count as scientific domains. In his book on the visual culture of spectroscopy, historian of science Klaus Hentschel splits spectroscopy into several sub-units, or domains, defined mostly by certain preferences for visualization and interpretation, and directed to specific application fields.[7] In NMR and mass spectrometry, visualization played a certain role, though not as dominating as in Hentschel's account of spectroscopy in the last third of the nineteenth and the first third of the twentieth centuries. We have seen that Biemann's element mapping demanded a different visual feeling than the interpretation of a traditional mass spectrum. The same applies to the contour plots in 2D NMR. We can thus speak of the visual cultures of NMR, and mass spectrometry, respectively. Also historian Jody Roberts, emphasizing the fluid character of the boundaries of NMR, refers to NMR as a domain of knowledge, constituting a "dynamic network of individual, group, and material agents."[8] For Roberts, the NMR spectrometer provided access to a knowledge domain, its boundaries with other analytical techniques being defined by its scientific content. Basically, in Roberts's view, a domain is an actor-network, and the instrument the embodiment of the domain.[9] Certainly, one can refer to NMR and mass spectrometry as domains, both in the sense of actor-network theory and of a visual culture. But there is no need to create a second denominator if the notion accepted and used in the scientific community itself, whether method or technique, is sufficient. The term method directly refers to the meaning that is of highest interest here, its use as a pathway in scientific research. Thus it seems better to refer to NMR and mass spectrometry as highly elaborated and standardized methods. Each of those techniques encompassed much more than a

[7] Hentschel encircles his understanding of "spectro-scopic domains" on a "sphere of activity," and a "department of knowledge." In his view, "each of these domains is characterized by specific visual practices, preferences for certain recording and printing techniques, pattern recognition skills, manipulations of specialized scientific instruments, and other tacit knowledge, none easily transferable from one domain to the next." Hentschel, *Mapping the Spectrum*, 435–436.

[8] Roberts, "Instruments and Domains of Knowledge," 19.

[9] Roberts, "Instruments and Domains of Knowledge," 21, 60–65.

single spectroscopic domain, because they referred to many application fields, and entailed many representation styles. In addition, neither spectroscopy, nor its sub-units such as NMR, mass spectrometry, infrared spectroscopy and the like, were scientific disciplines. Neither NMR nor mass spectrometry was a fully developed subject in university programs. But there was a core of common knowledge, further sub-divided into more fundamental (quantum physics), technical (electronics), and applied (chemistry, biomedicine, etc.) parts. There was certainly agreement on what counted as an NMR experiment, and agreement on how to interpret a spectrum, depending on the question in hand. Moreover, there were textbooks, societies, and prizes. All this contributed to the formation of a scientific community, centered around a method.

Philosopher of science Henk Zandvoort refers to NMR as a scientific research program. He explains its development with intrinsic and extrinsic success, intrinsic being the theory and conceptualization of NMR itself, while research problems of other scientific fields as well as social needs fall under extrinsic influence. Building on Imre Lakatos's philosophy of scientific research programs, and on the finalization concept of the Starnberg school, Zandvoort describes NMR as a cooperatively developing research program, in close contact with other fields, e.g., chemistry, biochemistry and molecular biology. The latter count as "guide programs" for the "supply program" of NMR. In these cases, NMR supplied experimental devices to tackle problems and research questions in the respective guide programs. As further examples of supply programs, Zandvoort mentions X-ray crystallography and photoelectron spectroscopy, being of an experimental type, and quantum chemistry and molecular dynamics, as more theoretically oriented.[10] Zandvoort has not offered a mechanism for the interaction of guide and supply programs. How does chemistry, for example, guide NMR? Do we see instances where the supply program suddenly directs research in the guide program, thus does NMR guide chemistry? And what kinds of development trigger advances and applications?

In a similar vein, and in the context of 1960s science policy, physicist Alvin M. Weinberg raised up the issue of external criteria. He opined that scientific fields should be evaluated by their benefit to other disciplines, technologies, and society, and made a plea for a new ethical principle in science. Internal issues were relevant only to judge the exploitability and available competence:

> Not only must science seek truth, it must seek *relatedness*. The value of chemistry as a field can hardly be decided by the chemists alone; they must ask the biologists who need the results of structural chemistry to elucidate the genetic mechanisms of the cell; or they must ask the physicist who cannot probe

[10] Zandvoort, *Models of Scientific Development*, 231–241.

> nuclear magnetic resonances unless they understand how the chemical environment affects the details of their NMR signals; or they must ask the reactor technologist who needs the chemistry of protactinium to design a continuous purification system in a thorium breeder. And so it is with the rest of science. The scientific merit of a field must be judged in large part by the contribution it makes, by the illumination it affords, and by the cohesion it produces in the neighboring fields.[11]

Weinberg's aim was to regulate and guide science, and for that he needed criteria. But it does not make sense to compare the contributions of separate blocks to each other, at least if one does not have a fixed point. For Weinberg, seemingly, genetics, NMR, and nuclear power possessed merits in their own right.

I argue that the research programs of NMR, mass spectrometry, and probably other spectroscopies, were not homogeneous blocks, neither cognitively nor socially. NMR for example was influenced by many scientific fields, both in synchronic and diachronic perspective. The research projects of scientists such as Roberts, Ernst, Biemann and McLafferty all showed a breadth that ranged from physical and organic chemistry to biomedical applications. Socially, they were approached not only by academic colleagues in need of novel experimental techniques, and funding agencies responding to this need, but also by the chemical industry. Because of their diversity and the many resources they could tap, NMR and mass spectrometry showed a relatively great measure of autonomy, as well as a self-generated thrust. How did this interaction work? The complexity of NMR, mass spectrometry and other techniques called for specialization; while at the same time, scientific norms asked the chemist to be able to understand the techniques in question. Thus, a network of specialists close to the discipline in need of support emerged: close enough to understand the chemistry, but specialized enough to allow the development of sophisticated methods. The work of those experts decided on the type of technique to be used, and its scale and scope. With experts' work I do not just mean research, or instrument development, but also the advertising and marketing parts of the scientific endeavor. We have seen that experts originated mostly in intermediate fields between the classical disciplines, Gutowsky and Ernst coming from a more physical side than Roberts, McLafferty, Biemann, and Djerassi. All of them had, in varying degrees, understanding of the opportunities and principles of both chemistry and spectroscopy. The flexibility of the experimental technique decided on the choice of and variation in applications, ranging from small molecules in physical organic chemistry to large proteins in molecular biology. With the exception of Djerassi, all scientists presented in this book relied on and remained with the same method dur-

[11] Weinberg, *Reflections on Big Science*, 116. See Zandvoort, *Models of Scientific Development*, 251.

ing the largest parts of their careers, showing versatility and breadth in fields of application.

Thus, can we, in the case of NMR and mass spectrometry from the 1950s to the 1970s, really distinguish between intrinsic and extrinsic success? I tend to say no. The work of the experts, by creation of methods, secured the blending, or partial merger, of hitherto divided fields. It was their task to make the boundaries invisible, to add to a new chemistry and a new technique at the same time. NMR and mass spectrometry became a part of chemistry, though chemistry changed a lot in this process. The differentiation of intrinsic and extrinsic success does not hold because the goal of the "supply programs" was to act from the inside of the "guide programs." Methods development was a core task of science, and no scientific discipline worthy of its name could—in the long run—afford to rely to a large extent on outside developments.

The focus on methods seems to be trivial. Of course, every scientist thought about methods in his work, and most, if not all, created new methods in the process of research. This is not new in the history of science, nor is it a surprising finding that scientists cooperated with instrument makers in this endeavor. What I claim to be still relatively unknown is the activity of scientists in the making of methods for their constituency. To put it in different words: The argument that methods stand at the beginning of an experiment, and continue to be important during its course is a commonplace. But that methods constitute the final outcome of experiment seems to me not just a trifle. Methods decide about what is worth pursuing in scientific research, and the ways of how to reach it. We can approach methods as being standardized experiments, technical objects in the terms of historian Hans-Jörg Rheinberger.[12] Standards constitute an agreement of a techno-scientific community, and they decide on the state of the art in scientific research. The creation and diffusion of methods thus entails a considerable share of power in a discipline. As a consequence, the making of methods leads to scientific reputation and becomes an occupation in its own right. The importance that I assign to methods should not lead to an argument of technological determinism. Though science heavily depends on technical objects, such as tools, instruments, and infrastructure, the necessity to adapt technical instruments to diverse application fields defers a one-sided dependence on technical, or, in the case of physical methods in chemistry, also on a physical foundation. Moreover, as Rheinberger has shown, both research objects (in his terminology, epistemic things) and experimental conditions, including methods (technical objects) are inseparable entities. Only their interaction, and interconversion, leads to unexpected findings and to new scientific knowledge. However, methods are both objects of research and research means. In my view, methods connect the technical (certain, fixed) with the scientific

---

[12] Rheinberger, *Toward a History of Epistemic Things*, 24–37.

(uncertain, fluid) parts of scientific work. The scientists presented in the case studies covered by this book consciously and energetically worked on projects aimed at transforming their research objects into methods for other scientists. They only achieved this with research in the same areas as their prospective followers, and through the design and development of instrumental techniques. Thus, the result of the method makers' work was both science and technology.

In the introduction, we reviewed different dimensions of scientific methods, and the work leading to them. This historiographic taxonomy is not thought to be complete, and serves more as an ordering factor than as an outline for further research.[13] The first dimension, laboratory work, is most closely connected to studies in scientific practice. The second, adaptation to chemical concepts, brings forward the importance of hybrid fields and intermediate concepts. Standardization and teaching, the third dimension, embraces the public activities of method makers, as well as communitary aspects. Fourth, the university-industry nexus, arguably, constitutes a *sine qua non* for a method-oriented chemist in the second half of the twentieth century. Finally, the social organization in centers and facilities, as well as departmental laboratories, is the result of a peculiar political economy of chemical research during this period.

Out of the Laboratory

Physical instrumentation had its first impact on chemical research in chemical physics and its experimental culture, molecular spectroscopy. Following World War II, chemical physicists developed research instrumentation based on novel spectroscopic techniques in the radiofrequency and microwave regions of the spectrum, among them NMR. NMR became a successful and acknowledged technique for tackling questions of chemical structure and molecular dynamics. The different techniques of molecular spectroscopy closely interacted with quantum chemical theories. In the laboratory world of molecular spectroscopy, phenomena were accessible only with special kinds of instrumentation. Theoretical rules, molecular models, material means, and objects of inquiry merged to afford a specific experimental style. Its major features were the rooting in quantum mechanical concepts, the control over design, construction, and use of instruments, and the application of methods to chemical research problems. Because of theoretical limitations in explaining and predicting chemical phenomena, an empirical approach prospered. This attitude was assisted by a pragmatic outlook, in which physical phenomena were defined by the operations

[13] For a more encompassing taxonomy of instrument and experiment see Ian Hacking, "Self-vindication of the Laboratory Sciences"; and Hentschel, "Historiographische Anmerkungen." See Laszlo, "Tools, Instruments and Concepts," for a list of historiographic questions.

needed to produce, observe, and measure them. In this context, and at Harvard University, the physical chemist Herbert S. Gutowsky first used NMR to chemical ends.

When Gutowsky arrived at the University of Illinois in Urbana-Champaign in September 1948, his aim was to emulate the laboratory he had left at Harvard. Piece by piece, Gutowksy assembled the instrument he had been familiar with, and he aimed to continue his research along the accustomed lines. What he did not (and could not) take into account were unexpected difficulties in constructing the apparatus, and the equally unexpected discoveries that he, and others, were about to make. Together with his change of institutional environment, this laid the foundations from which Gutowsky, and a few other spectrometrists, transformed NMR from an esoteric sub-specialty of molecular spectroscopy into a major research method of organic chemists. Before NMR data left the laboratory, control of the method should have been achieved. For that, Gutowsky applied the trusted arsenal of molecular spectroscopists: calibration, convergence, calculation, experimental characterization, and connection to chemical concepts.[14] Control measurements with a known sample, and the correlation of a new instrument's data to values obtained by older, similar ones, enabled the calibration of an instrument. Gutowsky's early struggle to achieve a reliable calibration points to the pitfalls of this type of experimental work, even during a time when standardized equipment was available. In his quest to learn how to handle his complex instrumentation, Gutowsky compared his results with other scientists' data, thus establishing convergence with outside work. Convergence and calibration largely remained inside the boundaries of the community of molecular spectroscopists. In contrast, the reduction of chemical effects to fundamental physical theories—calculation—involved cooperation between experimentalists and theoreticians. Gutowsky's interaction with the Harvard physicist Norman Ramsey in tackling the experimental confirmation of the latter's theory of the chemical shift shows the difficulties of such an approach. In the early 1950s, NMR instrumentation reached such a stage of precision that small shifts caused by the electronic structure of the molecules became a matter of concern to the physicists involved. These shifts were disturbing effects in the physicists' quest for precision measurements of nuclear moments. With "mild dismay" on the physicists' part, they were called chemical shifts, because of their dependence on the chemical nature of the sample. Although Ramsey achieved a satisfying explanation of the effect in qualitative terms, he was not able to account for a quantitative treatment. He was not even searching for an experimental confirmation of his theory, about which he seemed to be sure; instead, he asked only for experiments that allowed the correlation of his theoretical result with as many experimental

[14] For calibration, convergence, and calculation see Rasmussen, *Picture Control,* 12–14.

data as possible. With that as his major goal, the validation of the precision of ongoing measurements in nuclear physics could be achieved. Thus, for the theoretical physicist, the NMR experiment was a routine process. In general, attempts to base the chemical shift on sound quantum mechanical calculations failed. Though the underlying theory was deemed sufficient in qualitative terms, the empirically found chemical shift values could not be rigorously compared to a theoretical standard. The breakthrough of the chemical shift came with its correlation to chemical bond theories, and with its applications in structural elucidation of molecules. This happened because Gutowsky and his colleagues connected the regularities that had been found in NMR with established theoretical entities and experimental methods of chemistry. The practical use of NMR in chemistry, and not its theoretical foundation in physics, decided on the success of the technique. In the case of spin-spin coupling, the second "chemical" effect in NMR discovered around 1950, a similar story can be told. Here, a careful experimental characterization of the new phenomenon took place in Gutowsky's group. Its confirmation involved four research teams, working with different techniques at Stanford and Urbana. This discovery was a slow encircling of a natural-technical phenomenon. Again it was Ramsey who proposed a qualitative theory. Again, he left it at that. The background for the "theories of measurement"[15] based on the chemical shift and spin-spin coupling remained largely empirical. Experimental chemists and theoretical physicists had different interests in the follow-up experiments in NMR. Soon Gutowsky, by correlating chemical shift data to chemical concepts and parameters, went his own, independent way. This strategy set NMR on the track that made it a chemical method.

Correlation to chemical concepts had its inner-laboratory counterpart in relating NMR to chemical practice. In the case of Gutowsky's research, the process of bringing an apparatus under control was a matter of group-internal teamwork, combining chemical, physical, and electronic expertise. Thus, knowledge of reaction pathways and impurities decided on the interpretation of NMR experiments as did physical and electronic knowledge. This interpretation technique kept being used by chemists working in NMR in the years to come. It had its pitfalls, though. In the process leading to the discovery of spin-spin coupling, the knowledge of the chemical synthesis of the sample compound first pointed to an impurity. Thus, the effect seemed to be an artifact. Even after the group could exclude this interpretation by different synthetic routes, they did not pursue the phenomenon further. For a while, it simply was an anomaly. Only after other researchers recognized similar effects did Gutowsky realize that they had found something new. In Gutowsky's research program, the instrument became a machine for constructing novel knowledge.

[15] Zandvoort, *Models of Scientific Development*, 227.

Through NMR, new questions arose that were within reach of answers. This searching for the unexpected took place at the borderline between physics and chemistry. Anomalies constituted the foundations of new effects if they were shown to be independent from previous experience and familiar theoretical explanations. In the case of spin-spin coupling, this realization was a difficult one. The temptation to interpret the splittings of resonance lines as instrumental artifacts or as effects of known chemical origin was great, and the complexity of the phenomenon prevented a smooth unraveling. The processes that took place during the discovery of the indirect coupling of nuclear spins were sometimes described as serendipitous and lucky by their protagonists. For the historian, they represent a pattern of gaining trust in the reproducibility of observables, the exclusion of instrumental malperformance, the constraints of empirical evidence on theoretical explanations, and the interrelating and cross-stabilizing of different areas of experimental science. The experimental characterization of the new effect of spin-spin coupling was a step in a series of attempts to define its meaning, followed by the convergence with different techniques, and its qualitative theoretical foundation.

With the building of his own instrument, Gutowsky acquired the freedom and status to decide about the direction of research, and he followed an important feature of the experimental style of chemical physics: the ability to design, to control, and to improve the research instrumentation. From 1949 to 1953, Gutowsky and his co-workers at the University of Illinois played a crucial role in the determination and the subsequent process of stabilizing the effects that formed the basis of chemical NMR. The experimental culture of chemical physics provided Gutowsky's questions, methods, and theories, among them bond distances, the unraveling of molecular and crystal structure, and the rates and energies of dynamic processes. In this way, Gutowsky and other scientists at Oxford, Stanford, Harvard, and MIT established the usefulness and credibility of NMR in the field of molecular spectroscopy. The experimental style of chemical physics dictated both the opportunities and the limitations of Gutowsky's work. His aim was the construction of apparatus and the establishment of methods through a complex interplay of experiment and theory. Routine use, as it was made possible with commercial NMR spectrometers available from 1953 on, did not belong to his area of interest.

The same applied to the Swiss physical chemist Richard R. Ernst. During his Ph.D. studies with Hans Primas at ETH Zurich, Ernst became acquainted with NMR. As with Gutowsky, Ernst was imbued with a specific research style that focused on the construction of an instrument, or part thereof, as the crucial sign that one had become a member of the community: "That is the *Gesellenstück* that you had to perform before you were accepted."[16] To a certain

[16] "Das ist das Gesellenstück, das man zu leisten hatte, bevor man akzeptiert wurde." Richard R. Ernst, interview by Carsten Reinhardt, 25 February 2002.

extent, NMR research at ETH was research for the sake of the instrumentation. This focus had two important constraints: First, the physical chemistry division at ETH Zurich had close contacts to the dominant organic chemists there, and was obliged to set up service instrumentation. Second, Primas himself arranged for commercial production of his instruments by a Swiss instrument manufacturer. Thus, Ernst shared with Gutowsky the close connection with organic chemists, and was greatly aware of their needs. In contrast to Gutowsky, Ernst from early on collaborated with instrument manufacturers. In the mid-1960s, he worked for Varian Associates at Palo Alto, California. There, the ruling topic of the decade was the enhancement of sensitivity, the bottleneck of NMR when compared to other spectroscopic methods. Following up the needs of a community of users, coupled with a footing in the quantum mechanical foundation of NMR, nuclear spin, became Ernst's specialty. During his tenure at Varian Associates, he laid the groundwork for FT NMR, a technique that soon thereafter revolutionized the impact of NMR in the chemical and biochemical sciences. Back in Zurich, Ernst, building on his capabilities of manipulating the complex response pattern of nuclear spins when subjected to electromagnetic radiation, invented a method that finally made NMR competitive with X-ray crystallography in the structural elucidation of large biomolecules. This was just a part of his encompassing research in all major features of NMR. His aim was to use NMR as an example of what could be achieved in spectroscopy in general. Many of his methods, their theory at least, were also applicable in other spectroscopic techniques. The endeavors of Ernst and his colleagues made NMR a specialty that enjoyed a large measure of independence, and a lead in refinement of theory and applications. Like Gutowsky, Ernst worked in the tradition of molecular spectroscopy. Unlike his earlier counterpart, Ernst merged the industrial with the academic laboratory.

Adapted to Chemistry

The natural product chemist Carl Djerassi once compared his science to the situation of a person who entered a dark room and had to figure out its contents. Djerassi related the touching of the pieces of furniture to find out about their form, size, and material to the reactions that were used by the chemists in degradative, analytical work to determine the structure of an unknown compound. What flashlights were for the person in the dark, physical methods were for the chemist. Each "flashlight," such as ultraviolet spectroscopy, infrared spectroscopy, NMR, and mass spectrometry, allowed the illumination of certain parts of the molecule only. Because the spectroscopic methods and mass spectrometry could not easily disentangle the whole puzzle of a complex structure, they retained parts of the intellectual excitement of the unraveling of an unknown molecule. The information gained by these physical methods had to be combined and judged, and supplemented with additional information obtained by chemical means. In contrast, the "flash camera" of X-ray crystallography had the capability of showing the whole three-dimensional structure of a crystalline sample molecule at once. This made the elucidation of natural product structures a much less fascinating endeavor. The chemist was "reduced to growing the crystal and pushing the camera's button."[17] Djerassi resented the X-ray technique, mainly because he did not want to be relegated to the role of a technician. But he was eager to employ the other flashlights. Moreover, his most active time in the structural elucidation of natural compounds, the 1950s to the 1970s, happened to be the most exciting years for the establishment of physical methods. This was an intellectual challenge in its own right.

According to Carl Djerassi, chemists active in the field of natural products had to use every possible assistance in the elucidation of their unknown compounds, which often were available only in very small amounts. For this reason, physical methods entered organic chemistry through the efforts of the natural product chemists, and the success of these methods was such that they literally obliterated the traditional methods in this field:

> I know of no area where organic chemistry can be taught better than in natural product chemistry. All bets are off—you are dealing with an unknown compound where you've got to use every help you possibly can to learn something about it . . . . Yet because we wanted help from everything, a natural product chemist was more receptive than anyone else. Why is it that all the advances in UV, IR, NMR, chiroptical methods, and mass spectrometry, entered into organic chemistry invariably through the natural product chemist? In fact, in the process it killed the traditional structure elucidation natural products chemistry. Now, it is so sophisticated that we don't do any more chemistry with the natural products. We're isolating new compounds, and we can establish their structure, but we don't do any chemistry with them.[18]

Djerassi's verdict applies to the classical style of structure elucidation, using chemical reactions in the test tube. Certainly, this type of work has vanished as a result of the inroads of high-technology instrumentation. But exactly because of the introduction of chemical concepts into mass spectrometry, natural product chemistry remained a chemical undertaking.

In many ways, and apart from any intrinsic intellectual pride, the extent to which the new physical instruments could be adapted to chemical paradigms decided on their initial success. In the case of mass spectrometry, chemical

[17] Djerassi, *The Pill, Pygmy Chimps, and Degas' Horse*, 84.

[18] Carl Djerassi, interview by Jeffrey L. Sturchio and Arnold Thackray, 31 July 1985, p. 39, Chemical Heritage Foundation, Philadelphia, Pennsylvania.

concepts explained the working mode of a part of the instrument itself. Concepts such as reaction mechanisms, the electronic effects of atomic groupings, the courses of rearrangements of molecular parts—all part and parcel of the well-established subdiscipline of physical organic chemistry—explained what happened to a molecule inside a mass spectrometer. Moreover, and most important, knowledge of physical organic chemistry enabled the mass spectrometrist to interpret mass spectra in a way that would have been otherwise impossible. Chemists connected the previously separated spheres of natural product chemistry, physical organic chemistry, and mass spectrometry, and in the 1960s merged them into a new field of activity, organic mass spectrometry.

During World War II, mass spectrometry had become an established branch of instrumentation used in the chemical and petroleum industries. Instrument manufacturers produced reliable, though costly, mass spectrometers that were well equipped for the investigation of organic chemicals. Dozens of industrial chemists built up and acquired the know-how for using the technique in analytical chemistry. But unlike ultraviolet and infrared spectroscopy, mass spectrometry did not enter chemical research in the 1940s. The overwhelming pressure of wartime needs restricted mass spectrometry's applications almost completely to hydrocarbons, and these molecules were not of much interest to academic research. The high prices of mass spectrometers, both in capital investment and operational costs, prevented easy access by academic chemists. Most important, though, was the lack of applicability of mass spectrometry in a crucial domain of chemical research, the structural elucidation of unknown molecules. In this field, a huge resistance on the part of academic organic chemists had to be overcome. This changed in the late 1950s, when mass spectrometrists at industrial companies and universities built up a knowledge base that rested on accepted chemical theories of the time: reaction mechanisms described in the terms of physical organic chemistry.

The analytical chemist Fred Warren McLafferty, who started out at the industrial enterprise Dow Chemical of Midland, Michigan, and in 1964 moved to Purdue University at Lafayette, Indiana, pushed forward the use of physical organic reasoning for the interpretation of mass spectra. In the early 1960s, McLafferty clearly formulated the dogma of mass spectrometry in chemical terms. Moreover, and in a pragmatic stance, McLafferty made no attempt to postulate a theoretical foundation of this dogma:

> The author would like to propose as a basis for further discussion . . . that *the unimolecular degradation reactions of the energetic ions demonstrated in the mass spectrum are similar to, and controlled through much the same energy effects as ordinary, thermal-energy, chemical reactions* . . . . The author will attempt no theoretical justification for applying the principles of the reactions of low-energy molecules to the decompositions of high-energy cations, except that

> for many cases to be illustrated it allows a basis for explanation and prediction of the observed results.[19]

Carl Djerassi belonged, together with McLafferty and Klaus Biemann, to the pioneering group of organic chemists that developed mass spectrometry as a tool for the structural elucidation of unknown compounds. Djerassi was convinced that "much of our success has been due to the fact that we approached the field through the eyes of the organic chemist in the context of concrete experimental problems." The tremendous success of mass spectrometry in the 1960s becomes visible through a glance at journals publishing in organic chemistry, when it became "probably impossible to find a journal . . . in which there are not numerous references to the use of this technique."[20] Of greatest importance, the necessary knowledge about "the ground rules of organic mass spectrometry" had to be generated. Organic chemistry provided for untold possible combinations. It was the established order system of this science that Djerassi put into place to give structure to findings in organic mass spectrometry. This pedagogical trick made possible the classification and presentation of the huge amount of data, as well as appealing to the majority of users, organic chemists with no special knowledge about mass spectrometry. This fashion was not relegated to the forms of presentation. It led to specific assumptions about the molecular events after initial ionization of the molecule in the mass spectrometer. Although mass spectrometrists' opinions were based on sound scientific knowledge, and their conclusions were shared by many, there was no unequivocal evidence as to their correctness. The appeal was mainly through parallels with established concepts of physical organic chemistry. Moreover, the successful use of the theories and rules of physical organic chemistry in mass spectrometry strengthened the position of the former even more. In the realms of biochemistry and natural product chemistry, mass spectrometrists concentrated on model systems and exemplary compounds, where they could show the advantages of their technique. In such a vein, mass spectrometry contributed to the chemistry of alkaloids, steroids, and peptides. Also, the chemistry of these substance classes decisively shaped the method. Especially in the 1950s and 1960s, when many physical methods still had to prove their worth, most organic chemists applied a combination of classical chemical knowledge derived from reactions, and the modern chemical knowledge made available by the physical methods. The puzzle of the structure of an unknown compound had to be solved, no matter how, and the diverse methods were reduced to something astonishingly compatible, in terms of the type of reasoning familiar to organic chemists.

---

[19] McLafferty, "Mass Spectrometry," 113–114. Emphasis in the original.

[20] Carl Djerassi, NSF grant application, "Mass spectrometry in organic chemistry and biochemistry," 1969, p. 4, Djerassi Papers, box 22a.

Nevertheless, mass spectrometry experienced resistance. Mass spectrometrists had to argue for acceptance by a chemists' community engrossed in traditional methods, and at the same time mass spectrometry competed with other ascending methods for efficient problem solving. The extraordinary capability of the mass spectrometer to detect and analyze minute amounts of a sample, one of its greatest advantages, was also the reason for initial disbelief and doubt. With the claim that elemental compositions and molecular weights were determined far better with mass spectrometry than by traditional microanalytical methods, mass spectrometrists were a threat for an established intellectual, practical, and social system. Mass spectrometry was on the verge of replacing the traditional methods of organic chemists employed in the determination of the empirical formula of a compound. These methods, well established for about 130 years, required isolation of the unknown compound, preferably in a crystalline state to establish its purity. Then it was subjected to combustion analysis, to determine its elemental composition. For Klaus Biemann at MIT, and his fellow mass spectrometrists, this was an old-fashioned procedure, in most cases replaceable by more precise mass spectrometric techniques. But the characterization of a new compound entirely by mass spectrometric and spectroscopic means was a battle that had to be fought. With the changing means used to establish substance identity, this identity itself changed. Instead of a proper crystal, mixtures of products were deemed sufficient for identification. Moreover, in the traditional system, vocationally trained laboratory workers performed these tasks. Though well paid, they had a relatively low social status, and did not pose a challenge to the supremacy of academic organic chemists who could use their results without any conflicts on rights of scientific interpretation. Now, high-tech instruments, whose results could be interpreted only by highly-qualified academic and industrial experts, were about to take over. For the leading group in the hierarchical system of academic chemistry, the synthetic organic chemists, this posed a threat. They feared that they had to give away a crucial part of their competence to their academic counterparts in analytical chemistry. Characteristically, Biemann emphasized the responsibility that mass spectrometrists had for the correct application of the technique and the interpretation of its results. Thus, in Biemann's view, traditional, labor-intensive methods of analysis would not be replaced by automated procedures. Still, a human interpreter would rule the process. In Biemann's system, the academic mass spectrometrist should fulfill this role. In the long run, the organic chemists countered this challenge to their status by taming mass spectrometry (and other spectroscopic techniques) in the form of service laboratories run by subordinated technicians. In response, experts, such as Biemann, focused on the development of methods for structural elucidation of very complex molecules. This more demanding intellectual field guaranteed their independence from the organic chemists, and enabled them to establish a scientific specialty in its own right. However, they depended on the applicability of their methods by biochemists and natural product chemists.

At first glance, NMR did not experience the same inroads by chemical concepts as did mass spectrometry. But in the long run, because NMR was, and still is, much more widely used in chemistry than mass spectrometry, NMR spectrometrists made their method perhaps even more chemical than the mass spectrometrists. In the beginning, two issues were important: the concepts of molecular structure and chemical bonding; and correlation analysis, an interpretative technique that enabled chemists to point to analogies among different parameters. The first point made available many of the research problems of early chemical NMR in the form of hypotheses that could be tested. The second allowed for building relations between chemical entities and parameters in NMR. Both strategies had an explanatory function in the theory of NMR, next to their heuristic role. In this regard, the attitudes of physicists and chemists, even when working on the same topic, showed characteristic differences: In commenting on their first article that appeared in the *Journal of Chemical Physics*, Gutowsky proposed to his collaborator, the physicist George Pake, "to make life a little simpler for the average chemist who might be more interested in the method and results than in the quantum mechanics."[21] Utility, not theory, was important for the chemist. The pragmatic outlook proved to be most important for future uses of NMR in organic chemistry, especially in structural analysis. This involved a change in perspective of interpreting the results. With Gutowsky's contributions, and those of his co-workers, NMR became a tool of distinct interest for chemical work. The work that related chemical shifts to accepted notions of physical organic chemistry built up the credibility of the technique, and also pointed to refinements in the theory of NMR itself. Correlation of chemical shifts to functional groups in molecules made clear the tremendous utility of NMR for analytical and structural work. Gutowsky's small team of chemists and electronic engineers skillfully transferred their results from the small community of molecular spectroscopists to the wide-ranging discipline of organic chemistry.

For chemists, the applicability of methods that had to be used with the physical instruments decided on the success of the latter. Naturally, such methods could have been designed by physicists, and especially by instrument manufacturers. But differences in paradigmatic outlooks of these communities decided in favor of method makers with strong roots inside chemistry. Moreover, because the use of classical, chemical methods in structural elucidation was becoming more and more out of date, intellectual capacities of organic

[21] Herbert S. Gutowsky to George Pake, 14 October 1948, Gutowsky Papers, folder "N.M. I correspondence."

chemists moved on to other fields. Method makers filled this void. Still being chemists, though now members of a specialized sub-group, they decided on the methods to be used in chemical research. In so doing, they rescued the methodological autonomy of chemistry.

Standardized and Taught

As standardized procedures, methods decided what counted as state of the art in science, technology, and even many social fields. Thus, control over standardization represented an important share of the power exerted in scientific disciplines. In the case of physical methods, four spheres battled over influence: the chemical industry, partially through industrial associations; governmental agencies; the instrument manufacturers; and specialized scientists. In the end, after many compromises, a balance was reached between users on the one side (including academic as well as industrial chemists), and suppliers of instruments on the other. Neither could be dominant, because their respective contributions to the research system were too important. The fields in which standardization played a role were diverse. Representation modes, choice of parameters, data files, reference substances, and instrumental hardware were among the disputed issues. Instrument manufacturers enjoyed the fact that, with the in-built standards of their instruments, their status was hard to dispute. Especially in the beginning, industry had a powerful position through the design of instruments, and accessories. An early manufacturer of mass spectrometers, CEC of Pasadena, for example, decided on certification of spectra, and it took the combined efforts of mass spectrometrists to set up an independent system of "uncertified spectra" that catered to their own needs. Manufacturers added the publication of data files for reference, research, and teaching to their activities. So did the user scientists, and they joined forces in associations. In mass spectrometry, the American Society for Testing Materials (ASTM) E-14 subcommittee evolved into the American Society of Mass Spectrometry. Originally, the ASTM committee had emerged out of user meetings arranged for by the manufacturers of mass spectrometers. Thus, the society was an example of the continuing emancipation of scientists from their suppliers of instrumental hardware. In NMR, the Experimental NMR conferences, and especially the MELLONMR newsletter, shaped the public activities of companies and academic scientists alike. Clearly, the topic of standardization had an important impact on the foundation of method-oriented societies. The union of scientists was a necessary strategy in their dealing with instrument manufacturers. Many scientists had close contacts to the suppliers of their research instrumentation, and in this way sought to exert influence over design, and standards. The same applies to the kind of service, and construction of hardware, where scientists had to make their needs, and arguments, visible. Thus, one should not forget that a clear-cut division between user-scientists on one side, and supplier-industrialists on the other side, never existed. Despite the great dependence of scientists on industry and government in the area of standardization, the original measure of setting and controlling scientific standards, that is, peer review, was kept firmly in place. Most methods were published, and subjected to review. Thus, although many techniques were patented, or made available as software, the scientific power system continued to be intact. But it had to contend with competitive interests of industry. In order to launch a successful method, the innovative scientist had to arrange for its compatibility both with similar methods and especially with existing instruments. No method came without a standard. Some were not disputed, and even decided on in the academic sphere alone, but most had to be discussed with a multitude of players from diverse fields. No wonder that participants attempted to conquer influential positions, mainly by taking over functions that originally belonged to the other side. This is especially clear in the case of the first manufacturer of NMR spectrometers, Varian Associates, which aggressively pursued a strategy of setting standards in any dimension possible: Standards of achievement were published side by side with standards of representation and standards of procedure in the company-owned bulletins, spectra catalogues and advertisements. In the 1960s, Varian Associates was tremendously successful, but soon had to accept an increased independence of users, who relied on the products of competing firms that did not come with so many constraints and restrictions.

While standardization often seemed to be a struggle over power and influence between users and manufacturers, teaching more often remained a complementary task. Though it happened that company scientists published textbooks, they more often stuck to their traditional measures in diffusion of methods: workshops, user clinics, instrument fairs; and bulletins, catalogues and instrument documentation. In contrast, though academic scientists participated as lecturers at workshops organized by companies, they normally chose the traditional channels for their educational endeavors: textbooks, lectures, seminars, exchange of research personnel, and short-term stays at their colleagues' laboratories. Some academic scientists regarded the authoring of textbooks as an intellectually worthless and boring task that they avoided. Not so in the method-making community. A major part of the influence that scientists and historians assign to Roberts, Biemann, Djerassi and McLafferty stems from their textbooks. An exception is Gutowsky, who never wrote an academic textbook. Ernst co-authored an influential textbook on NMR, but it seemed to be too technically demanding to really have an impact on a wide community of chemists. Clearly, the authoring of textbooks constituted an important part of a scientist's strategy to spread his methods. Biemann even made it explicit, when he replied to criticisms with respect to possible misinterpretation of mass spectrometric data, that this was the reason he had written his book. Thus, scientific standards of interpretation were transported by textbooks, along with scientific articles. In science, reputation is bound to citation of scientific work. In this respect, method makers enjoyed an advantage over other scientists, because articles describing methods had the potential to be cited with increasing frequency. But with entirely new methods, a critical number of knowledgeable users had to be generated, and in this challenge, textbooks helped. Thus, method-oriented scientists engaged in the preparation of teaching, and in teaching itself, also because this would contribute indirectly to their scientific reputation.

Textbooks, with their terminologies, notations, and representation systems, decisively influenced the standards in a scientific field. Moreover, the textbook afforded a systematic treatment under the headings of interest for the user-scientist. Djerassi and co-authors, for example, explicitly chose the ordering system of organic chemistry to present mass spectrometry. Biemann and McLafferty, in contrast, opted for the more method-oriented approach. Roberts, in his first book, applied the teaching-by-example method; in his second he opted for a thorough mathematical treatment; and in his third textbook, on general chemistry, he and his co-author introduced spectroscopic methods as the first and correct way to approach chemistry. Thus, Pauling's chemistry of substances and their reactions in 1947 had changed to a chemistry of substances and their spectra in the mid-1960s.

Of course, instrument manufacturers engaged in the spread of instrumental methods, too. Most successful in this respect were workshops, often organized in cooperation with academic scientists and chemical societies. In the mid-1960s, such workshops attracted many hundreds of attendants. The training of experienced mass spectrometrists, for example, was deemed to be the only practical solution in dealing with the bottleneck of the technique, the time-consuming interpretation of spectra. Notably, the chemical and pharmaceutical industries felt this need when they decided to expand their activities in this field. Many of these teaching jobs were done in a face-to-face manner. Naturally, the education of graduate students followed this way; but also, more remarkably, so did the teaching of industrial scientists and academic colleagues. Biemann, for example, spent several weeks in the laboratory of his soon-to-be competitor Djerassi to get him started in organic mass spectrometry. In addition, he engaged in long-term cooperations with chemical and pharmaceutical companies that involved the education of company scientists and the enabling of them to reaching an independent standing in mass spectrometry. It was not so much the tacit-knowledge component of technology transfer that made this necessary, but more the continuous training over a rather long period that it took to become a reliable and efficient mass spectrometrist.

Connecting University and Industry

Recent scholarship in the history of science has shown that instrument manufacturers entered academic and industrial research laboratories both with mass-produced equipment and with cutting-edge instrumentation. The standardization of instrumentation played an important part in the stabilization and dissemination of local knowledge, while the specialized market of instruments exerted considerable influence in science.[22] In the second half of the twentieth century, chemists who specialized in the development and pioneering use of instrument-based methods could hardly avoid entertaining close relationships with instrument manufacturers. First of all, the complexity of the research instrumentation and the needed expertise in electronics and physics made in-house construction an insurmountable hurdle for most chemists. Second, the cost of equipment contributed to a drain on resources. Third, the need to have one's own methods spread as widely as possible in the scientific community made collaboration with an efficient manufacturer attractive, and almost essential, since the sales of the instrument contributed to the use of the method, and vice versa. Fourth, the opportunity to influence the design process for advanced instrumentation in the direction of an individual's own wishes and preferences, and possibly the access to one-of-a-kind instruments, constituted a huge advantage in competition with other scientists. As an outcome of this scenario, scientists, via the development of methods, were as much present in the innovation and diffusion of scientific instruments as instrument manufacturers were in the production of scientific knowledge.

With their contacts to instrument manufacturers, method makers added another type of academic-industrial relationship to the already long list of chemists' collaborations with industry.[23] In contrast to the familiar consultancies with the chemical industry, chemists suddenly relied on industry not for cooperation with regard to their research objects, nor for absorption of their research results, but for access to their research means. As was the case with the older type of relationship, the main factor was not money, but access to knowledge and information. Knowledge, codified or non-codified, in the form of substances, reaction pathways, methods, instruments and their parts, included market and political, and not only scientific, knowledge. Its flow went in both directions, the academic and the industrial partners being both in the giving and the receiving positions. It is important to note that not only knowledge but also information crossed the border between the industrial and the academic spheres. Information took the form of data and their interpretation, and this

---

[22] Gaudillière and Löwy, *The Invisible Industrialist*, 6, passim. See also Joerges and Shinn, *Instrumentation Between Science, State and Industry*; Bud and Cozzens, *Invisible Connections*.

[23] For an overview see Reinhardt and Schröter, "Academia and Industry in Chemistry."

often brought about long-term dependencies. In contrast, transfer of knowledge enabled the partner to achieve independence in the main questions at stake, because it entailed a training part. Particularly for the use and development of instrumentation, training was an indispensable condition of technology transfer.

The leitmotiv of the relationship between instrument manufacturer and academic chemist was access to advanced instrumentation for the latter and the gain of user and market knowledge for the former. This brought about new collaborative forms. Users contributed to design, and sometimes even construction, of instrumentation, and played roles in the marketing of instruments and the diffusion of methods.[24] Such lead users took over important parts of industrial activities. In contrast, industrial employees, such as James N. Shoolery at Varian Associates, acted sometimes more like academic scientists than as industrial scientists. It is with such cases that the new character of the industrial-academic relationship involving physical methods and instruments is most clearly visible. For the academic scientist, exclusive access to cutting-edge instruments was a necessary part of his research program in developing new methods. Without such instrumentation, competition by other scientists would have made his results much harder to obtain. But this exclusivity bore fruit only when it was a transitory phenomenon. In contrast to other instrument-dependent scientists, the method makers' reputation relied on the use of their methods, and not simply on the acceptance of their research results. In the end, they had to enable scientific colleagues to imitate, and to expand, their work. For using methods such as carbon-13 NMR or high resolution mass spectrometry, instruments had to be made available to the scientific community at large, and as a consequence the original lead of the method maker disappeared. Lead users worked most efficiently during a time period when they enjoyed an advantage with regard to instrumentation over their competitors. In order to keep this position during their academic career, they needed a continuous and close cooperation with an instrument manufacturer who was willing to supply them with the newest state-of-the-art or even one-of-a-kind instrumentation. Thus, lead users had to move all the time in front of a majority of users. If they were able to act as trend-setters, their methods being widely appreciated and used, this would enhance their chances both for procuring grant money to develop improved instrumentation and for an ongoing satisfying cooperation with the instrument manufacturer. In order to achieve such a role, they needed status in the respective scientific communities, including chemistry, biology and medicine between 1950 and 1975. Lead users actually had to show that their methods would solve the problems of their scientific colleagues, and these problems included new and even future ones. If a method had proven (or disproven) its

[24] See von Hippel, *Sources of Innovation*.

worth in a special field, its designer moved on to a new method, and left the field to the non-specialists. This feature is the core of the economy of a method maker's activities in general, and explains the sometimes urgent pleas directed to colleagues for communicating solvable problems. Thus, lead users depended both on cooperation with instrument manufacturers and on participation in the respective scientific fields, which involved cooperation with other academic and industrial scientists.

The role of a lead user of scientific instruments entailed many functions. Next to development of methods, it was their dissemination in the scientific community, achieved via the channels of scientific cooperation, publication, and teaching. It involved the counseling of the instrument manufacturer with respect to acceptance, usability, and performance of instrumentation. Both were to the advantage of both scientist and company, and their relation can best be compared to a symbiosis. Another task of a lead user consisted in the anticipation of future research fields, best performed when he contributed to directing their course. Possibilities for influencing future science involved his own successful research, which led to imitation; participation in agencies for research funding, which involved the setting of incentives; and teaching and other public activities. Thus, both sides benefited most if the lead user was an acknowledged and experienced scientist, his performance in the development of methods for use with soon-to-be commercial equipment deciding on this issue.

Conflicts between the industrial manufacturer and academic lead user arose if the former undertook an aggressive scientific strategy, that is, if he acted as competitor and not as supporter in chemical research. This was a thin line that Shoolery crossed several times. Unproblematic was in-house research by the company in the fields of electronics and other hardware-related fields. However, the chemical community did not accept non-cooperative chemical research, and exerted its power as customers to block such ventures. After a transient period of learning, this no longer posed problems. A harmonious long-term development of business and scientific strategies was more difficult to achieve. When Varian Associates decided to concentrate on routine instrumentation, and was not willing to follow the direction towards advanced research instruments as recommended by Roberts, the scientist decided to change to a competitive company. This alludes to the importance for a method maker to engage in a close technological symbiosis with an instrument manufacturer, as seen also in the example of Richard Ernst, who changed his industrial partner in the late 1970s. During this decade, Varian Associates lost its near monopoly in the NMR spectrometer market. In the long run, the continuous development of advanced research instrumentation, and the cooperation with lead users, proved to be decisive also for the routine, "analytical" market, probably not only for technical but also for prestige reasons.

For the lead user, the specification of advanced instruments continued to constitute a central part of his research program. Their conceptualization and commissioning counted as scientific achievement in itself, sometimes a major one. Their marketing and sales, even if in altered form, served similar purposes as textbooks and lectures: inbuilt in the instruments were methods and procedures, which were spread with the instrument in a scientific community. Instruments, in this sense, were not only industrial, but also scientific commodities. The necessary investment led to the inclusion of a third player, governmental agencies, supplying large parts of development funds in the form of grant money. When the development of research instrumentation began to be regarded as a scientific activity in its own right, it justified special grant applications at the National Institutes of Health (NIH), the National Science Foundation (NSF), and other institutions of this kind. In the late 1950s, NSF started an instrument-related program, joined in the early 1960s by NIH. These organizations decisively shaped the direction of instrument-related research and the institutional forms of research instrumentation.

While instrument manufacturers were a new type of partner for academic chemists, their expertise in instruments soon added a new style of contacts to the chemical and pharmaceutical industries. In the beginning, many instruments, and related methods, originated in the chemical and petroleum business, and academic scientists had to accept that industrial scientists often were more advanced in the art of developing and using research instrumentation. Most academic chemists in this book first learned about their later method of choice from industrial scientists. Thus, scientists active in the chemical industry were the peers, and teachers, for their counterparts at universities as well as for the fledgling instrument industry. Soon, academic method makers occupied positions as experts for industry, and in this regard acted as teachers and multipliers. Transfer of knowledge was often accompanied by the sharing of the instrument, as the case of Klaus Biemann's cooperation with the Swiss Firmenich & Cie. has shown. The same happened in cooperations with academic chemists. The sharing of the instrument actually became a transfer of knowledge, and not just information, through the training that accompanied its joint use. If this did not take place, as in Biemann's strained relationship with Firmenich & Cie., it led to dependence. In Carl Djerassi's career, his positions as research director and chairman of several industrial enterprises during his academic tenure led to conflicts of interest. Djerassi attempted to separate his industrial activities from his academic duties, partially through his decision to separate the studies on the compound classes he worked on. Steroids were tackled in industry, while alkaloids, terpenes, and others were investigated in academia. Another distinction was Djerassi's decision to concentrate on synthetic work in his industrial projects, and to focus on analytical research in his academic investigations. However, mass spectrometry soon merged these two spheres. First, Djerassi chose steroids as model compounds for his academic, analytical mass spectrometric program, and, second, he was forced to include synthetic work in order to generate the variety of substances needed. In an initial step, his research style led to knowledge transfer from the company Syntex to the university, but also, in a second step, back to industry, when the know-how gained at Stanford moved back into the entrepreneurial realm. For some, including the "father of Silicon Valley" and Stanford's provost, Frederick E. Terman, this constituted an attractive feature of Djerassi's research. For others, especially NIH officials, it came close to a misuse of tax money for private interests.

With the exception of Gutowsky, all scientists described in this study entertained close and long-lasting contacts to instrument manufacturers, and other industrial enterprises. Almost all of these contacts were built around access to instruments and training of methods. At least for Roberts and Ernst, and to a lesser extent for McLafferty and Biemann, the symbiosis with an instrument firm represented the constitutive basis of their work. All of them engaged in construction of instruments, but not to such an extent that they got enamored with instrumental design and forgot to take care of their major task, the development of methods, both for a market of instruments, and, especially, for a scientific community.

At Research Centers and Facilities

When compared to other disciplines, most notably physics, chemical research was considered as less of a teamwork affair. It did not require the spectacular, large research facilities of particle accelerators, vessels for oceanographic research, and radio telescopes in astronomy. In contrast to physics, the research culture of chemistry did not experience a shift towards Big Science. Chemistry continued to be a largely decentralized endeavor, but adjustments had to be made to accommodate the new requirements of instrumentation.

With the advent of Big Science, the funding of science shifted its focus, to the disadvantage of chemistry. The percentage of funding for chemistry within the total subsidy for the mathematical, physical, and engineering sciences by NSF decreased from 23.4 percent in 1953 to 8.6 percent in 1963. Large projects were easier to lobby for in the public sphere; they had a greater visibility, and were often administered by special budgets. The chemists tried to invert this trend with surveys specifying their needs for so-called community equipment, that is, instrumentation shared by two or more researchers. In 1962, the total need for new instrumentation of this kind was estimated at $15 million, for more than 110 chemistry departments in the United States. In 1964, an estimated accumulated investment of $55 million in 125 chemistry departments was reported, at an annual increase rate of 20–25 percent.[25] This was a little more than half of the projected investment for the Stanford Linear Accelerator alone. Thus, John D. Roberts could write to Leland Haworth, director of NSF, that "support for such instruments would have the advantage of being relatively cheap, very broad-based and practically 100% of direct aid to education."[26] Roberts wanted to prevent the exclusion of small flexible grants compared to large installations. Moreover, he was sure that the pace of new developments in chemistry depended largely on broad access to novel types of instrumentation, and he feared the opening of a gap between a few large institutions that could afford the new instruments and the many who could not. Heads of chemistry departments described the need for modern instrumentation as one of the most critical financial difficulties they had to face. Chemists' preference to apply for small, individual grants made it especially hard for younger scientists to afford the cost of instrumentation. For example, the average annual grant available to a faculty member in chemistry was reported to be $17,000 in 1964, while the average cost of an instrument was about $14,000. Thus, the typical grant could barely cover the cost of a single instrument.

Because of their complexity and their cost, the sharing of the instruments by several scientists became an imperative. Basically, two different types of organization can be distinguished: the departmental laboratory, and the regional or national facility. Most often, and because of their decentralized needs and traditions of work style, chemists preferred the departmental solution. But the enormous cost of the development of cutting-edge instrumentation made regional centers necessary, especially in the biomedical sciences. There were intellectual reasons, too. Synthetic organic chemists, the biggest and most powerful group in United States chemistry, resisted dependence on their colleagues in physical and analytical chemistry in vital research questions. Herbert S. Gutowsky, who had substantial experience in infrared spectroscopy service work, recalled some of the difficulties he experienced as a physical chemist in a department that was dominated by organic chemistry:

> At least in the early days of using infrared which was when I came here it was frustrating to the organic chemist who ruled the roost to be dependent on a whippersnapper of a physical chemist who had just gotten a PhD. The thing is they found it very frustrating to depend on the interpretation of somebody else. It's probably the way it should be. If you have too much dependence by a researcher on the work of somebody else you miss too much, you see.[27]

---

[25] National Academy of Sciences, *Chemistry: Opportunities and Needs*, 96–99.

[26] Roberts to Haworth, 13 August 1963, NSF office of the director subject files, folder "MPS chemistry," 307-75-051, box 3.

[27] Gutowsky, interview by Reinhardt, 1 and 2 December 1998.

Consequently, the major instruments were pooled, and operated on a centralized basis in order that a broad base of people could have access to them. With respect to support at the departmental level, NSF started a Chemical Research Instrumentation Program in 1957.[28] The money was given on a cost-matching basis. There was a clear correlation between the reputation of the department and the amount of funding it received.[29] At first, typical departments allowed research workers free access to the research instruments, with the benefits of high educational value and use on a seven-days-a-week, 24-hours basis, though there was also the risk of damage to the equipment. Very often, faculty members who were experienced in these techniques were assigned to supervise these open departmental facilities, informally helping other faculty members. As demand for assistance increased, it became common to set up service centers with professional staff members, instrumentation experts who split their work time between service work and instrument-related research. The example of the chemistry department at the University of Illinois shows how much instrument facilities changed departmental structure. Around 1940, only three small service units existed at the department level: machine shop, microanalytical laboratory, and glassblowing shop. In 1957, an NMR service laboratory was established, followed by an electronic shop. In the mid-1960s, nearly a dozen such facilities were necessary, ranging from mass spectrometry to a computer center. A service laboratories committee supervised this organization, and in 1962 the hiring of competent technical staff was regarded as an urgent problem of highest importance. No less than nine positions had to be filled, among them electronic technicians, programmers, and a full-time spectrometrist. These social changes were regarded as revolutionary by the contemporary scientists themselves.[30] Chemists regarded the availability of instrumentation as essential for research and education, but availability no longer meant the simple presence of equipment in the laboratory. The complexity and the high cost made absolutely necessary the establishment of service laboratories with trained technical personnel to operate and maintain the instrumentation.

As a result of the rapidly increasing need for research instrumentation, chemistry departments had increasing difficulties in raising the funds. In 1970,

---

[28] See Coulter, "Research Instrument Sharing"; and Stine, "Scientific Instrumentation."

[29] Table viii in Arthur Findeis, memorandum "Staff report on chemistry research instruments program," 10 October 1967, NSF office of the director subject files, folder "MPS chemistry," 307-75-051, box 1.

[30] Department of Chemistry and Chemical Engineering, annual report 1961–1962, Gunsalus Papers, 15/5/40, box 8, folder "Dept. reports—annual 1963/64." See also Nelson J. Leonard to H. E. Carter, 5 October 1955, Leonard to Carter, 25 February 1956, and an anonymous memo, 15 May 1956, all Gunsalus Papers, 15/5/40, box 8, folder "Instrument rooms 1956;" and Gutowsky to Meyer, 2 April 1957, Meyer to Gutowsky, 21 April 1957, Gutowsky Papers, folder "Meyer, L.H. 3."

the individual start-up costs of a new faculty member in synthetic organic chemistry were estimated to be $8,000 with access needed to departmental service facilities worth more than $100,000; in 1979 the amount for individuals was raised to $44,000, while access to instrumentation that was valued at nearly $750,000 was judged necessary by the Association of American Universities.[31] Certainly, this was an extreme example, presented in order to justify demands for increased funding. Nevertheless, with the move from a bench-top and labor-intensive science to an equipment-intensive discipline, chemistry went through many changes, similar to those earlier experienced in physics. Some chemistry departments could no longer afford to pursue expensive lines of research. Thus, under tight financial constraints, a solution seemed to be the development of regional instrument facilities, on a shared-access basis. In chemistry, NSF began with such a program in 1978, while the National Institutes of Health (NIH) had started such a program already in the 1960s. But still, the traditional funding patterns of the chemical sciences resisted this move. This was a problem of the sociology of the chemists, who preferred to work as small-time entrepreneurs. Nevertheless, instrumentation required chemists to rethink this behavior and shift to more collaborative modes of operation. Specialists who focused on the development of instrumentation and methodology became important mediators between the instrument manufacturers and chemists interested in routine use.

NIH became a major investor in the advancement of research instrumentation, as well as an important contributor to organizational innovations of how to use it. In addition to its intra-mural research laboratories in Bethesda, Maryland, NIH supported health-related research projects all over the United States, in the early 1960s starting a program to set up instrumentation in centers. These Special Research Resources resided in universities and other nonprofit institutions, and provided service to the academic community in a certain geographical area. The personnel of the facility were expected to pursue "core research" in instrumentation science and methods development, thought to assist in continuously modernizing the facility. In addition, training constituted an important task of a special research resource. The sharing of the instrument became essential, especially in fields where research was still in flux. At such centers, scientists gained access to instruments and expertise that would be too time- and resource-consuming for them to develop and operate on their own. Instrument facilities became meeting grounds for funding agencies, instrument makers, and user-scientists, with instrument specialists such as Biemann and McLafferty in the middle.

Not all, even not the majority, of method makers continuously worked in such instrument centers. Nevertheless, facilities and research resources became

---

[31] See Berlowitz et al., "Instrumentation Needs of Research Universities," 1015.

centers for methods development. They provided crucial contacts to a scientific constituency, and supplied additional resources. Seemingly, such service centers worked best if they did not constitute the only type of funding of a researcher. For example, although Biemann and Ernst ran service facilities during periods in their careers, both oversaw laboratories of their own, with different funding. This secured their independence, and left space to pursue special scientific interests. In essence, instrument centers and facilities institutionalized the established feature of cooperation between scientists, and combined it with the style of service work done by university departments and industrial analytical laboratories. Because of their community-wide importance, method makers were enabled to add their own research activities in instrument and methods development.

MAKING METHODS

The first impact that physical instrumentation had on chemistry was the alienation of chemists from the design of their research means, the chemical reaction no longer the major method. Physics, electronics, and other technologies threatened to decide the nature and the direction of chemical research. In response, and from many areas inside the chemical community, scientists transformed their research programs towards the development of methods that were accepted by a majority of their colleagues. As users, innovators, and mediators, chemists adapted physical instrumentation to the reigning paradigms and the habitus of their scientific community. Arguably, their contributions rescued chemistry's autonomy. In any case, competence in physical methods became an acknowledged chemical qualification, and a scientific merit in itself. In this process, the intermediate fields of chemical physics and the chemical industry were supplemented by novel communities of instrument experts, the method makers. At universities, research and service centers, and instrument manufacturers these experts pursued their careers. For them, machines formerly serving for the discovery of natural effects became objects of inquiry in their own right. Chemistry expanded to the biomedical sciences through the funds offered, the instruments' capabilities, and the intellectual challenges posed. Although all this seems to be a history of progress and achievement, something was lost. In this regard, the chemist Robert Robinson and the mountaineer Albert F. Mummery shared some insights: They disliked the inroads of physicists and guides, spectrometers and theodolites, charts and maps, and they regretted the losses of skill, adventure, and excitement that their fields experienced during this process.

The access to chemical phenomena via physical instrumentation was at odds with traditional methods of chemists. Nevertheless, during the 1950s and 1960s, scientists connected physical evidence to chemical ideas and theories.

This correlation was essential for the success of physical methods in chemistry. In the case of mass spectrometry, chemical theories decisively shaped the interpretation of the instrument's working mode. In NMR, rules simplified the interpretation of spectra and enabled clear-cut correlations between physical spectra and chemical structures. A pattern of incorporation by adaptation is also recognized if we take note of the organizational changes within chemical research. Practical concerns, and traditional modes of funding and laboratory organization, prevented a sudden change of chemistry to the Big Science practiced in some parts of physics. Chemists tried to make use of the new methods, without giving up too much of their own. In this process, a splitting of the new techniques occurred: Chemists incorporated fully the knowledge of how to interpret and use the data for their own purposes. They subordinated to a service role, but kept under control, the practical operations necessary to gain these data. In the beginning, they relied on scientific colleagues, who were specialists in the new techniques. But when the dependence on them became too great, chemists arranged for a service system. The instrument specialists set out to continue to work in their own domain, with practice and interpretation unified. Their research style made possible the innovations necessary for cutting-edge instrument research. Together, method makers and instrument manufacturers innovated novel research techniques, pushed their diffusion in the scientific community, and transformed complex research spectrometers into standardized instruments for routine use.

The scientists described in this book developed special styles for making and using methods in their respective work. Biemann, and especially McLafferty, undertook research programs that included improvements of instrumentation. However, Djerassi was not involved in the development of hardware technology in his own laboratory. McLafferty, an analytical chemist, pursued mass spectrometry with a small group of specialists, while the organic chemist Djerassi used a variety of physical methods for chemical work in a factory-like style. With the lead user in NMR, Roberts, the mass spectrometrists shared a leaning towards developing methods for bigger and bigger molecules, thus shifting their focus from chemistry to biochemistry and finally molecular biology. Chemistry expanded, but also lost its homogeneity, in this process.

In contrast, Gutowsky and Ernst focused on the science of the instrument itself. Gutowsky's example shows how the first inroads of physical instrumentation into chemistry were made in chemical physics. I have argued that this went hand in hand with the adoption of a pragmatic experimental style, and I have emphasized the crucial role of controlling the instrument in this style. The successful correlation of parameters in NMR to chemical concepts led to the creation of an instrument-focused community of chemists, fostering the innovation of techniques. Moreover, the pragmatic style of chemical physics led to a relatively smooth transfer of methods to other chemical fields. Ernst's story shed light on the establishment of a community of NMR spectrometrists at research universities and service centers. Ernst worked in the context of specialists who designed methods for the manipulation of spin systems, the conceptual basis of NMR, and brought the manipulative aspects of modern NMR techniques to the fore. In their work, the scientists presented in this book connected the theoretical to the experimental and instrumental sides of their science, and in addition took care of the needs of users. In doing so, they made methods and brought them to the scientific community.

Appendix

Archival Sources

Biemann Papers: Klaus Biemann Papers (private)
Djerassi Papers: Carl Djerassi Papers, Department of Special Collections, Stanford University Libraries, Stanford, California
Ernst Papers: Richard R. Ernst Papers (private)
Gunsalus Papers: Irwin C. Gunsalus Papers, University of Illinois, Archives, Urbana-Champaign, Illinois
Gutowsky Papers: Herbert S. Gutowsky Papers (private)
Jardetzky Papers: Oleg Jardetzky Papers (private)
Kistiakowsky Papers: Papers of George B. Kistiakowsky, Harvard University Archives, Cambridge, Massachusetts
McLafferty Papers: Fred Warren McLafferty Papers (private)
MIT Archives: Institute Archives, Massachusetts Institute of Technology, Cambridge, Massachusetts
NARA: National Archives and Records Administration, College Park, Maryland
NSF: National Science Foundation, Arlington, Virginia
Roberts Papers: John D. Roberts Papers (private)
Terman Papers: Frederick Terman Papers, Department of Special Collections, Stanford University Libraries, Stanford, California
Varian Associates Papers: Varian Associates, SC 345, series Varian Associates, Department of Special Collections, Stanford University Libraries, Stanford, California
Varian Associates Oral History: Varian Associates Inc. Oral History Project, Department of Special Collections, Stanford University Libraries, Stanford, California
Yost Papers: Don Yost Papers, Institute Archives, California Institute of Technology, Pasadena, California

Interviews

Weston Anderson, interview by Sharon Mercer, February 1990, Varian Associates Inc. Oral History Project, Stanford University Archives, SC M 708, box 1.
Klaus Biemann, interview by Carsten Reinhardt, 10 December 1998.
Nicolaas Bloembergen, interview by Joan Bromberg and Paul L. Kelley, 27 June 1983, Niels Bohr Library, American Institute of Physics, College Park, MD, USA.
Mildred Cohn, interview by Carsten Reinhardt, 25 November 1998.

Carl Djerassi, interview by Jeffrey L. Sturchio and Arnold Thackray, 31 July 1985, Chemical Heritage Foundation, Philadelphia, Pennsylvania, USA.

Carl Djerassi, interview by Carsten Reinhardt, 20 January 1999.

Richard R. Ernst, interview by Carsten Reinhardt, 25 February 2002.

Herbert S. Gutowsky, interview by Carsten Reinhardt, 1 and 2 December 1998.

Ralph Kane, interview by Sharon Mercer, November 1989, Varian Associates Inc. Oral History Project, Stanford University Archives, SC M 708, box 1.

Fred W. McLafferty, interview by Carsten Reinhardt, 16 and 17 December 1998.

Martin Packard, interview by Sharon Mercer, December 1989, Varian Associates Inc. Oral History Project, Stanford University Archives, SC M 708, box 1.

Edward M. Purcell, interview by Katherine R. Sopka, 8 and 14 June 1977, Niels Bohr Library, American Institute of Physics, College Park, MD, USA.

Norman F. Ramsey, interview by Paul Forman, 12 July 1983, Niels Bohr Library, American Institute of Physics, College Park, MD, USA.

John D. Roberts, interview by Rachel Prud'homme, February to May 1985, Caltech Archives, Oral History Collection.

John D. Roberts, interview by Carsten Reinhardt, 28 January 1999.

James N. Shoolery, interview by Sharon Mercer, April 1990, Varian Associates Inc. Oral History Project, Stanford University Archives, SC M 708, box 1.

Bibliography

Abir-Am, Pnina. "The Discourse of Physical Power and Biological Knowledge in the 1930s: A Reappraisal of the Rockefeller Foundation's 'Policy' in Molecular Biology." *Social Studies of Science* 12 (1982): 341–82.

Abrahamsson, Sixten, Einar Stenhagen, and Fred W. McLafferty. *Atlas of Mass Spectral Data*. New York: Wiley 1969.

Allen, James S. "The Detection of Single Positive Ions, Electrons, and Photons by a Secondary Electron Multiplier." *Physical Review* 55 (1939): 966–71.

Anderson, D. H. "Early NMR Experiences and Experiments." In *Encyclopedia of Nuclear Magnetic Resonance*, vol. 1, *Historical Perspectives*, edited by D. M. Grant and R. K. Harris. Chichester: John Wiley & Sons 1996, 168–76.

Anderson, Herbert L. "Precise Measurement of the Gyromagnetic Ratio of He-3." *Physical Review* 76 (1949): 1460–70.

Anderson, P. W. "John Hasbrouck Van Vleck." *Biographical Memoirs of the National Academy of Sciences* 56 (1987): 501–40.

Anderson, R. G. W., J. A. Bennett, and W. F. Ryan, eds. *Making Instruments Count. Essays on Historical Scientific Instruments Presented to Gerald L'Estrange Turner*. Aldershot: Variorum 1993.

Anderson, Weston A. "Fourier Transform Spectroscopy." In *Encyclopedia of Nuclear Magnetic Resonance*, edited by D. M. Grant and R. K. Harris. Chichester: John Wiley & Sons 1996, 2126–36.

Andrew, E. Raymond. "Nuclear Magnetic Resonance Absorption in $NaSbF_6$." *Physical Review* 82 (1951): 443–44.

Andrew, E. Raymond. *Nuclear Magnetic Resonance*. Cambridge: Cambridge University Press 1955.

———. "Spinning the Spins: A Lifetime in NMR." In *Encyclopedia of Nuclear Magnetic Resonance*, vol. 1, *Historical Perspectives*, edited by D. M. Grant and R. K. Harris. Chichester: John Wiley & Sons 1996, 180–87.

Andrew, E. Raymond, and Richard Bersohn. "Nuclear Magnetic Resonance Line Shape for a Triangular Configuration of Nuclei." *Journal of Chemical Physics* 18 (1950): 159–61.

Andrew, E. Raymond, and E. Szczesniak. "A Historical Account of NMR in the Solid State." *Progress in Nuclear Magnetic Resonance Spectroscopy* 28 (1995): 11–36.

Anonymous. *About Lead User Concepts* [cited 20 September 2003]. Available from <http://www.leaduser.com>.

———. "Element Maps Depict Complete Mass Spectra. MIT Scientists Use All Lines in High-Resolution Mass Spectra to Determine Element Distribution in Large Molecules." *Chemical and Engineering News* 6 July (1964): 42–44.

———. "Norman Wright (1906–1994): Co-Founder of the Coblentz Society." *Applied Spectroscopy* 49 (1995): 688.

Arnold Engineering Company. *Group Arnold History* [cited 29 August 2001]. Available from <www.grouparnold.com/corp/history.htm>.

Assmus, Alexi. "The Americanization of Molecular Physics." *Historical Studies in the Physical and Biological Sciences* 23 (1992): 1–34.

———."The Molecular Tradition in Early Quantum Theory." *Historical Studies in the Physical and Biological Sciences* 22 (1992): 209–31.

Bachelard, Gaston. "Le Problème Philosophique des Méthodes Scientifiques" (1951). In id., *L'Engagement Rationaliste*, Paris: Presses Universitaires de France 1972, 35–44.

Bachovchin, William W., and John D. Roberts. "Nitrogen-15 Nuclear Magnetic Resonance Spectroscopy. The State of Histidine in the Catalytic Triad of α-Lytic Protease. Implications for the Charge-Relay Mechanism of Peptide-Bond Cleavage by Serine Proteases." *Journal of the American Chemical Society* 100 (1978): 8041–47.

Baird, Davis. "Analytical Chemistry and the 'Big' Scientific Instrumentation Revolution." *Annals of Science* 50 (1993): 267–90.

———. "Encapsulating Knowledge: The Direct Reading Spectrometer." *Foundations of Chemistry* 2 (2000): 5–46.

Barrie, James Matthew. *What Every Woman Knows, and Other Plays*, vol. 12, *The Works of J. M. Barrie*. New York: AMS Press 1975. Original edition Charles Scribner's Sons 1930.

Bartky, W., and Arthur J. Dempster. "Paths of Charged Particles in Electric and Magnetic Fields." *Physical Review* 33 (1929): 1019–22.

Bartlett, M. F., R. Sklar, and W. I. Taylor. "Rauwolfia Alkaloids XXXIII. The Structure and Stereochemistry of Sarpagine." *Journal of the American Chemical Society* 82 (1960): 3790.

Bartlett, Paul D. *Nonclassical Ions. Reprints and Commentary*. New York: Benjamin 1965.

Barton, Derek H. R. "Some Reflections on the Present Status of Organic Chemistry." In *Science and Human Progress. Addresses at the Celebration of the 50th Anniversary of Mellon Institute*, edited by Mellon Institute of Industrial Research. Pittsburgh 1964, 85–100.

Beard, C., J. M. Wilson, Herbert Budzikiewicz, and Carl Djerassi. "Mass Spectrometry in Structural and Stereochemical Problems. XXXV." *Journal of the American Chemical Society* 86 (1964): 269–84.

Becker, Edwin D., Cherie Fisk, and C. L. Khetrapal. "The Development of NMR." In *Encyclopedia of Nuclear Magnetic Resonance*, vol. 1, *Historical Perspectives*, edited by D. M. Grant and R. K. Harris. Chichester: John Wiley & Sons 1996, 1–158.

Benfey, Theodor. "Teaching Chemistry Embedded in History: Reflections on C. K. Ingold's Influence as Historian and Educator." *Bulletin of the History of Chemistry* 19 (1996): 19–24.

Benfey, Theodor, and Peter J. T. Morris, eds. *Robert Burns Woodward. Architect and Artist in the World of Molecules.* Philadelphia: Chemical Heritage Foundation 2001.

Berchtold, Glenn A., and Louise H. Foley. "George Hermann Büchi, August 1, 1921–August 28, 1998." *Biographical Memoirs of the National Academy of Sciences* 78 (2000). Available from <http://stills.nap.edu/html/biomems/gbuchi.html>.

Berlowitz, Laurence, Richard A. Zdanis, John C. Crowley, and John C. Vaughn. "Instrumentation Needs of Research Universities." *Science* 211 (1981): 1013–18.

Bersohn, Richard, and Herbert S. Gutowsky. "Proton Magnetic Resonance in an Ammonium Chloride Single Crystal." *Journal of Chemical Physics* 22 (1954): 651–58.

Beynon, John H. "The Eternal Triangle: Research, Universities and Industry. John D Rose Memorial Lecture." *Chemistry and Industry* (1979): 175–82.

———. "High Resolution Mass Spectrometry of Organic Materials." In *Advances in Mass Spectrometry*, edited by J. D. Waldron. New York: Pergamon Press 1959, 328–54.

———. "Qualitative Analysis of Organic Compounds by Mass Spectrometry." *Nature* 174 (1954): 735–37.

Beynon, John H., and R. P. Morgan. "The Development of Mass Spectrometry. An Historical Account." *International Journal of Mass Spectrometry and Ion Processes* 27 (1978): 1–30.

Bhacca, Norman S., and Dudley H. Williams. *Applications of NMR Spectroscopy in Organic Chemistry. Illustrations from the Steroid Field.* San Francisco: Holden-Day 1964.

Bible, Roy H. *Guide to the NMR Empirical Method. A Workbook.* New York: Plenum Press 1967.

———. *Interpretation of NMR Spectra. An Empirical Approach.* New York: Plenum Press 1965.

Biemann, Klaus. "The Application of Mass Spectrometry in Amino Acid and Peptide Chemistry." *Chimia* 14 (1960): 393–401.

———. "The Determination of Carbon Skeleton of Sarpagine by Mass Spectrometry." *Tetrahedron Letters* 15 (1960): 9–14.

———. "The Application of Mass Spectrometry in Organic Chemistry. Determination of the Structure of Natural Products." *Angewandte Chemie, International Edition* 1 (1962): 98–111.

———. "Applications of Mass Spectrometry." In *Elucidation of Structures by Physical and Chemical Methods*, edited by K. W. Bentley. New York: Interscience Publisher 1963, 259–316.

———. "The Coming of Age of Mass Spectrometry in Peptide and Protein Chemistry." *Protein Science* 4 (1995): 1920–27.

———. "Four Decades of Structure Determination by Mass Spectrometry: From Alkaloids to Heparin." *Journal of the American Society for Mass Spectrometry* 13 (2002): 1254–72.

———. "Determination of the Structure of Alkaloids by Mass Spectrometry." In *Advances in Mass Spectrometry*, edited by R. M. Elliott. New York: Macmillan 1963, 408–15.

———. "High Resolution Mass Spectrometry." *Topics in Organic Mass Spectrometry* 8 (1970): 185 221.

———. "High Resolution Mass Spectrometry of Natural Products." *Journal of Pure and Applied Chemistry* 9 (1964): 95–118.

———. *Mass Spectrometry, Organic Chemical Applications*. New York: McGraw-Hill 1962.

———. "The Massachusetts Institute of Technology Mass Spectrometry School." *Journal of the American Society for Mass Spectrometry* 5 (1994): 332–38.

Biemann, Klaus, et al. "On the Structure of Lysopine, a New Amino Acid Isolated from Crown Gall Tissue." *Biochimica et Biophysica Acta* 40 (1960): 369–70.

Biemann, Klaus, Peter Bommer, and Dominic M. Desiderio. "Element-Mapping. A New Approach to the Interpretation of High Resolution Mass Spectra." *Tetrahedron Letters* 26 (1964): 1725–31.

Biemann, Klaus, George Büchi, and B. H. Walker. "Structure and Synthesis of Muscopyridine." *Journal of the American Chemical Society* 79 (1957): 5558–64.

Biemann, Klaus, C. Cone, and B. R. Webster. "Computer-Aided Interpretation of High-Resolution Mass Spectra. II. Amino Acid Sequence of Peptides." *Journal of the American Chemical Society* 88 (1966): 2597–98.

Biemann, Klaus, and Paul V. Fennessey. "Problems and Progress in High Resolution Mass Spectrometry." *Chimia* 21 (1967): 226–35.

Biemann, Klaus, and Margot Friedmann-Spiteller. "Application of Mass Spectrometry to Structure Problems. V. Iboga Alkaloids." *Journal of the American Chemical Society* 83 (1961): 4805–10.

Biemann, Klaus, Margot Friedmann-Spiteller, and Gerhard Spiteller. "Application of Mass Spectrometry to Structure Problems. X. Alkaloids of the Bark of *Aspidosperma Quebracho Blanco*." *Journal of the American Chemical Society* 85 (1963): 631–38.

Biemann, Klaus, Fritz Gapp, and Josef Seibl. "Application of Mass Spectrometry to Structure Problems. I. Amino Acid Sequence in Peptides." *Journal of the American Chemical Society* 81 (1959): 2274–75.

Biemann, Klaus, and Walter J. McMurray. "Computer-Aided Interpretation of High Resolution Mass Spectra." *Tetrahedron Letters* 11 (1965): 647–53.

Biemann, Klaus, Josef Seibl, and Fritz Gapp. "Mass Spectra of Organic Molecules. I. Ethyl Esters of Amino Acids." *Journal of the American Chemical Society* 83 (1961): 3795–804.

Biemann, Klaus, and W. Vetter. "Quantitative Amino Acid Analysis by Mass Spectrometry." *Biochemical and Biophysical Research Communications* 2 (1960): 93–96.

Birch, Arthur J. *To See the Obvious.* Washington DC: American Chemical Society 1995.

Bitter, Francis. *Magnets: The Education of a Physicist.* Garden City: Doubleday Anchor Books 1959.

Bloch, Felix. "Nuclear Induction." *Physical Review* 70 (1946): 460–74.

Bloembergen, Nicolaas. *Nuclear Magnetic Relaxation. A Reprint Volume.* New York: W.A. Benjamin 1961.

Bloembergen, N., Edward M. Purcell, and Robert V. Pound. "Relaxation Effects in Nuclear Magnetic Resonance Absorption." *Physical Review* 73 (1948): 679–712.

Bloom, Evelyn G., et al. "Mass Spectra of Octanes." *Journal of Research of the NBS* 41 (1948): 129–33.

Bodenhausen, Geoffrey, and Ray Freeman. "Correlation of Proton and Carbon-13 NMR Spectra by Heteronuclear Two-Dimensional Spectroscopy." *Journal of Magnetic Resonance* 28 (1977): 471–76.

Bommer, P., W. J. McMurray, and Klaus Biemann. "High Resolution Mass Spectra of Natural Products. Vinblastine and Derivatives." *Journal of the American Chemical Society* 86 (1964): 1439–40.

Bonacich, Edna. "A Theory of Middleman Minorities." *American Sociological Review* 38 (1973): 583–94.

Booth, Harold Simmons, and Roscoe Bozarth. "Fluorination of Phosphorus Trichloride." *Journal of the American Chemical Society* 61 (1939): 2927–34.

Boreham, G. R., F. R. Goss, and G. J. Minkoff. "The Structure of Feist's Acid." *Chemistry and Industry* (1955): 1354–55.

Borman, Stu. *A Brief History of Mass Spectrometry Instrumentation* [cited 20 November 1998]. Available from <http://masspec.scripps.edu/Hist-ms.htm>.

Bothner-By, Aksel A., and S. M. Castellano. "LAOCN 3." In *Computer Programs in Chemistry*, edited by D. F. DeTar. New York: Benjamin 1968, 10–53.

Bottini, Albert T., and John D. Roberts. "The Nitrogen Inversion Frequency in Cyclic Amines." *Journal of the American Chemical Society* 78 (1956): 5126.

———. "Nuclear Magnetic Resonance Spectra. Nitrogen Inversion Rates of N-Substituted Aziridines (Ethylenimines)." *Journal of the American Chemical Society* 80 (1958): 5203–08.

———."The Nuclear Magnetic Resonance Spectrum of Feist's Acid." *Journal of Organic Chemistry* 21 (1956): 1169–70.

Bowden, Mary Ellen. *Chemical Achievers. The Human Face of the Chemical Sciences.* Philadelphia: Chemical Heritage Foundation 1997.

Brewer, Keith A., and Vernon H. Dibeler. "Mass Spectrometric Analyses of Hydrocarbon Mixtures." *Journal of Research of the National Bureau of Standards* 35 (1945): 125–39.

Bridgman, Percy W. *The Logic of Modern Physics.* New York: Macmillan 1927.

Brock, William H. *The Fontana History of Chemistry.* Hammersmith: Fontana Press 1992.

Brown, Russell H., and Seymour Meyerson. "Cyclic Sulfides in a Petroleum Distillate." *Industrial and Engineering Chemistry* 44 (1952): 2620–23.

Brush, Stephen. "Dynamics of Theory Change in Chemistry: Part 2. Benzene and Molecular Orbitals, 1945–1980." *Studies in the History and Philosophy of Science* 30 (1999): 263–302.

Buchanan, Bruce G., Georgia L. Sutherland, and Edward A. Feigenbaum. "Heuristic Dendral. A Program for Generating Explanatory Hypotheses in Organic Chemistry." *Machine Intelligence* 4 (1969): 209–54.

———. "Rediscovering Some Problems of Artificial Intelligence in the Context of Organic Chemistry." *Machine Intelligence* 5 (1969): 253–80.

Büchi, George, and Klaus Biemann. "Conversion of Sclareol to Manool." *Croat. Chem. Acta* 29 (1957): 163.

Büchi, George, Klaus Biemann, B. Vittimberga, and Max Stoll. "Terpenes IV. The Acid-Induced Cyclization of Dihydro-Alpha-Ionone." *Journal of the American Chemical Society* 78 (1956): 2622–25.

Büchi, George, R. E. Erickson, and Nobel Wakabayashi. "Terpenes. XVI. Constitution of Patchouli Alcohol and Absolute Configuration of Cedrene." *Journal of the American Chemical Society* 83 (1961): 927–38.

Büchi, George, and W. MacLeod. "Synthesis of Patchouli Alcohol." *Journal of the American Chemical Society* 84 (1962): 3205–06.

Buchwald, Jed Z., ed. *Scientific Practice. Theories and Stories of Doing Physics.* Chicago: University of Chicago Press 1995.

Bud, Robert, and Susan E. Cozzens, eds. *Invisible Connections. Instruments, Institutions and Science.* Bellingham: SPIE Optical Engineering Press 1992.

Budzikiewicz, Herbert, and Carl Djerassi. "Mass Spectrometry in Structural and Stereochemical Problems. I. Steroid Ketones." *Journal of the American Chemical Society* 84 (1962): 1430–39.

Budzikiewicz, Herbert, Carl Djerassi, and Dudley H. Williams. *Interpretation of Mass Spectra of Organic Compounds.* 2nd ed. San Francisco: Holden-Day 1965.

———. *Mass Spectrometry of Organic Compounds.* San Francisco: Holden-Day 1967.

———. *Structure Elucidation of Natural Products by Mass Spectrometry.* 2 vols. San Francisco: Holden-Day 1964.

Bursey, Joan T., Maurice M. Bursey, and David G. I. Kingston. "Intramolecular Hydrogen Transfer in Mass Spectra. I. Rearrangements in Aliphatic Hydrocarbons and Aromatic Compounds." *Chemical Reviews* 73 (1973): 191–234.

Bursey, Maurice M. "F. W. McLafferty: An Appreciation." *Organic Mass Spectrometry* 23 (1988): 297–98.

———. "An Appreciation of Fred W. McLafferty." *Journal of the American Society for Mass Spectrometry* 6 (1995): 992.

Caldecourt, Victor J. "Heated Sample Inlet System for Mass Spectrometry." *Analytical Chemistry* 27 (1955): 1670.

———. "A Mass Indicator." *Review of Scientific Instruments* 21 (1950): 772.

California Institute of Technology. *Chemistry and Chemical Engineering at the California Institute of Technology, 1952–1953. A Report of the Academic Year and Other Activities of the Division of Chemistry and Chemical Engineering.* Pasadena: Caltech 1953.

Cameron, Frank. *Cottrell: Samaritan of Science.* Garden City: Doubleday 1952.

Carhart, Raymond E., et al. "Applications of Artificial Intelligence for Chemical Inference. XVII. An Approach to Computer-Assisted Elucidation of Molecular Structure." *Journal of the American Chemical Society* 97 (1975): 5755–62.

Carr, Herman Y. "Early Years of Free Precession Revisited." In *Encyclopedia of Nuclear Magnetic Resonance*, vol. 1, *Historical Perspectives*, edited by D. M. Grant and R. K. Harris. Chichester: John Wiley & Sons, 253–60.

Carrier, Martin. "New Experimentalism and the Changing Significance of Experiments: On the Short-Comings of an Equipment-Centered Guide to History." In *Experimental Essays—Versuche zum Experiment*, edited by M. Heidelberger and F. Steinle. Baden-Baden: Nomos Verlagsgesellschaft 1998.

Cartwright, Hugh M. *Applications of Artificial Intelligence in Chemistry*. Oxford: Oxford University Press 1993.

Chapman, D., and P. D. Magnus. *Introduction to Practical High Resolution Nuclear Magnetic Resonance Spectroscopy*. London: Academic Press 1966.

Cohen, H. Floris. *The Scientific Revolution. A Historiographical Inquiry*. Chicago: University of Chicago Press 1994.

Cohen, I. Bernard. *Revolution in Science*. Cambridge, Mass.: Belknap Press 1985.

Collins, Harry. *Artificial Experts. Social Knowledge and Intelligent Machines*. Cambridge, Mass.: MIT Press 1990.

———. *Changing Order. Replication and Induction in Scientific Practice*. London: SAGE 1985.

Colthup, Norman B. *The Origins of the IR Spectra-Structure Correlations Chart* [cited 21 May 2002]. Available from <http://www.s-a-s.org/epstein/colthup/index.html>.

———. "Spectra-Structure Correlations in the Infra-Red Region." *Journal of the Optical Society of America* 40 (1950): 397–400.

Conant, James B. "Equilibria and Rates of Some Organic Reactions." *Industrial and Engineering Chemistry* 24 (1932): 466–72.

Coulson, Charles A. "Representation of Simple Molecules by Molecular Orbitals." *Quarterly Reviews* 1 (1947): 144–78.

Coulter, Charles C. "Research Instrument Sharing." *Science* 201 (1978): 415–20.

Crabbé, Pierre. "A Dynamic Philosophy in Industrial Research." *Chemistry and Industry* (1967): 21–24.

Craig, R. D., and G. A. Errock. "Design and Performance of a Double-Focusing Mass Spectrometer for Analytical Work." In *Advances in Mass Spectrometry*, edited by J. D. Waldron. New York: Pergamon Press 1959, 66–85.

Crevier, Daniel. *AI. The Tumultuous History of the Search for Artificial Intelligence*. New York: Basic Books 1993.

Crombie, Alistair C. *Styles of Scientific Thinking in the European Tradition. The History of Argument and Explanation Especially in the Mathematical and Biomedical Sciences and Arts*. 3 vols. London: Duckworth 1994.

Dainton, Frederick. "George Bogdan Kistiakowsky." *Biographical Memoirs of Fellows of the Royal Society* 31 (1985): 377–408.

Davenport, Derek. "On the Comparative Unimportance of the Invective Effect." *Chemtech* 17 (1987): 526–31.

Dervan, Peter B. "John D. Roberts." *Aldrichimica Acta* 21 (1988): 71–77.

Dewar, Michael J. S. *The Electronic Theory of Organic Chemistry*. Oxford: Clarendon Press 1949.

Dicke, Robert. "The Measurement of Thermal Radiation at Microwave Frequencies." *Review of Scientific Instruments* 17 (1946): 268–75.

Dickinson, William C. "Dependence of the F^{19} Nuclear Resonance Position on Chemical Compound." *Physical Review* 77 (1950): 736–37.

———. "Factors Influencing the Positions of Nuclear Magnetic Resonances." *Physical Review* 78 (1950): 339.

———. "The Time Average Magnetic Field at the Nucleus in Nuclear Magnetic Resonance Experiments." *Physical Review* 81 (1951): 717–31.

Division of Research Resources. *Biotechnology Resources. A Research Resources Directory*, DHEW Publication No. (NIH) 77-1430. Bethesda: U.S. Department of Health, Education, and Welfare 1977.

Djerassi, Carl. "Foreword." In *Applications of Artificial Intelligence for Organic Chemistry. The Dendral Project*, edited by R. K. Lindsay, B. G. Buchanan, E. A. Feigenbaum and J. Lederberg. New York: McGraw-Hill 1980, ix–x.

———. "Isotope Labelling and Mass Spectrometry of Natural Products." *Pure and Applied Chemistry* 9 (1964): 159–78.

———. "Mass Spectrometric Investigations in the Steroid, Terpenoid and Alkaloid Fields." *Pure and Applied Chemistry* 6 (1963): 575–99.

———. "Mass Spectrometry. Organic Chemical Applications." *Journal of the American Chemical Society* 85 (1963): 2190–92.

———. "Natural Products Chemistry 1950 to 1980: A Personal View." *Pure and Applied Chemistry* 41 (1975): 113–44.

———. *Optical Rotatory Dispersion. Applications to Organic Chemistry*. New York: McGraw-Hill 1960.

———. *The Pill, Pygmy Chimps, and Degas' Horse. The Autobiography of Carl Djerassi*. New York: Basic Books 1992.

———. *The Politics of Contraception*. 2 vols. Stanford: Stanford Alumni Association 1979.

———. "Recent Advances in the Mass Spectrometry of Steroids." *Pure and Applied Chemistry* 50 (1978): 171–84.

———. "A Steroid Autobiography." *Steroids* 43 (1984): 351–61.

———. "Steroid Research at Syntex: 'The Pill' and Cortisone." *Steroids* 57 (1992): 631–41.

———. *Steroids Made It Possible*. Washington, DC: American Chemical Society 1990.

Djerassi, Carl, et al. "The Dendral Project: Computational Aids to Natural Products Structure Elucidation." *Pure and Applied Chemistry* 54 (1982): 2425–42.

Djerassi, Carl, et al. "Mass Spectrometry in Structural and Stereochemical Problems. XIII. Echitamidine." *Tetrahedron Letters* (1962): 653–59.

Djerassi, Carl, et al. "The Natural Constituents of the Cactus Lophocereus Schottii." *Journal of the American Chemical Society* 80 (1958): 6284–92.

Djerassi, Carl, B. Gilbert, J. N. Shoolery, L. F. Johnson, and K. Biemann. "Alkaloid Studies XXVI. The Constitution of Pyrifolidine." *Experientia* 17 (1961): 162–66.

Djerassi, Carl, J. P. Kutney, M. Shamma, J. N. Shoolery, and L. F. Johnson. "Alkaloid Studies. XXVII. The Structure of Skytanthine." *Chemistry and Industry* (1961): 210–11.

Djerassi, Carl, J. S. Mills, and Riccardo Villotti. "The Structure of the Cactus Sterol Lophenol. A Link in Sterol Biogenesis." *Journal of the American Chemical Society* 80 (1958): 1005–06.

Djerassi, Carl, G. von Mutzenbecher, J. Fajkos, Dudley H. Williams, and Herbert Budzikiewicz. "Mass Spectrometry in Structural and Stereochemical Problems. LXV." *Journal of the American Chemical Society* 87 (1965): 817–26.

Djerassi, Carl, Dennis H. Smith, and Tomas H. Varkony. "A Novel Role of Computers in the Natural Products Field." *Naturwissenschaften* 66 (1979): 9–21.

Doering, William von Eggers, and Lawrence H. Knox. "The Cycloheptatrienylium (Tropylium) Ion." *Journal of the American Chemical Society* 76 (1954): 3203–06.

Drysdale, John J., and William D. Phillips. "Restricted Rotation in Substituted Ethanes as Evidenced by Nuclear Magnetic Resonance." *Journal of the American Chemical Society* 79 (1957): 319–22.

Dunitz, Jack D. "Linus Carl Pauling." *Biographical Memoirs of Fellows of the Royal Society* 42 (1996): 317–38.

Eliel, Ernest L. *Stereochemistry of Carbon Compounds*. New York: McGraw-Hill 1962.

Emsley, James W., and James Feeney. "Milestones in the First Fifty Years of NMR." *Progress in Nuclear Magnetic Spectroscopy* 28 (1995): 1–9.

Emsley, James W., James Feeney, and Leslie H. Sutcliffe. *High Resolution Nuclear Magnetic Resonance Spectroscopy*. 2 vols. Oxford: Pergamon Press 1965.

Ernst, Richard R. "Arts and Sciences. A Personal Perspective of Tibetan Painting." *Chimia* 55 (2001): 900–14.

———. "ENC, a Motor of Progress in NMR." *Concepts in Magnetic Resonance* 6 (1994): 201–29.

———. "Measurement and Control of Magnetic Field Homogeneity." *Review of Scientific Instruments* 39 (1968): 998–1012.

———. "Nuclear Magnetic Double Resonance with an Incoherent Radio-Frequency Field." *Journal of Chemical Physics* 45 (1966): 3845–61.

———. "Nuclear Magnetic Resonance Fourier Transform Spectroscopy." In *Nobel Lectures Chemistry 1991–1995*, edited by B. G. Malmström. Singapore: World Scientific 1997, 12–57.

———. "Richard R. Ernst." In *Nobel Lectures Chemistry 1991–1995*, edited by B. G. Malmström. Singapore: World Scientific 1997, 7–11.

———. "Sensitivity Enhancement in Magnetic Resonance. I. Analysis of the Method of Time Averaging." *Review of Scientific Instruments* 36 (1965): 1689–95.

———. "Sensitivity Enhancement in Magnetic Resonance. II. Investigation of Intermediate Passage Conditions." *Review of Scientific Instruments* 36 (1965): 1696–706.

———. "The Success Story of Fourier Transformation in NMR." In *Encyclopedia of Nuclear Magnetic Resonance*, vol. 1, *Historical Perspectives*, edited by D. M. Grant and R. K. Harris. Chichester: John Wiley & Sons 1996, 293–306.

———. "Two-Dimensional Spectroscopy." *Chimia* 29 (1973): 179–83.

———. "Without Computers—No Modern NMR." In *Computational Aspects of the Study of Biological Macromolecules by Nuclear Magnetic Resonance Spectroscopy*, edited by J. C. Hoch et al. New York: Plenum Press 1991, 1–25.

Ernst, Richard R., and Weston A. Anderson. "Application of Fourier Transform Spectroscopy to Magnetic Resonance." *Review of Scientific Instruments* 37 (1966): 93–102.

Ernst, Richard R., Geoffrey Bodenhausen, and Alexander Wokaun. *Principles of Nuclear Magnetic Resonance in One and Two Dimensions.* Oxford: Clarendon Press 1987.

Ernst, Richard R., and Hans Primas. "Gegenwärtiger Stand und Entwicklungstendenzen in der Instrumentierung hochauflösender Kernresonanz-Spektrometer." *Berichte der Bunsengesellschaft* 67 (1963): 261–67.

Ettlinger, Martin G. "Structure of Feist's Methylcyclopropenedicarboxylic Acid." *Journal of the American Chemical Society* 74 (1952): 5805–06.

Ettlinger, Martin G., and Flynt Kennedy. "The Structure of Feist's Acid and Esters." *Chemistry and Industry* (1956): 166–67.

Ettre, Leslie S. "Gas Chromatography." In *A History of Analytical Chemistry*, edited by H. A. Laitinen and G. W. Ewing. Washington DC: American Chemical Society 1977, 296–306.

Eyring, Henry, John Walter, and George E. Kimball. *Quantum Chemistry.* New York: Wiley 1944.

Feigenbaum, Edward A., Bruce G. Buchanan, and Joshua Lederberg. "On Generality and Problem Solving: A Case Study Using the Dendral Program." *Machine Intelligence* 6 (1971): 165–90.

Feigenbaum, Edward A., and Pamela McCorduck. *The Fifth Generation. Artificial Intelligence and Japan's Computer Challenge to the World.* London: Pan Books 1984.

Feist, Franz. "Ueber den Abbau des Cumalinringes." *Berichte der Deutschen Chemischen Gesellschaft* 26 (1893): 747–64.

———. "Über 3-Methyl-Cyclo-Propen-1,2-Dicarbonsäure." *Annalen der Chemie* 436 (1924): 125–53.

Ferguson, Raymond C. "William D. Phillips and Nuclear Magnetic Resonance Spectroscopy at Du Pont." In *Encyclopedia of Nuclear Magnetic Resonance*, vol. 1, *Historical Perspectives*, edited by D. M. Grant and R. K. Harris. Chichester: John Wiley & Sons 1996, 309–13.

Finnigan, Robert E. "Quadrupole Mass Spectrometers." *Analytical Chemistry* 66 (1994): 969A–75A.

Flett, M. St. C. *Physical Aids to the Organic Chemist.* Amsterdam: Elsevier 1962.

Forman, Paul. "Behind Quantum Electronics: National Security as Basis for Physical Research in the United States, 1940–1960." *Historical Studies in the Physical Sciences* 18 (1) (1987): 149–229.

———. "Lock-in Detection/Amplifier." In *Instruments of Science. An Historical Encyclopedia*, edited by R. Bud and D. J. Warner. New York: Garland Publishing 1998, 359–61.

———. "Molecular Beam Measurements of Nuclear Moments before Magnetic Resonance, Part I." *Annals of Science* 55 (1998): 111–60.

———. "Swords to Ploughshares: Breaking New Ground with Radar Hardware and Technique in Physical Research after World War II." *Reviews of Modern Physics* 67 (1995): 397–455.

Forsén, Sture. "The Nobel Prize in Chemistry." In *Nobel Lectures Chemistry 1991–1995*, edited by B. G. Malmström. Singapore: World Scientific 1997, 3–5.

Foster, Harlan. "Application of Nuclear Magnetic Resonance Spectroscopy to Organic Analysis." In *Organic Analysis*, vol. 4, edited by J. Mitchell. New York: Interscience 1960, 229–91.

Freeman, Ray. "The Fourier Transform Revolution in NMR Spectroscopy." *Analytical Chemistry* 65 (1993): 743A–53A.

———. *A Handbook of Nuclear Magnetic Resonance*. 2nd ed. Harlow: Longman 1997.

Freeman, Ray, and Weston A. Anderson. "Use of Weak Perturbing Radio-Frequency Field in Nuclear Magnetic Double Resonance." *Journal of Chemical Physics* 37 (1962): 2053–73.

Freeman, Ray, and Gareth A. Morris. "Two-Dimensional Fourier Transformation in NMR." *Bulletin of Magnetic Resonance* 1 (1979): 5–26.

Friedman, Lewis, and Franklin A. Long. "Mass Spectra of Six Lactones." *Journal of the American Chemical Society* 75 (1953): 2832–36.

Friedman, Lewis, and John Turkevich. "The Mass Spectra of Some Deuterated Isopropyl Alcohols." *Journal of the American Chemical Society* 74 (1952): 1666–68.

Galison, Peter. *Image and Logic. A Material Culture of Microphysics*. Chicago: University of Chicago Press 1997.

Gardner, J. H., and Edward M. Purcell. "A Precise Determination of the Proton Magnetic Moment in Bohr Magnetons." *Physical Review* 76 (1949): 1262–63.

Gaudillière, Jean-Paul, and Ilana Löwy, eds. *The Invisible Industrialist. Manufactures and the Production of Scientific Knowledge*. Houndmills: Macmillan 1998.

Gavroglu, Kostas, and Ana Simões. "The Americans, the Germans, and the Beginnings of Quantum Chemistry: The Confluence of Diverging Traditions." *Historical Studies in the Physical Sciences* 25 (1994): 47–110.

———. "One Face or Many? The Role of Textbooks in Building the New Discipline of Quantum Chemistry." In *Communicating Chemistry. Textbooks and Their Audiences*, edited by A. Lundgren and B. Bensaude-Vincent. Canton, Mass.: Science History Publications 2000, 415–49.

Gershinowitz, Harold. "The First Infrared Spectrometer." *Journal of Physical Chemistry* 83 (1979): 1363–65.

Gerstein, Mark. "Purcell's Role in the Discovery of Nuclear Magnetic Resonance: Contingency Versus Inevitability." *American Journal of Physics* 62 (1994): 596–601.

Gilpin, Jo Ann. "Mass Spectra Rearrangements of 2-Phenyl Alcohols." *Journal of Chemical Physics* 28 (1958): 521–22.

Gilpin, Jo Ann, and Fred W. McLafferty. "Mass Spectrometric Analysis. Aliphatic Aldehydes." *Analytical Chemistry* 29 (1957): 990–94.

Gohlke, Roland S., and Fred W. McLafferty. "Early Gas Chromatography/Mass Spectrometry." *Journal of the American Society for Mass Spectrometry* 4 (1993): 367–71.

Golinski, Jan. *Making Natural Knowledge. Constructivism and the History of Science*. Cambridge: Cambridge University Press 1998.

———. *Science as Public Culture: Chemistry and Enlightenment in Britain, 1760–1820*. Cambridge: Cambridge University Press 1992.

Gooding, David. "Putting Agency Back into Experiment." In *Science as Practice and Culture*, edited by A. Pickering. Chicago: University of Chicago Press 1992, 65–112.

Gortler, L. "The Physical Organic Chemistry Community in the United States, 1925–1950." *Journal of Chemical Education* 62 (1985): 753–57.

Goss, Frank Robert, Christopher Kelk Ingold, and Jocelyn Field Thorpe. "The Chemistry of the Glutaconic Acids, Part XIV. Three-Carbon Tautomerism in the Cyclo-Propane Series." *Journal of the Chemical Society* 123 (1923): 327–61.

Grayson, Michael A., ed. *Measuring Mass. From Positive Rays to Proteins.* Philadelphia: CHF 2002.

Gust, Devens, Richard B. Moon, and John D. Roberts. "Applications of Natural-Abundance Nitrogen-15 Nuclear Magnetic Resonance to Large Biochemically Important Molecules." *Proceedings of the National Academy of Sciences, USA* 72 (1975): 4696–5000.

Gutowsky, Herbert S. "Chemical Aspects of Nuclear Magnetic Resonance." *Journal of Magnetic Resonance* 17 (1975): 281–94.

———. "The Coupling of Chemical and Nuclear Magnetic Phenomena." In *Encyclopedia of Nuclear Magnetic Resonance*, vol. 1, *Historical Perspectives*, edited by D. M. Grant and R. K. Harris. Chichester: John Wiley & Sons 1996, 360–68.

———. "Effects of Ring Size on Electron Distribution in Saturated Heterocyclic Compounds." *Journal of the American Chemical Society* 76 (1954): 4242.

———. "General Discussion." *Discussions of the Faraday Society* 19 (1955): 246–47.

———. "NMR Boon and Boom." *Chemical and Engineering News* 37, 25 May (1959): 96.

———."Nuclear Magnetic Resonance." *Annual Review of Physical Chemistry* 5 (1954): 333–56.

Gutowsky, Herbert S., and Charles J. Hoffman. "Chemical Shifts in the Magnetic Resonance of F^{19}." *Physical Review* 80 (1950): 110–11.

———. "Nuclear Magnetic Shielding in Fluorine and Hydrogen Compounds." *Journal of Chemical Physics* 19 (1951): 1259–67.

Gutowsky, Herbert S., and Charles H. Holm. "Rate Processes and Nuclear Magnetic Resonance Spectra II. Hindered Internal Rotation of Amides." *Journal of Chemical Physics* 25 (1956): 1228–34.

Gutowsky, Herbert S., Martin Karplus, and David M. Grant. "Angular Dependence of Electron-Coupled Interactions in CH_2 Groups." *Journal of Chemical Physics* 31 (1959): 1278–89.

Gutowsky, Herbert S., George B. Kistiakowsky, and George E. Pake. "Structural Investigations by Means of Nuclear Magnetism, I. Rigid Crystal Lattices." *Journal of Chemical Physics* 17 (1949): 972–81.

Gutowsky, Herbert S., and David W. McCall. "Nuclear Magnetic Resonance Fine Structure in Liquids." *Physical Review* 82 (1951): 748–49.

———. "Electron Distribution in Molecules IV. Phosphorus Magnetic Resonance Shifts." *Journal of Chemical Physics* 22 (1954): 162–64.

Gutowsky, Herbert S., David W. McCall, Bruce R. McGarvey, and Leon H. Meyer. "Electron Distribution in Molecules I. F^{19} Nuclear Magnetic Shielding and Substituent Effects in Some Benzene Derivatives." *Journal of the American Chemical Society* 74 (1952): 4809–17.

———. "A Nuclear Magnetic Parameter Related to Electron Distribution in Molecules." *Journal of Chemical Physics* 19 (1951): 1328–29.

Gutowsky, Herbert S., David W. McCall, and Charles P. Slichter. "Coupling among Nuclear Magnetic Dipoles in Molecules." *Physical Review* 84 (1951): 589–90.

———. "Nuclear Magnetic Resonance Multiplets in Liquids." *Journal of Chemical Physics* 21 (1953): 279–92.

Gutowsky, Herbert S., and Robert E. McClure. "Magnetic Shielding of the Proton Resonance in H_2, H_2O, and Mineral Oil." *Physical Review* 81 (1951): 276–77.

Gutowsky, Herbert S., and Bruce R. McGarvey. "Nuclear Magnetic Resonance in Metals, I. Broadening of Absorption Lines by Spin Lattice Interactions." *Journal of Chemical Physics* 20 (1952): 1472–77.

———. "Nuclear Magnetic Resonance in Metals, II. Temperature Dependence of the Resonance Shifts." *Journal of Chemical Physics* 21 (1953): 2114–19.

Gutowsky, Herbert S., and Leon H. Meyer. "The Proton Magnetic Resonance in Natural Rubber." *Journal of Chemical Physics* 21 (1953): 2122–26.

Gutowsky, Herbert S., Leon H. Meyer, and Robert E. McClure. "Apparatus for Nuclear Magnetic Resonance." *The Review of Scientific Instruments* 24 (1953): 644–52.

Gutowsky, Herbert S., and George E. Pake. "Structural Investigations by Means of Nuclear Magnetism, II. Hindered Rotation in Solids." *Journal of Chemical Physics* 18 (1950): 162–70.

Gutowsky, Herbert S., George E. Pake, and Richard Bersohn. "Structural Investigations by Means of Nuclear Magnetism, III. Ammonium Halides." *Journal of Chemical Physics* 22 (1954): 643–50.

Gutowsky, Herbert S., and Apollo Saika. "Dissociation, Chemical Exchange, and the Proton Magnetic Resonances in Some Aqueous Electrolytes." *Journal of Chemical Physics* 21 (1953): 1688–94.

Hacking, Ian. *Representing and Intervening. Introductory Topics in the Philosophy of Natural Science*. Cambridge: Cambridge University Press 1983.

———. "The Self-vindication of the Laboratory Sciences." In *Science as Practice and Culture*, edited by A. Pickering. Chicago: University of Chicago Press 1992, 29–64.

———. "Style for Historians and Philosophers." *Studies in the History and Philosophy of Science* 23 (1992): 1–20.

Hahn, Erwin L. "Pulsed NMR—A Personal History." In *Encyclopedia of Nuclear Magnetic Resonance*, vol. 1, *Historical Perspectives*, edited by D. M. Grant and R. K. Harris. Chichester: John Wiley & Sons 1996, 373–78.

———. "Spin Echoes." *Physical Review* 80 (1950): 580–94.

Hahn, Erwin L., and D. E. Maxwell. "Chemical Shift and Field Independent Frequency Modulation of the Spin Echo Envelope." *Physical Review* 84 (1951): 1246–47.

Hammett, Louis P. "The Effect of Structure Upon the Reactions of Organic Compounds. Benzene Derivatives." *Journal of the American Chemical Society* 59 (1937): 96–103.

———. "Physical Organic Chemistry in Retrospect." *Journal of Chemical Education* 43 (1966): 464–69.

———. *Physical Organic Chemistry. Reaction Rates, Equilibria, and Mechanisms*. New York: McGraw-Hill 1940.

Hammond, George S. "Foreword." In: John D. Roberts, *Collected Works*. New York: Benjamin 1970, v–vii.

Happ, Glenn P., and D. W. Stewart. "Rearrangement Peaks in the Mass Spectra of Certain Aliphatic Acids." *Journal of the American Chemical Society* 74 (1952): 4404–08.

Hargittai, István. "Interview with John D. Roberts." *The Chemical Intelligencer* 4 (1998): 29–39.

Harmon, Paul, and David King. *Expert Systems. Artificial Intelligence in Business.* New York: Wiley 1985.

Heidelberger, Michael, and Friedrich Steinle, eds. *Experimental Essays—Versuche zum Experiment.* Baden-Baden: Nomos Verlagsgesellschaft 1998.

Hein, Claudia. "Über die Anfänge der Kernspinresonanzforschung." Staatsexamensarbeit, Universität Regensburg 2000.

Heller, Stephen R., John M. McGuire, and William L. Budde. "Trace Organics by GC/MS." *Environmental Science & Technology* 9 (1975): 210–13.

Henahan, John F. "Klaus Biemann. A Renaissance Man in Mass Spectrometry. The Chemical Innovators, 5." *Chemical and Engineering News* 1970: 50–54.

Hentschel, Klaus. "Historiographische Anmerkungen zum Verhältnis von Experiment, Instrumentation und Theorie." In *Instrument-Experiment. Historische Studien,* edited by C. Meinel. Berlin: GNT-Verlag 2000, 13–51.

———. *Mapping the Spectrum. Techniques of Visual Representation in Research and Teaching.* Oxford: Oxford University Press 2002.

———. *Zum Zusammenspiel von Instrument, Experiment und Theorie. Rotverschiebung im Sonnenspektrum und verwandte spektrale Verschiebungseffekte von 1880 bis 1960.* 2 vols. Hamburg: Kovac 1998.

Hertz, H. S., D. A. Evans, and Klaus Biemann. "A User-Oriented Computer-Searchable Library of Mass Spectrometry Literature References." *Organic Mass Spectrometry* 4 (1970): 453–60.

Herzberg, Gerhard. "Molecular Spectroscopy: A Personal History." *Annual Review of Physical Chemistry* 36 (1985): 1–30.

———. *Zweiatomige Moleküle,* vol. I, *Molekülspektren und Molekülstruktur.* Dresden: Steinkopff 1939.

Hilne, S. R., and G. W. A. Milne. *EPA/NIH Mass Spectral Data Base.* Washington DC: U.S. GPO 1978.

Hine, Jack. *Physical Organic Chemistry.* 2nd ed. New York: McGraw-Hill 1962.

Hintenberger, Heinrich. "Anwendung der Massenspektroskopie in der analytischen Chemie." *Mikrochimica Acta* (1956): 71–90.

Hippel, Eric von. *The Sources of Innovation.* New York: Oxford University Press 1995. Original edition 1988.

Hipple, John A. "Gas Analysis with the Mass Spectrometer." *Journal of Applied Physics* 13 (1942): 551–60.

———. "Peak Contours and Half Life of Metastable Ions Appearing in Mass Spectra." *Physical Review* 71 (1947): 594–99.

Hipple, John A., and Edward U. Condon. "Detection of Metastable Ions with the Mass Spectrometer." *Physical Review* 68 (1945): 54–55.

Hipple, John A., R. E. Fox, and Edward U. Condon. "Metastable Ions Formed by Electron Impact in Hydrocarbon Gases." *Physical Review* 69 (1946): 347–56.

Hites, R. A., and Klaus Biemann. "Computer Evaluation of Continuously Scanned Mass Spectra of Gas Chromatographic Effluents." *Analytical Chemistry* 42 (1970): 855–60.

———. "Mass Spectrometer-Computer System Particularly Suited for Gas Chromatography of Complex Mixtures." *Analytical Chemistry* 40 (1968): 1217–21.

Holmes, Frederic L. "Do We Understand Historically How Experimental Knowledge Is Acquired?" *History of Science* 30 (1992): 119–36.

Honig, Richard E. "On the Isomerization of Hydrocarbons by Electron Impact." *Physical Review* 75A (1949): 1319–20.

Hutchinson, Eric. *The Department of Chemistry, Stanford University, 1891–1976. A Brief Account of the First Eighty-Five Years*. Stanford: Stanford University, Department of Chemistry 1977.

Ingram, D. J. E. *Spectroscopy at Radio and Microwave Frequencies*. London: Butterworth 1967. Original edition 1955.

Isenhour, Thomas L., and Peter C. Jurs. "Chemical Applications of Machine Intelligence." *Analytical Chemistry* 43 (1971): 20A–21A, 23A–26A, 29A, 31A, 33A–35A.

Jackman, Lloyd M. *Applications of Nuclear Magnetic Resonance Spectroscopy in Organic Chemistry*. London: Pergamon Press 1959.

———. "Magnetic Resonance Spectroscopy." In *Physical Methods in Organic Chemistry*, edited by J. C. P. Schwarz. San Francisco: Holden-Day 1964, 168–209.

Jackman, Lloyd M., and Sever Sternhell. *Applications of Nuclear Magnetic Resonance Spectroscopy in Organic Chemistry*. 2nd ed. Oxford: Pergamon Press 1969.

Jaffé, Hans H. "Theoretical Considerations Concerning Hammett's Equation. II. Correlation of Sigma-Values for Toluene and Naphthalene." *Journal of Chemical Physics* 20 (1952): 778–80.

Jeener, Jean. "Reminiscences About the Early Days of 2D NMR." In *Encyclopedia of Nuclear Magnetic Resonance*, vol. 1, *Historical Perspectives*, edited by D. M. Grant and R. K. Harris. Chichester: John Wiley & Sons 1996, 409–10.

Joerges, Bernward, and Terry Shinn, "A Fresh Look at Instrumentation: An Introduction." In *Instrumentation between Science, State and Industry*, edited by B. Joerges and T. Shinn. Dordrecht: Kluwer Academic Publishers 2001, 1–13.

Joerges, Bernward, and Terry Shinn, eds. *Instrumentation between Science, State and Industry*. Dordrecht: Kluwer Academic Publishers 2001.

Johnson, William S. *A Fifty-Year Love Affair with Organic Chemistry*. Washington DC: American Chemical Society 1997.

Johnston, Herrick L. "A Symposium on Molecular Structure: Introduction to the Symposium." *Journal of Physical Chemistry* 41 (1937): 1–4.

Johnston, Sean. "In Search of Space: Fourier Spectroscopy 1950–1970." In *Instrumentation between Science, State and Industry*, edited by B. Joerges and T. Shinn. Dordrecht: Kluwer 2001, 121–41.

Jonas, Jiri, and Herbert S. Gutowsky. "NMR—An Evergreen." *Annual Review of Physical Chemistry* 31 (1980): 1–27.

Jordan, E. B., and Louis B. Young. "A Short History of Isotopes and the Measurement of Their Abundances." *Journal of Applied Physics* 13 (1942): 526–38.

Jurs, Peter C., Bruce R. Kowalski, and Thomas L. Isenhour. "Computerized Learning Machines Applied to Chemical Problems. Molecular Formula Determination from Low Resolution Mass Spectrometry." *Analytical Chemistry* 41 (1969): 21–27.

Jurs, Peter C., Bruce R. Kowalski, Thomas L. Isenhour, and Charles N. Reilly. "Computerized Learning Machines Applied to Chemical Problems. Molecular Structure Parameters from Low Resolution Mass Spectrometry." *Analytical Chemistry* 42 (1970): 1387–94.

Kanamori, Keiko, and John D. Roberts. "^{15}N NMR Studies of Biological Systems." *Accounts of Chemical Research* 16 (1983): 35–41.

Karplus, Martin, D. H. Anderson, T. C. Farrar, and H. S. Gutowsky. "Valence-Bond Interpretation of Electron Coupled Proton-Proton Magnetic Interactions Measured Via Deuterium Substitution." *Journal of Chemical Physics* 27 (1958): 597–98.

Kende, Andrew S. "Nuclear Magnetic Resonance Spectrum of Feist's Acid." *Chemistry and Industry* (1956): 437, 544.

King, A. Bruce, and F. A. Long. "Mass Spectra of Some Simple Esters and Their Interpretation by Quasi-Equilibrium Theory." *Journal of Chemical Physics* 29 (1958): 374–82.

King, W. H., and William Priestley. "Spectrometric Analysis Employing Punch Card Calculators." *Analytical Chemistry* 23 (1951): 1418–21.

Kingston, David G. I., Joan T. Bursey, and Maurice M. Bursey. "Intramolecular Hydrogen Transfer in Mass Spectra. II. The McLafferty Rearrangement and Related Reactions." *Chemical Reviews* 74 (1974): 215–42.

Kinney, I. W., and G. L. Cook. "Identification of Thiophene and Benzene Homologs." *Analytical Chemistry* 24 (1952): 1391–96.

Kistiakowsky, George B. "Edgar Bright Wilson, Jr. Theodore William Richards Professor of Chemistry." *Journal of Physical Chemistry* 83 (1979): 5A–7A.

Knight, David. "Then . . . And Now." In *From Classical to Modern Chemistry. The Instrumental Revolution*, edited by P. J. T. Morris. Cambridge: Royal Society of Chemistry 2002, 87–94.

Knight, Walter D. "Nuclear Magnetic Resonance Shifts in Metals." *Physical Review* 76 (1949): 1259–60.

———. "The Knight Shift." In *Encyclopedia of Nuclear Magnetic Resonance*, vol. 1, *Historical Perspectives*, edited by D. M. Grant and R. K. Harris. Chichester: John Wiley & Sons 1996, 431–34.

Kohler, Robert E. *Lords of the Fly. Drosophila Genetics and the Experimental Life*. Chicago: University of Chicago Press 1994.

———. *Partners in Science. Foundations and Natural Scientists 1900–1945*. Chicago: University of Chicago Press 1991.

———. "Systems of Production: Drosophila, Neurospora, and Biochemical Genetics." *Historical Studies in the Physical and Biological Sciences* 22 (1991/1992): 87–130.

Kosover, Edward M. *Molecular Biochemistry*. New York: McGraw-Hill 1962.

Kuhn, Thomas. *The Structure of Scientific Revolutions*. Chicago: University of Chicago Press 1970. Original edition 1962.

Kumar, Anil, Dieter Welti, and Richard R. Ernst. "NMR Fourier Zeugmatography." *Journal of Magnetic Resonance* 18 (1975): 69–83.

Kwok, Kain-Sze, Rengachari Venkataraghavan, and Fred W. McLafferty. “Computer-Aided Interpretation of Mass Spectra. III. A Self-Training Interpretive and Retrieval System.” *Journal of the American Chemical Society* 95 (1973): 4185–94.

Laitinen, Herbert A., and Galen W. Ewing, eds. *A History of Analytical Chemistry.* Washington, DC: Division of Analytical Chemistry, American Chemical Society 1977.

Langer, Alois J. “Rearrangement Peaks Observed in Some Mass Spectra.” *Journal of Physical and Colloid Chemistry* 54 (1950): 618–29.

Lassman, Thomas C. “Government Science in Postwar America. Henry A. Wallace, Edward U. Condon, and the Transformation of the National Bureau of Standards, 1945–1951.” *Isis* 96 (2005): 25–51.

Laszlo, Pierre. “Eine Geschichte des Diborans.” *Angewandte Chemie* 112 (2000): 2151–52.

———. “Tools, Instruments and Concepts: The Influence of the Second Chemical Revolution.” In *From Classical to Modern Chemistry*, edited by P. J. T. Morris. Cambridge: Royal Society of Chemistry 2002, 171–87.

Latour, Bruno. *Science in Action. How to Follow Scientists and Engineers through Society.* Cambridge: Harvard University Press 1987.

Latour, Bruno, and Steve Woolgar. “The Cycle of Credibility.” In *Science in Context. Readings in the Sociology of Science*, edited by B. Barnes and D. Edge. Cambridge, Mass.: MIT Press 1982, 35–43.

Lécuyer, Christophe. “The Making of a Science Based Technology University: Karl Compton, James Killian, and the Reform of MIT, 1930–1957.” *Historical Studies in the Physical and Biological Sciences* 23 (1992): 153–80.

Lederberg, Joshua. *Computation of Molecular Formulas for Mass Spectrometry.* San Francisco: Holden-Day 1964.

Lederberg, Joshua. “Topological Mapping of Organic Molecules.” *Proceedings of the National Academy of Sciences (US)* 53 (1965): 134–39.

———. “Topology of Molecules.” In *The Mathematical Sciences. A Collection of Essays*, edited by the Committee on Support of Research on the Mathematical Sciences; National Research Council. Cambridge, Mass.: MIT Press 1969, 37–51.

Lederberg, Joshua, and Edward A. Feigenbaum. “Mechanization of Inductive Inference in Organic Chemistry.” In *Formal Representation of Human Judgment*, edited by B. Kleinmuntz. New York: Wiley 1968, 187–218.

Lee, Laurance Lem. “Nuclear Magnetic Resonance from Nuclear Physics to Organic Chemistry.” B.A. Thesis, Harvard University, Cambridge, Mass. 1992.

Lenoir, Timothy, and Christophe Lécuyer. “Instrument Makers and Discipline Builders: The Case of Nuclear Magnetic Resonance.” *Perspectives on Science* 3 (1995): 276–345.

Leslie, Stuart W., and Robert H. Kargon. “Selling Silicon Valley: Frederick Terman’s Model for Regional Advantage.” *Business History Review* 70 (1996): 435–72.

Levere, Trevor H., and Frederic L. Holmes. “Introduction: A Practical Science.” In *Instruments and Experimentation in the History of Chemistry*, edited by F. L. Holmes and T. H. Levere. Cambridge, Mass.: MIT Press 2000, vii–xvii.

Levine, Samuel G. “A Short History of the Chemical Shift.” *Journal of Chemical Education* 78 (2001): 133.

Linder, Ernest G. "Mass-Spectrographic Study of the Ionization and Dissociation by Electron Impact of Benzene and Carbon Disulfide." *Physical Review* 41 (1932): 149–53.

Lindsay, Robert K., Bruce G. Buchanan, Edward A. Feigenbaum, and Joshua Lederberg. *Applications of Artificial Intelligence for Organic Chemistry. The Dendral Project.* New York: McGraw-Hill 1980.

Lloyd, Douglas, T. C. Downie, and J. C. Speakman. "Structure of Feist's Acid." *Chemistry and Industry* (1954): 222–23, 492.

Long, Franklin A., and Lewis Friedman. "Mass Spectra and Appearance Potentials of Ketene Monomer and Dimer. Relation to Structure of Dimer." *Journal of the American Chemical Society* 75 (1953): 2837–40.

Lowe, Irving J. "My Life in the Rotating Frame." In *Encyclopedia of Nuclear Magnetic Resonance*, vol. 1, *Historical Perspectives*, edited by D. M. Grant and R. K. Harris. Chichester: John Wiley & Sons 1996, 457–61.

Lowe, Irving J., and Richard E. Norberg. "Free-Induction Decays in Solids." *Physical Review* 107 (1957): 46–61.

Lowen, Rebecca S. *Creating the Cold War University. The Transformation of Stanford.* Berkeley: University of California Press 1997.

Lucas, Howard J. *Organic Chemistry.* New York: American Book Company 1935.

MacDowell, Charles A., ed. *Mass Spectrometry.* New York: McGraw-Hill 1963.

Mack, J. E. "A Table of Nuclear Moments, January 1950." *Reviews of Modern Physics* 22 (1950): 64–76.

MacKenzie, Donald. "Negotiating Arithmetic, Constructing Proof." In *Knowing Machines: Essays on Technical Change*, edited by D. MacKenzie. Cambridge, Mass.: MIT Press 1996, 165–83.

Maienschein, Jane. "Epistemic Styles in German and American Embryology." *Science in Context* 4 (1991): 407–27.

Manning, P. P. "The Photolysis of Saturated Aldehydes and Ketones." *Journal of the American Chemical Society* 79 (1957): 5151–53.

March, Jerry. *Advanced Organic Chemistry. Reactions, Mechanisms, and Structure.* 4th ed. New York: Wiley-Interscience 1992.

Markley, John L., and Stanley J. Opella, eds. *Biological NMR Spectroscopy.* New York: Oxford University Press 1997.

Maudsley, A. A., and Richard R. Ernst. "Indirect Detection of Magnetic Resonance by Heteronuclear Two-Dimensional Spectroscopy." *Chemical Physics Letters* 50 (1977): 368–72.

Maushart, Marie-Ann. *"Um mich nicht zu vergessen." Hertha Sponer—Ein Frauenleben für die Physik im 20. Jahrhundert.* Bassum: GNT-Verlag 1997.

Mayo, Paul de, and Rowland Ivor Reed. "The Application of the Mass Spectrometer to Steroid and Terpenoid Chemistry." *Chemistry and Industry* (1956): 1481–82.

McConnell, Harden M., A. D. MacLean, and C. A. Reilly. "Analysis of Spin-Spin Multiplets in Nuclear Magnetic Resonance Spectra." *Journal of Chemical Physics* 23 (1955): 1152–59.

McFadden, W. H., and A. L. Wahrhaftig. "The Mass Spectra of Four Deuterated Butanes." *Journal of the American Chemical Society* 78 (1956): 1572–77.

McLafferty, Fred W. "Billionfold Data Increase from Mass Spectrometry Instrumentation." *Journal of the American Society for Mass Spectrometry* 8 (1997): 1–7.

———. *Interpretation of Mass Spectra. An Introduction.* New York: Benjamin 1967.

———. *Mass Spectral Correlations.* Washington DC: American Chemical Society 1963.

———. "Mass Spectrometric Analysis. Broad Applicability to Chemical Research." *Analytical Chemistry* 28 (1956): 306–16.

———. "Mass Spectrometric Analysis. Molecular Rearrangements." *Analytical Chemistry* 31 (1959): 82–87.

———. "Mass Spectrometry." In *Determination of Organic Structures by Physical Methods*, edited by F. C. Nachod and W. D. Phillips. New York: Academic Press 1962, 93–179.

———. "Mass Spectrometry and Analytical Chemistry." *Journal of the American Society for Mass Spectrometry* 6 (1995): 993–94.

McLafferty, Fred W., ed. *Mass Spectrometry of Organic Ions.* New York: Academic Press 1963.

McLafferty, Fred W., et al. "A Self-Training Interpretive and Retrieval System for Mass Spectra. The Data Base." In *Mass Spectrometry and NMR Spectroscopy in Pesticide Chemistry*, edited by R. Haque and F. J. Biros. New York: Plenum Press 1973, 49–60.

McLafferty, Fred W., et al. "Substituent Effects in Unimolecular Ion Decompositions. XV." *Journal of the American Chemical Society* 92 (1970): 6867–80.

McLafferty, Fred W., H. E. Dayringer, and Rengachari Venkataraghavan. "Computerizing the Spectra Puzzle." *Industrial Research* (1976): 78–83.

McLafferty, Fred W., and Roland S. Gohlke. "Mass Spectrometric Analysis. Spectral Data File Utilizing Machine Filing and Manual Searching." *Analytical Chemistry* 31 (1959): 1160–63.

McLafferty, Fred W., and Mynard C. Hamming. "Mechanism of Rearrangements in Mass Spectra." *Chemistry and Industry* (1958): 1366–67.

McLafferty, Fred W., R. H. Hertel, and R. D. Villwock. "Probability Based Matching of Mass Spectra. Rapid Identification of Specific Compounds in Mixtures." *Organic Mass Spectrometry* 9 (1974): 690–702.

McLafferty, Fred W., Douglas Stauffer, Stanton Y. Loh, and Chrysotomos Wesdemiotis. "Unknown Identification Using Reference Mass Spectra. Quality Evaluation of Databases." *Journal of the American Society for Mass Spectrometry* 10 (1999): 1229–40.

McNeil, E. B., C. P. Slichter, and H. S. Gutowsky. "Slow Beats in F^{19} Nuclear Spin Echoes." *Physical Review* 84 (1951): 1245–46.

Meinel, Christoph, ed. *Instrument-Experiment. Historische Studien.* Berlin: GNT-Verlag 2001.

Meyer, Leon H. "NMR Some Fifty Years Later. Personal Recollections." *Chemical Heritage Magazine* 15 (1998): 47–48.

Meyer, Leon H., and Herbert S. Gutowsky. "Electron Distribution in Molecules II. Proton and Fluorine Magnetic Resonance Shifts in the Halomethanes." *Journal of Physical Chemistry* 57 (1953): 481–86.

Meyer, Leon H., Apollo Saika, and Herbert S. Gutowsky. "Electron Distribution in Molecules III. The Proton Magnetic Spectra of Simple Organic Groups." *Journal of the American Chemical Society* 75 (1953): 4567–73.

Meyerson, Seymour. "Cationated Cyclopropanes as Reaction Intermediates in Mass Spectra: An Earlier Incarnation of Ion-Neutral Complexes." *Organic Mass Spectrometry* 24 (1989): 267–70.

———. "From Black Magic to Chemistry. The Metamorphosis of Organic Mass Spectrometry." *Analytical Chemistry* 66 (1994): 960A–64A.

———. "Reminiscences of the Early Days of Mass Spectrometry in the Petroleum Industry." *Organic Mass Spectrometry* 21 (1986): 197–208.

———. "Tropylium, Chlorine Isotopic Abundances, Monomeric Metaphosphate Anion, and Conestoga Wagon Theory." *Journal of the American Society for Mass Spectrometry* 4 (1993): 761–68.

Miller, Foil A. "The Infrastructure of IR Spectrometry: Reminiscences of Pioneers and Early Commercial IR Instruments." *Analytical Chemistry* 64 (1992): 824A–31A.

Morris, Peter J. T. "From Basel to Austin: A Brief History of Ozonolysis." *Special Publications of the Royal Society of Chemistry* 170 (1995): 170–90.

Morris, Peter J. T., ed. *From Classical to Modern Chemistry. The Instrumental Revolution.* Cambridge: Royal Society of Chemistry 2002.

Morris, Peter J. T., and Anthony S. Travis. "The Role of Physical Instrumentation in Structural Organic Chemistry." In *Science in the Twentieth Century*, edited by J. Krige and D. Pestre. Amsterdam: Harwood Academic Publishers 1997, 715–39.

Morris, Peter J. T., Anthony S. Travis, and Carsten Reinhardt. "Research Fields and Boundaries in Twentieth-Century Organic Chemistry." In *Chemical Sciences in the 20th Century. Bridging Boundaries*, edited by C. Reinhardt. Weinheim: Wiley-VCH 2001, 14–42.

Mors, Walter B. "Editorial." *Journal of the Brazilian Chemical Society* 12 (2001). Available from <http://jbcs.sbq.org.br/>.

Müller, Eugen, ed. *Physikalische Forschungsmethoden*, vol. 3, *Methoden der Organischen Chemie*. Stuttgart: Thieme 1955.

Müller, Luciano, Anil Kumar, and Richard R. Ernst. "Two-Dimensional Carbon-13 NMR Spectroscopy." *Journal of Chemical Physics* 63 (1975): 5490–91.

Mulliken, Robert S. "Interview by Thomas Kuhn, February 1964." In *Selected Papers of Robert S. Mulliken*, edited by D. A. Ramsay and J. Hinze. Chicago: University of Chicago Press 1975, 5–10.

———. *Life of a Scientist. An Autobiographical Account of the Development of Molecular Orbital Theory with an Introductory Memoir by Friedrich Hund.* Berlin: Springer 1989.

———. "Molekülspektren und ihre Anwendungen auf chemische Probleme." *Review of Scientific Instruments* 7 (1936): 171–72.

———. "Spectroscopy, Molecular Orbitals, and Chemical Bonding." *Science* 157 (1967): 13–24.

Mummery, Albert Frederick. *My Climbs in the Alps and Caucasus.* 2nd ed. London: T. Fisher Unwin 1895.

Mutzenbecher, Gerhard von, Z. Pelah, Dudley H. Williams, Herbert Budzikiewicz, and Carl Djerassi. "Mass Spectrometry in Structural and Stereochemical Problems. XLVI." *Steroids* 2 (1963): 475–84.

Nagayama, Kuniaki. "The First Protein Two-Dimensional (2D) NMR." In *Encyclopedia of Nuclear Magnetic Resonance*, vol. 1, *Historical Perspectives*, edited by D. M. Grant and R. K. Harris. Chichester: John Wiley & Sons 1996, 500–02.

Nair, P. Madhavan, and John D. Roberts. "Nuclear Magnetic Resonance Spectra. Hindered Rotation and Molecular Asymmetry." *Journal of the American Chemical Society* 79 (1957): 4565–66.

National Academy of Sciences, and National Research Council. *Chemistry: Opportunities and Needs. A Report on Basic Research in US Chemistry by the Committee for the Survey of Chemistry*. Washington DC: National Academy of Sciences 1965.

Neuert, H. "Gasanalyse mit dem Massenspektrometer." *Angewandte Chemie* 61 (1949): 369–78.

Neuss, Norbert. "Alkaloids from Apocyanaceae. II. Ibogaline, a New Alkaloid from Tabernanthe Iboga Baill." *Journal of Organic Chemistry* 24 (1959): 2047–48.

Neuss, Norbert, et al. "Vinca Alkaloids. XXI. The Structures of the Oncolytic Alkaloids Vinblastine (Vlb) and Vincristine (Vcr)." *Journal of the American Chemical Society* 86 (1964): 1440–42.

Nicholson, A. J. C. "The Photochemical Decomposition of the Aliphatic Methyl Ketones." *Transactions of the Faraday Society* 50 (1954): 1067–73.

Norberg, Richard E. "NMR in Urbana and St. Louis." In *Encyclopedia of Nuclear Magnetic Resonance*, vol. 1, *Historical Perspectives*, edited by D. M. Grant and R. K. Harris. Chichester: John Wiley & Sons 1996, 504–407.

Nowicki, Henry. "A Tribute to Foil Miller. A Career Focus on Infrared and Raman Spectroscopy." *Spectroscopy* 15 (11) (2000): 16–18.

Nye, Mary Jo. *From Chemical Philosophy to Theoretical Chemistry. Dynamics of Matter and Dynamics of Disciplines, 1800–1950*. Berkeley: University of California Press 1993.

———. "From Student to Teacher. Linus Pauling and the Reformulation of the Principles of Chemistry in the 1930s." In *Communicating Chemistry. Textbooks and Their Audiences*, edited by A. Lundgren and B. Bensaude-Vincent. Canton, Mass.: Science History Publications 2000, 397–414.

———. "Physics and Chemistry: Commensurate or Incommensurate Sciences?" In *The Invention of Physical Science. Intersections of Mathematics, Theology and Natural Philosophy since the Seventeenth Century*, edited by M. J. Nye, J. L. Richards and R. H. Stuewer. Dordrecht: Kluwer 1992, 205–24.

Ogata, Koreichi, and Teruo Hayakawa, eds. *Recent Developments in Mass Spectroscopy*. Baltimore: University Park Press 1970.

Ohloff, Günther. "Ein fragmentarischer Beitrag zur Geschichte der Firmenich-Forschung." *Riechstoffe, Aromen, Körperpflegemittel* 22 (1972): 242–49.

———. "In Place of a Foreword. A Fragmentary Contribution to a History of Firmenich Research." *Helvetica Chimica Acta* 54 (1971): v–xvii.

O'Neal, Milburn J., and Thomas P. Wier. "Mass Spectrometry of Heavy Hydrocarbons." *Analytical Chemistry* 23 (1951): 840–43.

Oster, Gerald, and Arthur W. Pollister, eds. *Physical Techniques in Biological Research*. New York: Academic Press 1955–56.

Packard, Martin E. "Nuclear Induction at Stanford and the Transition to Varian." In *Encyclopedia of Nuclear Magnetic Resonance*, vol. 1, *Historical Perspectives*, edited by D. M. Grant and R. K. Harris. Chichester: John Wiley & Sons 1996, 516–25.

Packard, Martin E., and J. T. Arnold. "A Fine Structure in Nuclear Induction Signals from Ethyl Alcohol." *Physical Review* 83 (1951): 210.

Pake, George E. "Fundamentals of Nuclear Magnetic Resonance Absorption. I, II." *American Journal of Physics* 18 (1950): 438–52, 473–86.

———. "Nuclear Resonance Absorption in Hydrated Crystals: Fine Structure of the Proton Line." *Journal of Chemical Physics* 16 (1948): 327–36.

———. *Paramagnetic Resonance. An Introductory Monograph.* New York: Benjamin 1962.

Pake, George E., and Edward M. Purcell. "Line Shapes in Nuclear Paramagnetism." *Physical Review* 74 (1948): 1184–88.

Pauling, Linus. *General Chemistry. An Introduction to Descriptive Chemistry and Modern Chemical Theory.* San Francisco: Freeman 1947.

———. *The Nature of the Chemical Bond and the Structure of Molecules and Crystals.* 2nd ed. Ithaca: Cornell University Press 1948. Original edition 1939.

Pauling, Linus, and Robert E. Corey. "Atomic Coordinates and Structure Factors for Two Helical Configurations of Polypeptide Chains." *Proceedings of the National Academy of Sciences (US)* 37 (1951): 235–40.

Pauling, Linus, and E. Bright Wilson. *Introduction to Quantum Mechanics with Applications to Chemistry.* New York: McGraw-Hill 1935.

Pesyna, G. M., and Fred W. McLafferty. "Computerized Structure Retrieval and Interpretation of Mass Spectra." In *Determination of Organic Structures by Physical Methods*, edited by F. C. Nachod, J. J. Zuckerman and E. W. Randall. New York: Academic Press 1976, 91–155.

Pesyna, G. M., Fred W. McLafferty, Rengachari Venkataraghavan, and H. E. Dayringer. "Probability Based Matching System Using a Large Collection of Reference Mass Spectra." *Analytical Chemistry* 48 (1976): 1362–68.

———. "Statistical Occurrence of Mass and Abundance Values in Mass Spectra." *Analytical Chemistry* 47 (1975): 1161–64.

Petersen, D. R. "X-Ray Investigation of the Structure of Feist's Acid." *Chemistry and Industry* (1956): 904–05.

Phillips, William D. "Restricted Rotation in Amides as Evidenced by Nuclear Magnetic Resonance." *Journal of Chemical Physics* 23 (1955): 1363–64.

Physics Survey Committee, National Research Council. *Physics in Perspective*, vol. II, part B. Washington DC: National Academy of Sciences 1973.

Pickering, Andrew. "Beyond Constraint. The Temporality of Practice and the Historicity of Knowledge." In *Scientific Practice. Theories and Stories of Doing Physics*, edited by J. Z. Buchwald. Chicago: University of Chicago Press 1995, 42–55.

———. "Living in the Material World: On Realism and Experimental Practice." In *The Uses of Experiment. Studies in the Natural Sciences*, edited by D. Gooding, T. Pinch and S. Schaffer. Cambridge: Cambridge University Press 1989, 275–96.

Pickering, Andrew, ed. *Science as Practice and Culture.* Chicago: University of Chicago Press 1992.

Pitzer, Kenneth S. "Of Physical Chemistry and Other Activities." *Annual Review of Physical Chemistry* 38 (1987): 1–25.

———. *Quantum Chemistry.* New York: Prentice-Hall 1953.

———. "Repulsive Forces in Relation to Bond Energies, Distances and Other Properties." *Journal of the American Chemical Society* 70 (1948): 2140–45.

Pitzer, Kenneth S., and Herbert S. Gutowsky. "Electron Deficient Molecules II. Aluminum Alkyles." *Journal of the American Chemical Society* 68 (1946): 2204–09.

Pollard, C. B., and R. F. Parcell. "Synthesis of N-Allylidene-Alkylamines." *Journal of the American Chemical Society* 73 (1951): 2925–27.

Pople, J. A., W. G. Schneider, and H. J. Bernstein. *High-Resolution Nuclear Magnetic Resonance.* New York: McGraw-Hill 1959.

Pound, Robert V. "Early Days in NMR." In *Encyclopedia of Nuclear Magnetic Resonance*, vol. 1, *Historical Perspectives*, edited by D. M. Grant and R. K. Harris. Chichester: John Wiley & Sons 1996, 541–46.

Powles, J. G., and H. S. Gutowsky. "Proton Magnetic Resonance of the CH_3 Group, III. Reorientation Mechanism in Solids." *Journal of Chemical Physics* 23 (1955): 1692–99.

Primas, Hans. "Anwendungen der magnetischen Kernresonanz in der Chemie." *Chimia* 13 (1959): 15–23.

———. *Chemistry, Quantum Mechanics and Reductionism. Perspectives in Theoretical Chemistry.* Berlin: Springer 1981.

———. "Foundations of Theoretical Chemistry." In *Quantum Dynamics of Molecules. The New Experimental Challenge to Theorists*, edited by R. G. Woolley. New York: Plenum Press 1980, 39–113.

Proctor, Warren G. "When You and I Were Young, Magnet." In *Encyclopedia of Nuclear Magnetic Resonance*, vol. 1, *Historical Perspectives*, edited by D. M. Grant and R. K. Harris. Chichester: John Wiley & Sons 1996, 548–51.

Proctor, Warren G., and Fu Chun Yu. "The Dependence of a Nuclear Magnetic Resonance Frequency Upon Chemical Compound." *Physical Review* 77 (1950): 717.

———. "On the Nuclear Magnetic Moments of Several Stable Isotopes." *Physical Review* 81 (1951): 20–31.

Purcell, Edward M. "Nuclear Magnetism in Relation to Problems of the Liquid and Solid States." *Science* 107 (1948): 433–40.

———. "Nuclear Resonance in Crystals." *Physica* 17 (1951): 282–302.

Purcell, Edward M., Nicolaas Bloembergen, and Robert V. Pound. "Resonance Absorption by Nuclear Magnetic Moments in a Single Crystal of CaF_2." *Physical Review* 70 (1946): 988.

Rabkin, Yakov M. "Technological Innovation in Science. The Adoption of Infrared Spectroscopy by Chemists." *Isis* 78 (1987): 31–54.

———. "Uses and Images of Instruments in Chemistry." In *Chemical Sciences in the Modern World*, edited by S. H. Mauskopf. Philadelphia: University of Pennsylvania Press 1993, 25–42.

Ramsey, Norman F. "Early History of Magnetic Resonance." *Physics in Perspective* 1 (1999): 123–35.

———. "Electron Coupled Interactions between Nuclear Spins in Molecules." *Physical Review* 91 (1953): 303–07.

———. "The Internal Diamagnetic Field Correction in Measurements of the Proton Magnetic Moment." *Physical Review* 77 (1950): 567.

———. "Magnetic Shielding of Nuclei in Molecules." *Physical Review* 78 (1950): 699–703.

———. "The Rotational Magnetic Moments of H_2, D_2, and HD Molecules." *Physical Review* 58 (1940): 226–36.

Ramsey, Norman F., and Edward M. Purcell. "Interactions between Nuclear Spins in Molecules." *Physical Review* 85 (1952): 143–45.

Rasmussen, Nicolas. "Innovation in Chemical Separation and Detection Instruments: Reflections on the Role of Research-Technology in the History of Science." In *From Classical to Modern Chemistry*, edited by P. J. T. Morris. Cambridge: Royal Society of Chemistry 2002, 251–58.

———. *Picture Control. The Electron Microscope and the Transformation of Biology in America, 1940–1960*. Stanford: Stanford University Press 1997.

Ravetz, Jerome R. *Scientific Knowledge and Its Social Problems*. Oxford: Clarendon Press 1972.

Reed, Rowland Ivor. "Some Problems in Organic Mass Spectrometry." In *Recent Developments in Mass Spectroscopy*, edited by K. Ogata and T. Hayakawa. Baltimore: University Park Press 1970, 1214–17.

Reich, Hans J., Manfred Jautelat, Mark T. Messe, Frank J. Weigert, and John D. Roberts. "Nuclear Magnetic Resonance Spectroscopy. Carbon-13 Spectra of Steroids." *Journal of the American Chemical Society* 91 (1969): 7445–54.

Reilly, Joseph, and William Norman Rae. *Physico-Chemical Methods*. London: Methuen 1926.

Reinhardt, Carsten. "Chemistry in a Physical Mode: Molecular Spectroscopy and the Emergence of NMR." *Annals of Science* 61 (2004): 1–32.

Reinhardt, Carsten, ed. *Chemical Sciences in the 20th Century. Bridging Boundaries*. Weinheim: Wiley-VCH 2001.

Reinhardt, Carsten, and Harm G. Schröter. "Academia and Industry in Chemistry. The Impact of State Intervention and the Effects of Cultural Values." *Ambix* 51 (2004): 99–106.

Remane, Horst. "Zur Entwicklung der Massenspektroskopie von den Anfängen bis zur Strukturaufklärung organischer Verbindungen." *NTM* 24 (1987): 93–106.

Research Corporation. *Research Corporation* [cited 26 February 2002]. Available from <http://www.rescorp.org/history.htm>.

Rheinberger, Hans-Jörg. *Toward a History of Epistemic Things. Synthesizing Proteins in the Test Tube*. Stanford: Stanford University Press 1997.

Richards, Rex. "Development of NMR in Oxford." In *Encyclopedia of Nuclear Magnetic Resonance*, vol. 1, *Historical Perspectives*, edited by D. M. Grant and R. K. Harris. Chichester: John Wiley & Sons 1996, 580–85.

Ridley, R. D. "Mass Spectrometry Data Center." In *Modern Aspects of Mass Spectrometry*, edited by R. I. Reed. New York: Plenum Press 1968, 361–66.

Rigden, John S. "The Birth of the Magnetic Resonance Method." In *Observation, Experiment, and Hypothesis in Modern Physical Science*, edited by P. Achinstein and O. Hannaway. Cambridge, Mass. 1985, 205–37.

———. "Quantum States and Precession: The Two Discoveries of NMR." *Reviews of Modern Physics* 58 (1986): 433–48.

Rittenberg, D. "Some Applications of Mass Spectrometric Analysis to Chemistry." *Journal of Applied Physics* 13 (1942): 561–69.

Roberts, Jody. "Instruments and Domains of Knowledge: The Case of Nuclear Magnetic Resonance Spectroscopy, 1956–1969." MSc thesis, Blacksburg: Virginia Polytechnic Institute, 2002.

Roberts, John D. "Applications of Nuclear Magnetic Resonance Spectroscopy in Organic Chemistry." *Journal of the American Chemical Society* 82 (1960): 5767.

———. "The Beginnings of Physical Organic Chemistry in the United States (1)." *Bulletin of the History of Chemistry* 19 (1996): 48–56.

———. *Collected Works.* New York: W. A. Benjamin 1970.

———. "Interview by István Hargittai." *The Chemical Intelligencer* 4 (1998): 29–39.

———. *An Introduction to the Analysis of Spin-Spin Splitting in High-Resolution Nuclear Magnetic Resonance Spectra.* New York: Benjamin 1961.

———. *Notes on Molecular Orbital Calculations.* New York: Benjamin 1962.

———. *Nuclear Magnetic Resonance. Applications to Organic Chemistry.* New York: McGraw-Hill 1959.

———. *The Right Place at the Right Time.* Washington DC: ACS 1990.

Roberts, John D., and Marjorie C. Caserio. *Basic Principles of Organic Chemistry.* New York: Benjamin 1964.

———. *Modern Organic Chemistry.* New York: Benjamin 1967.

Roberts, John D., and C. C. Lee. "The Nature of the Intermediate in the Solvolysis of Norbornyl Derivatives." *Journal of the American Chemical Society* 73 (1951): 5009.

Robinson, Robert. "Introductory Remarks." In *Some Newer Physical Methods in Structural Chemistry. Mass Spectrometry, Optical Rotatory Dispersion, and Circular Dichroism,* edited by R. Bonnett and J. G. Davis. London: United Trade Press 1967.

———. *The Structural Relations of Natural Products.* Oxford: Clarendon Press 1955.

Rock, Sybil M. "Qualitative Analysis from Mass Spectra." *Analytical Chemistry* 23 (1951): 261–68.

Rosenkranz, George. "From Ruzicka's Terpenes in Zurich to Mexican Steroids Via Cuba." *Steroids* 57 (1992): 409–18.

Rosenstock, Henry M., and Morris Krauss. "Quasi-Equilibrium Theory of Mass Spectra." In *Mass Spectrometry of Organic Ions,* edited by F. W. McLafferty. New York: Academic Press 1963, 2–64.

Rosenstock, Henry M., Merrill B. Wallenstein, Austin L. Wahrhaftig, and Henry Eyring. "Absolute Rate Theory for Isolated Systems and the Mass Spectra of Polyatomic Molecules." *Proceedings of the National Academy of Sciences (US)* 38 (1952): 667–78.

Rossini, Frederick D. "American Petroleum Institute. Spectral Data and Standard Samples." *Applied Spectroscopy* 6 (1951): 3–13.

Russell, Colin A. "The Changing Role of Synthesis in Organic Chemistry." *Ambix* 34 (1987): 169–80.

Russell, N. C., E. M. Tansey, and P. V. Lear. "Missing Links in the History and Practice of Science: Teams, Technicians and Technical Work." *History of Science* 38 (2000): 237–41.

Ruzicka, Leopold. "Rolle der Riechstoffe in meinem chemischen Lebenswerk." *Riechstoffe, Aromen, Körperpflegemittel* 22 (1972): 1753–59.

Ryhage, Ragnar. "The Mass Spectrometry Laboratory at the Karolinska Institute 1944–1987." *Mass Spectrometry Reviews* 12 (1993): 1–49.

Rylander, Paul N., and Seymour Meyerson. "Organic Ions in the Gas Phase. I. The Cationated Cyclopropane Ring." *Journal of the American Chemical Society* 78 (1956): 5799–802.

Rylander, Paul N., Seymour Meyerson, and Henry M. Grubb. "Organic Ions in the Gas Phase. II. The Tropylium Ion." *Journal of the American Chemical Society* 79 (1957): 842–46.

Saika, Apollo, and Charles P. Slichter. "A Note on the Fluorine Resonance Shifts." *Journal of Chemical Physics* 22 (1954): 26–28.

Salton, Gerald. *Automatic Information Organization and Retrieval.* New York: McGraw-Hill 1968.

Saltzman, M. D. "The Development of Physical Organic Chemistry in the United States and the United Kingdom: 1919–1939, Parallels and Contrasts." *Journal of Chemical Education* 63 (1986): 588–93.

Sanger, Frederick. "Free Amino Groups of Insulin." *Biochemical Journal* 39 (1945): 507–15.

Schaffer, Simon. "Glass Works: Newton's Prisms and the Uses of Experiment." In *The Uses of Experiment. Studies in the Natural Sciences*, edited by D. Gooding, T. Pinch and S. Schaffer. Cambridge: Cambridge University Press 1989, 67–104.

Schissler, D. O., S. O. Thompson, and John Turkevich. "Behaviour of Paraffin Hydrocarbons on Electron Impact. Synthesis and Mass Spectra of Some Deuterated Paraffin Hydrocarbons." *Discussions of the Faraday Society* 10 (1951): 46–53.

Schönberger, Angela, ed. *Raymond Loewy. Pionier des amerikanischen Industriedesigns.* München: Prestel 1990.

Schummer, Joachim. "The Impact of Instrumentation on Chemical Species Identity." In *From Classical to Modern Chemistry: The Instrumental Revolution*, edited by P. J. T. Morris. Cambridge: Royal Society of Chemistry 2002, 188–211.

Schuster, John A., and Richard R. Yeo, eds. *The Politics and Rhetoric of Scientific Method.* Dordrecht: Reidel 1986.

Schwarz, Johann C. P., ed. *Physical Methods in Organic Chemistry.* San Francisco: Holden-Day 1964.

Schweber, Sylvan S. "The Empiricist Temper Regnant: Theoretical Physics in the United States 1920–1950." *Historical Studies in the Physical Sciences* 17 (1986): 55–98.

———. "Physics, Community and the Crisis in Physical Theory." In *Physics, Philosophy, and the Scientific Community. Essays in the Philosophy and History of the Natural Sciences and Mathematics in Honor of Robert S. Cohen*, edited by K. Gavroglu, J. Stachel and M. W. Wartofsky. Dordrecht: Kluwer 1995, 125–52.

———. "The Young John Clarke Slater." *Historical Studies in the Physical and Biological Sciences* 20 (1992): 339–406.

Seibl, Josef. *Massenspektrometrie. Studienbuch für Studierende der Chemie nach dem Vordiplom.* Frankfurt am Main: Akademische Verlagsgesellschaft 1970.

Seidel, Casimir, Dorothee Felix, Albert Eschenmoser, Klaus Biemann, Edouard Palluy, and Max Stoll. "Zur Kenntnis des Rosenöls. 2. Mitteilung. Die Konstitution des Oxyds $C_{10}H_{18}O$ aus bulgarischem Rosenöl." *Helvetica Chimica Acta* 44 (1961): 598–606.

Servos, John W. *Physical Chemistry from Ostwald to Pauling. The Making of a Science in America.* Princeton: Princeton University Press 1990.

Shapin, Steven. "Cordelia's Love: Credibility and the Social Studies of Science." *Perspectives on Science* 3 (1995): 255–75.

Shapin, Steven, and Simon Schaffer. *Leviathan and the Air-Pump. Hobbes, Boyle, and the Experimental Life.* Princeton: Princeton University Press 1985.

Shapiro, Bernard L. "The NMR Newsletter." In *Encyclopedia of Nuclear Magnetic Resonance*, vol. 1, *Historical Perspectives*, edited by D. M. Grant and R. K. Harris. Chichester: John Wiley & Sons 1996, 621–23.

Shapiro, Robert, and Carl Djerassi. "Mass Spectrometry in Structural and Stereochemical Problems. L." *Journal of the American Chemical Society* 86 (1964): 2825–32.

Sharkey, A. G., J. L. Shultz, and R. A. Friedel. "Mass Spectra of Ketones." *Analytical Chemistry* 28 (1956): 934–44.

Shepherd, Martin, Sybil M. Rock, Royce Howard, and John Stormes. "Isolation, Identification, and Estimation of Gaseous Pollutants of Air." *Analytical Chemistry* 23 (1951): 1431–40.

Shoolery, James N. "NMR Spectroscopy in the Beginning." *Analytical Chemistry* 65 (1993): 731A–41A.

Shoolery, James N., and John D. Roberts. "High Resolution Nuclear Magnetic Resonance Spectroscopy at Elevated Temperatures." *Review of Scientific Instruments* 28 (1957): 61–62.

Shoolery, James N., and Max T. Rogers. "Nuclear Magnetic Resonance Spectra of Steroids." *Journal of the American Chemical Society* 80 (1958): 5121–35.

Shorter, John. "Die Hammett-Gleichung—und was daraus in fünfzig Jahren wurde." *Chemie in unserer Zeit* 19 (1985): 197–208.

———. "Hammett Memorial Lecture." *Progress in Physical Organic Chemistry* 17 (1990): 1–29.

Shriner, Ralph L., Reynold C. Fuson, and David Y. Curtin. *The Systematic Identification of Organic Compounds. A Laboratory Manual.* 4th ed. New York: Wiley 1956.

Shulman, Robert. "My Years in NMR." In *Encyclopedia of Nuclear Magnetic Resonance*, vol. 1, *Historical Perspectives*, edited by D. M. Grant and R. K. Harris. Chichester: John Wiley & Sons 1996, 635–42.

Simões, Ana, and Kostas Gavroglu. "Issues in the History of Theoretical and Quantum Chemistry." In *Chemical Sciences in the 20th Century. Bridging Boundaries*, edited by C. Reinhardt. Weinheim: Wiley-VCH 2001, 51–74.

Slater, John Clarke. "The Current State of Solid-State and Molecular Theory." *International Journal of Quantum Chemistry* 1 (1967): 37–102.

———. *Introduction to Chemical Physics.* New York: McGraw-Hill 1939.

———. *Solid-State and Molecular Theory: A Scientific Biography.* New York: Wiley 1975.

Slater, Leo B. "Instruments and Rules: R. B. Woodward and the Tools of Twentieth-Century Organic Chemistry." *Studies in History and Philosophy of Science* 33 (2002): 1–33.

———. "Organic Chemistry and Instrumentation: R. B. Woodward and the Reification of Chemical Structures." In *From Classical to Modern Chemistry: The Instrumental Revolution*, edited by P. J. T. Morris. Cambridge: Royal Society of Chemistry 2002, 212–28.

———. "Woodward, Robinson, and Strychnine: Chemical Structure and Chemists' Challenge." *Ambix* 48 (2001): 161–89.

Slichter, Charles P. "Some Scientific Contributions of Herbert S. Gutowsky." *Journal of Magnetic Resonance* 17 (1975): 274–80.

Smith, Dennis H., et al. "Applications of Artificial Intelligence for Chemical Inference. IX. Analysis of Mixtures without Prior Separation as Illustrated for Estrogens." *Journal of the American Chemical Society* 95 (1973): 6078–84.

Smith, J. A. S. "Nuclear Magnetic Resonance Absorption." *Quarterly Reviews* 7 (1953): 279–306.

Smith, Lincoln G. "Ionization and Dissociation of Polyatomic Molecules by Electron Impact. I. Methane." *Physical Review* 51 (1937): 263–75.

Smyth, H. D. "Products and Processes of Ionization by Low Speed Electrons." *Reviews of Modern Physics* 3 (1931): 347–91.

Solomon, Miriam. "The Pragmatic Turn in Naturalistic Philosophy of Science." *Perspectives on Science* 3 (1995): 206–28.

Sparkman, O. David. "Mass Spectrometry. Overview and History." In *Encyclopedia of Analytical Chemistry. Applications, Theory and Instrumentation*, edited by R. A. Meyers. Chichester: Wiley 2000, 11501–58.

Spiteller, Gerhard, and Margot Spiteller-Friedmann. "Zur Umlagerung aliphatischer Verbindungen im Massenspektrometer." *Monatshefte für Chemie* 95 (1964): 257–64.

Stadler, P. A. "Gaschromatographische Untersuchung der Ketone des Lavandinöls." *Helvetica Chimica Acta* 43 (1960): 1601–12.

Starr, C. E. Jr., and Trent Lane. "Accuracy and Precision of Analysis of Light Hydrocarbon Mixtures." *Analytical Chemistry* 21 (1949): 572–82.

Stenhagen, Einar, Sixten Abrahamsson, and Fred W. McLafferty. *Registry of Mass Spectral Data*. New York: Wiley-Interscience 1974.

Stevenson, David P. "On the Mass Spectra of Propanes and Butanes Containing C-13." *Journal of Chemical Physics* 19 (1951): 17–21.

———. "On the Strength of Carbon-Hydrogen and Carbon-Carbon Bonds." *Journal of Chemical Physics* 10 (1942): 291–94.

Stevenson, David P. "Ionization and Dissociation by Electronic Impact. Ionization Potentials and Energies of Formation of Sec-Propyl and Tert-Butyl Radicals. Some Limitations of the Method." *Discussions of the Faraday Society* 10 (1951): 35–45.

Stevenson, David P., and J. A. Hipple. "Ionization and Dissociation by Electron Impact. Normal Butane, Isobutane, and Ethane." *Journal of the American Chemical Society* 64 (1942): 1588–94, 2766–72.

———. "Ionization and Dissociation by Electron Impact. The Methyl and Ethyl Radicals." *Physical Review* 63 (1943): 121–26.

Stevenson, David P., and C. D. Wagner. "The Mass Spectra of C1–C4 Monodeutero Paraffins." *Journal of Chemical Physics* 19 (1951): 11–16.

———. "Mass Spectrometric Analysis of Low-Molecular-Weight Monodeutero Paraffins." *Journal of the American Chemical Society* 72 (1950): 5612–17.

Stewart, David W. "Mass Spectrometry." In *Physical Methods of Organic Chemistry*, edited by A. Weissberger. New York: Interscience Publishers 1946, 1291–314.

Stewart, H. R., and A. R. Olson. "The Decomposition of Hydrocarbons in the Positive Ray Tube." *Journal of the American Chemical Society* 53 (1931): 1236–44.

Stine, Jeffrey K. "Scientific Instrumentation as an Element of US Science Policy: National Science Foundation Support of Chemistry Instrumentation." In *Invisible Connections. Instruments, Institutions, and Science*, edited by R. Bud and S. E. Cozzens. Bellingham: SPIE Optical Engineering Press 1992, 238–63.

Stout, J. W. "The Journal of Chemical Physics: The First 50 Years." *Annual Review of Physical Chemistry* 37 (1986): 1–23.

Streitwieser, Andrew. *Molecular Orbital Theory for Organic Chemists*. New York: Wiley 1961.

Stroke, H. H. "The First Magnet Laboratory at M.I.T." In *Francis Bitter. Selected Papers and Commentaries*, edited by T. Erber and C. M. Fowler. Cambridge, Mass.: MIT Press 1969, 282–301.

Sutton, M. A. "Spectroscopy and the Chemists: A Neglected Opportunity?" *Ambix* 23 (1976): 16–26.

Swalen, J. D., and C. A. Reilly. "Analysis of Complex NMR Spectra. An Iterative Method." *Journal of Chemical Physics* 37 (1962): 21–29.

Tarbell, Dean Stanley, and Ann Tracy Tarbell. "The Instrumental Revolution, 1930–1955." In *Essays on the History of Organic Chemistry in the United States, 1875–1955*, edited by D. S. Tarbell and A. T. Tarbell. Nashville, Tenn.: Folio Press 1986, 335–52.

Taylor, Daniel D. "A Modified Aston-Type Mass Spectrometer and Some Preliminary Results." *Physical Review* 47 (1935): 666–71.

Taylor, John K. "The Impact of Instrumentation on Analytical Chemistry." In *The History and Preservation of Chemical Instrumentation. Proceedings of the ACS Division of History of Chemistry Symposium Held in Chicago, Ill., Sept. 9–10, 1985*, edited by J. T. Stock and M. V. Orna. Dordrecht: Reidel 1985, 1–10.

Taylor, R. C., et al. "The Mass Spectrometer in Organic Chemical Analysis." *Analytical Chemistry* 20 (1948): 396–401.

Thomas, B. W., and W. D. Seyfried. "Mass Spectrometer Analyses of Oxygenated Compounds." *Analytical Chemistry* 21 (1949): 1022–26.

Thomas, Harold A., Raymond L. Driscoll, and John A. Hipple. "Measurement of the Proton Moment in Absolute Units." *Physical Review* 78 (1950): 787–90.

Tiers, George V. D. "Proton Nuclear Resonance Spectroscopy. I. Reliable Shielding Values by 'Internal Referencing' with Tetramethylsilane." *Journal of the American Chemical Society* 62 (1958): 1151–52.

Todd, Alexander. *A Time to Remember*. Cambridge: Cambridge University Press 1983.

Townes, C. H., and B. P. Dailey. "Determination of Electronic Structure of Molecules from Nuclear Quadrupole Effects." *Journal of Chemical Physics* 17 (1949): 782–96.

Turkevich, John, et al. "Determination of Position of Tracer Atom in a Molecule. Mass Spectra of Some Deuteriated Hydrocarbons." *Journal of the American Chemical Society* 70 (1948): 2638–43.

Twiss, Richard Q., and Yardley Beers. "Minimal Noise Circuits." In *Vacuum Tube Amplifiers*, vol. 18, *Radiation Laboratory Series*, edited by G. E. Valley and H. Wallman. Lexington, Mass.: Boston Technical Publishers 1964, 615–94. Original edition 1946.

Van Meter, R. A., et al. "Oxygen and Nitrogen Compounds in Shale-Oil Naphtha." *Analytical Chemistry* 24 (1952): 1758–63.

Van Vleck, John H. *Electric and Magnetic Susceptibilities*. London: Oxford University Press 1932.

———. "A Third of a Century of Paramagnetic Relaxation and Resonance." In *Magnetic Resonance. Proceedings of the International Symposium on Electron and Mag-*

netic Resonance, Held in Melbourne, August 1969, edited by C. K. Coogan et al. New York, London: Plenum Press 1970.

Varian Associates. *High Resolution NMR Spectra Catalog*. Palo Alto: Varian Associates 1962–63.

Venkataraghavan, Rengachari, Richard J. Klimowski, and Fred W. McLafferty. "On-Line Computers in Research: High-Resolution Mass Spectrometry." *Accounts of Chemical Research* 3 (1970): 158–65.

Vicedo, Marga. "Scientific Styles: Toward Some Common Ground in the History, Philosophy, and Sociology of Science." *Perspectives on Science* 3 (1995): 231–54.

Voorhies, H. G., et al. "Theoretical and Experimental Study of High-Mass High-Resolution Mass Spectrometers." In *Advances in Mass Spectrometry*, edited by J. D. Waldron. New York: Pergamon Press 1959, 44–65.

Wallenstein, Merrill B., Austin L. Wahrhaftig, Henry M. Rosenstock, and Henry Eyring. "Chemical Reactions in the Gas Phase Connected with Ionization." In *Symposium on Radiobiology: The Basic Aspects of Radiation Effects on Living Systems, Oberlin College 1950*, edited by J. J. Nickson. New York: Wiley 1952, 70–96.

Wang, Jessica. "Science, Security, and the Cold War. The Case of E. U. Condon." *Isis* 83 (1992): 238–69.

Wangen, L. E., W. S. Woodward, and Thomas L. Isenhour. "Small Computer, Magnetic Tape Oriented, Rapid Search System Applied to Mass Spectrometry." *Analytical Chemistry* 43 (1971): 1605–14.

Warner, Mary, ed. *Milestones in Analytical Chemistry*. Washington DC: American Chemical Society 1994.

Washburn, H., H. F. Wiley, and S. M. Rock. "The Mass Spectrometer as an Analytical Tool." *Industrial and Engineering Chemistry, Analytical Edition* 15 (1943): 541–47.

Washburn, H., H. F. Wiley, S. M. Rock, and C. E. Berry. "Mass Spectrometry." *Industrial and Engineering Chemistry, Analytical Edition* 17 (1945): 74–81.

Watson, James T., and Klaus Biemann. "Direct Recording of High Resolution Mass Spectra of Gas Chromatic Effluents." *Analytical Chemistry* 37 (1965): 844–51.

Waugh, John S. "NMR Spectroscopy in Solids: A Historical Review." *Analytical Chemistry* 65 (1993): 725A–29A.

Weinberg, Alvin M. *Reflections on Big Science*. Cambridge, Mass.: MIT Press 1968.

Weininger, Stephen J. " 'What's in a Name?' From Designation to Denunciation—the Nonclassical Cation Controversy." *Bulletin for the History of Chemistry* 25 (2000): 123–31.

Weissberger, Arnold, ed. *Physical Methods of Organic Chemistry*. 2 vols. New York: Interscience 1945–46.

Wertz, John E. "Nuclear and Electronic Spin Magnetic Resonance." *Chemical Reviews* 5 (1955): 829–955.

Wheland, George W. *Resonance in Organic Chemistry*. New York: Wiley 1955.

———. *The Theory of Resonance and Its Application to Organic Chemistry*. New York: Wiley 1944.

———. *The Theory of Resonance and Its Application to Organic Chemistry*. 5th ed. New York: Wiley 1949.

Wiberg, Kenneth B., and Bernard J. Nist. *The Interpretation of NMR Spectra*. New York: Benjamin 1962.

Wilson, E. Bright. *An Introduction to Scientific Research*. New York: McGraw-Hill 1952.

Wiswesser, William J. "107 Years of Line-Formula Notations (1861–1968)." *Journal of Chemical Documentation* 8 (1968): 146–50.

Woessner, Don. "Early Days of NMR at Mobil in Dallas." *TAMU NMR* 449 (1995): 33–37.

Woodward, Robert B. "Synthesis." In *Perspectives in Organic Chemistry*, edited by A. Todd. New York: Interscience Publishers 1956, 155–84.

Wüthrich, Kurt. "NMR Structures of Biological Macromolecules." In *Encyclopedia of Nuclear Magnetic Resonance*, vol. 1, *Historical Perspectives*, edited by D. M. Grant and R. K. Harris. Chichester: John Wiley & Sons 1996, 710–19.

———. "Pancreatictrypsin Inhibitor." In *Encyclopedia of Nuclear Magnetic Resonance*, edited by D. M. Grant and R. K. Harris. Chichester: John Wiley & Sons 1996, 3449–56.

Yost, Don M., and Thomas F. Anderson. "The Raman Spectra and Molecular Constants of Phosphorus Trifluoride and Phosphine." *Journal of Chemical Physics* 2 (1934): 624–27.

Zaffaroni, Alejandro. "From Paper Chromatography to Drug Discovery: Zaffaroni." *Steroids* 57 (1992): 642–48.

Zandvoort, Henk. "Concepts of Interdisciplinarity and Environmental Science." In *Cognitive Patterns in Science and Common Sense. Groningen Studies in Philosophy of Science, Logic and Epistemology*, edited by T. A. F. Kuipers and A. R. Mackor. Amsterdam: Rodopi 1995, 45–68.

———. *Models of Scientific Development and the Case of Nuclear Magnetic Resonance*. Dordrecht: Reidel 1986.

———. "Nuclear Magnetic Resonance and the Acceptability of Guiding Assumptions." In *Scrutinizing Science. Empirical Studies of Scientific Change*, edited by A. Donovan et al. Dordrecht: Kluwer 1988, 337–58.

Zemany, Paul D. "Punched-Card Catalog of Mass Spectra Useful in Qualitative Analysis." *Analytical Chemistry* 22 (1950): 920–22.

Zenner, Walter P. *Minorities in the Middle. A Cross-Cultural Analysis*. Albany: SUNY Press 1991.

Company Index

Name Index